Saab 90, 99 & 900
Service and Repair Manual

A K Legg LAE MIMI and Spencer Drayton

Models covered

Saab 90, 99 and 900 Saloon, Hatchback & Coupe models to October 1993, including Turbo, 16-valve and limited edition versions; 1985 cc
Covers most mechanical features of Convertible
Does NOT cover revised 900 range introduced October 1993

(0765-296-4AA2)

© Haynes Publishing 1997

A book in the Haynes Service and Repair Manual Series

All rights reserved. No part of this book may be reproduced or transmitted in any form or by any means, electronic or mechanical, including photocopying, recording or by any information storage or retrieval system, without permission in writing from the copyright holder.

ISBN 1 85960 064 6

British Library Cataloguing in Publication Data
A catalogue record for this book is available from the British Library.

Printed in the USA

Haynes Publishing
Sparkford, Nr Yeovil, Somerset BA22 7JJ, England

Haynes North America, Inc
861 Lawrence Drive, Newbury Park, California 91320, USA

Editions Haynes S.A.
Tour Aurore - IBC, 18 Place des Reflets,
92975 Paris La Défense 2, Cedex, France

Haynes Publishing Nordiska AB
Box 1504, 751 45 Uppsala, Sverige

Contents

LIVING WITH YOUR SAAB

Introduction	Page 0•4
Safety First!	Page 0•5

Roadside Repairs

If your car won't start	Page 0•6
Jump starting	Page 0•7
Wheel changing	Page 0•8
Identifying leaks	Page 0•9
Towing	Page 0•9

Weekly Checks

Introduction	Page 0•10
Underbonnet check points	Page 0•10
Engine oil level	Page 0•11
Coolant level	Page 0•11
Screen washer fluid level	Page 0•12
Brake and clutch fluid level	Page 0•12
Tyre condition and pressure	Page 0•13
Power steering fluid level	Page 0•14
Electrical systems	Page 0•14
Battery	Page 0•15
Wiper blades	Page 0•15

Lubricants, fluids and tyre pressures	Page 0•16

MAINTENANCE

Routine Maintenance and Servicing

Models from 1979 to 1984	Page 1A•1
Maintenance schedule	Page 1A•2
Maintenance procedures	Page 1A•5
Models from 1985 onwards	Page 1B•1
Maintenance schedule	Page 1B•2
Maintenance procedures	Page 1B•5

Contents

REPAIRS & OVERHAUL

Engine and Associated Systems

Engine in-car repair procedures	Page **2A•1**
Engine removal and general overhaul procedures	Page **2B•1**
Cooling, heating and ventilation systems	Page **3•1**
Fuel/exhaust systems - carburettor models	Page **4A•1**
Fuel/exhaust systems - fuel injection models	Page **4B•1**
Emission control systems	Page **4C•1**
Engine electrical systems	Page **5•1**

Transmission

Clutch	Page **6•1**
Manual transmission	Page **7A•1**
Automatic transmission	Page **7B•1**
Driveshafts	Page **8•1**

Brakes and Suspension

Braking system	Page **9•1**
Suspension and steering systems	Page **10•1**

Body Equipment

Bodywork and fittings	Page **11•1**
Body electrical system	Page **12•1**

Wiring Diagrams

	Page **12•12**

REFERENCE

Dimensions and Weights	Page **REF•1**
Conversion Factors	Page **REF•2**
Buying Spare Parts and Vehicle Identification	Page **REF•3**
General Repair Procedures	Page **REF•4**
Jacking and Vehicle Support	Page **REF•5**
Radio/cassette Anti-theft System	Page **REF•5**
Tools and Working Facilities	Page **REF•6**
MOT Test Checks	Page **REF•8**
Fault Finding	Page **REF•12**
Glossary of Technical Terms	Page **REF•20**
Index	Page **REF•25**

Introduction

The Saab 900 was first introduced in early 1979 and was similar to the existing 99 Combi Coupe, but with a longer more sloping bonnet, front spoiler and deeper windscreen. The body is of rigid construction employing safety beams in the doors and a reinforced roof.

The in-line mounted engine is located over the gearbox or automatic transmission providing drive to the front wheels. Unlike the more conventional layout, the engine is positioned with the flywheel at the front of the car, but this facilitates removal of the clutch without having to remove either the engine or gearbox. Both 8-valve and 16-valve engines are covered by this manual.

Being of an up-market design, the car incorporates many extras to add to comfort and driveability.

Although a number of minor changes have been made to the Saab 99 and 900 models the most significant revisions occurred in 1984 with the introduction of the Saab 90 to replace the 99 model, and the introduction of the Saab 900 Turbo 16. The Saab 90 is essentially a two-door model comprising the front half of a 99 and rear half of a 900, powered by a single carburettor version of the 1985 cc engine. Apart from various minor modifications, the 90 should be regarded mechanically as a 99 with 900 rear suspension. The Saab 90 was discontinued in October 1985 and replaced by the 900 2-door. At the same time the Turbo 16-valve 4-door Saloon was discontinued and the 900 2-door introduced. October 1988 saw the introduction of the 900 i 16-valve models as 2 or 4-door Saloons or 3 or 5-door Hatchbacks with a 5-speed manual transmission as standard and automatic transmission, a catalytic converter and ABS optional. All Saab 900 models covered by this manual were discontinued in October 1993 in favour of the new range with new body and transverse engine.

All models have fully-independent front suspension. The rear suspension incorporates leading and trailing arms, shock absorbers and a Panhard rod.

A wide range of standard and optional equipment is available within the Saab 900 range to suit most tastes, including central locking and electric windows. An anti-lock braking system, traction control system and air conditioning system are available on certain models.

Provided that regular servicing is carried out in accordance with the manufacturer's recommendations, the Saab 900 should prove reliable and economical.

Saab 90 2-door Saloon

Saab 900XS 3-door Hatchback

The Saab 90, 99 & 900 Team

Haynes manuals are produced by dedicated and enthusiastic people working in close co-operation. The team responsible for the creation of this book included:

Authors	Andy Legg
	Spencer Drayton
Editor & Page Make-up	Bob Jex
Workshop manager	Paul Buckland
Photo Scans	John Martin
	Paul Tanswell
Cover illustration	Phillip Cox
Line art	Roger Healing
Wiring diagrams	Matthew Marke

We hope the book will help you to get the maximum enjoyment from your car. By carrying out routine maintenance as described you will ensure your car's reliability and preserve its resale value.

Your Saab Manual

The aim of this manual is to help you get the best value from your vehicle. It can do so in several ways. It can help you decide what work must be done (even should you choose to get it done by a garage), provide information on routine maintenance and servicing, and give a logical course of action and diagnosis when random faults occur. However, it is hoped that you will use the manual by tackling the work yourself. On simpler jobs, it may even be quicker than booking the car into a garage and going there twice, to leave and collect it. Perhaps most important, a lot of money can be saved by avoiding the costs a garage must charge to cover its labour and overheads.

The manual has drawings and descriptions to show the function of the various components, so that their layout can be understood. Then the tasks are described and photographed in a clear step-by-step sequence.

Acknowledgements

Thanks are due to Champion Spark Plug, who supplied the illustrations showing spark plug conditions. Thanks are also due to Sykes-Pickavant Limited, who provided some of the workshop tools, and to all those people at Sparkford and Newbury Park who helped in the production of this manual.

We take great pride in the accuracy of information given in this manual, but vehicle manufacturers make alterations and design changes during the production run of a particular vehicle of which they do not inform us. No liability can be accepted by the authors or publishers for loss, damage or injury caused by any errors in, or omissions from, the information given.

Safety First!

Working on your car can be dangerous. This page shows just some of the potential risks and hazards, with the aim of creating a safety-conscious attitude.

General hazards

Scalding
• Don't remove the radiator or expansion tank cap while the engine is hot.
• Engine oil, automatic transmission fluid or power steering fluid may also be dangerously hot if the engine has recently been running.

Burning
• Beware of burns from the exhaust system and from any part of the engine. Brake discs and drums can also be extremely hot immediately after use.

Crushing

• When working under or near a raised vehicle, always supplement the jack with axle stands, or use drive-on ramps. *Never venture under a car which is only supported by a jack.*
• Take care if loosening or tightening high-torque nuts when the vehicle is on stands. Initial loosening and final tightening should be done with the wheels on the ground.

Fire
• Fuel is highly flammable; fuel vapour is explosive.
• Don't let fuel spill onto a hot engine.
• Do not smoke or allow naked lights (including pilot lights) anywhere near a vehicle being worked on. Also beware of creating sparks (electrically or by use of tools).
• Fuel vapour is heavier than air, so don't work on the fuel system with the vehicle over an inspection pit.
• Another cause of fire is an electrical overload or short-circuit. Take care when repairing or modifying the vehicle wiring.
• Keep a fire extinguisher handy, of a type suitable for use on fuel and electrical fires.

Electric shock
• Ignition HT voltage can be dangerous, especially to people with heart problems or a pacemaker. Don't work on or near the ignition system with the engine running or the ignition switched on.
• Mains voltage is also dangerous. Make sure that any mains-operated equipment is correctly earthed. Mains power points should be protected by a residual current device (RCD) circuit breaker.

Fume or gas intoxication

• Exhaust fumes are poisonous; they often contain carbon monoxide, which is rapidly fatal if inhaled. Never run the engine in a confined space such as a garage with the doors shut.
• Fuel vapour is also poisonous, as are the vapours from some cleaning solvents and paint thinners.

Poisonous or irritant substances
• Avoid skin contact with battery acid and with any fuel, fluid or lubricant, especially antifreeze, brake hydraulic fluid and Diesel fuel. Don't syphon them by mouth. If such a substance is swallowed or gets into the eyes, seek medical advice.
• Prolonged contact with used engine oil can cause skin cancer. Wear gloves or use a barrier cream if necessary. Change out of oil-soaked clothes and do not keep oily rags in your pocket.
• Air conditioning refrigerant forms a poisonous gas if exposed to a naked flame (including a cigarette). It can also cause skin burns on contact.

Asbestos
• Asbestos dust can cause cancer if inhaled or swallowed. Asbestos may be found in gaskets and in brake and clutch linings. When dealing with such components it is safest to assume that they contain asbestos.

Special hazards

Hydrofluoric acid
• This extremely corrosive acid is formed when certain types of synthetic rubber, found in some O-rings, oil seals, fuel hoses etc, are exposed to temperatures above 400°C. The rubber changes into a charred or sticky substance containing the acid. *Once formed, the acid remains dangerous for years. If it gets onto the skin, it may be necessary to amputate the limb concerned.*
• When dealing with a vehicle which has suffered a fire, or with components salvaged from such a vehicle, wear protective gloves and discard them after use.

The battery
• Batteries contain sulphuric acid, which attacks clothing, eyes and skin. Take care when topping-up or carrying the battery.
• The hydrogen gas given off by the battery is highly explosive. Never cause a spark or allow a naked light nearby. Be careful when connecting and disconnecting battery chargers or jump leads.

Air bags
• Air bags can cause injury if they go off accidentally. Take care when removing the steering wheel and/or facia. Special storage instructions may apply.

Diesel injection equipment
• Diesel injection pumps supply fuel at very high pressure. Take care when working on the fuel injectors and fuel pipes.

⚠️ *Warning: Never expose the hands, face or any other part of the body to injector spray; the fuel can penetrate the skin with potentially fatal results.*

Remember...

DO
• Do use eye protection when using power tools, and when working under the vehicle.
• Do wear gloves or use barrier cream to protect your hands when necessary.
• Do get someone to check periodically that all is well when working alone on the vehicle.
• Do keep loose clothing and long hair well out of the way of moving mechanical parts.
• Do remove rings, wristwatch etc, before working on the vehicle – especially the electrical system.
• Do ensure that any lifting or jacking equipment has a safe working load rating adequate for the job.

DON'T
• Don't attempt to lift a heavy component which may be beyond your capability – get assistance.
• Don't rush to finish a job, or take unverified short cuts.
• Don't use ill-fitting tools which may slip and cause injury.
• Don't leave tools or parts lying around where someone can trip over them. Mop up oil and fuel spills at once.
• Don't allow children or pets to play in or near a vehicle being worked on.

0•6 Roadside Repairs

The following pages are intended to help in dealing with common roadside emergencies and breakdowns. You will find more detailed fault finding information at the back of the manual, and repair information in the main chapters.

If your car won't start and the starter motor doesn't turn

☐ If it's a model with automatic transmission, make sure the selector is in 'P' or 'N'.

☐ Open the bonnet and make sure that the battery terminals are clean and tight.

☐ Switch on the headlights and try to start the engine. If the headlights go very dim when you're trying to start, the battery is probably flat. Get out of trouble by jump starting (see next page) using a friend's car.

If your car won't start even though the starter motor turns as normal

☐ Is there fuel in the tank?

☐ Is there moisture on electrical components under the bonnet? Switch off the ignition, then wipe off any obvious dampness with a dry cloth. Spray a water-repellent aerosol product (WD-40 or equivalent) on ignition and fuel system electrical connectors like those shown in the photos. Pay special attention to the ignition coil and distributor wiring connectors and HT leads.

A Check that the spark plug HT leads (hidden below a cover on some models) are secure by pushing them home. Check the lead connection at both the plug end (seen here) and the distributor.

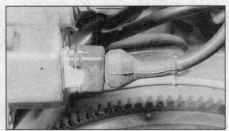

B The fuel injection wiring plugs may cause problems if not connected securely (typical plug shown).

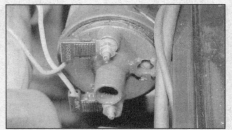

C Check the ignition coil wiring plugs for security (note that the central HT lead has been disconnected in this photo).

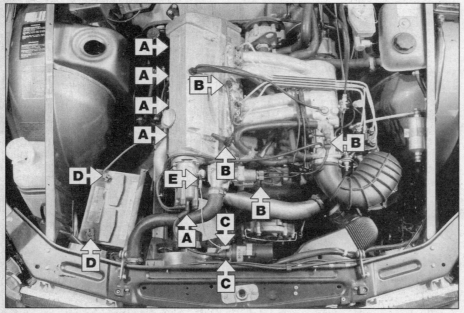

Check that electrical connections are secure (with the ignition switched off) and spray them with a water-dispersing spray like WD-40 if you suspect a problem due to damp.

D Check the security and condition of the battery connections.

E Check that the distributor wiring plug is secure, and spray with water-dispersant if necessary.

Roadside Repairs 0•7

Jump starting

Haynes Hint

Jump starting will get you out of trouble, but you must correct whatever made the battery go flat in the first place. There are three possibilities:

1 The battery has been drained by repeated attempts to start, or by leaving the lights on.

2 The charging system is not working properly (alternator drivebelt slack or broken, alternator wiring fault or alternator itself faulty).

3 The battery itself is at fault (electrolyte low, or battery worn out).

When jump-starting a car using a booster battery, observe the following precautions:

A) Before connecting the booster battery, make sure that the ignition is switched off.

B) Ensure that all electrical equipment (lights, heater, wipers, etc) is switched off.

C) Make sure that the booster battery is the same voltage as the discharged one in the vehicle.

D) If the battery is being jump-started from the battery in another vehicle, the two vehcles MUST NOT TOUCH each other.

E) Make sure that the transmission is in neutral (or PARK, in the case of automatic transmission).

1 Connect one end of the red jump lead to the positive (+) terminal of the flat battery

2 Connect the other end of the red lead to the positive (+) terminal of the booster battery.

3 Connect one end of the black jump lead to the negative (-) terminal of the booster battery

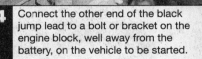

4 Connect the other end of the black jump lead to a bolt or bracket on the engine block, well away from the battery, on the vehicle to be started.

5 Make sure that the jump leads will not come into contact with the fan, drive-belts or other moving parts of the engine.

6 Start the engine using the booster battery, then with the engine running at idle speed, disconnect the jump leads in the reverse order of connection.

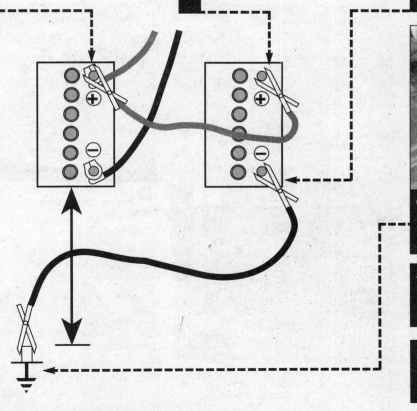

Roadside Repairs

Wheel changing

Some of the details shown here will vary according to model. For instance, the location of the spare wheel and jack is not the same on all cars. However, the basic principles apply to all vehicles.

Warning: *Do not change a wheel in a situation where you risk being hit by other traffic. On busy roads, try to stop in a lay-by or a gateway. Be wary of passing traffic while changing the wheel – it is easy to become distracted by the job in hand.*

Preparation

☐ When a puncture occurs, stop as soon as it is safe to do so.
☐ Park on firm level ground, if possible, and well out of the way of other traffic.
☐ Use hazard warning lights if necessary.
☐ If you have one, use a warning triangle to alert other drivers of your presence.
☐ Apply the handbrake and engage first or reverse gear.
☐ Chock the wheel diagonally opposite the one being removed – a couple of large stones will do for this.
☐ If the ground is soft, use a flat piece of wood to spread the load under the foot of the jack.

Changing the wheel

1 The spare wheel and jack are located beneath the rear floor panel

2 Prise off the wheel trim . . .

3 . . . then loosen all four wheel bolts by half a turn

4 Locate the jacking point closest to the punctured wheel

5 Then locate the jack securely into the jacking point . . .

6 . . . until the jack head is fully engaged as shown

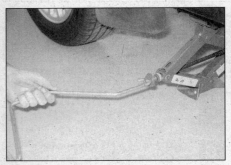

7 Turn the jack handle to raise the car until the wheel is clear of the ground. Remove the wheel bolts, and lift off the wheel.

8 Fit the spare wheel, insert the bolts, and tighten them securely by hand. Lower the car, then tighten the wheel bolts fully.

Finally...

☐ Refit the wheel trim.
☐ Remove the wheel chocks. Stow the jack and tools, and the punctured wheel, in the correct locations in the car.
☐ Check the tyre pressure on the wheel just fitted. If it is low, or if you don't have a pressure gauge with you, drive slowly to the nearest garage and inflate the tyre to the correct pressure.
☐ If a compact spare wheel has been fitted, drive at reduced speed. Have the damaged tyre or wheel repaired as soon as possible.

Roadside Repairs 0•9

Identifying leaks

Puddles on the garage floor or drive, or obvious wetness under the bonnet or underneath the car, suggest a leak that needs investigating. It can sometimes be difficult to decide where the leak is coming from, especially if the engine bay is very dirty already. Leaking oil or fluid can also be blown rearwards by the passage of air under the car, giving a false impression of where the problem lies.

 Warning: Most automotive oils and fluids are poisonous. Wash them off skin, and change out of contaminated clothing, without delay.

HAYNES HINT The smell of a fluid leaking from the car may provide a clue to what's leaking. Some fluids are distictively coloured. It may help to clean the car carefully and to park it over some clean paper overnight as an aid to locating the source of the leak.
Remember that some leaks may only occur while the engine is running.

Sump oil

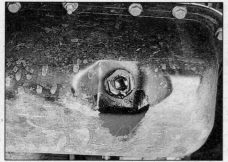

Engine oil may leak from the drain plug...

Oil from filter

...or from the base of the oil filter.

Gearbox oil

Gearbox oil can leak from the seals at the inboard ends of the driveshafts.

Antifreeze

Leaking antifreeze often leaves a crystalline deposit like this.

Brake fluid

A leak occurring at a wheel is almost certainly brake fluid.

Power steering fluid

Power steering fluid may leak from the pipe connectors on the steering rack.

Towing

When all else fails, you may find yourself having to get a tow home – or of course you may be helping somebody else. Long-distance recovery should only be done by a garage or breakdown service. For shorter distances, DIY towing using another car is easy enough, but observe the following points:

☐ Use a proper tow-rope – they are not expensive. The vehicle being towed must display an 'ON TOW' sign in its rear window.
☐ Always turn the ignition key to the 'on' position when the vehicle is being towed, so that the steering lock is released, and that the direction indicator and brake lights will work.
☐ Only attach the tow-rope to the towing eyes provided.
☐ Before being towed, release the handbrake and select neutral on the transmission.
☐ Note that greater-than-usual pedal pressure will be required to operate the brakes, since the vacuum servo unit is only operational with the engine running.
☐ On models with power steering, greater-than-usual steering effort will also be required.
☐ The driver of the car being towed must keep the tow-rope taut at all times to avoid snatching.
☐ Make sure that both drivers know the route before setting off.
☐ Only drive at moderate speeds and keep the distance towed to a minimum. Drive smoothly and allow plenty of time for slowing down at junctions.
☐ On models with automatic transmission, special precautions apply. If in doubt, do not tow, or transmission damage may result.

0•10 Weekly Checks

Introduction

There are some very simple checks which need only take a few minutes to carry out, but which could save you a lot of inconvenience and expense.

These "Weekly checks" require no great skill or special tools, and the small amount of time they take to perform could prove to be very well spent, for example;

☐ Keeping an eye on tyre condition and pressures, will not only help to stop them wearing out prematurely, but could also save your life.

☐ Many breakdowns are caused by electrical problems. Battery-related faults are particularly common, and a quick check on a regular basis will often prevent the majority of these.

☐ If your car develops a brake fluid leak, the first time you might know about it is when your brakes don't work properly. Checking the level regularly will give advance warning of this kind of problem.

☐ If the oil or coolant levels run low, the cost of repairing any engine damage will be far greater than fixing the leak, for example.

Underbonnet check points

◀ **Early models**

A *Engine oil level dipstick*
B *Engine oil filler cap*
C *Coolant expansion tank*
D *Brake fluid reservoir*
E *Screen washer fluid reservoir*
F *Power steering fluid reservoir*
G *Battery*

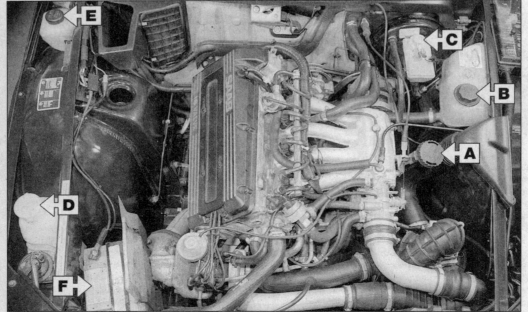

◀ **Later models**

A *Engine filler cap and oil level dipstick*
B *Coolant expansion tank*
C *Brake fluid reservoir*
D *Screen washer fluid reservoir*
E *Power steering fluid reservoir*
F *Battery*

Weekly Checks 0•11

Engine oil level

Before you start

✔ Make sure that your car is on level ground.
✔ Check the oil level before the car is driven, or at least 5 minutes after the engine has been switched off.

 HAYNES HiNT *If the oil is checked immediately after driving the vehicle, some of the oil will remain in the upper engine components, resulting in an inaccurate reading on the dipstick!*

The correct oil
Modern engines place great demands on their oil. It is very important that the correct oil for your car is used (See "Lubricants and Fluids" on page 0•17).

Car Care

● If you have to add oil frequently, you should check whether you have any oil leaks. Place some clean paper under the car overnight, and check for stains in the morning. If there are no leaks, the engine may be burning oil (see "Fault Finding").

● Always maintain the level between the upper and lower dipstick marks (see photo 3). If the level is too low severe engine damage may occur. Oil seal failure may result if the engine is overfilled by adding too much oil.

1 The oil filler/dipstick top location varies according to model (see "Underbonnet check points" on page 0•10 for exact location). Withdraw the dipstick.

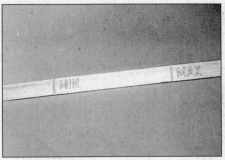
3 Note the oil level on the end of the dipstick, which should be between the upper ("MAX") mark and lower ("MIN") mark. Approximately 1.0 litre of oil will raise the level from the lower mark to the upper mark.

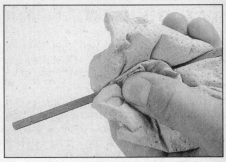
2 Using a clean rag or paper towel remove all oil from the dipstick. Insert the clean dipstick into the tube as far as it will go, then withdraw it again.

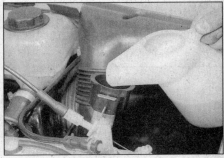
4 Oil is added through the filler cap. Unscrew the cap and top-up the level; a funnel may help to reduce spillage. Add the oil slowly, checking the level on the dipstick often. Don't overfill (see "Car Care" left).

Coolant level

 Warning: *DO NOT attempt to remove the expansion tank pressure cap when the engine is hot, as there is a very great risk of scalding. Do not leave open containers of coolant about, as it is poisonous.*

Car Care

● With a sealed-type cooling system, adding coolant should not be necessary on a regular basis. If frequent topping-up is required, it is likely there is a leak. Check the radiator, all hoses and joint faces for signs of staining or wetness, and rectify as necessary.

● It is important that antifreeze is used in the cooling system all year round, not just during the winter months. Don't top-up with water alone, as the antifreeze will become too diluted.

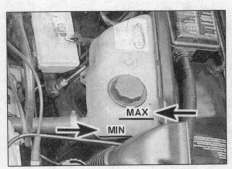
1 The coolant level varies with the temperature of the engine. When the engine is cold, the coolant level should be between the two marks. When the engine is hot, the level may rise slightly above the "MAX" mark.

2 If topping-up is necessary, **wait until the engine is cold**. Turn the expansion tank cap anti-clockwise until it reaches the first stop. Once any pressure is released, push the cap down, turn it anti-clockwise to the second stop and lift it off.

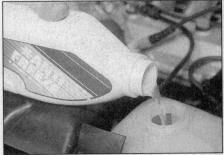
3 Add a mixture of water and antifreeze through the expansion tank filler neck until the coolant is halfway between the two level marks. Refit the cap, turning it clockwise as far as it will go until it is secure.

0•12 Weekly Checks

Screen washer fluid level

Screenwash additives not only keep the winscreen clean during foul weather, they also prevent the washer system freezing in cold weather - which is when you are likely to need it most.

Don't top up using plain water as the screenwash will become too diluted, and will freeze during cold weather. On no account use coolant antifreeze in the washer system - this could discolour or damage paintwork.

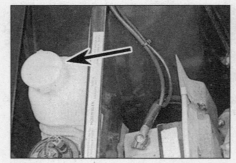

1 The windscreen/tailgate washer fluid reservoir filler is located at the front right-hand side of the engine compartment.

2 When topping-up the reservoir(s), a screenwash additive should be added in the quantities recommended on the bottle.

Brake and clutch fluid level

Warning:

● Brake hydraulic fluid can harm your eyes and damage painted surfaces, so use extreme caution when handling and pouring it.

● Do not use fluid that has been standing open for some time, as it absorbs moisture from the air which can cause a dangerous loss of braking effectiveness.

 ● Make sure that your car is on level ground.
● The fluid level in the master cylinder reservoir will drop slightly as the brake pads wear down, but the fluid level must never be allowed to drop below the 'MIN' mark.

Safety first

● If the reservoir requires repeated topping-up this is an indication of a fluid leak somewhere in the system, which should be investigated immediately.

● If a leak is suspected, the car should not be driven until the braking system has been checked. Never take risks with brakes.

1 The brake and clutch fluid reservoir is located at the rear left-hand side of the engine compartment. The MAX and MIN marks are indicated on the side of the reservoir. The fluid level must be kept between the marks.

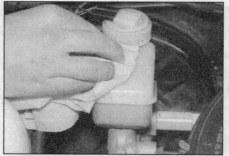

2 If topping-up is necessary, first wipe the area around the filler cap with a clean rag before removing the cap. When adding fluid, it's a good idea to inspect the reservoir. The system should be drained and refilled if dirt is seen in the fluid (see Chapter 9 for details).

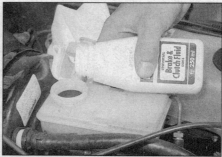

3 Carefully add fluid, avoiding spilling it on surrounding paintwork. Use only the specified hydraulic fluid; mixing different types of fluid can cause damage to the system. After filling to the correct level, refit the cap securely, to prevent leaks and the entry of foreign matter. Wipe off any spilt fluid.

Weekly Checks 0•13

Tyre condition and pressure

It is very important that tyres are in good condition, and at the correct pressure - having a tyre failure at any speed is highly dangerous. Tyre wear is influenced by driving style - harsh braking and acceleration, or fast cornering, will all produce more rapid tyre wear. As a general rule, the front tyres wear out faster than the rears. Interchanging the tyres from front to rear ("rotating" the tyres) may result in more even wear. However, if this is completely effective, you may have the expense of replacing all four tyres at once!

Remove any nails or stones embedded in the tread before they penetrate the tyre to cause deflation. If removal of a nail does reveal that the tyre has been punctured, refit the nail so that its point of penetration is marked. Then immediately change the wheel, and have the tyre repaired by a tyre dealer.

Regularly check the tyres for damage in the form of cuts or bulges, especially in the sidewalls. Periodically remove the wheels, and clean any dirt or mud from the inside and outside surfaces. Examine the wheel rims for signs of rusting, corrosion or other damage. Light alloy wheels are easily damaged by "kerbing" whilst parking; steel wheels may also become dented or buckled. A new wheel is very often the only way to overcome severe damage.

New tyres should be balanced when they are fitted, but it may become necessary to re-balance them as they wear, or if the balance weights fitted to the wheel rim should fall off. Unbalanced tyres will wear more quickly, as will the steering and suspension components. Wheel imbalance is normally signified by vibration, particularly at a certain speed (typically around 50 mph). If this vibration is felt only through the steering, then it is likely that just the front wheels need balancing. If, however, the vibration is felt through the whole car, the rear wheels could be out of balance. Wheel balancing should be carried out by a tyre dealer or garage.

1 Tread Depth - visual check
The original tyres have tread wear safety bands (B), which will appear when the tread depth reaches approximately 1.6 mm. The band positions are indicated by a triangular mark on the tyre sidewall (A).

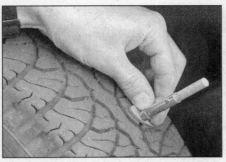

2 Tread Depth - manual check
Alternatively, tread wear can be monitored with a simple, inexpensive device known as a tread depth indicator gauge.

3 Tyre Pressure Check
Check the tyre pressures regularly with the tyres cold. Do not adjust the tyre pressures immediately after the vehicle has been used, or an inaccurate setting will result.

4 Tyre tread wear patterns

Shoulder Wear

Underinflation (wear on both sides)
Under-inflation will cause overheating of the tyre, because the tyre will flex too much, and the tread will not sit correctly on the road surface. This will cause a loss of grip and excessive wear, not to mention the danger of sudden tyre failure due to heat build-up.
Check and adjust pressures
Incorrect wheel camber (wear on one side)
Repair or renew suspension parts
Hard cornering
Reduce speed!

Centre Wear

Overinflation
Over-inflation will cause rapid wear of the centre part of the tyre tread, coupled with reduced grip, harsher ride, and the danger of shock damage occurring in the tyre casing.
Check and adjust pressures

If you sometimes have to inflate your car's tyres to the higher pressures specified for maximum load or sustained high speed, don't forget to reduce the pressures to normal afterwards.

Uneven Wear

Front tyres may wear unevenly as a result of wheel misalignment. Most tyre dealers and garages can check and adjust the wheel alignment (or "tracking") for a modest charge.
Incorrect camber or castor
Repair or renew suspension parts
Malfunctioning suspension
Repair or renew suspension parts
Unbalanced wheel
Balance tyres
Incorrect toe setting
Adjust front wheel alignment
Note: *The feathered edge of the tread which typifies toe wear is best checked by feel.*

0•14 Weekly Checks

Power steering fluid level

Before you start:
- ✓ Park the vehicle on level ground.
- ✓ Set the steering wheel pointing straight-ahead.
- ✓ The engine should be turned off.

 HAYNES HiNT *For the check to be accurate, the steering must not be turned once the engine has been stopped.*

Safety First:
- ● The need for frequent topping-up indicates a leak, which should be investigated immediately.

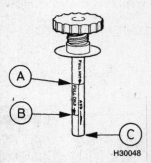

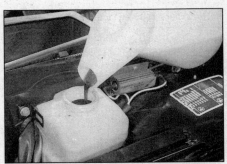

1 The power steering fluid reservoir is located at the rear right-hand side of the engine compartment. To check the level on early models, first unscrew and remove the filler cap, which incorporates its own dipstick. On later models, the fluid level is marked on the reservoir body, and can be seen without removing the cap.

2 The fluid level markings have the same meaning, whether on a dipstick or on the reservoir body. If the car has just been running and the fluid is hot, the level should be up to the "FULL HOT" mark (A). If the car has not been in recent use, and the fluid is cold, the level should be at or just above the "FULL COLD" mark (B) - this is the minimum permitted fluid level. If the fluid level is in the "ADD" zone, fluid should be added to at least the "FULL COLD" mark without delay.

3 If topping-up is necessary, use only the specified type of fluid (see *"Lubricants and Fluids"*). Take great care not to allow any dirt into the hydraulic system, and do not overfill the reservoir. When the level is correct, refit the cap.

Electrical systems

✓ Check all external lights and the horn. Refer to the appropriate Sections of Chapter 12 for details if any of the circuits are found to be inoperative.

✓ Visually check all accessible wiring connectors, harnesses and retaining clips for security, and for signs of chafing or damage.

 HAYNES HiNT *If you need to check your brake lights and indicators unaided, back up to a wall or garage door and operate the lights. The reflected light should show if they are working properly.*

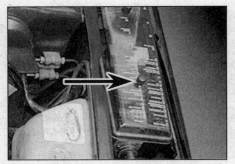

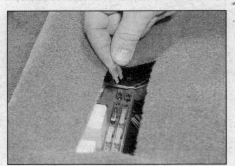

1 If a single indicator light, brake light or headlight has failed, it is likely that a bulb has blown and will need to be replaced. Refer to Chapter 12 for details.
If both brake lights have failed, it is possible that the brake light switch needs adjusting. Refer to Chapter 9 for details.

2 If more than one indicator light or headlight has failed it is likely that either a fuse has blown or that there is a fault in the circuit (see *"Electrical fault-finding"* in Chapter 12).
The fuses are mounted in a lidded box at the left rear corner of the engine compartment. Later models may have additional fuses under the rear seat (see next photo)

3 To replace a blown fuse, simply prise it out. Fit a new fuse of the same rating, available from car accessory shops.
It is important that you find the reason that the fuse blew - a complete checking procedure is given in Chapter 12.

Weekly Checks 0•15

Battery

Caution: *Before carrying out any work on the vehicle battery, read the precautions given in "Safety first" at the start of this manual.*

✔ Make sure that the battery tray is in good condition, and that the clamp is tight. Corrosion on the tray, retaining clamp and the battery itself can be removed with a solution of water and baking soda. Thoroughly rinse all cleaned areas with water. Any metal parts damaged by corrosion should be covered with a zinc-based primer, then painted.

✔ Periodically (approximately every three months), check the charge condition of the battery as described in Chapter 5.

✔ If the battery is flat, and you need to jump start your vehicle, see *"Roadside Repairs"*.

1 The battery is located on the left-hand side of the engine compartment. The exterior of the battery should be inspected periodically for damage such as a cracked case or cover.

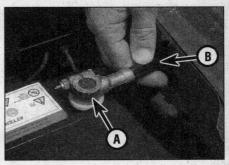

2 Check the tightness of battery clamps (A) to ensure good electrical connections. You should not be able to move them. Also check each cable (B) for cracks and frayed conductors.

HAYNES HiNT

Battery corrosion can be kept to a minimum by applying a layer of petroleum jelly to the clamps and terminals after they are reconnected.

3 If corrosion (white, fluffy deposits) is evident, remove the cables from the battery terminals, clean them with a small wire brush, then refit them. Automotive stores sell a useful tool for cleaning the battery post...

4 ...as well as the battery cable clamps

Wiper blades

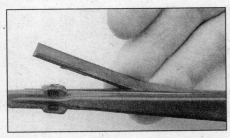

1 Check the condition of the wiper blades; if they are cracked or show any signs of deterioration, or if the glass swept area is smeared, renew them. For maximum clarity of vision, wiper blades should be renewed annually, as a matter of course.

2 To remove a windscreen wiper blade, pull the arm fully away from the glass until it locks. Swivel the blade through 90°, press the locking tab(s) with your fingers, and slide the blade out of the arm's hooked end. Don't forget to check the tailgate wiper blade as well, where applicable.

0•16 Lubricants, fluids and tyre pressures

Lubricants and fluids

Engine	Multigrade engine oil, viscosity SAE 10W/30, 10W/40, or 5W/30, to API SG/CD or SH/CD
Cooling system	Ethylene glycol-based antifreeze
Manual transmission	Multigrade engine oil, viscosity 10W/30 or 10W/40, or gear oil SAE EP75 API-GL4 or API-GL5
Automatic transmission	Automatic transmission fluid to Ford Specification M2C.33G
Automatic transmission final drive	EP oil SAE 80 W or 75 API-GL-4 or APT-GL-5
Driveshaft CV joints	Outer driveshaft: Molycote VN 2461C Inner driveshaft: Mobil grease GS 57C
Brake and clutch hydraulic systems	Brake fluid to DOT 4
Hub/wheel bearings	Multi-purpose lithium based grease
Steering rack	Multi-purpose lithium based grease
Power steering	Saab or Texaco 4634 power steering fluid

Tyre pressures

Tyre size	Up to 3 occupants		Full load	
	Front	Rear	Front	Rear
175/70 R15 87T/86T	31 (2.1)	32 (2.2)	33 (2.3)	35 (2.4)
185/65 R15 87T	29 (2.0)	31 (2.1)	32 (2.2)	33 (2.3)
185/65 R15 87H	29 (2.0)	31 (2.1)	35 (2.4)	36 (2.5)
195/60 R15 86H	31 (2.1)	32 (2.2)	35 (2.4)	36 (2.5)
195/60 VR15 - up to 130 mph	31 (2.1)	32 (2.2)	35 (2.4)	36 (2.5)
195/60 VR15 - up to 130 mph	35 (2.4)	36 (2.5)	-	-
T95/110 R15 (Compact spare wheel)	80 (5.5)			
T115/70 D15 (Compact spare wheel)	61 (4.2)			

Note: *Pressures apply only to original-equipment tyres, and may vary if any other make or type is fitted; check with the tyre manufacturer or supplier for correct pressures if necessary.*

Chapter 1 Part A:
Routine maintenance and servicing - 1979 to 1984 models

Contents

Air cleaner element renewal/cleaning 12/39	Exhaust system check 9
Antifreeze concentration check 6	Facia security check 33
Automatic transmission fluid and filter renewal, and cable adjustment 35	Front wheel alignment check 36
	Fuel filter renewal (fuel injection models) 41
Automatic transmission fluid and final drive oil level check 28	Fuel pump filter cleaning (carburettor models) 4
Auxiliary drivebelt check 20	Headlight beam alignment 26
Battery terminal check 7	Hinge and lock lubrication 10
Brake fluid renewal 43	Hose and fluid leak check 15
Brake line and flexible hose check 30	Idle speed and mixture check and adjustment 16
Brake pad, disc and handbrake check .29	Ignition timing check 24
Choke fast idling speed check (carburettor models) 14	Intensive maintenance 2
Choke mechanism check (carburettor engines) 14	Introduction 1
Contact breaker points check, renewal and adjustment 23	Manual transmission oil level check 27
Coolant renewal 44	Pollen filter renewal 37
Deceleration device (carburettor engines) check and adjustment 17	Power steering fluid level check 32
	Road test 38
Delay valve and fuel enrichment device renewal 34	Spark plug renewal 5/21
Delay valve check and renewal 25	Steering, suspension and shock absorber check 31
Distributor and ignition HT lead check .22	Throttle control lubrication 8
EGR system check 13/40	Turbocharger - check 19
Engine oil and filter renewal 3/11	Valve clearances (B201 engine) - checking and adjustment 18/42

Degrees of difficulty

Easy, suitable for novice with little experience	Fairly easy, suitable for beginner with some experience	Fairly difficult, suitable for competent DIY mechanic	Difficult, suitable for experienced DIY mechanic	Very difficult, suitable for expert DIY or professional

Maintenance & Servicing – '79 to '84 models

Maintenance schedule

The maintenance intervals in this manual are provided with the assumption that you will be carrying out the work yourself. These are the minimum maintenance intervals recommended by the manufacturer for vehicles driven daily. If you wish to keep your vehicle in peak condition at all times, you may wish to perform some of these procedures more often. We encourage frequent maintenance, because it enhances the efficiency, performance and resale value of your vehicle.

If the vehicle is driven in dusty areas, used to tow a trailer, or driven frequently at slow speeds (idling in traffic) or on short journeys, more frequent maintenance intervals are recommended.

When the vehicle is new, it should be serviced by a factory-authorised dealer service department, in order to preserve the factory warranty.

Every 250 miles or Weekly
☐ Refer to *Weekly Checks*.

Every 5000 miles (7500 km)
☐ Change the engine oil - Turbo models - and non-Turbo models in heavy use (Section 3).
☐ Renew the engine oil filter - non-Turbo models in heavy use (Section 3).
☐ Clean the fuel pump filter on carburettor engines (Section 4).
☐ Renew the spark plugs - Turbo models only (Section 5).

Every 10 000 miles (15 000 km)
☐ Check the antifreeze concentration (Section 6).
☐ Check and lubricate the battery terminals (Section 7).
☐ Lubricate the throttle control - but not the cable (Section 8).
☐ Check the exhaust system for leaks and condition (Section 9).
☐ Check the exhaust mounting on the transmission for tightness (Section 9).
☐ Lubricate all door stops, hinges and bonnet locks (Section 10).
☐ Change the engine oil and filter (Section 11).
☐ Clean the air cleaner element - non-Turbo models (Section 12).
☐ Change the air cleaner element - Turbo models (Section 12).
☐ Check the EGR system (Section 13).
☐ Clean the EGR system - B20 engines (Section 13).
☐ Check choke fast idling setting - H engine (Section 14).
☐ Check choke mechanism free play - B20 engine (Section 14).
☐ Cooling system pressure test and check hoses (Section 15).
☐ Check all vacuum lines and connections (Section 15).
☐ Check fuel lines in the engine compartment (Section 15).
☐ Check twin carburettor synchronisation (Section 16).
☐ Check and adjust the idling speed and CO content (Section 16).

Every 10 000 miles (15 000 km) - continued
☐ Check and top up the carburettor damper (Section 16).
☐ Check and adjust the deceleration valve (Section 17).
☐ Check and adjust the valve clearances - Turbo models only (Section 18).
☐ Check and adjust the Turbo charge pressure (Section 19).
☐ Check the Turbo overpressure switch (Section 19).
☐ Check the Turbo charge pressure regulator seal (Section 19).
☐ Check the Turbo APC control unit security seal - where applicable (Section 19).
☐ Check the Turbo fuel boosting device (Section 19).
☐ Check the auxiliary drivebelt for condition and tension (Section 20).
☐ Renew the spark plugs (Section 21).
☐ Distributor and ignition HT lead check (Section 22).
☐ Renew the contact breaker points. Lubricate the rotor arm and lubricating pad (Section 23).
☐ Check/adjust ignition timing (Section 24).
☐ Check the operation of delay valve (Section 25).
☐ Check/adjust headlight main beam (Section 26).
☐ Check and top up the manual transmission oil level (Section 27).
☐ Check and top up the automatic transmission fluid level (Section 28).
☐ Check and top up the automatic transmission final drive oil level (Section 28).
☐ Check all brake pads for wear - wheels removed (Section 29).
☐ Lubricate the front wheel brake yokes (Section 29).
☐ Check the handbrake lever operation on the front calipers (Section 29).
☐ Check all brake pipes and hoses (Section 30).
☐ Check all rubber boots and bellows on universal joints, ball joints and steering joints (Section 31).
☐ Check steering gear inner joints for wear (Section 31).
☐ Check steering track rod ends for wear (Section 31).
☐ Check and top up the power steering fluid level (Section 32).
☐ Check the facia for security (Section 33).

Maintenance & Servicing - '79 to '84 models

Every 20 000 miles (30 000 km)
- [ ] Renew the delay valve and fuel enrichment device - Turbo and non-Turbo models (Section 34).
- [ ] Check the automatic transmission, change the oil, clean the filter, and adjust gear selector and kickdown cables (Section 35).
- [ ] Check and adjust all front wheel alignment angles (Section 36).
- [ ] Renew the pollen filter (Section 37).
- [ ] Road test (Section 38).

Every 30 000 miles (45 000 km)
- [ ] Renew the air cleaner element (Section 39).
- [ ] Clean the EGR system - H engines (Section 40).
- [ ] Renew the fuel filter - fuel injection engines (Section 41).
- [ ] Check and adjust the valve clearances (Section 42).
- [ ] Renew the brake fluid (every 2 years maximum) (Section 43).

Every 2 years (regardless of mileage)
- [ ] Renew the coolant/antifreeze (Section 44).

Underbonnet view of Saab 900 with B201 Bosch CI fuel injection engine

1. Chassis number plate
2. Interior air filter location
3. Power steering pump
4. Engine oil filler cap
5. Distributor
6. Thermostat housing
7. Injector
8. Water pump
9. Inlet manifold
10. Brake/clutch fluid reservoir
11. Coolant expansion tank
12. Fuse box
13. Fuel filter
14. Fuel distributor
15. Transmission and primary gear housing
16. Ignition coil
17. Warm-up regulator
18. Electric cooling fan
19. Battery
20. Washer fluid reservoir

1A•4 Maintenance & Servicing – '79 to '84 models

Front underbody view of Saáb 900 with B201 Bosch CI fuel injection engine

1 Exhaust system
2 Steering track rod
3 Shock absorber
4 Lower control arm
5 Brake caliper
6 Crossmember
7 Engine oil drain plug
8 Transmission oil drain plug
9 Steering gear
10 Jacking point

Rear underbody view of Saab 900 with B201 Bosch CI fuel injection engine

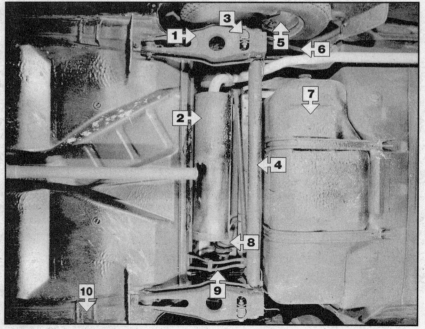

1 Trailing arm
2 Exhaust system
3 Shock absorber
4 Rear axle tube
5 Brake caliper
6 Link
7 Fuel tank
8 Fuel accumulator
9 Brake hydraulic hoses
10 Jacking point

Every 5000 miles (7500 km) - '79 to '84 models

Maintenance procedures

1 Introduction

This Chapter is designed to help the home mechanic maintain his/her vehicle for safety, economy, long life and peak performance.

The Chapter contains a master maintenance schedule, followed by Sections dealing specifically with each task in the schedule. Visual checks, adjustments, component renewal and other helpful items are included. Refer to the accompanying illustrations of the engine compartment and the underside of the vehicle for the locations of the various components.

Servicing your vehicle in accordance with the mileage/time maintenance schedule and the following Sections will provide a planned maintenance programme, which should result in a long and reliable service life. This is a comprehensive plan, so maintaining some items but not others at the specified service intervals, will not produce the same results.

As you service your vehicle, you will discover that many of the procedures can - and should - be grouped together, because of the particular procedure being performed, or because of the close proximity of two otherwise-unrelated components to one another. For example, if the vehicle is raised for any reason, the exhaust can be inspected at the same time as the suspension and steering components.

The first step in this maintenance programme is to prepare yourself before the actual work begins. Read through all the Sections relevant to the work to be carried out, then make a list and gather together all the parts and tools required. If a problem is encountered, seek advice from a parts specialist, or a dealer service department.

2 Intensive maintenance

If, from the time the vehicle is new, the routine maintenance schedule is followed closely, and frequent checks are made of fluid levels and high-wear items, as suggested throughout this manual, the engine will be kept in relatively good running condition, and the need for additional work will be minimised.

It is possible that there will be times when the engine is running poorly due to the lack of regular maintenance. This is even more likely if a used vehicle, which has not received regular and frequent maintenance checks, is purchased. In such cases, additional work may need to be carried out, outside of the regular maintenance intervals.

If engine wear is suspected, a compression test (refer to the relevant Part of Chapter 2) will provide valuable information regarding the overall performance of the main internal components. Such a test can be used as a basis to decide on the extent of the work to be carried out. If, for example, a compression test indicates serious internal engine wear, conventional maintenance as described in this Chapter will not greatly improve the performance of the engine, and may prove a waste of time and money, unless extensive overhaul work (Chapter 2B) is carried out first.

The following series of operations are those most often required to improve the performance of a generally poor-running engine:

Primary operations

a) Clean, inspect and test the battery (see "Weekly Checks" and Chapter 5).
b) Check all the engine-related fluids (See "Weekly Checks").
c) Check the condition and tension of the auxiliary drivebelt (Section 20).
d) Renew the spark plugs (Section 5).
e) Inspect the distributor cap, rotor arm and HT leads - as applicable (Section 22).
f) Check the condition of the air cleaner filter element, and renew if necessary (Section 12).
g) Clean/renew the fuel filter (Section 4 or 41).
h) Check the condition of all hoses, and check for fluid leaks (Section 15).
i) Check the idle speed and mixture settings - as applicable (Section 16).

If the above operations do not prove fully effective, carry out the following secondary operations:

Secondary operations

a) Check the charging system (Chapter 5).
b) Check the ignition system (Chapter 5).
c) Check the fuel system (Chapter 4).
d) Renew the distributor cap and rotor arm - as applicable (Chapter 5).
f) Renew the ignition HT leads - as applicable (Chapter 5).

Every 5000 miles (7500 km)

3 Engine oil and filter renewal

Note: *Owners of high-mileage vehicles, or those who do a lot of stop-start driving, may prefer to carry out engine oil and filter renewal more frequently.*

1 Frequent oil and filter changes are the most important preventative maintenance procedures which can be undertaken by the DIY owner. As engine oil ages, it becomes diluted and contaminated, which leads to premature engine wear.
2 Before starting this procedure, gather together all the necessary tools and materials. Also make sure that you have plenty of clean rags and newspapers handy, to mop up any spills. Ideally, the engine oil should be warm, as it will drain better, and more built-up sludge will be removed with it. Take care, however, not to touch the exhaust or any other hot parts of the engine when working under the vehicle. To avoid any possibility of scalding, and to protect yourself from possible skin irritants and other harmful contaminants in used engine oils, it is advisable to wear gloves when carrying out this work. Access to the underside of the vehicle will be greatly improved if it can be raised on a lift, driven onto ramps, or jacked up and supported on axle stands (see *"Jacking and Vehicle Support"*). Whichever method is chosen, make sure that the vehicle remains level, or if it is at an angle, that the drain plug is at the lowest point. **Do not** confuse the transmission drain plug with the engine drain plug - the engine drain plug is towards the front left-hand side of the transmission.

3 Slacken the drain plug about half a turn **(see illustration)**. Position the draining container under the drain plug, then remove the plug completely **(see Haynes Hint overleaf)**. Recover the sealing washer from the drain plug.

3.3 Unscrewing engine oil drain plug

1A•6 Every 5000 miles (7500 km) – '79 to '84 models

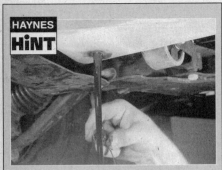

If possible, try to keep the plug pressed into the transmission casing while unscrewing it by hand the last couple of turns. As the plug releases from the threads, move it away sharply so the stream of oil runs into the container, not up your sleeve!

3.6 Oil filter mounting on B201 engine

3.7 Removing the oil filter from a B202 engine

4 Allow some time for the old oil to drain, noting that it may be necessary to reposition the container as the oil flow slows to a trickle.
5 After all the oil has drained, wipe off the drain plug with a clean rag. Check the sealing washer for condition and renew it if necessary. Clean the area around the drain plug opening, and refit the plug. Tighten the plug to the specified torque.
6 Move the container into position under the oil filter. The filter is located on an adaptor housing bolted to the left-hand front of the cylinder block on B201 engines and further back on the left-hand side just in front of the alternator on B202 engines **(see illustration)**.
7 Using an oil filter removal tool if necessary, slacken the filter initially, then unscrew it by hand the rest of the way **(see illustration)**. Empty the oil from the old filter into the container and discard the filter.
8 Use a clean rag to remove all oil, dirt and sludge from the filter sealing area. Check the old filter to make sure that the rubber sealing ring hasn't stuck to the engine. If it has, carefully remove it.
9 Apply a light coating of clean engine oil to the sealing ring on the new filter, then screw it into position on the engine until it just touches the adaptor housing. Now tighten the filter a further half a turn. **Do not** use any tools to tighten the filter since overtightening will distort the seal. Wipe clean the filter and drain plug.
10 Remove the old oil and all tools from under the car then lower the car to the ground.
11 Remove the oil filler cap and fill the engine with the correct grade and type of oil (see Specifications at the end of this Chapter). An oil can spout or funnel may help to reduce spillage. Pour in half the specified quantity of oil first, then wait a few minutes for the oil to drain to the bottom of the engine. Continue adding oil a small quantity at a time until the level is up to the lower mark on the dipstick. Adding a further 1.0 litre will bring the level up to the upper mark on the dipstick. Refit the filler cap.
12 Start the engine and run it for a few minutes; check for leaks around the oil filter seal and the drain plug. Note that there may be a delay of a few seconds before the oil pressure warning light goes out when the engine is first started, as the oil circulates through the engine oil galleries and the new oil filter before the pressure builds up.
13 Switch off the engine, and wait a few minutes for the oil to settle once more. With the new oil circulated and the filter completely full, recheck the level on the dipstick, and add more oil as necessary.
14 Dispose of the used engine oil safely, with reference to "General repair procedures" in the reference Sections of this manual.

4 Fuel pump filter cleaning (carburettor models)

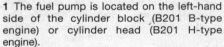

1 The fuel pump is located on the left-hand side of the cylinder block (B201 B-type engine) or cylinder head (B201 H-type engine).
2 Mark the fuel pump cover in relation to the body then remove the central screw and lift off the cover.
3 Remove the gauze filter and rubber seal and clean them in fuel. Also clean the cover and body. Check the seal for condition and renew it if necessary.
4 Refit the gauze filter and cover together with the seal, and insert and tighten the screw.

5 Spark plug renewal

1 The correct functioning of the spark plugs is vital for the correct running and efficiency of the engine. It is essential that the plugs fitted are appropriate for the engine (a suitable type is specified at the end of this Chapter). If this type is used and the engine is in good condition, the spark plugs should not need attention between scheduled replacement intervals. Spark plug cleaning is rarely necessary, and should not be attempted unless specialised equipment is available, as damage can be caused to the firing ends.
2 To remove the spark plugs, first on B202 engines remove the screws and lift the inspection cover from the centre of the camshaft cover **(see illustrations)**.
3 If the marks on the spark plug (HT) leads cannot be seen, mark the leads '1' to '4', to correspond to the cylinder the lead serves (No 1 cylinder is at the timing chain end of the engine) **(see illustration)**.
4 Pull the leads from the plugs by gripping the end fitting, not the lead, otherwise the lead connection may be fractured. On B202 engines, release the leads from the retainers which are located in the valve cover slots **(see illustrations)**.
5 It is advisable to remove the dirt from the spark plug recesses using a clean brush, vacuum cleaner or compressed air before removing the plugs, to prevent dirt dropping into the cylinders.

5.2a Remove the inspection cover for access to the spark plugs

5.2b HT spark plug leads with the inspection cover removed

Every 5000 miles (7500 km) - '79 to '84 models

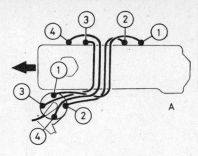
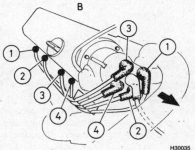

5.3 Showing spark plug HT lead positions
A B-type engine
B H-type engine
Arrow indicates flywheel end of the engine

5.4a Releasing the HT leads from the retainers located in the valve cover slots

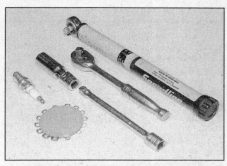

5.6a Tools required for spark plug removal, gap adjustment and refitting

5.4b Disconnecting the HT leads from the plugs

5.6b Removing a spark plug

6 Unscrew the plugs using a spark plug spanner, suitable box spanner or a deep socket and extension bar **(see illustrations)**. Keep the socket aligned with the spark plug - if it is forcibly moved to one side, the ceramic insulator may be broken off. As each plug is removed, examine it as follows.

7 Examination of the spark plugs will give a good indication of the condition of the engine. If the insulator nose of the spark plug is clean and white, with no deposits, this is indicative of a weak mixture or too hot a plug (a hot plug transfers heat away from the electrode slowly, a cold plug transfers heat away quickly).

8 If the tip and insulator nose are covered with hard black-looking deposits, then this is indicative that the mixture is too rich. Should the plug be black and oily, then it is likely that the engine is fairly worn, as well as the mixture being too rich.

9 If the insulator nose is covered with light tan to greyish-brown deposits, then the mixture is correct and it is likely that the engine is in good condition.

10 The spark plug electrode gap is of considerable importance as, if it is too large or too small, the size of the spark and its efficiency will be seriously impaired. The gap should be set to the value given in the Specifications at the end of this Chapter.

11 To set the gap, measure it with a feeler blade or wire gauge and then bend open, or close, the outer plug electrode until the correct gap is achieved **(see illustrations)**. The centre electrode should never be bent, as this will crack the insulator and cause plug failure, if nothing worse. If using feeler blades, the gap is correct when the appropriate-size blade is a firm sliding fit.

12 Special spark plug electrode gap adjusting tools are available from most motor accessory shops, or from some spark plug manufacturers.

13 Before fitting the spark plugs, check that the threaded connector sleeves are tight and the threads are clean. It is often difficult to insert spark plugs into their holes without cross-threading them. To avoid this possibility, fit a short length of 5/16 inch internal diameter rubber hose over the end of the spark plug **(see Haynes Hint)**.

HAYNES HINT

To avoid cross-threading the plugs, fit a short length of rubber hose over the end of the spark plug. The flexible hose acts as a universal joint to help align the plug with the plug hole. Should the plug begin to cross-thread, the hose will slip on the spark plug, preventing thread damage to the aluminium cylinder head.

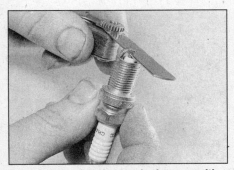

5.11a Measuring the spark plug gap with a feeler blade . . .

5.11b . . . and adjusting the gap using a special adjusting tool

1A•8 Every 5000 miles (7500 km) - '79 to '84 models

14 Remove the rubber hose, and tighten the plug to the specified torque using the spark plug socket and a torque wrench **(see illustration)**. Fit the remaining spark plugs in the same manner.
15 Connect the HT leads in their correct order.
16 On the B202 engine refit the inspection cover and tighten the retaining screws.

5.14 Tightening a spark plug

Every 10 000 miles (15 000 km)

6 Antifreeze concentration check

1 The antifreeze should always be maintained at the specified concentration. This is necessary not only to maintain the antifreeze properties, but also to prevent corrosion which would otherwise occur as the corrosion inhibitors become progressively less effective.
2 The check should be made with the engine cold, and it will be necessary to obtain an antifreeze tester from a car accessory shop.
3 Slowly unscrew the cap from the top of the coolant expansion tank, then draw coolant into the tester **(see illustration)**. Check the concentration of the antifreeze according to the tester manufacturer's instructions. The most common tester consists of three coloured balls of varying density - a high concentration will cause all three balls to float whereas a low concentration may cause only one ball to float.
4 If the concentration is incorrect, slight adjustments may be made by drawing some of the coolant out of the expansion tank and replacing it with undiluted antifreeze. If the concentration is excessively out, it will be necessary to completely drain the system and renew the solution - see Section 44.
5 Tighten the cap onto the expansion tank on completion.

7 Battery terminal check

Note: *Before carrying out any work on the vehicle battery, read through the precautions given in "Safety first!" at the beginning of this manual.*
1 The battery is located on the right-hand side of the engine compartment. The exterior of the battery should be inspected periodically for damage such as a cracked case or cover.
2 Check the tightness of the battery cable clamps to ensure good electrical connections, and check the entire length of each cable for cracks and frayed conductors **(see illustration)**. Check the positive cable between the battery and the starter motor.
3 If corrosion (visible as white, fluffy deposits) is evident, remove the cables from the battery terminals, clean them with a small wire brush, then refit them. Corrosion can be kept to a minimum by applying a layer of petroleum jelly to the clamps and terminals after they are reconnected.
4 Make sure that the battery retaining clamp is secure and the nuts are tight.
5 Corrosion on the retaining clamp and the battery terminals can be removed with a solution of water and baking soda. Thoroughly rinse all cleaned areas with plain water.
6 Any metal parts of the vehicle damaged by corrosion should be covered with a zinc-based primer, then painted.
7 Periodically (approximately every three months), check the charge condition of the battery as described in Chapter 5.
8 Further information on the battery, charging and jump starting can be found in Chapter 5 and in the preliminary sections of this manual.

8 Throttle control lubrication

1 Apply a little engine oil to the accelerator pedal and to the throttle control at the carburettor (carburettor models) or throttle valve housing (fuel injection models). **Do not** lubricate the accelerator cable.

9 Exhaust system check

 Warning: *If the engine has just been running, take care not to touch the exhaust system, especially the front section, as it may still be hot.*

1 Position the car over an inspection pit or on car ramps. Alternatively raise the front and rear of the car and support on axle stands (see *"Jacking and Vehicle Support"*).
2 Examine the exhaust system over its entire length checking for any damaged, broken or missing mountings **(see illustration)**, the security of the pipe retaining clamps, and the

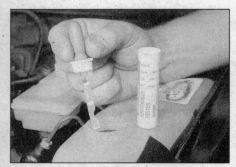

6.3 Checking the antifreeze concentration with a special tester

7.2 Check the tightness of the battery cable clamps

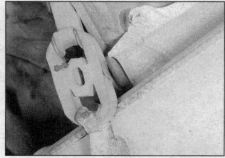

9.2 Typical exhaust rubber mounting

Every 10 000 miles (15 000 km) - '79 to '84 models

12.2a Air cleaner and cover on a B201 fuel injection engine

12.2b Removing the air cleaner element on a B201 fuel injection engine

12.2c Releasing the air cleaner cover clips on a B202 fuel injection engine

12.2d Removing the cover . . .

12.2e . . . and filter element on a B202 fuel injection engine

12.2f The insert locates on the bottom of the air cleaner filter element

condition of the system with regard to rust and corrosion.
3 Check the exhaust mounting on the transmission for tightness.
4 Lower the vehicle to the ground on completion.

10 Hinge and lock lubrication

1 Work around the vehicle, clean and then lubricate the hinges of the doors and tailgate with a light machine oil.
2 Lubricate the bonnet release mechanism, hinges, guide rails and safety locks with a smear of petroleum jelly.
3 Check carefully the security and operation of all hinges, latches and locks, adjusting them where required. Check the operation of the central locking system (if fitted).
4 Check the condition and operation of the tailgate struts, renewing them if either is leaking or no longer able to support the tailgate securely when raised.
5 On completion check the operation of all door locks, tailgate/boot locks and the fuel filler flap. Check that the child safety catches on the rear doors operate correctly.

11 Engine oil and filter renewal

Refer to the information given in Section 3.

12 Air cleaner element renewal/cleaning

1 The air cleaner is located on the left-hand side of the engine compartment.
2 To remove the element either release the clips or remove the screws from the air cleaner cover, move the cover to one side, and withdraw the element. Where fitted, remove the insert from the bottom of the body (see illustrations).
3 If cleaning the element, tap it to release the accumulated dust and if available use an air line from the inside of the element outwards. Do not attempt to wash the element.
4 Wipe clean the inside surfaces of the air cleaner and insert.
5 Refit the element (and insert where necessary) using a reversal of the removal procedure, but check all associated hoses and air ducts for condition and security.

13 EGR system check

Refer to Chapter 4, Part C, Section 2.

14 Choke mechanism check (carburettor engines)

1 On twin carburettor models check that both choke controls touch their stops at the same time when the choke is operated. If not, adjust the spindle linkages.

2 On B type engines check that the distance between the adjusting screw on the throttle lever and the choke cam is 1.0 mm with the choke knob fully inserted (see illustration). Where twin carburettors are fitted check the front carburettor only.
3 On H type engines check the fast idling speed with the engine warm by inserting an 8.0 mm twist drill behind the choke cam. Check that with the engine idling, the fast idle speed is as given in the Specifications. If not, loosen the locknut and adjust the screw as necessary. Tighten the locknut and remove the drill.
4 If any adjustments have been made, check the idling speed and mixture as described in Section 16, then repeat the checks described in the previous paragraphs.

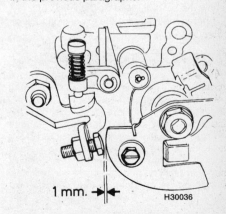

14.2 Basic choke setting clearance on the carburettor on B-type engines

Every 10 000 miles (15 000 km) – '79 to '84 models

16.4 Using the special tool to adjust the mixture on Zenith/Solex carburettors

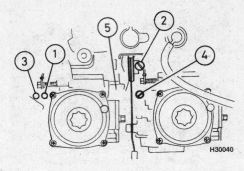

16.7 Zenith/Solex carburettor adjusting screw locations

Pre-1984 models
1. Vent valve (front carburettor)
2. Synchronising adjustment
3. Engine idling speed
4. Vent valve (rear carburettor)

1984 models
1. Redundant tapped hole
2. Synchronising adjustment
3. Vent valve (front carburettor)
4. Vent valve (rear carburettor)
5. Engine idling speed

15 Hose and fluid leak check

1 Visually inspect the engine joint faces, gaskets and seals for any signs of water or oil leaks. Pay particular attention to the areas around the camshaft cover, cylinder head, oil filter and transmission joint faces. Bear in mind that, over a period of time, some very slight seepage from these areas is to be expected - what you are really looking for is any indication of a serious leak. Should a leak be found, renew the offending gasket or oil seal by referring to the appropriate Chapters in this manual.

2 Also check the security and condition of all the engine-related pipes and hoses including vacuum hoses. Ensure that all cable ties or securing clips are in place and in good condition. Clips which are broken or missing can lead to chafing of the hoses, pipes or wiring, which could cause more serious problems in the future.

3 Carefully check the radiator hoses and heater hoses along their entire length. Renew any hose which is cracked, swollen or deteriorated. Cracks will show up better if the hose is squeezed. Pay close attention to the hose clips that secure the hoses to the cooling system components. Hose clips can pinch and puncture hoses, resulting in cooling system leaks.

4 Inspect all the cooling system components (hoses, joint faces etc.) for leaks. A leak in the cooling system will usually show up as white or rust-coloured deposits on the area adjoining the leak. Where any problems of this nature are found on system components, renew the component or gasket with reference to Chapter 3.

5 Check that the pressure cap on the expansion tank is fully tightened and shows no sign of coolant leakage. If possible have a Saab dealer pressure test the cooling system.

6 With the car raised, inspect the fuel tank and filler neck for punctures, cracks and other damage. The connection between the filler neck and tank is especially critical. Sometimes a rubber filler neck or connecting hose will leak due to loose retaining clamps or deteriorated rubber.

7 Carefully check all rubber hoses and metal fuel lines leading away from the fuel tank. Check for loose connections, deteriorated hoses, crimped lines, and other damage. Pay particular attention to the vent pipes and hoses, which often loop up around the filler neck and can become blocked or crimped. Follow the lines to the front of the vehicle, carefully inspecting them all the way. Renew damaged sections as necessary.

8 From within the engine compartment, check the security of all fuel hose attachments and pipe unions, and inspect the fuel hoses and vacuum hoses for kinks, chafing and deterioration.

16 Idle speed and mixture check and adjustment

Zenith/Solex carburettor

Note: *On the Zenith carburettor if any adjustment is made to the synchronisation, it will also be necessary to adjust the idling speed and rear carburettor vent valve IN THIS ORDER. If any adjustment is made to the idling speed it will also be necessary to adjust the rear carburettor vent valve.*

1 Before checking the idle speed and mixture make sure that the basic choke setting is correct as described in Section 14, and check that the damper oil level is as described in Chapter 4, Part A, Section 11.

Basic setting - metering needle

2 Unscrew the damper and cap.
3 Mark the vacuum chamber cover in relation to the carburettor body then remove the screws and withdraw the cover and spring.
4 Lift out the piston and diaphragm then using Saab tool 83 93 035 adjust the metering needle so that its shoulder is level with the bottom of the piston **(see illustration)**.
5 Refit the piston, diaphragm, and cover together with the spring making sure the diaphragm tab engages the cut-out. Top up the damper oil to within 10.0 mm of the top.

Twin carburettors - synchronisation and idling

6 Run the engine to normal operating temperature and allow it to idle.

7 Adjust the idle speed screw on the front carburettor so that the engine is running at the specified idling speed **(see illustration)**.
8 If an air flow balancer is available check that the air flow through both carburettors is identical. Alternatively a rough check can be made using a short length of plastic tube positioned in each carburettor inlet in turn - the amount of hiss should be identical. If necessary loosen the locknut and adjust the screw on the intermediate linkage. Tighten the locknut afterwards.

CO content (mixture) setting

9 Run the engine to normal operating temperature and make sure that the choke control is fully off.
10 Where the setting speed is 2000 rpm clamp the distributor vacuum hose, and disconnect the crankcase ventilation hoses.
11 Connect a CO meter and tachometer to the engine and check that the readings are as given in the Specifications with the engine idling.
12 If adjustment is necessary unscrew and remove the damper and cap and use tool 83 93 035 to raise or lower the metering needle as required. To use the tool hold the piston and diaphragm stationary with the sleeve and turn the spindle clockwise to richen the mixture or anticlockwise to weaken the mixture **(see illustration)**.
13 Stop the engine and refit the hoses as necessary. Refit the damper and cap. Remove the CO meter and tachometer.

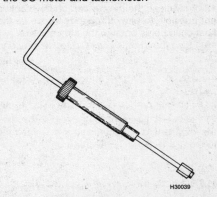

16.12 Carburettor mixture adjusting tool

Every 10 000 miles (15 000 km) – '79 to '84 models

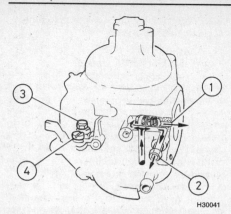

16.17 Carburettor float chamber ventilation on the Zenith carburettor

1 To the air cleaner
2 To the atmosphere
3 Vent valve adjustment (pre-1984 front carburettor)
4 Idle speed adjustment (pre-1984 front carburettor) OR vent valve adjustment (rear carburettor) OR vent valve adjustment (1984 front carburettor)

Vent valve adjustment (Zenith carburettor)

14 Refer to the illustration 16.7. The float chamber vent valve on the front carburettor of twin carburettor models is set by the manufacturers and will not normally require adjustment. If it is altered it will affect the other slow running adjustments.
15 Similarly, the vent valve on the rear carburettor will not normally require adjustment. However, if necessary the following check and adjustment may be made.
16 The vent valve facilitates good starting when the engine is hot by preventing vaporised fuel entering the inlet manifold. On twin carburettor engines it also prevents 'run-on' after the ignition has been switched off.
17 To check the valve first connect a hose to the atmosphere air aperture (see illustration). If the check is to be made with the carburettor removed from the engine and the fuel inlet pipe disconnected, plug the fuel inlet.
18 Fully close the throttle and check that it is not possible to blow through the hose (ie the float chamber is open to the atmosphere).
19 Open the throttle lever 0.5 to 1.0 mm, and check that it is now possible to blow through the hose proving that the valve is half open.
20 Further opening of the throttle should prevent blowing through the hose as the valve will only allow venting to the air cleaner.
21 If necessary adjust the valve by loosening the locknut and turning the screw. Tighten the locknut afterwards. If adjustment has been made recheck the slow running adjustment as described previously.

Pierburg 175 CDUS carburettor

22 The engine should be at its normal operating temperature, the cooling fan having cut in at least once, and the choke fully disengaged.
23 Before checking the idle speed and mixture make sure that the damper oil level is as described in Chapter 4, Part A, Section 11.
24 Connect up a tachometer to the engine in accordance with the manufacturer's instructions, then check that the idle speed is as specified. Adjust if necessary via the idle speed adjustment screw.
25 Idle mixture adjustment should only be carried out if an exhaust gas analyser is available. Detach the crankcase ventilation hose nipple from the camshaft cover and plug the hose. Connect up a temporary fume extraction hose to the camshaft cover and direct it to atmosphere (downstream of the CO meter).
26 A CO extraction hose with an open-type coupling must be used to prevent a vacuum build-up in the exhaust system.
27 Detach the EGR vacuum hose and plug the hose to the EGR valve (where equipped). Detach the vacuum hose from the distributor and plug it. Switch on the driving lights.
28 Start the engine and run it at a steady 2000 rpm. When the radiator cooling fan cuts in, check the CO content reading. If the CO content is not within the specified amount, adjustment is necessary.
29 Prise free the seal plug on the float chamber cover then, using an 8 mm socket, turn the adjuster as necessary (see illustration). Turning clockwise will reduce the CO content. If a swivel type spanner is used to make the adjustment, take care not to short it onto the starter motor or alternator connections.
30 If a CO adjustment was made, the fast idle speed should be checked as follows.
31 With the engine at its normal operating temperature, detach the distributor advance hose and pull out the choke control so that the line on the lever aligns with the fast idle screw. Check the engine speed and compare it with the fast idle speed specified. If adjustment is required, turn the screw in the required direction.
32 Pull out the choke control knob and ensure that the lever at the carburettor fully deflects. Now push the control knob fully in and check that the lever moves back to the

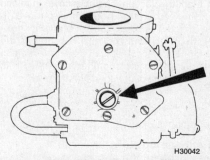

16.29 CO content (mixture) setting screw (arrowed) on the Pierburg carburettor

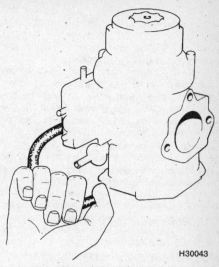

16.33 Checking the modulator valve on the Pierburg carburettor

lowest travel limit and that the fast idle screw is not in contact with it. Adjust if necessary.
33 To check the modulator valve, disconnect its hose and plug it with the engine idling (see illustration). The CO level should rise and the idle speed may fall slightly. If not, the valve is faulty or the hose is leaking.
34 When all checks and adjustments are complete, disconnect the test gear and remake the original connections.
35 Fit a new seal plug to the CO adjuster screw when required. In some EEC countries (though not yet in the UK) this is required by law.

Fuel injection models

Note: The checking of idle speed and mixture is possible on all models by using a tachometer and exhaust gas analyser. However it is not possible to adjust the idling speed and mixture on models with the Lucas CU fuel injection system or on models with idle control and a catalytic converter. On all other models continue as follows.

36 Before checking and adjusting the idle speed or mixture settings, always check the following first:
  a) Check the ignition timing (Section 24).
  b) Check that the spark plugs are in good condition and correctly gapped (Section 5).
  c) Check that the accelerator cable is correctly adjusted (Chapter 4, Part A or B).
  d) Check that the crankcase breather hoses are secure, with no leaks or kinks (Chapter 2, Part A).
  e) Check that the air cleaner filter element is clean (Section 12).
  f) Check that the exhaust system is in good condition (Section 9).
  g) If the engine is running very roughly, check the compression pressures (Chapter 2, Part A).

Every 10 000 miles (15 000 km) – '79 to '84 models

37 The engine should be at normal temperature, the cooling fan having cut in at least once.
Note: *Checking/adjustment should be completed as quickly as possible so that the engine is still at its normal operating temperature. If the radiator electric cooling fan operates again, wait for it to stop before continuing. Clear any excess fuel from the inlet manifold by racing the engine two or three times to between 2000 and 3000 rpm, then allow it to idle again.*
38 Ensure all electrical loads are switched off, then stop the engine and connect a tachometer to it following its manufacturer's instructions. Where a tachometer is fitted to the instrument panel, this may be used instead. If the idle mixture is to be checked, connect an exhaust gas analyser in accordance with its manufacturer's instructions.

Models with Bosch CI (K-Jetronic) fuel injection (except models with catalytic converter and modulating valve)

39 The idle speed adjustment screw is located on the throttle housing. Start the engine and allow it to idle, then check that the idle speed is as given in the Specifications. If adjustment is necessary, loosen the locknut and adjust the screw in to reduce the speed or out to increase the speed. Tighten the locknut after making the adjustment.
40 The idle mixture adjustment screw is located on the top of the airflow meter next to the fuel distributor and a long Allen key will be required to turn it **(see illustration)**. The screw is fitted with a tamperproof plug which must be removed in order to make any adjustments. **Note:** *In some countries adjustment is only allowed by qualified persons and it is also a legal requirement to renew the plug after making an adjustment.*
41 With the engine idling at the correct idle speed, check that the CO level is as given in the Specifications. If adjustment is necessary, use an Allen key or fine screwdriver to turn the mixture adjustment screw in or out (in very small increments) until the CO level is as given in the Specifications. Turning the screw in (clockwise) richens the mixture and increases the CO level, turning it out will weaken the mixture and reduce the CO level.
42 If necessary readjust the idling speed.
43 Temporarily increase the engine speed, then allow it to idle and recheck the settings.
44 When adjustments are complete, stop the engine and disconnect the test equipment. Where necessary fit a new tamperproof plug to the mixture adjustment screw.

Models with Bosch LH-Jetronic fuel injection (except models with idle air control and catalytic converter)

45 The idle speed adjustment screw is located on the throttle housing. Start the engine and allow it to idle, then check that the idle speed is as given in the Specifications. If adjustment is necessary, loosen the locknut and adjust the screw in to reduce the speed or out to increase the speed. Tighten the locknut after making the adjustment.
46 The idle mixture adjustment screw is located on the airflow meter and may be hidden under a tamperproof cap. First remove the cap. With the engine idling at the correct idling speed, check that the CO level is as given in the Specifications. If adjustment is necessary, turn the mixture adjustment screw in or out (in very small increments) until the CO level is as given in the Specifications. Turning the screw in (clockwise) richens the mixture and increases the CO level, turning it out will weaken the mixture and reduce the CO level.
47 If the CO percentage is higher than 6% and it is not possible to adjust it within limits, stop the engine and disconnect the wiring plug from the airflow meter. Connect a digital multimeter across terminals 3 and 6 of the airflow meter, then turn the adjustment screw as necessary to obtain 380 ohms. If it is not possible to do this the airflow meter is faulty and should be renewed, however if it is possible reconnect the wiring and recheck the CO level as described in paragraph 46.
48 If necessary readjust the idling speed.
49 Temporarily increase the engine speed, then allow it to idle and recheck the settings.
50 When adjustments are complete, stop the engine and disconnect the test equipment. Where necessary, fit a new tamperproof plug to the mixture adjustment screw.

Other models with LH-Jetronic and Lucas CU14 systems

51 The idle speed is controlled by an idle air control valve in conjunction with the system ECU. With the engine at normal operating temperature, check that the idle speed is as specified. No adjustment is possible.
52 The idle mixture is controlled by the Lambda sensor in conjunction with the system ECU. With the engine at normal operating temperature, check that the idle mixture is as given in the Specifications. No adjustment is possible.
53 If incorrect readings are obtained, take the car to a Saab dealer and have the system checked with the special tester.

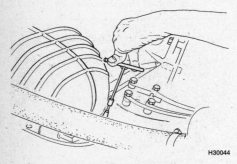

16.40 Adjusting the idling mixture on engines with the Bosch CI (K-Jetronic) fuel injection system

17 Deceleration device (carburettor engines) - check and adjustment

1 The deceleration device assists combustion during engine overrun in order to prevent the emission of unburned hydrocarbons. There are several types of device fitted - prior to 1984 a vacuum-controlled device and a dashpot were fitted, and later models had a disc valve incorporated in the throttle butterfly. An electric type was also fitted.
2 The vacuum type incorporates a spring-loaded valve actuated by inlet manifold vacuum and the device effectively by-passes the throttle valve until the engine speed approaches normal idling speed.
3 The dashpot system prevents the throttle shutting too quickly which would otherwise result in unburnt hydrocarbons.
4 The disc valve system consists of a spring tensioned valve mounted on the throttle valve. When the throttle is shut and the engine is in overrun, the valve allows additional air to by-pass the throttle until the engine speed approaches idling speed.
5 The electric type incorporates an engine speed transmitter and a solenoid operated idling stop. During engine overrun, the idling speed is increased to around 1550 rpm if the speed of the car exceeds 18 mph (30 kph).
6 To check the vacuum controlled type run the engine to normal operating temperature and connect a tachometer. Adjust the idling speed to 875 rpm then rev up the engine to 3000 rpm, release the throttle, and check that the time for the engine to return to idling speed is 4 to 6 seconds. Unscrew the unit screw until the valve closes then adjust the idling speed of the engine in the normal way. Now screw in the unit screw until the engine speed is 1600 rpm and back off the screw two complete turns. Finally adjust the idling speed again. Note that the check should be completed without the radiator fan cutting in.
7 To check the dashpot system, run the engine to normal operating temperature, then connect a tachometer and adjust the idling speed to 875 rpm. Rev up the engine to 3000 rpm, release the throttle, and check that the time for the engine to return to idling speed is 3 to 6 seconds. If not, loosen the locknut and reposition the unit.
8 To check the electric type, connect a tachometer and run the engine to normal operating temperature. Disconnect the positive solenoid cable and connect battery voltage. Rev the engine then release the throttle and check that the idling speed is maintained at 1550 rpm. If necessary adjust the solenoid with the screw. Connect a test lamp between the positive solenoid wire and earth, then drive the car at about 25 mph (40 kph) and declutch. Brake the car and check that the test lamp goes out at a speed of 18.6 mph (30 kph).

Every 10 000 miles (15 000 km) – '79 to '84 models

9 The vacuum valve can be removed by disconnecting the hose and unscrewing the unit from the inlet manifold. The electric speed transmitter can be removed by lowering the facia panel and unplugging the unit. The solenoid can be removed by disconnecting the wiring and unscrewing the unit.

18 Valve clearances (B201 engine) – checking and adjustment

Note: *This Section applies to the B201 engine only since the B202 engine is fitted with hydraulic tappets.*

1 The valve clearances must be adjusted with the engine cold. First remove the cylinder head cover as described in Chapter 2A, Section 4.
2 Turn the engine using a spanner on the crankshaft pulley bolt until No 1 cam lobe peak (timing chain end of the engine) is pointing away from the cam follower (tappet).
3 Using a feeler gauge check that the clearance between the heel of the cam and the cam follower is as given in the Specifications. The feeler blade should be a firm sliding fit **(see illustration)**. If not, record the actual clearance.

Draw the valve positions on a piece of paper and record the actual clearances as you check them.

18.3 Checking the valve clearances

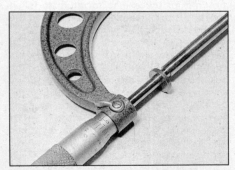
18.5 Using a micrometer to check the cam follower shim thickness

4 Check the remaining clearances using the same procedure and record them as necessary.
5 If adjustment is required remove the camshaft with reference to Chapter 2, Part A, Section 5, then remove the cam followers and shims from the relevant valves, keeping them identified. Measure the thickness of the existing shims and by comparison with the correct clearances determine the thickness of the new shims. Use a micrometer to measure the shims **(see illustration)**.
6 Fit the correct shims followed by the cam followers and camshaft (Chapter 2, Part A, Section 5).
7 Rotate the engine several times then recheck the clearances.
8 Refit the cylinder head cover (see Chapter 2).

19 Turbocharger – check

The following work should be carried out by your Saab dealer as special equipment is required to make the check, however refer to Chapter 4, Part B, Sections 16 and 17 for more information.
 a) Check and adjust the Turbo charge pressure.
 b) Check the Turbo overpressure switch.
 c) Check the Turbo charge pressure regulator seal.
 d) Check the Turbo APC control unit security seal - where applicable.
 e) Check the Turbo fuel boosting device.

20 Auxiliary drivebelt check

Checking the condition of the auxiliary drivebelts

1 The main drivebelt is used to drive the alternator and on the H-type B201 engine the water pump. On later models two drivebelts are fitted instead of the one. The secondary drivebelt is used to drive the power steering pump and where fitted a further drivebelt drives the air conditioning compressor.
2 Access to the drivebelts is restricted by their position at the rear of the engine and the use of a mirror will help when checking them. Using a spanner on the crankshaft pulley bolt, rotate the crankshaft so that the entire length of the drivebelts can be examined. Examine the drivebelt for cracks, splitting, fraying, or other damage. Check also for signs of glazing (shiny patches) and for separation of the belt plies. Renew the belt if worn or damaged.

Air conditioning compressor drivebelt - removal, refitting and tensioning

Removal

3 On models with the B201 engine, loosen the adjusting and pivot bolts and move the compressor to one side to release the tension on the drivebelt.
4 On models with the B202 engine loosen the tensioner bolt and swivel the tensioner to release the tension on the drivebelt.
5 Slip the drivebelt from the pulleys on the crankshaft, compressor and on the B202 engine the tensioner.

Refitting and tensioning

6 Locate the drivebelt on the pulleys then move the compressor or tensioner until it is possible to depress the belt approximately 15 mm under firm thumb pressure midway between the crankshaft and compressor pulleys. Tighten the adjustment and pivot bolts as applicable. Note that Saab technicians use a special tensioning tool to set the drivebelt tension - if any doubt exists about the tension of the belt, it should be checked by a Saab dealer.

Power steering pump drivebelt - removal, refitting and tensioning

Removal

7 Where fitted remove the air conditioning compressor drivebelt as described earlier.
8 Loosen the power steering pump pivot and adjustment bolts and move the pump to release the tension on the drivebelt **(see illustration)**.
9 Slip the drivebelt from the pulleys on the crankshaft and power steering pump pulleys **(see illustration)**.

Refitting and tensioning

10 Locate the drivebelt on the pulleys, then

20.8 Loosening the power steering pump drivebelt adjustment bolts

20.9 Removing steering pump drivebelt from crankshaft and pump pulleys

1A•14 Every 10 000 miles (15 000 km) – '79 to '84 models

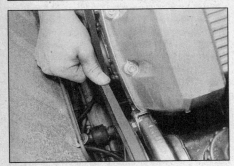

20.10 Checking the power steering pump drivebelt tension

20.15 Loosening the pivot and adjustment bolts

20.16 Removing the alternator/water pump drivebelts

move the pump until it is possible to depress the belt between 10 and 15 mm under firm thumb pressure midway between the crankshaft and pump pulleys **(see illustration)**. Tighten the pivot and adjustment bolts.
11 Refit and tension the air conditioning compressor drivebelt as described earlier.
12 Note that Saab technicians use a special tensioning tool to set the drivebelt tension - if any doubt exists about the tension of the belt, it should be checked by a Saab dealer.

Alternator/water pump drivebelt(s) - removal, refitting and tensioning

Removal
13 Remove the air conditioning compressor drivebelt and the power steering pump drivebelt as described earlier.
14 Disconnect the battery negative lead.
15 Loosen the alternator pivot and adjustment bolts and swivel the alternator in toward the cylinder block **(see illustration)**.
16 Slip the drivebelt(s) from the alternator, crankshaft and where applicable the water pump pulleys **(see illustration)**.

Refitting and tensioning
17 Locate the drivebelt on the pulleys, then lever the alternator away from the cylinder block until it is possible to depress the belt(s) between 10 and 15 mm under firm thumb pressure midway between the longest run of the belt. Lever the alternator on the drive end bracket to prevent straining the brackets. It is helpful to semi-tighten the adjustment link bolt before tensioning the drivebelt.
18 Fully tighten the alternator pivot and adjustment bolts.
19 Reconnect the battery negative lead.
20 Refit and tension the power steering pump drivebelt and the air conditioning compressor drivebelt.
21 Note that Saab technicians use a special tensioning tool to set the drivebelt tension - if any doubt exists about the tension of the belt, it should be checked by a Saab dealer.

21 Spark plug renewal

Refer to the information given in Section 5.

22 Distributor and ignition HT lead check

1 On B202 engines remove the inspection cover from the top of the cylinder head cover.
2 Wipe clean all of the ignition HT leads using a dry cloth. Check that the leads are attached to the spark plugs, distributor cap and ignition coil securely and are located in their fasteners correctly. Make sure that the leads are held clear of other components to prevent the possibility of arcing.
3 If necessary the HT leads may be checked as follows. Remove the leads together with the distributor cap, then connect an ohmmeter to each end of the leads and the appropriate terminal within the cap in turn. If the resistance is greater than that given in the Specifications, check that the lead connection in the cap is good before renewing the lead.
4 With the distributor cap removed, check the cap and rotor for hairline cracks and signs of arcing. If evident renew the component **(see illustration)**.

23 Contact breaker points check, renewal and adjustment

Checking
1 Disconnect the ignition coil HT lead from the distributor cap, then prise back the spring clips, remove the distributor cap and place it to one side **(see illustration)**.
2 Pull off the rotor arm and on H-type engines remove the plastic dust cover **(see illustration)**.
3 Prise open the contact points and examine the condition of their faces. If they are discoloured or slightly pitted, remove them as described later in this Section and dress them using emery tape or a grindstone. If they are worn excessively, renew them.
4 If the contact points are in good condition, adjust them as described later in this Section.
5 Apply a smear of oil to the distributor cam which operates the contact points. Where applicable apply one or two drops of oil to the cam lubricating pad and the felt pad in the top of the driveshaft.
6 Refit the plastic dust cover (if applicable), the rotor arm, and the distributor cap.

22.4 Removing the distributor cap for checking

23.1 Disconnecting the ignition coil HT lead from the distributor cap

23.2 Removing the plastic dust cover from the distributor on the H-type engine

Every 10 000 miles (15 000 km) – '79 to '84 models 1A•15

23.9 Removing the distributor bearing plate on the H-type engine

23.10 Contact breaker points location on the H-type engine

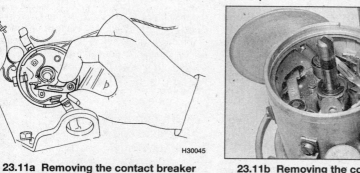

23.11a Removing the contact breaker points on the B-type engine

23.11b Removing the contact breaker points retaining screw

Renewal and adjustment

7 Disconnect the ignition HT lead from the distributor cap, then prise back the spring clips, remove the distributor cap and place it to one side.
8 Carefully pull off the rotor arm.
9 On H-type engines remove the plastic dust cover, then mark the bearing plate for position, loosen the screws and withdraw the plate (see illustration).
10 Disconnect the LT lead from the terminal inside the distributor (see illustration).
11 Unscrew the fixed contact retaining screw and remove the points (see illustrations).
12 Wipe clean the distributor base plate, the distributor cap and the HT leads. Check that the carbon brush moves freely in the distributor cap and that the metal HT segments are clean.
13 Apply one or two drops of oil to the felt pad at the top of the driveshaft.
14 Fit the new contact breaker points and tighten the fixed contact screw finger tight.
15 Connect the LT lead to the terminal.
16 Turn the engine with a spanner on the crankshaft pulley bolt until the heel of the moving contact is on the high point of one of the cam lobes.
17 Using a feeler blade check that the gap between the two points is as given in the Specifications (see illustration). If not, with the fixed contact screw finger tight use a screwdriver in the baseplate slot to reposition the fixed contact until the feeler blade is a firm sliding fit between the two points. When correct tighten the screw.
18 Connect a dwellmeter to the engine and turn the engine on the starter - the dwell angle should be within the limits given in the Specifications but if not, adjust the contact points gap as necessary. Reduce the gap to increase the angle or increase the gap to reduce the angle.
19 On H-type engines refit the bearing plate and tighten the screws, and fit the plastic dust cover (see illustration). Note that adjustment of the contact points gap is possible with the bearing plate fitted.
20 Refit the rotor arm and distributor cap.

24 Ignition timing check

1 Before adjusting the ignition timing on models equipped with a conventional ignition system, check and if necessary adjust the contact points dwell angle as described in Section 23. The initial setting method should be used in order to start the engine or for emergency roadside repairs, but the final setting must always be made using a stroboscopic timing light. The clutch (or torque converter) housing cover incorporates an attachment point for a special instrument which gives an instant ignition timing read-out, but this instrument will not normally be available to the home mechanic.

Initial setting

2 Remove No 1 spark plug (timing chain end of the engine). Put a finger over the plug hole.
3 Turn the engine in the running direction (clockwise viewed from the timing chain end of engine, anti-clockwise from the front of the car) until pressure is felt in No 1 cylinder, indicating the piston is on its compression stroke. Use a spanner on the crankshaft pulley bolt or engage top gear and pull the car forwards (except automatic transmission models).
4 Continue turning the engine until the correct ignition timing mark appears opposite the mark on the clutch housing cover timing hole (see illustration).
5 Remove the distributor cap and check that the rotor arm is pointing in the direction of the No 1 terminal of the cap. The rotor arm should also be pointing to the timing groove in the rim of the distributor body (see illustration).
6 On conventional ignition models connect a 12 volt test lamp and leads between the coil terminal 1 (with blue wire attached) and a suitable earthing point on the engine. Loosen the distributor retaining bolt(s) and switch on the ignition. If the bulb is already lit turn the distributor body slightly anti-clockwise until

23.17 Using a feeler blade when adjusting the contact breaker points gap

23.19 Showing bearing plate location tab on the H-type engine distributor

24.4 Ignition timing marks on the flywheel

1A•16 Every 10 000 miles (15 000 km) - '79 to '84 models

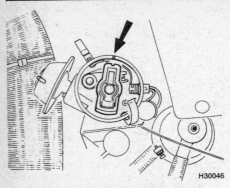

24.5 Distributor rotor arm on the B-type engine pointing to the ignition timing groove

the bulb goes out. Turn the distributor body clockwise until the bulb just lights up indicating that the points have just opened, then tighten the bolt(s). Switch off the ignition and remove the test lamp.

7 On electronic ignition models remove the rotor arm and dust cover and check that the rotor arms are aligned with the stator posts on the inductive transmitter type, or one of the rotor slots is aligned with the transmitter on the Hall effect type. If not, loosen the distributor retaining bolt(s) and turn the distributor body as necessary. Make sure that the rotor arm is still pointing in the direction of the No 1 terminal of the cap. Refit the dust cover and rotor arm.

8 Refit the distributor cap and No 1 spark plug and HT lead.

9 Once the engine has been started check the timing stroboscopically as follows.

Final setting

10 Disconnect and plug the vacuum hose at the distributor. Where a delay valve is fitted in the vacuum line do not disconnect the hose with the engine running otherwise foreign matter may cause the valve to be inoperative.

11 Connect a timing light and tachometer to the engine, following their manufacturer's instructions.

12 Start the engine and run it at the speed given in the Specifications.

13 Point the timing light at the timing hole in the clutch housing cover. The correct timing mark on the flywheel should appear to be stationary and aligned with the mark on the cover. If not, loosen the distributor retaining bolt(s) and turn it clockwise to advance or anticlockwise to retard the ignition. Tighten the bolt(s) when the setting is correct.

14 Gradually increase the engine speed while still pointing the timing light at the timing marks. The flywheel marks should appear to move clockwise when viewed from the front of the car proving that the distributor centrifugal weights are operating.

15 Run the engine at about 4000 rpm and note the ignition timing, then reconnect the vacuum hose. On non-Turbo models the timing should advance a few degrees, but on Turbo models it should retard a few degrees, proving that the vacuum unit is operating.

16 Switch off the engine and disconnect the timing light and tachometer. Check that the vacuum hose is secure.

25 Delay valve check and renewal

1 The delay valve is located in the distributor vacuum advance pipe and its purpose is to delay ignition advance during acceleration so reducing nitrous oxide emissions.

Checking

2 Connect a tachometer and timing light to the engine, following their manufacturers instructions, and run the engine to operating temperature. Have an assistant open the throttle quickly so that the engine runs at 3000 rpm. Using the timing light check that the time from when the throttle was opened to when the ignition advances is between 4 and 8 seconds. If not, renew the delay valve.

Removal and refitting

3 To remove the delay valve disconnect the hoses and remove the valve.

4 Refitting is a reversal of removal, but note that the white end of the valve must be towards the distributor.

26 Headlight beam alignment

1 It is recommended that the headlamp alignment is carried out by a Saab dealer using beam setting equipment. However in an emergency the following procedure will provide an acceptable light pattern, however the alignment should be checked as soon as possible by a Saab dealer.

2 Position the car on a level surface with tyres correctly inflated, about 5.0 metres in front of, and at right-angles to, a wall or garage door.

3 Draw a vertical line on the wall corresponding to the centre line of the car. The position of the line can be ascertained by marking the centre of the front and rear screens with crayon then viewing the wall from the rear of the car.

4 Measure the distance between the headlamp centres and their height above the ground, then mark the positions on the wall.

5 Switch the headlamps on main beam and check that the areas of maximum illumination ("hot spots") coincide with the marks on the wall. On dipped beam, the hot spots should be 50 mm below the centre marks.

6 If adjustment is necessary, turn the adjustment screws on the headlamp rim (99 models) or on the rear of the headlamp (900 models) until the setting is correct. On 900 models insert a screwdriver in one of the bonnet hinge holes to hold the bonnet half open so that the screws can be reached when adjusting the headlamps.

27 Manual transmission oil level check

1 The manual transmission oil level plug is located midway along the right-hand side of the transmission (see illustration). On early models a separate level dipstick is fitted. **Note:** *The level plug at the front right-hand side of the transmission is for the primary gear housing, but is only used for adding some of the oil to the primary housing after overhauling the transmission. When the transmission is in motion oil is delivered to the primary gears via an oil catcher.*

2 On models with a dipstick remove the dipstick and wipe it clean with a piece of cloth, then re-insert it fully. Withdraw it again and check that the level is correct (see illustrations).

3 On models with an oil level plug, first wipe clean the area around the level plug, then

27.1 Manual transmission oil level plug

27.2a Removing the early manual transmission oil level dipstick

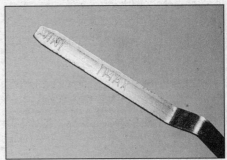

27.2b Manual transmission oil level dipstick markings

Every 10 000 miles (15 000 km) - '79 to '84 models 1A•17

unscrew and remove the plug which incorporates a dipstick and check that the transmission oil level is on the maximum mark.
4 If the oil level requires topping up, add oil as necessary using only good quality oil of the specified type as given in the Specifications at the end of this Chapter. On early models add oil through the dipstick tube and on later models add oil through the level plug hole. Frequent topping up indicates a leak which should be investigated and repaired.
5 With the level correct either insert the dipstick or wipe clean the level plug, refit and tighten it. Wipe off any spilt oil.

28 Automatic transmission fluid and final drive oil level check

Automatic transmission fluid

1 Three different types of oil are contained in the automatic transmission casing - the engine oil, final drive oil, and automatic transmission fluid. Check the automatic transmission fluid level as follows according to the type of transmission fitted.

Type 35 transmission

2 Run the engine at idling speed for a few minutes with the selector in P.
3 Switch off the engine and withdraw the transmission fluid dipstick located on the front right-hand side of the engine. Wipe it clean on a piece of non-fluffy rag and re-insert it. Withdraw it for the second time when the oil level should be within the limits of the cut-out on the 'cold' edge of the dipstick if the engine and transmission are cold. Use the 'hot' edge of the dipstick if the engine and transmission are at normal operating temperature (see illustration).
4 If necessary top up the fluid level through the dipstick/filler tube using only automatic transmission fluid as given in the Specifications.

Type 37 transmission

5 Run the engine at idling speed for at least 15 seconds in each of the following selector positions: D, R and then P.
6 With the engine still running, and the selector in P, check the fluid level as described in paragraph 3.
7 If necessary top up the fluid level through the dipstick tube using only the specified automatic transmission fluid quoted in the Specifications.
8 Allow the engine to idle for a few minutes then recheck the fluid level.

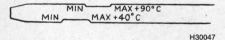

28.3 Automatic transmission fluid level dipstick markings

Final drive oil (automatic transmission models only)

9 The final drive oil plug is located on the rear left-hand side of the transmission, over the left-hand driveshaft. Access to the plug is best from under the car. First jack up the car and support on axle stands (see "Jacking and Vehicle Support") - make sure that the car is level.
10 Unscrew the level plug and check that the oil is level with the bottom edge of the plug hole.
11 If the level requires topping up add oil as necessary using only oil of the specified type as given in the Specifications at the end of this Chapter.
12 With the level correct, wipe clean the level plug then refit it and tighten it securely. Wipe off any spilt oil.
13 Lower the car to the ground.

29 Brake pad, disc and handbrake check

1 Jack up the front and rear of the car and support it securely on axle stands (see "Jacking and Vehicle Support"). Remove the front and rear roadwheels.
2 For a quick check, the thickness of friction material remaining on each brake pad can be measured through the aperture in the caliper body. If any pad's friction material is worn to the specified thickness or less, all four pads on that axle must be renewed as a set.
3 For a comprehensive check, the brake pads should be removed and cleaned. The operation of the caliper can then also be checked, and the condition of the brake discs checked on both sides. Refer to Chapter 9 for further information.
4 Lubricate the caliper guides of the front brake calipers only (refer to Chapter 9).
5 Have an assistant operate the handbrake while checking the operation of the levers on the front calipers. If there is any sign of the levers sticking, check the handbrake lever and cables, and the mechanism within the caliper with reference to Chapter 9.
6 On completion refit the wheels and lower the car to the ground.

30 Brake line and flexible hose check

1 Raise the front and rear of the car and securely support on axle stands (see "Jacking and Vehicle Support"). Remove all wheels.
2 Thoroughly examine all brake lines and brake flexible hoses for security and damage. To check the flexible hoses, bend them slightly in order to show up any cracking of the rubber.
3 Check the complete braking system for any signs of brake fluid leakage.
4 Where necessary carry out repairs to the braking system with reference to Chapter 9.

31 Steering, suspension and shock absorber check

1 Raise the front and rear of the car, and securely support it on axle stands (see "Jacking and Vehicle Support").
2 Visually inspect all balljoint dust covers and the steering rack-and-pinion gaiters for splits, chafing or deterioration. Any wear of these components will cause loss of lubricant, together with dirt and water entry, resulting in rapid deterioration of the balljoints or steering gear.
3 Grasp each front roadwheel in turn at the 12 o'clock and 6 o'clock positions, and try to rock it (see illustration). Very slight free play may be felt, but if the movement is appreciable, further investigation is necessary to determine the source. Continue rocking the wheel while an assistant depresses the footbrake. If the movement is now eliminated or significantly reduced, it is likely that the hub bearings are at fault. If the free play is still evident with the footbrake depressed, then there is wear in the suspension joints or mountings.
4 Now grasp the wheel at the 9 o'clock and 3 o'clock positions, and try to rock it as before. Any movement felt now may again be caused by wear in the hub bearings or the steering track-rod balljoints. If the inner or outer balljoint is worn, the visual movement will be obvious.
5 Using a large screwdriver or flat bar, check for wear in the suspension mounting bushes by levering between the relevant suspension component and its attachment point. Some movement is to be expected as the mountings are made of rubber, but excessive wear should be obvious. Also check the condition of any visible rubber bushes, looking for splits, cracks or contamination of the rubber.
6 Check for any signs of fluid leakage around the front suspension struts and rear shock absorbers. Should any fluid be noticed, the suspension strut/shock absorber is defective internally, and should be renewed. Note: Suspension struts/shock absorbers should always be renewed in pairs on the same axle.
7 Working from the front to the rear of the vehicle, check the security of all suspension

31.3 Rocking the roadwheel to check steering/suspension components

Every 10 000 miles (15 000 km) - '79 to '84 models

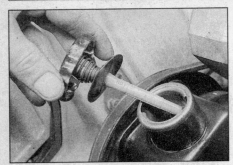

32.3 Unscrewing the power steering fluid reservoir filler cap from the top of the pump (early models)

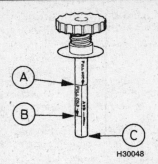

32.4 Filler cap and power steering fluid reservoir dipstick (early models)
A Full Hot   B Full Cold   C Add

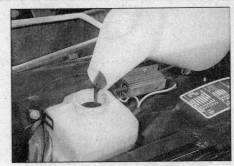

32.6 Topping-up the power steering fluid level (later models)

and steering nuts and bolts and if necessary tighten them to the specified torque as given in Chapter 10.
8 Lower the car to the ground.
9 The efficiency of the suspension struts and shock absorbers may be checked by depressing each corner of the car in turn. If the struts/shock absorbers are in good condition, the body will rise and then settle in its normal position. If it continues to rise and fall, the suspension strut or shock absorber is probably suspect. Examine also the suspension strut/shock absorber upper and lower mountings for any signs of wear.
10 With the car standing on its wheels, have an assistant turn the steering wheel back and forth about an eighth of a turn each way. There should be very little, if any, lost movement between the steering wheel and roadwheels. If this is not the case, closely observe the joints and mountings previously described, but in addition, check the steering column universal joints for wear, and the rack-and-pinion steering gear itself.

32 Power steering fluid level check

1 The power steering fluid reservoir is located either on the top of the power steering pump on the rear right-hand side of the engine, or in a separate reservoir attached to the rear right-hand side of the engine compartment.
2 For the check, the front wheels should be pointing straight-ahead and the engine should be stopped. The car should be positioned on level ground.
3 Before removing the filler cap use a clean rag to wipe the cap and the surrounding area to prevent any foreign matter from entering the reservoir. Unscrew and remove the filler cap **(see illustration)**.
4 Where the reservoir is located on top of the pump, the dipstick is incorporated into the filler cap. There are three markings **(see illustration)**. If the engine is hot, the level should be between the "Full Hot" and "Full Cold" marks, and if the engine is cold, the level should be between the "Full Cold" and "Add" marks.
5 Where the reservoir is located on the bulkhead, there are Hot, Cold and Add marks.
6 Top up if necessary with the specified power steering fluid **(see illustration)**. Be careful not to introduce dirt into the system, and do not overfill. Frequent topping up indicates a leak which should be investigated.

33 Facia security check

Check the facia panel mounting bolts for security and tighten as necessary. Refer to Chapter 12.

Every 20 000 miles (30 000 km)

34 Delay valve and fuel enrichment device renewal

Renewal of the delay valve is described in Section 25.
Renewal of the fuel enrichment device is described in Chapter 4, Part C, Section 2.

35 Automatic transmission fluid and filter renewal, and cable adjustment

Carry out the following work as described in Chapter 7, Part B.
 a) Visually check the automatic transmission for signs of fluid or oil leaks.
 b) Renew the automatic transmission fluid and where applicable clean the filter.
 c) Adjust the selector and kick-down cables.

36 Front wheel alignment check

Due to the special measuring equipment necessary to check the wheel alignment accurately, checking and adjustment is best left to a Saab dealer or similar expert. Note that most tyre-fitting shops now possess sophisticated checking equipment. Refer to Chapter 10 for more information.
Before having the front wheel alignment checked, all tyre pressures should be checked and if necessary adjusted.

37 Pollen filter renewal

1 The pollen filter is on the right-hand side of the bulkhead in the engine compartment.
2 On pre-1983 models unscrew the four screws and slide out the filter element **(see illustration)**.
3 On 1983-on models the cover is retained with plastic clips which are removed by turning them through 90° either clockwise or

37.2 Pollen filter (pre-1983 model)

Every 20 000 miles (30 000 km) - '79 to '84 models

37.3a Remove the cover . . .

37.3b . . . and slide out the pollen filter element

anticlockwise. With the cover off, slide out the filter element **(see illustrations)**.
4 Fit the new pollen filter using a reversal of the removal procedure.

38 Road test

Instruments and electrical equipment

1 Check the operation of all instruments and electrical equipment.
2 Make sure that all instruments read correctly, and switch on all electrical equipment in turn to check that it functions properly. Check the function of the heating, air conditioning and automatic climate control systems.

Steering and suspension

3 Check for any abnormalities in the steering, suspension, handling or road 'feel'.
4 Drive the vehicle, and check that there are no unusual vibrations or noises.

5 Check that the steering feels positive, with no excessive 'sloppiness', or roughness, and check for any suspension noises when cornering, or when driving over bumps. Check that the power steering system operates correctly. Check that the cruise control system (where fitted) operates correctly.

Drivetrain

6 Check the performance of the engine, clutch (manual transmission), transmission and driveshafts. On Turbo models, check that the boost pressure needle moves up to the red zone during brief acceleration.
7 Listen for any unusual noises from the engine, clutch (manual transmission) and transmission.
8 Make sure that the engine runs smoothly when idling, and that there is no hesitation when accelerating.
9 On manual transmission models, check that the clutch action is smooth and progressive, that the drive is taken up smoothly, and that the pedal travel is correct. Also listen for any noises when the clutch pedal is depressed. Check that all gears can be engaged smoothly, without noise, and that the gear lever action is not vague or 'notchy'.
10 On automatic transmission models, make sure that all gearchanges occur smoothly without snatching, and without an increase in engine speed between changes. Check that all the gear positions can be selected with the vehicle at rest. If any problems are found, they should be referred to a Saab dealer.
11 Listen for a metallic clicking sound from the front of the vehicle, as the vehicle is driven slowly in a circle with the steering on full lock. Carry out this check in both directions. If a clicking noise is heard, this indicates wear in a driveshaft joint, in which case, refer to Chapter 8.

Check the operation and performance of the braking system

12 Make sure that the vehicle does not pull to one side when braking, and that the wheels do not lock prematurely when braking hard.
13 Check that there is no vibration through the steering when braking.
14 Check that the handbrake operates correctly, without excessive movement of the lever, and that it holds the vehicle stationary on a slope.
15 Test the operation of the brake servo unit as follows. With the engine off, depress the footbrake four or five times to exhaust the vacuum. Start the engine, holding the brake pedal depressed. As the engine starts, there should be a noticeable 'give' in the brake pedal as vacuum builds up. Allow the engine to run for at least two minutes, and then switch it off. If the brake pedal is depressed now, it should be possible to detect a hiss from the servo as the pedal is depressed. After about four or five applications, no further hissing should be heard.

Every 30 000 miles (45 000 km)

39 Air cleaner element renewal/cleaning

Refer to the information given in Section 12.

40 EGR system check - H engines

Refer to the information given in Section 13.

41 Fuel filter renewal (fuel injection models)

 Warning: Before carrying out the following operation, refer to the precautions given in 'Safety first!' at the beginning of this manual, and follow them implicitly. Petrol is a highly-dangerous and volatile liquid - take extreme care when handling it.

1 The fuel filter is located in the engine compartment on the left-hand wheel arch, in the pressure line between the fuel pump and the fuel supply rail.
2 Depressurise the fuel system with reference to Chapter 4, Part B, Section 7.
3 Jack up the rear of the car and support on axle stands (see "Jacking and Vehicle Support").

4 Clean the areas around the fuel filter inlet and outlet unions.
5 Position a small container or cloth rags beneath the filter to catch spilt fuel.
6 Unscrew one of the banjo bolts while holding the flats on the filter with a further spanner, and disconnect the fuel line. Recover the sealing washers **(see illustrations)**.

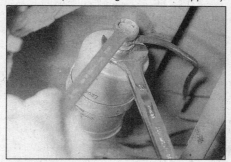

41.6a Unscrew the banjo bolts on the fuel filter . . .

41.6b . . . and remove the bolts and sealing washers

1A•20 Every 30 000 miles (45 000 km) – '79 to '84 models

 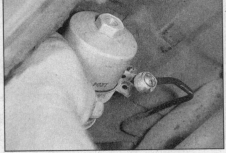 

41.8a Unscrew the retaining clip screw . . .   41.8b . . . and remove the fuel filter   41.9 The arrow on the filter body must point towards the outlet

7 Unscrew the remaining banjo bolt while holding the flats on the filter, disconnect the fuel line, and recover the sealing washers.
8 Noting the direction of the arrow marked on the filter body, unscrew the retaining clip screw and withdraw the filter (see illustrations).
9 Locate the new filter in the retaining clip and tighten the clip. Make sure the flow arrow on the filter body is pointing towards the outlet which leads to the fuel injection supply rail (see illustration).
10 Check the condition of the sealing washers and renew them if necessary.
11 Refit the banjo couplings and hoses to the top and bottom of the filter together with the sealing washers, and tighten the bolts securely while holding the filter with a further spanner.
12 Wipe away any excess fuel, then start the engine and check the filter union connections for leaks.
13 On completion, lower the car to the ground.
14 The old filter should be disposed of safely, bearing in mind that it will be highly inflammable.

42 Valve clearances check and adjustment

Refer to the information given in Section 18.

43 Brake fluid renewal

 Warning: Brake hydraulic fluid can harm your eyes and damage painted surfaces, so use extreme caution when handling and pouring it. Do not use fluid that has been standing open for some time, as it absorbs moisture from the air. Excess moisture can cause a dangerous loss of braking effectiveness.

1 The procedure is similar to that for the bleeding of the hydraulic system as described in Chapter 9, except that the brake fluid reservoir should be emptied by siphoning, using a clean poultry baster or similar before starting, and allowance should be made for the old fluid to be expelled when bleeding a section of the circuit. To prevent contamination from the fluid in the clutch hydraulic system (which uses the same reservoir) the latter system should be bled as described in Chapter 6 after bleeding the brake system.
2 Working as described in Chapter 9, open the first bleed screw in the sequence, and pump the brake pedal gently until nearly all the old fluid has been emptied from the fluid reservoir. Top-up to the 'MAX' level with new fluid, and continue pumping until only the new fluid remains in the reservoir, and new fluid can be seen emerging from the bleed screw. Tighten the screw, and top up the reservoir level to the 'MAX' level line.
3 Old hydraulic fluid is darker in colour than new, making it easy to distinguish the two.
4 Work through all the remaining bleed screws in the sequence until new fluid can be seen at all of them. Be careful to keep the master cylinder reservoir topped-up to above the 'MIN' level at all times, or air may enter the system and increase the length of the task.
5 When the operation is complete, check that all bleed screws are securely tightened, and that their dust caps are refitted. Wash off any spilt fluid, and recheck the reservoir fluid level.
6 Check the operation of the brakes before taking the car on the road.

Every 2 years (regardless of mileage)

44 Coolant renewal

Cooling system draining

 Warning: Wait until the engine is cold before starting this procedure. Do not allow antifreeze to come in contact with your skin, or with the painted surfaces of the vehicle. Rinse off spills immediately with plenty of water. Never leave antifreeze lying around in an open container - antifreeze can be fatal if ingested.

1 With the engine completely cold, unscrew and remove the expansion tank filler cap. Turn the cap a few turns anti-clockwise then wait until any pressure remaining in the system is released. Fully unscrew and remove the cap.
2 Remove the filler cap then position a suitable container beneath the radiator drain plug - on 99 models the plug is on the left-hand side, on 900 models it is on the right-hand side.
3 Set the heater control to maximum heat then unscrew the drain plug and drain the coolant.
4 Position another container beneath the right-hand side of the cylinder block, unscrew the drain plug, and drain the coolant from the block.
5 On 900 models loosen the clip and disconnect the bottom hose from the radiator in order to drain the remaining coolant (see illustration). Any remaining coolant on 99 models can only be drained after removing the radiator and inverting it, however this will only be necessary where severe contamination has occurred.

44.5 Radiator bottom hose connection on 900 models

Every 2 years - '79 to '84 models

44.19 On later models, the cooling system bleed screw is located on the thermostat cover

6 If the coolant has been drained for a reason other than renewal, then provided it is clean and less than 3 years old, it can be re-used.

Cooling system flushing

7 If coolant renewal has been neglected, or if the antifreeze mixture has become diluted, then in time, the cooling system may gradually lose efficiency, as the coolant passages become restricted due to rust, scale deposits, and other sediment. The cooling system efficiency can be restored by flushing the system clean.
8 The radiator should be flushed independently of the engine, to avoid unnecessary contamination.

Radiator flushing

9 To flush the radiator, first disconnect the top and bottom hoses from the radiator.
10 Insert a garden hose into the radiator top inlet. Direct a flow of clean water through the radiator, and continue flushing until clean water emerges from the radiator bottom outlet.
11 If after a reasonable period, the water still does not run clear, the radiator can be flushed with a good proprietary cleaning agent. It is important that their manufacturer's instructions are followed carefully. If the contamination is particularly bad, remove the radiator then insert the hose in the radiator bottom outlet, and reverse-flush the radiator.

Engine flushing

12 To flush the engine, first remove the thermostat as described in Chapter 3, then temporarily refit the thermostat cover.
13 With the top and bottom hoses disconnected from the radiator, insert a garden hose into the radiator top hose. Direct a clean flow of water through the engine, and continue flushing until clean water emerges from the radiator bottom hose. Also insert the hose into the expansion tank to flush away any sediment.
14 Disconnect the hose from the inlet manifold, insert the hose, and allow water to run through the heater and out of the bottom hose.
15 On completion of flushing, refit the thermostat and reconnect the hoses with reference to Chapter 3.
16 Remove the container from under the car, then refit and tighten the cylinder block and radiator drain plugs.

Cooling system filling

17 Before attempting to fill the cooling system, make sure that all hoses and clips are in good condition, and that the clips are tight. Note that an antifreeze mixture must be used all year round, to prevent corrosion of the engine components (see following sub-Section).
18 Pour the correct amount of antifreeze into a container, then add water until the total amount equals the cooling system capacity (see Specifications at the end of this Chapter).
19 Slowly fill the system through the expansion tank filler neck, allowing time for trapped air to escape. Where fitted, loosen the bleed screw on the thermostat cover, thermostat housing or heater temperature control valve and allow trapped air to escape. Tighten the screw when bubble-free coolant comes out **(see illustration)**.
20 Top up to the 'MAX' mark on the side of the expansion tank.
21 Refit and tighten the filler cap.
22 Start the engine, and run it at a fast idle speed until the cooling fan cuts in, and then cuts out. This will purge any remaining air from the system. Stop the engine.
23 Allow the engine to cool, then check the coolant level with reference to "Weekly Checks". Top-up the level if necessary. Saab recommend that the level be checked again after a few days and topped up as necessary.

Antifreeze mixture

24 The antifreeze should always be renewed at the specified intervals. This is necessary not only to maintain the antifreeze properties, but also to prevent corrosion which would otherwise occur as the corrosion inhibitors become progressively less effective.
25 Always use an ethylene-glycol based antifreeze which is suitable for use in mixed-metal cooling systems. The quantity of antifreeze and levels of protection are indicated in the Specifications.
26 Before adding antifreeze, the cooling system should be completely drained, preferably flushed, and all hoses checked for condition and security.
27 After filling with antifreeze, a label should be attached to the expansion tank filler neck, stating the type and concentration of antifreeze used, and the date installed. Any subsequent topping-up should be made with the same type and concentration of antifreeze.
28 Do not use engine antifreeze in the windscreen/tailgate washer system, as it will cause damage to the vehicle paintwork. A screenwash additive should be added to the washer system in the quantities stated on the bottle.

Servicing Specifications - '79 to '84 models

Lubricants and fluids
Refer to the end of "Weekly Checks"

Capacities

Engine oil
Including filter:
   B201 engine .... 3.8 litres
   B202 engine .... 4.0 litres
Difference between MAX and MIN dipstick marks .... 1.0 litre

Cooling system
   99 models .... 8.0 litres
   900 models .... 10.0 litres

Manual transmission
   4-speed .... 2.5 litres
   5-speed .... 3.0 litres

Automatic transmission
   Including torque converter and oil cooler .... 8.0 litres

Automatic transmission final drive
   Type 35 .... 1.25 litres
   Type 37 .... 1.4 litres

Fuel tank
   All models .... 63 litres

Braking system capacity
   All models .... 0.55 litres approx.

Power steering capacity
   All models .... 0.75 litre

Engine

Difference between MIN and MAX marks on dipstick .... 1.0 litre
Oil filter .... Champion C142
Valve clearances - cold:

	Checking value	Setting value
Inlet (all models)	0.15 to 0.30 mm (0.006 to 0.012 in)	0.20 to 0.25 mm (0.008 to 0.010 in)
Exhaust (except Turbo)	0.35 to 0.50 mm (0.014 to 0.020 in)	0.40 to 0.45 mm (0.016 to 0.018 in)
Exhaust (Turbo)	0.40 to 0.50 mm (0.016 to 0.020 in)	0.45 to 0.50 mm (0.018 to 0.020 in)

Cooling system

Antifreeze mixture:
   28% antifreeze .... Protection down to -15°C (5°F)
   50% antifreeze .... Protection down to -30°C (-22°F)

Note: Refer to antifreeze manufacturer for latest recommendations.

Fuel system

Air filter element:
   All models except 16V Turbo 1984-on .... Champion W198
Fuel filter:
   Saab 99i EMS to 1980 .... Champion L203
   Saab 900 16V Turbo .... Champion L204
   Saab 900 900i 16V 1981-on .... Champion L204
   Saab 900i EMS and GLE 1982-on .... Champion L203
Idling speed:
   Bosch CI (K-Jetronic) and LH-Jetronic fuel injection systems:
     Non-Turbo models up to 1981 .... 875 rpm
     Non-Turbo models from 1982 .... 850 rpm
     All Turbo models .... 850 rpm
Idling CO% .... 4.5% maximum

Ignition system

Distributor:
   Contact points gap .... 0.4 mm
   Dwell angle .... 50° ± 3°

Servicing Specifications - '79 to '84 models

Ignition system (continued)

Ignition timing:
UK models
- 99 models with conventional ignition .... 17° (B-type) or 18° (H-type) BTDC at maximum of 800 rpm (B-type) or 2000 rpm (H-type) and vacuum hose disconnected
- 99 models with electronic ignition .... 23° BTDC at 2000 rpm and vacuum hose disconnected and plugged
- Turbo models without APC (1979 to 1982) .... 23° BTDC at 2000 rpm and vacuum hose disconnected
- Turbo APC (B201) (1982) .... 20° BTDC at 2000 rpm and vacuum hose disconnected
- Turbo 16 (B202) (1984-on) .... 16° BTDC at 850 rpm and vacuum hose disconnected
- Carburettor models (1979 to 1980) .... 17° BTDC at 800 rpm and vacuum hose disconnected
- Non-Turbo fuel injection models (B201) (1981-on) .... 20° BTDC at 2000 rpm and vacuum hose disconnected

Sweden (1979-on) and Switzerland (1983-on) models
- Turbo (B201) (1979-on) .... 20° BTDC at 2000 rpm and vacuum hose disconnected
- Turbo 16 (B202) (1985-on) .... 16° BTDC at 850 rpm and vacuum hose disconnected
- Carburettor models (1979 to 1980):
    - Manual transmission .... 18° BTDC at 2000 rpm and vacuum hose disconnected
    - Automatic transmission .... 21° BTDC at 2000 rpm and vacuum hose disconnected
- Non-Turbo fuel injection models (B201) (1981-on) .... 18° BTDC at 2000 rpm and vacuum hose disconnected

Spark plugs:
- Saab 90 .... Champion N9YCC
- Saab 99 non-Turbo .... Champion N9YCC
- Saab 99 Turbo .... Champion N7YCC
- Saab 900 B201 non-Turbo .... Champion N9YCC
- Saab 900 B201 Turbo .... Champion N7YCC
- Saab 900 B202 non-Turbo .... Champion C9YCC
- Saab 900 B202 Turbo .... Champion C7YCC

Spark plug electrode gap* .... 0.8 mm

Ignition HT lead set:
- 90 .... Champion LS 04
- 99 models to 1981 .... Champion LS 08
- 99 models 1981-on .... Champion LS 04
- 900 Turbo to 1980 .... Champion LS 08
- 900 non-Turbo to 1980 .... Champion LS 04
- 900 Turbo 1981 to 1983 (H-type engine) .... Champion LS 23
- 900 Turbo 1983 to 1990 (H-type engine) .... Champion LS 13
- 900 GL/GLS/Turbo 1981-on .... Champion LS 13

Ignition HT lead resistance .... Approximately 600 ohms per 100 mm length

*The spark plug gap quoted is that recommended by Champion for their specified plugs listed above. If spark plugs of any other type are to be fitted, refer to their manufacturer's recommendations.

Brakes

Brake pad friction material minimum thickness .... 1.0 mm

Tyre pressures

Refer to the end of "Weekly Checks".

Wiper blades

Windscreen and tailgate:
- Saab 90 and 99 .... Champion X-3803
- Saab 900 .... Champion X-4103

Torque wrench settings

	Nm	lbf ft
Spark plugs	28	21
Cylinder block drain plug	55	41
Engine oil drain plug:		
Manual transmission models	50	37
Automatic transmission models	37	27
Cylinder block coolant drain plug	25	19
Manual transmission filler and drain plugs	49	36
Automatic transmission drain plug	7	5
Roadwheel nuts	98	72

Notes

Chapter 1 Part B:
Routine maintenance and servicing - 1985-on models

Contents

Air cleaner element renewal/cleaning . . . . .13 / 37	Front wheel alignment check . . . . .35 / 40
Airbag check 47	Fuel filter renewal (fuel injection models) . . . . .41 / 43
Automatic transmission fluid and filter renewal 44	Handbrake adjustment check . . . . .32 / 6
Automatic transmission fluid and final drive oil level check . . . . . .8 / 31	Headlight beam alignment 29
Auxiliary drivebelt check 24	Hose and fluid leak check 4
Brake fluid renewal 45	Idle speed and mixture check and adjustment 18
Brake line and flexible hose check 33	Ignition timing check 27
Brake pad, disc and handbrake check 6	Intensive maintenance 2
Choke fast idling speed (carburettor models) check 15	Introduction 1
Coolant renewal 46	Manual transmission oil level check 30
Cooling system pressure test 16	Pollen filter renewal 10
Crankcase ventilation system check 23	Power steering fluid level check 7
Deceleration device (carburettor engines) check and adjustment . . .19	Road test 11
Delay valve check and renewal . . . . .28 / 39	Seat belt check 9
Distributor and ignition HT lead check 26	Spark plug renewal or clean . . . . .5 / 25
EGR system check . . . . .14 / 38	Steering, suspension and shock absorber check 34
Engine oil and filter renewal . . . . .3 / 12	Turbocharger - check 21
Exhaust system check 22	Vacuum lines check 17
Facia security check 36	Valve clearances (B201 engine) - checking and adjustment . . .20 / 42

Degrees of difficulty

| Easy, suitable for novice with little experience | | Fairly easy, suitable for beginner with some experience | | Fairly difficult, suitable for competent DIY mechanic | | Difficult, suitable for experienced DIY mechanic | | Very difficult, suitable for expert DIY or professional | |

Maintenance & Servicing - '85-on models

Maintenance schedule

The maintenance intervals in this manual are provided with the assumption that you will be carrying out the work yourself. These are the minimum maintenance intervals recommended by the manufacturer for vehicles driven daily. If you wish to keep your vehicle in peak condition at all times, you may wish to perform some of these procedures more often. We encourage frequent maintenance, because it enhances the efficiency, performance and resale value of your vehicle.

If the vehicle is driven in dusty areas, used to tow a trailer, or driven frequently at slow speeds (idling in traffic) or on short journeys, more frequent maintenance intervals are recommended.

When the vehicle is new, it should be serviced by a factory-authorised dealer service department, in order to preserve the factory warranty.

Every 250 miles, or weekly
- ☐ Refer to *Weekly Checks*.

At first 6000 miles (10 000 km)*
*Note: Subsequent servicing of items in this Section must be carried out at 12 000 mile (20 000 km) intervals (ie at 18 000, 30 000, 42 000, 54 000, 66 000 miles et seq.) unless otherwise specified.
- ☐ Change the engine oil **every 6000 miles** (Section 3) - Turbo models only.
- ☐ Visually check the cooling system (Section 4).
- ☐ Check the fuel lines in the engine compartment **every 6000 miles** (Section 4).
- ☐ Check and adjust the spark plugs (Section 5).
- ☐ Check front and rear brake pads **every 6000 miles** (Section 6).
- ☐ Check power steering fluid level **every 6000 miles** (Section 7).
- ☐ Check and top-up automatic transmission fluid level (Section 8).
- ☐ Check the seat belts for operation and damage (Section 9).
- ☐ Renew the pollen filter (Section 10).
- ☐ Road test **every 6000 miles** (Section 11).

At 12 000 miles (20 000 km)*
*Note: Subsequent servicing of items in this Section must be carried out at 12 000 mile (20 000 km) intervals (ie at 24 000, 36 000, 48 000, 60 000 miles et seq.) unless otherwise specified.
- ☐ Change the engine oil and filter - all models (Section 12).
- ☐ Clean the air cleaner element **every 24 000 miles** (ie at 36 000, 60 000 miles et seq.) - non-Turbo models only (Section 13).
- ☐ Renew the air cleaner element - Turbo APC models only (Section 13).
- ☐ Check the EGR system (Section 14).
- ☐ Check the choke fast idling speed - carburettor engines only (Section 15).
- ☐ Pressure test the cooling system (Section 16).
- ☐ Check all vacuum lines (Section 17).
- ☐ Check oil level in carburettor (Section 18).
- ☐ Check idling speed and CO content (Section 18).

At 12 000 miles (20 000 km) - continued
- ☐ Check and adjust the deceleration valve (Section 19).
- ☐ Check and adjust the valve clearances - Turbo models (except B202) (Section 20).
- ☐ Check the Turbo overpressure switch (Section 21).
- ☐ Check the Turbo charge pressure regulator seal (Section 21).
- ☐ Check the Turbo APC control unit security seal - where applicable (Section 21).
- ☐ Check the Turbo fuel boosting device (Section 21).
- ☐ Check exhaust pipe mountings (Section 22).
- ☐ Check crankcase ventilation hoses and connections (Section 23).
- ☐ Check all drivebelts and adjust (Section 24).
- ☐ Renew the spark plugs (Section 25).
- ☐ Check the HT leads (Section 26).
- ☐ Check and adjust the ignition timing (Section 27).
- ☐ Check the delay valve **every 24 000 miles** (ie at 36 000 and 60 000 miles et seq.) (Section 28).
- ☐ Check and adjust the headlamp beam alignment (Section 29).
- ☐ Check the manual transmission oil level (Section 30).
- ☐ Check the final drive oil level - automatic transmission models only (Section 8).
- ☐ Check the handbrake adjustment (Section 32).
- ☐ Check all hydraulic brake lines (Section 33).
- ☐ Check all front suspension and steering rubber boots and gaiters (Section 34).
- ☐ Check steering gear ball joints and track rod end joints (Section 34).
- ☐ Check steering track (toe-in) **every 24 000 miles** (ie at 36 000 and 60 000 miles et seq.) (Section 34).
- ☐ Check facia for security (Section 36).

Every 24 000 miles (40 000 km)
- ☐ Renew the air cleaner element - except Turbo APC models (Section 37).
- ☐ Clean the EGR system (Section 38).
- ☐ Renew the delay valve (Section 39).
- ☐ Check all front wheel alignment angles (Section 40).

Maintenance & Servicing - '85-on models 1B•3

Every 30 000 miles (50 000 km)
☐ Renew the fuel filter - B201 engine only (Section 41).
☐ Check and adjust the valve clearances - non-Turbo models (except B202) (Section 42).

Every 54 000 miles (90 000 km)
☐ Renew the fuel filter - B202 engine only (Section 43).

Every 66 000 miles (110 000 km)
☐ Change the automatic transmission fluid (Section 44).

Every 2 years (regardless of mileage)
☐ Renew the brake fluid (Section 45).
☐ Renew the coolant/antifreeze (Section 46).

Every 10 years (regardless of mileage)
☐ Check the airbag system (Section 47).

Underbonnet view of 1992 Saab 900 Turbo

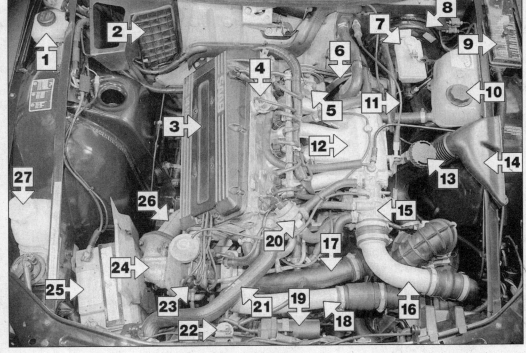

1 Power steering fluid reservoir
2 Heater inlet pollen filter
3 Spark plug inspection cover
4 Fuel pressure regulator
5 Water pump
6 Alternator
7 Brake fluid reservoir
8 Brake vacuum servo
9 Fusebox
10 Cooling system expansion tank
11 Accelerator cable
12 Inlet manifold
13 Engine oil filler cap/dipstick
14 Air inlet for air cleaner
15 Throttle housing
16 Air inlet tube (intercooler to throttle housing)
17 Air inlet tube (air cleaner to turbocharger)
18 Air inlet tube (turbocharger to intercooler)
19 Ignition coil
20 AIC (automatic idle control) valve
21 Radiator top hose
22 APC solenoid valve (controls turbocharger)
23 Distributor and cap
24 Turbocharger outlet elbow
25 Battery
26 Exhaust downpipe
27 Washer fluid reservoir

1B•4 Maintenance & Servicing - '85-on models

Front underbody view of 1992 Saab 900 Turbo

1. Catalytic converter and exhaust system
2. Flexible brake hydraulic hose
3. Steering track rod
4. Shock absorber
5. Front suspension lower control arm
6. Crossmember
7. Engine oil drain plug
8. Manual transmission (oil drain plug hole blanked off on later models)
9. Driveshaft inner joint
10. Power steering gear

Rear underbody view of 1992 Saab 900 Turbo

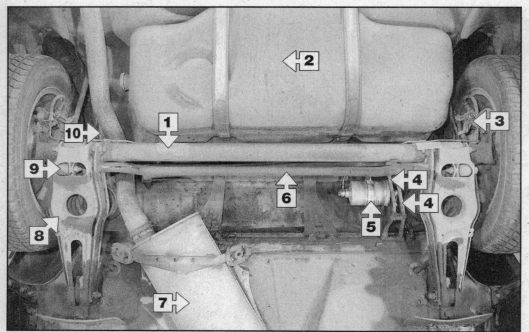

1. Rear axle tube
2. Fuel tank
3. Brake caliper
4. Brake hydraulic hoses
5. Fuel filter
6. Panhard rod
7. Rear silencer and tailpipe
8. Rear suspension trailing arm
9. Shock absorber
10. Rear suspension link

6000 miles (10 000 km) - '85-on models

Maintenance procedures

1 Introduction

This Chapter is designed to help the home mechanic maintain his/her vehicle for safety, economy, long life and peak performance.

The Chapter contains a master maintenance schedule, followed by Sections dealing specifically with each task in the schedule. Visual checks, adjustments, component renewal and other helpful items are included. Refer to the accompanying illustrations of the engine compartment and the underside of the vehicle for the locations of the various components.

Servicing your vehicle in accordance with the mileage/time maintenance schedule and the following Sections will provide a planned maintenance programme, which should result in a long and reliable service life. This is a comprehensive plan, so maintaining some items but not others at the specified service intervals, will not produce the same results.

As you service your vehicle, you will discover that many of the procedures can - and should - be grouped together, because of the particular procedure being performed, or because of the close proximity of two otherwise-unrelated components to one another. For example, if the vehicle is raised for any reason, the exhaust can be inspected at the same time as the suspension and steering components.

The first step in this maintenance programme is to prepare yourself before the actual work begins. Read through all the Sections relevant to the work to be carried out, then make a list and gather together all the parts and tools required. If a problem is encountered, seek advice from a parts specialist, or a dealer service department.

2 Intensive maintenance

If, from the time the vehicle is new, the routine maintenance schedule is followed closely, and frequent checks are made of fluid levels and high-wear items, as suggested throughout this manual, the engine will be kept in relatively good running condition, and the need for additional work will be minimised.

It is possible that there will be times when the engine is running poorly due to the lack of regular maintenance. This is even more likely if a used vehicle, which has not received regular and frequent maintenance checks, is purchased. In such cases, additional work may need to be carried out, outside of the regular maintenance intervals.

If engine wear is suspected, a compression test (refer to the relevant Part of Chapter 2) will provide valuable information regarding the overall performance of the main internal components. Such a test can be used as a basis to decide on the extent of the work to be carried out. If, for example, a compression test indicates serious internal engine wear, conventional maintenance as described in this Chapter will not greatly improve the performance of the engine, and may prove a waste of time and money, unless extensive overhaul work (Chapter 2B) is carried out first.

The following series of operations are those most often required to improve the performance of a generally poor-running engine:

Primary operations
a) Clean, inspect and test the battery (See "Weekly Checks" and Chapter 5).
b) Check all the engine-related fluids (See "Weekly Checks").
c) Check the condition and tension of the auxiliary drivebelt (Section 24).
d) Renew the spark plugs (Section 5).
e) Inspect the distributor cap, rotor arm and HT leads - as applicable (Section 26).
f) Check the condition of the air cleaner filter element, and renew if necessary (Section 13).
g) Renew the fuel filter (Section 41).
h) Check the condition of all hoses, and check for fluid leaks (Section 4).
i) Check the idle speed and mixture settings - as applicable (Section 18).

If the above operations do not prove fully effective, carry out the following secondary operations:

Secondary operations
a) Check the charging system (Chapter 5).
b) Check the ignition system (Chapter 5).
c) Check the fuel system (Chapter 4).
d) Renew the distributor cap and rotor arm - as applicable (Chapter 5).
f) Renew the ignition HT leads - as applicable (Chapter 5).

6000 miles (10 000 km)

3 Engine oil and filter renewal

Note: *Owners of high-mileage vehicles, or those who do a lot of stop-start driving, may prefer to carry out engine oil and filter renewal more frequently.*

1 Frequent oil and filter changes are the most important preventative maintenance procedures which can be undertaken by the DIY owner. As engine oil ages, it becomes diluted and contaminated, which leads to premature engine wear.

2 Before starting this procedure, gather together all the necessary tools and materials. Also make sure that you have plenty of clean rags and newspapers handy, to mop up any spills. Ideally, the engine oil should be warm, as it will drain better, and more built-up sludge will be removed with it. Take care, however, not to touch the exhaust or any other hot parts of the engine when working under the vehicle. To avoid any possibility of scalding, and to protect yourself from possible skin irritants and other harmful contaminants in used engine oils, it is advisable to wear gloves when carrying out this work. Access to the underside of the vehicle will be greatly improved if it can be raised on a lift, driven onto ramps, or jacked up and supported on axle stands (see "*Jacking and Vehicle Support*"). Whichever method is chosen, make sure that the vehicle remains level, or if it is at an angle, that the drain plug is at the lowest point. Do not confuse the transmission drain plug with the engine drain plug - the engine drain plug is towards the front left-hand side of the transmission.

3 Slacken the drain plug about half a turn **(see illustration)**. Position the draining container under the drain plug, then remove the plug completely **(see Haynes Hint overleaf)**. Recover the sealing washer from the drain plug.

3.3 Unscrewing engine oil drain plug

1B•6 6000 miles (10 000 km) - '85-on models

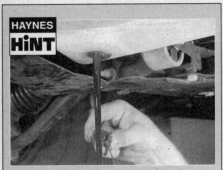

If possible, try to keep the plug pressed into the transmission casing while unscrewing it by hand the last couple of turns. As the plug releases from the threads, move it away sharply so the stream of oil runs into the container, not up your sleeve!

3.6 Oil filter mounting on B201 engine

3.7 Removing the oil filter from a B202 engine

4 Allow some time for the old oil to drain, noting that it may be necessary to reposition the container as the oil flow slows to a trickle.
5 After all the oil has drained, wipe off the drain plug with a clean rag. Check the sealing washer for condition and renew it if necessary. Clean the area around the drain plug opening, and refit the plug. Tighten the plug to the specified torque.
6 Move the container into position under the oil filter. The filter is located on an adapter housing bolted to the left-hand front of the cylinder block on B201 engines and further back on the left-hand side just in front of the alternator on B202 engines **(see illustration)**.
7 Using an oil filter removal tool if necessary, slacken the filter initially, then unscrew it by hand the rest of the way **(see illustration)**. Empty the oil from the old filter into the container and discard the filter.
8 Use a clean rag to remove all oil, dirt and sludge from the filter sealing area. Check the old filter to make sure that the rubber sealing ring hasn't stuck to the engine. If it has, carefully remove it.
9 Apply a light coating of clean engine oil to the sealing ring on the new filter, then screw it into position on the engine until it just touches the adapter housing. Now tighten the filter a further half a turn. Do not use any tools to tighten the filter since overtightening will distort the seal. Wipe clean the filter and drain plug.
10 Remove the old oil and all tools from under the car then lower the car to the ground.
11 Remove the oil filler cap and fill the engine with the correct grade and type of oil (see Specifications at the end of this Chapter). An oil can spout or funnel may help to reduce spillage. Pour in half the specified quantity of oil first, then wait a few minutes for the oil to drain to the bottom of the engine. Continue adding oil a small quantity at a time until the level is up to the lower mark on the dipstick. Adding a further 1.0 litre will bring the level up to the upper mark on the dipstick. Refit the filler cap.
12 Start the engine and run it for a few minutes; check for leaks around the oil filter seal and the drain plug. Note that there may be a delay of a few seconds before the oil pressure warning light goes out when the engine is first started, as the oil circulates through the engine oil galleries and the new oil filter before the pressure builds up.
13 Switch off the engine, and wait a few minutes for the oil to settle once more. With the new oil circulated and the filter completely full, recheck the level on the dipstick, and add more oil as necessary.
14 Dispose of the used engine oil safely, with reference to *"General repair procedures"* in the reference Section of this manual.

4 Hose and fluid leak check

1 Visually inspect the engine joint faces, gaskets and seals for any signs of water or oil leaks. Pay particular attention to the areas around the camshaft cover, cylinder head, oil filter and transmission joint faces. Bear in mind that, over a period of time, some very slight seepage from these areas is to be expected - what you are really looking for is any indication of a serious leak. Should a leak be found, renew the offending gasket or oil seal by referring to the appropriate Chapters in this manual.
2 Also check the security and condition of all the engine-related pipes and hoses including vacuum hoses. Ensure that all cable-ties or securing clips are in place and in good condition. Clips which are broken or missing can lead to chafing of the hoses, pipes or wiring, which could cause more serious problems in the future.
3 Carefully check the radiator hoses and heater hoses along their entire length. Renew any hose which is cracked, swollen or deteriorated. Cracks will show up better if the hose is squeezed. Pay close attention to the hose clips that secure the hoses to the cooling system components. Hose clips can pinch and puncture hoses, resulting in cooling system leaks.
4 Inspect all the cooling system components (hoses, joint faces etc.) for leaks **(see Haynes Hint below)**. Where any problems of this nature are found on system components, renew the component or gasket with reference to Chapter 3.
5 Check that the pressure cap on the expansion tank is fully tightened and shows no sign of coolant leakage. If possible have a Saab dealer pressure test the cooling system.
6 With the car raised, inspect the fuel tank and filler neck for punctures, cracks and other damage. The connection between the filler neck and tank is especially critical. Sometimes a rubber filler neck or connecting hose will leak due to loose retaining clamps or deteriorated rubber.
7 Carefully check all rubber hoses and metal fuel lines leading away from the fuel tank. Check for loose connections, deteriorated hoses, crimped lines, and other damage. Pay particular attention to the vent pipes and hoses, which often loop up around the filler neck and can become blocked or crimped. Follow the lines to the front of the vehicle, carefully inspecting them all the way. Renew damaged sections as necessary.
8 From within the engine compartment, check the security of all fuel hose attachments and pipe unions, and inspect the fuel hoses and vacuum hoses for kinks, chafing and deterioration.

A leak in the cooling system will usually show up as white or rust-coloured deposits on the area adjoining the leak.

6000 miles (10 000 km) - '85-on models

5.2a Removing the inspection cover for access to the spark plugs

5.2b HT spark plug leads with the inspection cover removed

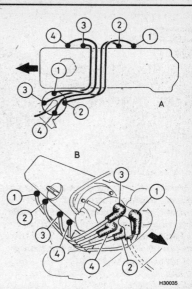

5.3 Showing spark plug HT lead positions

*A B-type engine   B H-type engine
Arrow indicates flywheel end of the engine*

5 Spark plug renewal

1 The correct functioning of the spark plugs is vital for the correct running and efficiency of the engine. It is essential that the plugs fitted are appropriate for the engine (a suitable type is specified at the end of this Chapter). If this type is used and the engine is in good condition, the spark plugs should not need attention between scheduled replacement intervals. Spark plug cleaning is rarely necessary, and should not be attempted unless specialised equipment is available, as damage can be caused to the firing ends.

2 To remove the spark plugs, first on B202 engines remove the screws and lift the inspection cover from the centre of the camshaft cover **(see illustrations)**.

3 If the marks on the spark plug (HT) leads cannot be seen, mark the leads '1' to '4', to correspond to the cylinder the lead serves (No 1 cylinder is at the timing chain end of the engine) **(see illustration)**.

4 Pull the leads from the plugs by gripping the end fitting, not the lead, otherwise the lead connection may be fractured. On B202 engines release the leads from the retainers which are located in the valve cover slots **(see illustrations)**.

5 It is advisable to remove the dirt from the spark plug recesses using a clean brush, vacuum cleaner or compressed air before removing the plugs, to prevent dirt dropping into the cylinders.

6 Unscrew the plugs using a spark plug spanner, suitable box spanner or a deep socket and extension bar **(see illustrations)**. Keep the socket aligned with the spark plug - if it is forcibly moved to one side, the ceramic insulator may be broken off. As each plug is removed, examine it as follows.

7 Examination of the spark plugs will give a good indication of the condition of the engine. If the insulator nose of the spark plug is clean and white, with no deposits, this is indicative of a weak mixture or too hot a plug (a hot plug transfers heat away from the electrode slowly, a cold plug transfers heat away quickly).

8 If the tip and insulator nose are covered with hard black-looking deposits, then this is indicative that the mixture is too rich. Should the plug be black and oily, then it is likely that the engine is fairly worn, as well as the mixture being too rich.

9 If the insulator nose is covered with light tan to greyish-brown deposits, then the mixture is correct and it is likely that the engine is in good condition.

10 The spark plug electrode gap is of considerable importance as, if it is too large or too small, the size of the spark and its efficiency will be seriously impaired. The gap should be set to the value given in the Specifications at the end of this Chapter.

11 To set the gap, measure it with a feeler blade or wire gauge and then bend open, or close, the outer plug electrode until the correct gap is achieved **(see illustration)**. The

5.4a Releasing the HT leads from the retainers located in the valve cover slots

5.4b Disconnecting the HT leads from the plugs

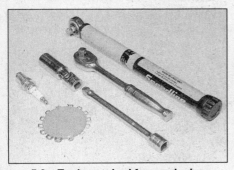

5.6a Tools required for spark plug removal, gap adjustment and refitting

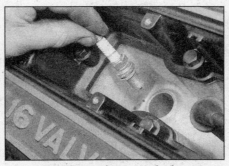

5.6b Removing a spark plug

5.11 Measuring the spark plug gap with a feeler blade

1B•8 6000 miles (10 000 km) – '85-on models

HAYNES HINT

To avoid cross-threading the plugs, fit a short length of rubber hose over the end of the spark plug. The flexible hose acts as a universal joint to help align the plug with the plug hole. Should the plug begin to cross-thread, the hose will slip on the spark plug, preventing thread damage to the aluminium cylinder head.

centre electrode should never be bent, as this will crack the insulator and cause plug failure, if nothing worse. If using feeler blades, the gap is correct when the appropriate-size blade is a firm sliding fit.
12 Special spark plug electrode gap adjusting tools are available from most motor accessory shops, or from some spark plug manufacturers.
13 Before fitting the spark plugs, check that the threaded connector sleeves are tight and the threads are clean. It is often difficult to insert spark plugs into their holes without cross-threading them. To avoid this possibility, fit a short length of 5/16 inch internal diameter rubber hose over the end of the spark plug **(see Haynes Hint)**.
14 Remove the rubber hose, and tighten the plug to the specified torque using the spark plug socket and a torque wrench **(see illustration)**. Fit the remaining spark plugs in the same manner.
15 Connect the HT leads in their correct order.
16 On the B202 engine refit the inspection cover and tighten the retaining screws.

7.3 Unscrewing the power steering fluid reservoir filler cap from the top of the pump (early models)

5.14 Tightening a spark plug

6 Brake pad, disc and handbrake check

1 Jack up the front and rear of the car and support it securely on axle stands (see *"Jacking and Vehicle Support"*). Remove the front and rear roadwheels.
2 For a quick check, the thickness of friction material remaining on each brake pad can be measured through the aperture in the caliper body. If any pad's friction material is worn to the specified thickness or less, all four pads on that axle must be renewed as a set.
3 For a comprehensive check, the brake pads should be removed and cleaned. The operation of the caliper can then also be checked, and the condition of the brake discs checked on both sides. Refer to Chapter 9 for further information.
4 Lubricate the caliper guides of the front brake calipers only (refer to Chapter 9).
5 Have an assistant operate the handbrake while checking the operation of the levers on the front calipers. If there is any sign of the levers sticking, check the handbrake lever and cables, and the mechanism within the caliper with reference to Chapter 9.
6 On completion refit the wheels and lower the car to the ground.

7 Power steering fluid level check

1 The power steering fluid reservoir is located either on the top of the power steering pump

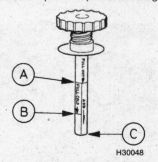

7.4 Filler cap and power steering fluid reservoir dipstick (early models)

A Full Hot   B Full Cold   C Add

on the rear right-hand side of the engine, or in a separate reservoir attached to the rear right-hand side of the engine compartment.
2 For the check, the front wheels should be pointing straight-ahead and the engine should be stopped. The car should be positioned on level ground.
3 Before removing the filler cap use a clean rag to wipe the cap and the surrounding area to prevent any foreign matter from entering the reservoir. Unscrew and remove the filler cap **(see illustration)**.
4 Where the reservoir is located on top of the pump, the dipstick is incorporated into the filler cap. There are three markings **(see illustration)**. If the engine is hot, the level should be between the "Full Hot" and "Full Cold" marks, and if the engine is cold, the level should be between the "Full Cold" and "Add" marks.
5 Where the reservoir is located on the bulkhead, there are Hot, Cold and Add marks.
6 Top-up if necessary with the specified power steering fluid **(see illustration)**. Be careful not to introduce dirt into the system, and do not overfill. Frequent topping up indicates a leak which should be investigated.

8 Automatic transmission fluid and final drive oil level check

Automatic transmission fluid

1 Three different types of oil are contained in the automatic transmission casing – the engine oil, final drive oil, and automatic transmission fluid. Check the automatic transmission fluid level as follows according to the type of transmission fitted.

Type 35 transmission

2 Run the engine at idling speed for a few minutes with the selector in P.
3 Switch off the engine and withdraw the transmission fluid dipstick located on the front right-hand side of the engine. Wipe it clean on a piece of non-fluffy rag and re-insert it. Withdraw it for the second time when the oil level should be within the limits of the cut-out on the 'cold' edge of the dipstick if the engine and transmission are cold. Use the 'hot' edge

7.6 Topping-up the power steering fluid level (late models)

6000 miles (10 000 km) – '85-on models

8.3 Automatic transmission fluid level dipstick markings

of the dipstick if the engine and transmission are at normal operating temperature **(see illustration)**.
4 If necessary top-up the fluid level through the dipstick/filler tube using only automatic transmission fluid as given in the Specifications at the end of this Chapter.

Type 37 transmission
5 Run the engine at idling speed for at least 15 seconds in each of the following selector positions: D, R and then P.
6 With the engine still running, and the selector in P, check the fluid level as described in paragraph 3.
7 If necessary top-up the fluid level through the dipstick tube using the specified fluid as given in the Specifications.
8 Allow the engine to idle for a few minutes then recheck the fluid level.

Final drive oil (automatic transmission models only)
9 The final drive oil level plug is located on the rear left-hand side of the transmission, over the left-hand driveshaft. Access to the plug is best from under the car. First jack up the car and support on axle stands (see *"Jacking and Vehicle Support"*) - make sure that the car is level.
10 Unscrew the level plug and check that the oil is level with the bottom edge of the plug hole.
11 If the level requires topping up add oil as necessary using only oil of the specified type as given in the Specifications at the end of this Chapter.
12 With the level correct, wipe clean the level plug then refit it and tighten securely. Wipe off any spilt oil.
13 Lower the car to the ground.

9 Seat belt check

1 Working on each seat belt in turn, carefully examine the seat belt webbing for cuts or any signs of serious fraying or deterioration. Pull the belt all the way out, and examine the full extent of the webbing.
2 Fasten and unfasten the belt, ensuring that the locking mechanism holds securely and releases properly when intended. Check also that the retracting mechanism operates correctly when the belt is released.
3 Check the security of all seat belt mountings and attachments which are accessible, without removing any trim or other components, from inside the vehicle.

10 Pollen filter renewal

1 The pollen filter is on the right-hand side of the bulkhead in the engine compartment.
2 On pre-1983 models, remove the four screws and slide out the element **(see illustration)**.
3 On 1983-on models the cover is retained with plastic clips which are removed by turning them through 90° either clockwise or anticlockwise. With the cover off, slide out the filter element **(see illustrations)**.
4 Fit the new pollen filter using a reversal of the removal procedure.

11 Road test

Instruments and electrical equipment
1 Check the operation of all instruments and electrical equipment.
2 Make sure all instruments read correctly, and switch on all electrical equipment in turn to check that it functions properly. Check the function of the heating, air conditioning and automatic climate control systems.

Steering and suspension
3 Check for any abnormalities in the steering, suspension, handling or road 'feel'.
4 Drive the vehicle, and check that there are no unusual vibrations or noises.
5 Check that the steering feels positive, with no excessive 'sloppiness', or roughness, and check for any suspension noises when cornering, or when driving over bumps. Check that the power steering system operates correctly. Check that the cruise control system (where fitted) operates correctly.

Drivetrain
6 Check the performance of the engine, clutch (manual transmission), transmission and driveshafts. On Turbo models, check that the boost pressure needle moves up to the red zone during brief acceleration.
7 Listen for any unusual noises from the engine, clutch (manual transmission) and transmission.
8 Make sure that the engine runs smoothly when idling, and that there is no hesitation when accelerating.
9 On manual transmission models, check that the clutch action is smooth and progressive, that the drive is taken up smoothly, and that the pedal travel is correct. Also listen for any noises when the clutch pedal is depressed. Check that all gears can be engaged smoothly, without noise, and that the gear lever action is not abnormally vague or 'notchy'.
10 On automatic transmission models, make sure that all gearchanges occur smoothly without snatching, and without an increase in engine speed between changes. Check that all the gear positions can be selected with the vehicle at rest. If any problems are found, they should be referred to a Saab dealer.
11 Listen for a metallic clicking sound from the front of the vehicle, as the vehicle is driven slowly in a circle with the steering on full lock. Carry out this check in both directions. If a clicking noise is heard, this indicates wear in a driveshaft joint (refer to Chapter 8).

Check the operation and performance of the braking system
12 Make sure that the vehicle does not pull to one side when braking, and that the wheels do not lock prematurely when braking hard.
13 Check that there is no vibration through the steering when braking.

10.2 Pollen filter (pre-1983 model)

10.3a Remove the cover . . .

10.3b . . . and slide out the pollen filter element

6000 miles (10 000 km) - '85-on models

14 Check that the handbrake operates correctly, without excessive movement of the lever, and that it holds the vehicle stationary on a slope.
15 Test the operation of the brake servo unit as follows. With the engine off, depress the footbrake four or five times to exhaust the vacuum. Start the engine, holding the brake pedal depressed. As the engine starts, there should be a noticeable 'give' in the brake pedal as vacuum builds up. Allow the engine to run for at least two minutes, and then switch it off. If the brake pedal is depressed now, it should be possible to detect a hiss from the servo as the pedal is depressed. After about four or five applications, no further hissing should be heard.

12 000 miles (20 000 km)

12 Engine oil and filter renewal

Refer to the information given in Section 3.

13 Air cleaner element renewal/cleaning

1 The air cleaner is located on the left-hand side of the engine compartment.
2 To remove the element either release the clips or remove the screws from the air cleaner cover, move the cover to one side, and withdraw the element. Where fitted, remove the insert from the bottom of the body **(see illustrations)**.
3 If cleaning the element, tap it to release the accumulated dust and if available use an air line from the inside of the element outwards. Do not attempt to wash the element.
4 Wipe clean the inside surfaces of the air cleaner and insert.
5 Refit the element (and insert where necessary) using a reversal of the removal procedure, but check all associated hoses and air ducts for condition and security.

14 EGR system check

Refer to Chapter 4, Part C, Section 2.

15 Choke fast idling speed (carburettor models) check

Refer to Chapter 4, Part A, Section 11.

16 Cooling system pressure test

Refer to the information given in Section 4.

17 Vacuum lines check

Refer to the information given in Section 4.

18 Idle speed and mixture check and adjustment

Zenith/Solex carburettor

Note: *On the Zenith carburettor if any adjustment is made to the synchronisation, it will also be necessary to adjust the idling speed and rear carburettor vent valve IN THIS ORDER. If any adjustment is made to the idling speed it will also be necessary to adjust the rear carburettor vent valve.*

1 Before checking the idle speed and mixture make sure that the damper oil level is as described in Chapter 4, Part A, Section 11.

Basic setting - metering needle

2 Unscrew the damper and cap.
3 Mark the vacuum chamber cover in relation to the carburettor body then remove the screws and withdraw the cover and spring.
4 Lift out the piston and diaphragm then using Saab tool 83 93 035 adjust the metering needle so that its shoulder is level with the bottom of the piston **(see illustration)**.

13.2a Air cleaner and cover on a B201 fuel injection engine

13.2b Removing the air cleaner element on a B201 fuel injection engine

13.2c Releasing the air cleaner cover clips on a B202 fuel injection engine

13.2d Removing the cover...

13.2e ...and filter element on a B202 fuel injection engine

13.2f The insert locates on the bottom of the air cleaner filter element

12 000 miles (10 000 km) – '85-on models

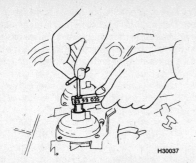

18.4 Using the special tool to adjust the mixture on Zenith/Solex carburettors

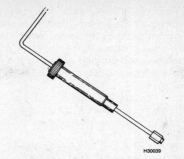

18.12 Carburettor mixture adjusting tool

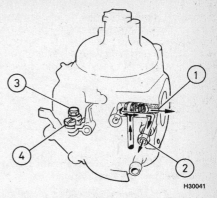

18.17 Carburettor float chamber ventilation on the Zenith carburettor

1 To the air cleaner   3 Vent valve adjustment
2 To the atmosphere   4 Not used

5 Refit the piston, diaphragm, and cover together with the spring making sure that the diaphragm tab engages the cut-out. Top-up the damper with oil to within 10.0 mm of the top.

Twin carburettors – synchronisation and idling

6 Run the engine to normal operating temperature and allow it to idle.
7 Adjust the idle speed screw on the front carburettor so that the engine is running at the specified idling speed (see illustration).
8 If an air flow balancer is available check that the air flow through both carburettors is identical. Alternatively a rough check can be made using a short length of plastic tube positioned in each carburettor inlet in turn – the amount of hiss should be identical. If necessary loosen the locknut and adjust the screw on the intermediate linkage. Tighten the locknut afterwards.

CO content (mixture) setting

9 Run the engine to normal operating temperature and make sure that the choke control is fully off.
10 Where the setting speed is 2000 rpm clamp the distributor vacuum hose, and disconnect the crankcase ventilation hoses.
11 Connect a CO meter and tachometer to the engine and check that the readings are as given in the Specifications with the engine idling.
12 If adjustment is necessary unscrew and remove the damper and cap and use tool 83 93 035 to raise or lower the metering needle as required. To use the tool hold the piston and diaphragm stationary with the sleeve and turn the spindle clockwise to richen the mixture or anticlockwise to weaken the mixture (see illustration).
13 Stop the engine and refit the hoses as necessary. Refit the damper and cap. Remove the CO meter and tachometer.

Vent valve adjustment (Zenith carburettor)

14 Refer to the illustration 18.7. The float chamber vent valve on the front carburettor of twin carburettor models is set by the manufacturers and will not normally require adjustment. If it is altered it will affect the other slow running adjustments.
15 Similarly the vent valve on the rear carburettor will not normally require adjustment, however if necessary the following check and adjustment may be made.
16 The vent valve facilitates good starting when the engine is hot by preventing vaporised fuel entering the inlet manifold. On twin carburettor engines it also prevents 'run-on' after the ignition has been switched off.
17 To check the valve first connect a hose to the atmosphere air aperture (see illustration). If the check is to be made with the carburettor removed from the engine and the fuel inlet pipe disconnected, plug the fuel inlet.
18 Fully close the throttle and check that it is not possible to blow through the hose (ie the float chamber is open to the atmosphere).
19 Open the throttle lever 0.5 to 1.0 mm and check that it is now possible to blow through the hose proving that the valve is half open.
20 Further opening of the throttle should prevent blowing through the hose as the valve will only allow venting to the air cleaner.
21 If necessary adjust the valve by loosening the locknut and turning the screw. Tighten the locknut afterwards. If adjustment has been made recheck the slow running adjustment as described previously.

Pierburg 175 CDUS carburettor

22 The engine should be at its normal operating temperature, the cooling fan having cut in at least once, and the choke fully disengaged.
23 Before checking the idle speed and mixture make sure that the damper oil level is as described in Chapter 4, Part A, Section 11.
24 Connect up a tachometer to the engine in accordance with the manufacturer's instructions, then check that the idle speed is as specified. Adjust if necessary via the idle speed adjustment screw.
25 Idle mixture adjustment should only be carried out if an exhaust gas analyser is available. Detach the crankcase ventilation hose nipple from the camshaft cover and plug the hose. Connect up a temporary fume extraction hose to the camshaft cover and direct it to atmosphere (downstream of the CO meter).
26 A CO extraction hose with an open-type coupling must be used to prevent a vacuum build-up in the exhaust system.
27 Detach the EGR vacuum hose and plug the hose to the EGR valve (where equipped). Detach the vacuum hose from the distributor and plug it. Switch on the driving lights.
28 Start the engine and run it at a steady 2000 rpm. When the radiator cooling fan cuts in, check the CO content reading. If the CO content is not within the specified amount, adjustment is necessary.
29 Prise free the seal plug on the float chamber cover then, using an 8 mm socket, turn the adjuster as necessary (see illustration). Turning clockwise will reduce the CO content. If a swivel type spanner is used to make the adjustment, take care not to short it onto the starter motor or alternator connections.
30 If a CO adjustment was made, the fast idle speed should be checked as follows.
31 With the engine at its normal operating temperature, detach the distributor advance hose and pull out the choke control so that

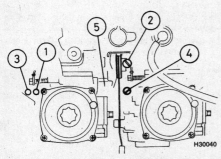

18.7 Zenith/Solex carburettor adjusting screw locations

1 Redundant tapped hole
2 Synchronising adjustment
3 Vent valve (front carburettor)
4 Vent valve (rear carburettor)
5 Engine idling speed

1B•12 12 000 miles (20 000 km) – '85-on models

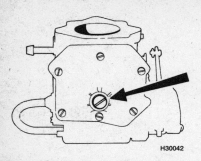

18.29 CO content (mixture) setting screw (arrowed) on the Pierburg carburettor

the line on the lever aligns with the fast idle screw. Check the engine speed and compare it with the fast idle speed specified. If adjustment is required, turn the screw in the required direction.

32 Pull out the choke control knob and ensure that the lever at the carburettor fully deflects. Now push the control knob fully in and check that the lever moves back to the lowest travel limit and that the fast idle screw is not in contact with it. Adjust if necessary.

33 To check the modulator valve, disconnect its hose and plug it with the engine idling **(see illustration)**. The CO level should rise and the idle speed may fall slightly. If not, the valve is faulty or the hose is leaking.

34 When all checks and adjustments are complete, disconnect the test gear and remake the original connections.

35 Fit a new seal plug to the CO adjuster screw when required. In some EEC countries (though not yet in the UK) this is required by law.

Fuel injection models

Note: *The checking of idle speed and mixture is possible on all models by using a tachometer and exhaust gas analyser. However it is not possible to adjust the idling*

speed and mixture on models with the Lucas CU fuel injection system or on models with idle control and a catalytic converter. On all other models continue as follows.

36 Before checking and adjusting the idle speed or mixture settings, always check the following first:
 a) Check the ignition timing (Section 27).
 b) Check that the spark plugs are in good condition and correctly gapped (Section 5).
 c) Check that the accelerator cable is correctly adjusted (Chapter 4, Part A or B).
 d) Check that the crankcase breather hoses are secure, with no leaks or kinks (Section 23).
 e) Check that the air cleaner filter element is clean (Section 13).
 f) Check that the exhaust system is in good condition (Section 22).
 g) If the engine is running very roughly, check the compression pressures (Chapter 2, Part A).

37 The engine should be at normal temperature, the cooling fan having cut in at least once.

Note: *Checking/adjustment should be completed as quickly as possible so that the engine is still at its normal operating temperature. If the radiator electric cooling fan operates again, wait for it to stop before continuing. Clear any excess fuel from the inlet manifold by racing the engine two or three times to between 2000 and 3000 rpm, then allow it to idle again.*

38 Ensure that all electrical loads are switched off, then stop the engine and connect a tachometer to it following the manufacturer's instructions. Where a tachometer is fitted to the instrument panel, this may be used instead. If the idle mixture is to be checked, connect an exhaust gas analyser in accordance with the manufacturer's instructions.

Models with Bosch CI (K-Jetronic) fuel injection (except models with catalytic converter and modulating valve)

39 The idle speed adjustment screw is located on the throttle housing. Start the engine and allow it to idle, then check that the idle speed is as given in the Specifications. If adjustment is necessary, loosen the locknut and adjust the screw in to reduce the speed or out to increase the speed. Tighten the locknut after making the adjustment.

40 The idle mixture adjustment screw is located on the top of the airflow meter next to the fuel distributor and a long Allen key will be required to turn it **(see illustration)**. The screw is fitted with a tamperproof plug which must be removed in order to make any adjustments. **Note:** *In some countries adjustment is only allowed by qualified persons and it is also a legal requirement to renew the plug after making an adjustment.*

41 With the engine idling at the correct idle speed, check that the CO level is as given in the Specifications. If adjustment is necessary, use an Allen key or fine screwdriver to turn the mixture adjustment screw in or out (in very small increments) until the CO level is as given in the Specifications. Turning the screw in (clockwise) richens the mixture and increases the CO level, turning it out will weaken the mixture and reduce the CO level.

42 If necessary readjust the idling speed.

43 Temporarily increase the engine speed, then allow it to idle and recheck the settings.

44 When adjustments are complete, stop the engine and disconnect the test equipment. Where necessary fit a new tamperproof plug to the mixture adjustment screw.

Models with Bosch LH-Jetronic fuel injection (except models with idle air control and catalytic converter)

45 The idle speed adjustment screw is located on the throttle housing. Start the engine and allow it to idle, then check that the idle speed is as given in the Specifications. If adjustment is necessary, loosen the locknut and adjust the screw in to reduce the speed or out to increase the speed. Tighten the locknut after making the adjustment.

46 The idle mixture adjustment screw is located on the airflow meter and may be hidden under a tamperproof cap. First remove the cap. With the engine idling at the correct idling speed, check that the CO level is as given in the Specifications. If adjustment is necessary, turn the mixture adjustment screw in or out (in very small increments) until the CO level is as given in the Specifications. Turning the screw in (clockwise) richens the mixture and increases the CO level, turning it out will weaken the mixture and reduce the CO level.

47 If the CO percentage is higher than 6% and it is not possible to adjust it within limits, stop the engine and disconnect the wiring plug from the airflow meter. Connect a digital multimeter across terminals 3 and 6 of the airflow meter, then turn the adjustment screw as necessary to obtain 380 ohms. If it is not possible to do this the airflow meter is faulty and should be renewed, however if it is possible reconnect the wiring and recheck the CO level as described in paragraph 46.

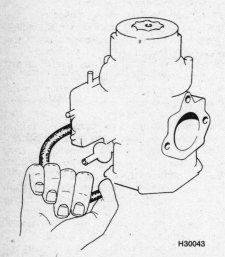

18.33 Checking the modulator valve on the Pierburg carburettor

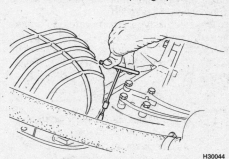

18.40 Adjusting the idling mixture on engines with the Bosch CI (K-Jetronic) fuel injection system

12 000 miles (20 000 km) – '85-on models

48 If necessary readjust the idling speed.
49 Temporarily increase the engine speed, then allow it to idle and recheck the settings.
50 When adjustments are complete, stop the engine and disconnect the test equipment. Where necessary, fit a new tamperproof plug to the mixture adjustment screw.

Other models with LH-Jetronic and Lucas CU14 systems

51 The idle speed is controlled by an idle air control valve in conjunction with the system ECU. With the engine at normal operating temperature, check that the idle speed is as given in the Specifications. No adjustment is possible.
52 The idle mixture is controlled by the Lambda sensor in conjunction with the system ECU. With the engine at normal operating temperature, check that the idle mixture is as given in the Specifications. No adjustment is possible.
53 If incorrect readings are obtained, take the car to a Saab dealer and have the system checked with the special tester.

19 Deceleration device (carburettor engines) check and adjustment

1 The deceleration device assists combustion during engine overrun in order to prevent the emission of unburned hydrocarbons. There are several types of device fitted - prior to 1984 a vacuum-controlled device and a dashpot were fitted, and later models had a disc valve incorporated in the throttle butterfly. An electric type was also fitted.
2 The vacuum type incorporates a spring-loaded valve actuated by inlet manifold vacuum and the device effectively by-passes the throttle valve until the engine speed approaches normal idling speed.
3 The dashpot system prevents the throttle shutting too quickly which would otherwise result in unburnt hydrocarbons.
4 The disc valve system consists of a spring tensioned valve mounted on the throttle valve. When the throttle is shut and the engine is in overrun, the valve allows additional air to by-pass the throttle until the engine speed approaches idling speed.
5 The electric type incorporates an engine speed transmitter and a solenoid operated idling stop. During engine overrun the idling speed is increased to about 1550 rpm if the speed of the car exceeds 18 mph (30 kph).
6 To check the vacuum controlled type run the engine to normal operating temperature and connect a tachometer. Adjust the idling speed to 875 rpm then rev up the engine to 3000 rpm, release the throttle, and check that the time for the engine to return to idling speed is 4 to 6 seconds. Unscrew the unit screw until the valve closes then adjust the idling speed of the engine in the normal way. Now screw in the unit screw until the engine speed is 1600 rpm and back off the screw two complete turns. Finally adjust the idling speed again. Note that the check should be completed without the radiator fan cutting in.
7 To check the dashpot system, run the engine to normal operating temperature, then connect a tachometer and adjust the idling speed to 875 rpm. Rev up the engine to 3000 rpm, release the throttle, and check that the time for the engine to return to idling speed is 3 to 6 seconds. If not, loosen the locknut and reposition the unit.
8 To check the electric type connect a tachometer and run the engine to normal operating temperature. Disconnect the positive solenoid cable and connect battery voltage. Rev the engine then release the throttle and check that the idling speed is maintained at 1550 rpm. If necessary, adjust the solenoid with the screw. Connect a test lamp between the positive solenoid wire and earth, then drive the car at about 25 mph (40 kph) and declutch. Brake the car and check that the test lamp goes out at a speed of 18.6 mph (30 kph).
9 The vacuum valve can be removed by disconnecting the hose and unscrewing the unit from the inlet manifold. The electric speed transmitter can be removed by lowering the facia panel and unplugging the unit. The solenoid can be removed by disconnecting the wiring and unscrewing the unit.

20 Valve clearances (Turbo models, except B202 engine) - checking and adjustment

Note: *This Section applies to the B201 engine only since the B202 engine is fitted with hydraulic tappets.*

1 The valve clearances must be adjusted with the engine cold. First remove the cylinder head cover as described in Chapter 2A, Section 1.
2 Turn the engine using a spanner on the crankshaft pulley bolt until No 1 cam lobe peak (timing chain end of the engine) is pointing away from the cam follower (tappet).
3 Using a feeler blade check that the clearance between the heel of the cam and the cam follower is as given in the Specifications. The feeler blade should be a firm sliding fit **(see illustration)**. If not, record the actual clearance.

> **HAYNES HiNT**: *Draw the valve positions on a piece of paper, and record the actual clearances as you check them.*

4 Check the remaining clearances using the same procedure and record them as necessary.
5 If adjustment is required remove the camshaft with reference to Chapter 2, Part A, Section 5, then remove the cam followers and shims from the relevant valves, keeping them identified. Measure the thickness of the existing shims and by comparison with the correct clearances determine the thickness of the new shims. Use a micrometer to measure the shims **(see illustration)**.
6 Fit the correct shims followed by the cam followers and camshaft (Chapter 2, Part A, Section 5).
7 Rotate the engine several times then recheck the clearances.
8 Refit the cylinder head cover with reference to Chapter 2.

21 Turbocharger - check

The following work should be carried out by your Saab dealer as special equipment is required to make the check, however refer to Chapter 4, Part B, Sections 16 and 17 for more information.
a) *Check and adjust the turbo charge pressure.*
b) *Check the turbo overpressure switch.*
c) *Check the turbo charge pressure regulator seal.*
d) *Check the turbo APC control unit security seal - where applicable.*
e) *Check the turbo fuel boosting device.*

20.3 Checking the valve clearances

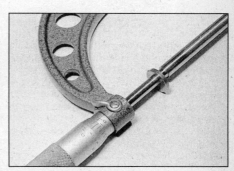

20.5 Using a micrometer to check the cam follower shim thickness

12 000 miles (20 000 km) – '85-on models

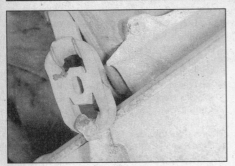

22.2 Typical exhaust rubber mounting

22 Exhaust system check

Warning: If the engine has just been running, take care not to touch the exhaust system, especially the front section, as it may still be hot.

1 Position the car over an inspection pit or on car ramps. Alternatively raise the front and rear of the car and support on axle stands (see *"Jacking and Vehicle Support"*).
2 Examine the exhaust system over its entire length checking for any damaged, broken or missing mountings **(see illustration)**, the security of the pipe clamps, and for excess rust and corrosion.
3 Check the exhaust mounting on the transmission for tightness.
4 Lower the car to the ground on completion.

23 Crankcase ventilation system check

Check all crankcase ventilation and vacuum hoses for damage and leakage (refer to Chapter 4). Where necessary remove the hoses and clear them of any blockage.

24 Auxiliary drivebelt check

Checking the condition of the auxiliary drivebelts

1 The main drivebelt is used to drive the alternator and on the H-type B201 engine the water pump. On later models two drivebelts are fitted instead of the one. The secondary drivebelt is used to drive the power steering pump and where fitted a further drivebelt drives the air conditioning compressor.
2 Access to the drivebelts is restricted by their position at the rear of the engine and the use of a mirror will help when checking them. Using a spanner on the crankshaft pulley bolt, rotate the crankshaft so that the entire length of the drivebelts can be examined. Examine the drivebelt for cracks, splitting, fraying, or other damage. Check also for signs of glazing (shiny patches) and for separation of the belt plies. Renew the belt if worn or damaged.

Air conditioning compressor drivebelt - removal, refitting and tensioning

Removal

3 On models with the B201 engine loosen the adjusting and pivot bolts and move the compressor to one side to release the tension on the drivebelt.
4 On models with the B202 engine loosen the tensioner bolt and swivel the tensioner to release the tension on the drivebelt.
5 Slip the drivebelt from the pulleys on the crankshaft, compressor and on the B202 engine, the tensioner.

Refitting and tensioning

6 Locate the drivebelt on the pulleys then move the compressor or tensioner until it is possible to depress the belt approximately 15 mm under firm thumb pressure midway between the crankshaft and compressor pulleys. Tighten the adjustment and pivot bolts as applicable. Note that Saab technicians use a special tensioning tool to set the drivebelt tension - if any doubt exists about the tension of the belt, it should be checked by a Saab dealer.

Power steering pump drivebelt - removal, refitting and tensioning

Removal

7 Where fitted, remove the air conditioning compressor drivebelt as described earlier.
8 Loosen the power steering pump pivot and adjustment bolts and move the pump to release the tension on the drivebelt **(see illustration)**.
9 Slip the drivebelt from the pulleys on the crankshaft and power steering pump pulleys **(see illustration)**.

Refitting and tensioning

10 Locate the drivebelt on the pulleys then move the pump until it is possible to depress the belt between 10 and 15 mm under firm thumb pressure midway between the crankshaft and pump pulleys. Tighten the pivot and adjustment bolts.
11 Refit and tension the air conditioning compressor drivebelt as described earlier.
12 Note that Saab technicians use a special tensioning tool to set the drivebelt tension - if any doubt exists about the tension of the belt, it should be checked by a Saab dealer.

Alternator/water pump drivebelt(s) - removal, refitting and tensioning

Removal

13 Remove the air conditioning compressor drivebelt and the power steering pump drivebelt as described earlier.
14 Disconnect the battery negative lead.
15 Loosen the alternator pivot and adjustment bolts and swivel the alternator in toward the cylinder block **(see illustration)**.
16 Slip the drivebelt(s) from the alternator, crankshaft and where applicable the water pump pulleys **(see illustration)**.

24.8 Loosening the power steering pump drivebelt adjustment bolts

24.9 Removing steering pump drivebelt from crankshaft and pump pulleys

24.15 Loosening the pivot and adjustment bolts

24.16 Removing the alternator/water pump drivebelts

12 000 miles (20 000 km) – '85-on models

Refitting and tensioning

17 Locate the drivebelt on the pulleys, then lever the alternator away from the cylinder block until it is possible to depress the belt(s) between 10 and 15 mm under firm thumb pressure midway between the longest run of the belt. Lever the alternator on the drive end bracket to prevent straining the brackets. It is helpful to semi-tighten the adjustment link bolt before tensioning the drivebelt.
18 Fully tighten the alternator pivot and adjustment bolts.
19 Reconnect the battery negative lead.
20 Refit and tension the power steering pump drivebelt and the air conditioning compressor drivebelt.
21 Note that Saab technicians use a special tensioning tool to set the drivebelt tension – if any doubt exists about the tension of the belt, it should be checked by a Saab dealer.

25 Spark plug renewal

Refer to the information given in Section 5.

26 Distributor and ignition HT lead check

1 On B202 engines remove the inspection cover from the top of the cylinder head cover.
2 Wipe clean all of the ignition HT leads using a dry cloth. Check that the leads are attached to the spark plugs, distributor cap and ignition coil securely and are located in their fasteners correctly. Make sure that the leads are held clear of other components to prevent the possibility of arcing.
3 If necessary the HT leads may be checked as follows. Remove the leads together with the distributor cap, then connect an ohmmeter to each end of the leads and the appropriate terminal within the cap in turn. If the resistance is greater than that given in the Specifications, check that the lead connection in the cap is good before renewing the lead.
4 With the distributor cap removed, check the cap and rotor for hairline cracks and signs of arcing. If evident renew the component **(see illustration)**.

26.4 Removing the distributor cap for checking

27 Ignition timing check

1 Before adjusting the ignition timing on models equipped with a conventional ignition system, check and if necessary adjust the contact points dwell angle as described in Chapter 1A, Section 23. The initial setting method should be used in order to start the engine or for emergency roadside repairs, but the final setting must always be made using a stroboscopic timing light. The clutch (or torque converter) housing cover incorporates an attachment point for a special instrument which gives an instant ignition timing read-out, but this instrument will not normally be available to the home mechanic.

Initial setting

2 Remove No 1 spark plug (timing chain end of the engine) and put a finger over the plug hole.
3 Turn the engine in the normal running direction (clockwise viewed from the timing chain end of engine, anti-clockwise from the front of the car) until pressure is felt in No 1 cylinder indicating that the piston is commencing its compression stroke. Use a spanner on the crankshaft pulley bolt or engage top gear and pull the car forwards (except automatic transmission models).
4 Continue turning the engine until the correct ignition timing mark appears opposite the mark on the clutch housing cover timing hole **(see illustration)**.
5 Remove the distributor cap and check that the rotor arm is pointing in the direction of the No 1 terminal of the cap. The rotor arm should also be pointing to the timing groove in the rim of the distributor body **(see illustration)**.
6 On conventional ignition models connect a 12 volt test lamp and leads between the coil terminal 1 (with blue wire attached) and a suitable earthing point on the engine. Loosen the distributor retaining bolt(s) and switch on the ignition. If the bulb is already lit turn the distributor body slightly anti-clockwise until the bulb goes out. Turn the distributor body clockwise until the bulb just lights up indicating that the points have just opened, then tighten the bolt(s). Switch off the ignition and remove the test lamp.

27.4 Ignition timing marks on the flywheel

7 On electronic ignition models remove the rotor arm and dust cover and check that the rotor arms are aligned with the stator posts on the inductive transmitter type, or one of the rotor slots is aligned with the transmitter on the Hall effect type. If not, loosen the distributor retaining bolt(s) and turn the distributor body as necessary. Make sure that the rotor arm is still pointing in the direction of the No 1 terminal of the cap. Refit the dust cover and rotor arm.
8 Refit the distributor cap and No 1 spark plug and HT lead.
9 Once the engine has been started check the timing stroboscopically as follows.

Final setting

10 Disconnect and plug the vacuum hose at the distributor. Where a delay valve is fitted in the vacuum line do not disconnect the hose with the engine running otherwise foreign matter may cause the valve to be inoperative.
11 Connect a timing light and tachometer to the engine in accordance with the manufacturer's instructions.
12 Start the engine and run it at the speed given in the Specifications.
13 Point the timing light at the timing hole in the clutch housing cover. The correct timing mark on the flywheel should appear to be stationary and aligned with the mark on the cover. If not, loosen the distributor retaining bolt(s) and turn it clockwise to advance or anticlockwise to retard the ignition. Tighten the bolt(s) when the setting is correct.
14 Gradually increase the engine speed while still pointing the timing light at the timing marks. The flywheel marks should appear to move clockwise when viewed from the front of the car proving that the distributor centrifugal weights are operating.
15 Run the engine at about 4000 rpm and note the ignition timing, then reconnect the vacuum hose. On non-Turbo models the timing should advance a few degrees, but on Turbo models it should retard a few degrees, proving that the vacuum unit is operating.
16 Switch off the engine and disconnect the timing light and tachometer. Check that the vacuum hose is secure.

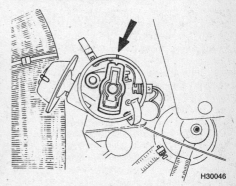

27.5 Distributor rotor arm on the B-type engine pointing to the ignition timing groove

12 000 miles (20 000 km) - '85-on models

30.1 The manual transmission oil level plug

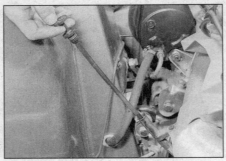

30.2a Removing the early manual transmission oil level dipstick

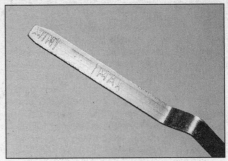

30.2b Manual transmission oil level dipstick markings

28 Delay valve check and renewal

1 The delay valve is located in the distributor vacuum advance pipe and its purpose is to delay ignition advance during acceleration so reducing nitrous oxide emissions.

Checking

2 Connect a tachometer and timing light to the engine according to the manufacturers instructions, and run the engine to normal operating temperature. Have an assistant open the throttle quickly so that the engine runs at 3000 rpm. Using the timing light check that the time from when the throttle was opened to when the ignition advances is between 4 and 8 seconds. If not, renew the delay valve.

Removal and refitting

3 To remove the delay valve disconnect the hoses and remove the valve.
4 Refitting is a reversal of removal, but note that the white end of the valve must be towards the distributor.

29 Headlight beam alignment

1 It is recommended that the headlamp alignment is carried out by a Saab dealer using modern beam setting equipment. However in an emergency the following procedure will provide an acceptable light pattern, however the alignment should be checked as soon as possible by a Saab dealer.
2 Position the car on a level surface with tyres correctly inflated approximately 5.0 metres in front of, and at right-angles to, a wall or garage door.
3 Draw a vertical line on the wall corresponding to the centre line of the car. The position of the line can be ascertained by marking the centre of the front and rear screens with crayon then viewing the wall from the rear of the car.
4 Measure the distance between the headlamp centres and their height above the ground, then mark the positions on the wall.

5 Switch the headlamps on main beam and check that the areas of maximum illumination coincide with the marks on the wall. On dipped beam the area of maximum illumination should be 50 mm below the centre marks.
6 If adjustment is necessary turn the adjustment screws on the headlamp rim (99 models) or on the rear of the headlamp (900 models) until the setting is correct. On 900 models insert a screwdriver in one of the bonnet hinge holes to hold the bonnet half open so that the screws can be reached when adjusting the headlamps.

30 Manual transmission oil level check

1 The manual transmission oil level plug is located midway along the right-hand side of the transmission (see illustration). On early models a separate level dipstick is fitted.
Note: *The level plug at the front right-hand side of the transmission is for the primary gear housing, but is only used for adding some of the oil to the primary housing after overhauling the transmission. When the transmission is in motion oil is delivered to the primary gears via an oil catcher.*

2 On models with a dipstick remove the dipstick and wipe it clean with a piece of cloth, then re-insert it fully. Withdraw it again and check that the level is correct **(see illustrations)**.
3 On models with an oil level plug first wipe clean the area around the level plug, then unscrew and remove the plug which incorporates a dipstick and check that the transmission oil level is on the maximum mark.
4 If the oil level requires topping up, add oil as necessary using only good quality oil of the specified type as given in the Specifications at the end of this Chapter. On early models add oil through the dipstick tube and on later models add oil through the level plug hole. Frequent topping up indicates a leak which should be investigated and repaired.
5 With the level correct either insert the dipstick or wipe clean the level plug, refit and tighten it. Wipe off any spilt oil

31 Automatic transmission fluid and final drive oil level check

Refer to the information given in Section 8.

32 Handbrake adjustment check

Refer to the information given in Section 6.

33 Brake line and flexible hose check

1 Raise the front and rear of the car and securely support on axle stands (see *"Jacking and Vehicle Support"*). Remove all wheels.
2 Thoroughly examine all brake lines and brake flexible hoses for security and damage. To check the flexible hoses, bend them slightly in order to show up any cracking of the rubber.
3 Check the complete braking system for any signs of brake fluid leakage.
4 Where necessary carry out repairs to the braking system with reference to Chapter 9.

34 Steering, suspension and shock absorber check

1 Raise the front and rear of the car, and securely support it on axle stands (see *"Jacking and Vehicle Support"*).
2 Visually inspect all balljoint dust covers and the steering rack-and-pinion gaiters for splits, chafing or deterioration. Any wear of these components will cause loss of lubricant, together with dirt and water entry, resulting in rapid deterioration of the balljoints or steering gear.
3 Grasp each front roadwheel in turn at the 12 o'clock and 6 o'clock positions, and try to rock it **(see illustration)**. Very slight free play may be felt, but if the movement is appreciable, further investigation is necessary to determine the source. Continue rocking the wheel while an assistant depresses the footbrake. If the movement is now eliminated or significantly reduced, it is likely that the hub

12 000 miles (20 000 km) – '85-on models 1B•17

34.3 Rocking the roadwheel to check steering/suspension components

bearings are at fault. If the free play is still evident with the footbrake depressed, then there is wear in the suspension joints or mountings.
4 Now grasp the wheel at the 9 o'clock and 3 o'clock positions, and try to rock it as before. Any movement felt now may again be caused by wear in the hub bearings or the steering track-rod balljoints. If the inner or outer balljoint is worn, the visual movement will be obvious.
5 Using a large screwdriver or flat bar, check for wear in the suspension mounting bushes by levering between the relevant suspension component and its attachment point. Some movement is to be expected as the mountings are made of rubber, but excessive wear should be obvious. Also check the condition of any visible rubber bushes, looking for splits, cracks or contamination of the rubber.
6 Check for any signs of fluid leakage around the front suspension struts and rear shock absorbers. Should any fluid be noticed, the suspension strut/shock absorber is defective internally, and should be renewed. **Note**: *Suspension struts/shock absorbers should always be renewed in pairs on the same axle.*
7 Working from the front to the rear of the vehicle, check the security of all suspension and steering nuts and bolts. If necessary, tighten them to the specified torque as given in Chapter 10.
8 Lower the car to the ground.
9 The efficiency of the suspension struts and shock absorbers may be checked by depressing each corner of the car in turn. If the struts/shock absorbers are in good condition, the body will rise and then settle in its normal position. If it continues to rise and fall, the suspension strut or shock absorber is probably suspect. Examine also the suspension strut/shock absorber upper and lower mountings for any signs of wear.
10 With the car standing on its wheels, have an assistant turn the steering wheel back and forth about an eighth of a turn each way. There should be very little, if any, lost movement between the steering wheel and roadwheels. If this is not the case, closely observe the joints and mountings previously described, but in addition, check the steering column universal joints for wear, and the rack-and-pinion steering gear itself.

35 Front wheel alignment check

Due to the special measuring equipment necessary to check the wheel alignment accurately, checking and adjustment is best left to a Saab dealer or similar expert. Note that most tyre-fitting shops now possess sophisticated checking equipment. Refer to Chapter 10 for more information.
Before having the front wheel alignment checked, all tyre pressures should be checked and if necessary adjusted (see *"Weekly Checks"*).

36 Facia security check

Check the facia panel mounting bolts for security and tighten as necessary. Refer to Chapter 12.

Every 24 000 miles (40 000 km)

37 Air cleaner element renewal – except Turbo APC models

Refer to the information given in Section 13.

38 EGR system check

Refer to the information given in Section 14.

39 Delay valve check and renewal

Refer to the information given in Section 28.

40 Front wheel alignment check

Refer to the information given in Section 35.

Every 30 000 miles (50 000 km)

41 Fuel filter renewal (fuel injection models)

⚠ **Warning: Before carrying out the following operation, refer to the precautions given in 'Safety first!' at the beginning of this manual, and follow them implicitly.**

1 On pre-1986 models the fuel filter is located in the engine compartment on the left-hand wheel arch, in the pressure line between the fuel pump and the fuel supply rail. On later models the filter is located beneath the rear of the car on the right-hand side just in front of the fuel tank.
2 Depressurise the fuel system with reference to Chapter 4, Part B, Section 7.
3 On 1986-on models jack up the rear of the car and support on axle stands (see *"Jacking and Vehicle Support"*).
4 Clean the areas around the fuel filter inlet and outlet unions.
5 Position a small container or cloth rags beneath the filter to catch spilt fuel.
6 Unscrew one of the banjo bolts while holding the flats on the filter with a further spanner, and disconnect the fuel line. Recover the sealing washers **(see illustrations)**.

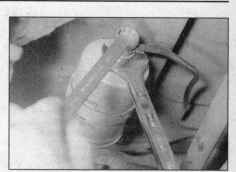

41.6a Unscrew the banjo bolts on the fuel filter . . .

1B•18 Every 30 000 miles (50 000 km) - '85-on models

7 Unscrew the remaining banjo bolt while holding the flats on the filter, disconnect the fuel line, and recover the sealing washers.
8 Noting the direction of the arrow marked on the filter body, unscrew the retaining clip screw and withdraw the filter (see illustrations).
9 Locate the new filter in the retaining clip and tighten the clip. Make sure the flow arrow on the filter body is pointing towards the outlet which leads to the fuel injection supply rail (see illustration).
10 Check the condition of the sealing washers and renew them if necessary.
11 Refit the banjo couplings and hoses to the top and bottom of the filter together with the washers, and tighten the bolts securely while holding the filter with a further spanner.
12 Wipe away any excess fuel, then start the engine and check the filter union connections for leaks.
13 If applicable, lower the car to the ground.
14 The old filter should be disposed of safely, bearing in mind that it will be highly inflammable.

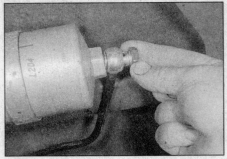

41.6b ... and remove the bolts and sealing washers

41.8a Unscrew the retaining clip screw ...

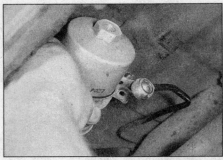

41.8b ... and remove the fuel filter

41.9 The arrow on the filter body must point towards the outlet

42 Valve clearances check and adjustment

Refer the information given in Section 20.

Every 54 000 miles (90 000 km)

43 Fuel filter renewal (B202 engine only)

Refer to the information given in Section 41.

Every 66 000 miles (110 000 km)

44 Automatic transmission fluid and filter renewal

Carry out the following work as described in Chapter 7, Part B.

a) Visually check the automatic transmission for signs of fluid or oil leaks.
b) Renew the automatic transmission fluid and where applicable clean the filter.

Every 2 years (regardless of mileage)

45 Brake fluid renewal

 Warning: *Brake hydraulic fluid can harm your eyes and damage painted surfaces, so use extreme caution when handling and pouring it. Do not use fluid that has been standing open for some time, as it absorbs moisture from the air. Excess moisture can cause a dangerous loss of braking effectiveness.*

1 The procedure is similar to that for the bleeding of the hydraulic system as described in Chapter 9, except that the brake fluid reservoir should be emptied by siphoning, using a clean poultry baster or similar before starting, and allowance should be made for the old fluid to be expelled when bleeding a section of the circuit. To prevent contamination from the fluid in the clutch hydraulic system (which uses the same reservoir) the latter system should be bled as described in Chapter 6 after bleeding the brake system.

2 Working as described in Chapter 9, open the first bleed screw in the sequence, and pump the brake pedal gently until nearly all the old fluid has been emptied from the fluid reservoir. Top-up to the 'MAX' level with new fluid, and continue pumping until only the new fluid remains in the reservoir, and new fluid can be seen emerging from the bleed screw. Tighten the screw, and top-up the reservoir level to the 'MAX' level line.

3 Old hydraulic fluid is invariably much darker in colour than the new, making it easy to distinguish the two.

Every 2 years - '85-on models

4 Work through all the remaining bleed screws in the sequence until new fluid can be seen at all of them. Be careful to keep the master cylinder reservoir topped-up to above the 'MIN' level at all times, or air may enter the system and greatly increase the length of the task.
5 When the operation is complete, check that all bleed screws are securely tightened, and that their dust caps are refitted. Wash off all traces of spilt fluid, and recheck the reservoir fluid level.
6 Check the operation of the brakes before taking the car on the road.

46 Coolant renewal

Cooling system draining

⚠ **Warning:** *Wait until the engine is cold before starting this procedure. Do not allow antifreeze to come in contact with your skin, or with the painted surfaces of the car. Rinse off spills with plenty of water. Never leave antifreeze lying around in an open container - antifreeze can be fatal if ingested.*

1 With the engine completely cold, unscrew and remove the expansion tank filler cap. Turn the cap a few turns anti-clockwise then wait until any pressure remaining in the system is released. Fully unscrew and remove the cap.
2 Remove the filler cap then position a suitable container beneath the radiator drain plug - on 99 models the plug is on the left-hand side, on 900 models it is on the right-hand side.
3 Set the heater control to maximum heat then unscrew the drain plug and drain the coolant.
4 Position another container beneath the right-hand side of the cylinder block, unscrew the drain plug, and drain the coolant from the block.
5 On 900 models, loosen the clip and disconnect the bottom hose from the radiator in order to drain the remaining coolant **(see illustration)**. Any remaining coolant on 99 models can only be drained after removing the radiator and inverting it, however this will only be necessary where severe contamination has occurred.
6 If the coolant has been drained for a reason other than renewal, then provided it is clean and less than 3 years old, it can be re-used.

Cooling system flushing

7 If coolant renewal has been neglected, or if the antifreeze mixture has become diluted, then in time, the cooling system may gradually lose efficiency, as the coolant passages become restricted due to rust, scale deposits, and other sediment. The cooling system efficiency can be restored by flushing the system clean.

46.5 Radiator bottom hose connection on 900 models

8 The radiator should be flushed independently of the engine, to avoid unnecessary contamination.

Radiator flushing

9 To flush the radiator, first disconnect the top and bottom hoses from the radiator.
10 Insert a garden hose into the radiator top inlet. Direct a flow of clean water through the radiator, and keep flushing until clean water emerges from the radiator bottom outlet.
11 If after a reasonable period, the water still does not run clear, the radiator can be flushed with a good proprietary cleaning agent. It is important that their manufacturer's instructions are followed carefully. If the contamination is particularly bad, remove the radiator then insert the hose in the radiator bottom outlet, and reverse-flush the radiator.

Engine flushing

12 To flush the engine, first remove the thermostat as described in Chapter 3, then temporarily refit the thermostat cover.
13 With the top and bottom hoses disconnected from the radiator, insert a garden hose into the radiator top hose. Direct a clean flow of water through the engine, and continue flushing until clean water emerges from the radiator bottom hose. Also insert the hose into the expansion tank to flush away any sediment.
14 Disconnect the hose from the inlet manifold, insert the hose, and allow water to run through the heater and out of the bottom hose.
15 On completion of flushing, refit the thermostat and reconnect the hoses with reference to Chapter 3.
16 Remove the container from under the car, then refit and tighten the cylinder block and radiator drain plugs.

Cooling system filling

17 Before attempting to fill the cooling system, make sure that all hoses and clips are in good condition, and that the clips are tight. Note that an antifreeze mixture must be used all year round, to prevent corrosion of the engine components (see following sub-Section).
18 Pour the correct amount of antifreeze into

46.19 On later models, the cooling system bleed screw is located on the thermostat cover

a container, then add water until the total amount equals the cooling system capacity (see *Specifications*).
19 Slowly fill the system through the expansion tank filler neck, allowing time for trapped air to escape. Where fitted, loosen the bleed screw on the thermostat cover, thermostat housing or heater temperature control valve and allow trapped air to escape. Tighten the screw when bubble-free coolant comes out **(see illustration)**.
20 Top-up to the 'MAX' mark on the side of the expansion tank.
21 Refit and tighten the filler cap.
22 Start the engine, and run it at a fast idle speed until the cooling fan cuts in, and then cuts out. This will purge any remaining air from the system. Stop the engine.
23 Allow the engine to cool, then check the coolant level with reference to *"Weekly Checks"*. Top-up the level if necessary. Saab recommend that the level be checked again after a few days and topped up as necessary.

Antifreeze mixture

24 The antifreeze should always be renewed at the specified intervals. This is necessary not only to maintain the antifreeze properties, but also to prevent corrosion which would otherwise occur as the corrosion inhibitors become progressively less effective.
25 Always use an ethylene-glycol based antifreeze which is suitable for use in mixed-metal cooling systems. The quantity of antifreeze and levels of protection are indicated in the Specifications.
26 Before adding antifreeze, the cooling system should be completely drained, preferably flushed, and all hoses checked for condition and security.
27 After filling with antifreeze, a label should be attached to the expansion tank filler neck, stating the type and concentration of antifreeze used, and the date installed. Any topping-up should be made with the same type and concentration of antifreeze.
28 Do not use engine antifreeze in the windscreen/tailgate washer system, as it will cause damage to the vehicle paintwork. A screenwash additive should be added to the washer system in the quantities stated on the bottle.

1B•20 Every 10 years - '85-on models

Every 10 years (regardless of mileage)

47 Airbag check

The airbag system should be checked every 10 years by a Saab dealer who will have the necessary equipment and expertise to ensure that the system is in good condition.

Servicing Specifications

Lubricants and fluids
Refer to the end of *"Weekly Checks"*

Capacities

Engine oil
Including filter:
- B201 engine .... 3.8 litres
- B202 engine .... 4.0 litres
- Difference between MAX and MIN dipstick marks .... 1.0 litre

Cooling system
- All 900 models .... 10.0 litres

Manual transmission
- 4-speed .... 2.5 litres
- 5-speed .... 3.0 litres

Automatic transmission
- Including torque converter and oil cooler .... 8.0 litres

Automatic transmission final drive
- Type 35 .... 1.25 litres
- Type 37 .... 1.4 litres

Fuel tank
- Up to 1989 models .... 63 litres
- 1990-on models .... 68 litres

Braking system capacity
- All models .... 0.55 litres approx.

Power steering capacity
- All models .... 0.75 litre

Engine
- Difference between MIN and MAX marks on dipstick .... 1.0 litre
- Oil filter .... Champion C142
- Valve clearances - cold: **Checking value**   **Setting value**
  - Inlet (all models) .... 0.15 to 0.30 mm   0.20 to 0.25 mm
                                                                (0.006 to 0.012 in)   (0.008 to 0.010 in)
  - Exhaust (except Turbo) .... 0.35 to 0.50 mm   0.40 to 0.45 mm
                                                                (0.014 to 0.020 in)   (0.016 to 0.018 in)
  - Exhaust (Turbo) .... 0.40 to 0.50 mm   0.45 to 0.50 mm
                                                                (0.016 to 0.020 in)   (0.018 to 0.020 in)

Cooling system
Antifreeze mixture:
- 28% antifreeze .... Protection down to -15ßC (5ßF)
- 50% antifreeze .... Protection down to -30ßC (-22ßF)

Note: *Refer to antifreeze manufacturer for latest recommendations.*

Servicing Specifications - '85-on models

Fuel system
Air filter element:
  All models except 16V Turbo .... Champion W198
Fuel filter:
  Saab 900 16V Turbo .... Champion L204
  Saab 900 900i 16V .... Champion L204
  Saab 900i EMS and GLE .... Champion L203
  Saab 900i S 1990-on .... Champion L204
  Saab 900i 16V 1991-on .... Champion L203
Idling speed:
  Bosch CI (K-Jetronic) and LH-Jetronic fuel injection systems:
    Non-Turbo models .... 850 rpm
    All Turbo models .... 850 rpm
Idling CO% .... 4.5% maximum

Ignition system
Distributor:
  Contact points gap .... 0.4 mm
  Dwell angle .... 50° ± 3°
Ignition timing:
UK models:
  Turbo APC (B201) .... 20° BTDC at 2000 rpm and vacuum hose disconnected
  Turbo 16 (B202) .... 16° BTDC at 850 rpm and vacuum hose disconnected
  Non-Turbo fuel injection models (B201) .... 20° BTDC at 2000 rpm and vacuum hose disconnected
  Non-Turbo fuel injection models (B202) (1986-on) 14° BTDC at 850 rpm and vacuum hose disconnected
Sweden and Switzerland models:
  Turbo (B201) .... 20° BTDC at 2000 rpm and vacuum hose disconnected
  Turbo 16 (B202) .... 16° BTDC at 850 rpm and vacuum hose disconnected
  Non-Turbo fuel injection models (B201) .... 18° BTDC at 2000 rpm and vacuum hose disconnected
  Non-Turbo fuel injection models (B202) (1986-on) 14° BTDC at 850 rpm and vacuum hose disconnected
Spark plugs:
  Saab 900 B201 non-Turbo .... Champion N9YCC
  Saab 900 B201 Turbo .... Champion N7YCC
  Saab 900 B202 non-Turbo .... Champion C9YCC
  Saab 900 B202 Turbo .... Champion C7YCC
Spark plug electrode gap* .... 0.8 mm
Ignition HT lead set:
  900 GL/GLS/Turbo .... Champion LS 13
Ignition HT lead resistance .... Approximately 600 ohms per 100 mm length

*The spark plug gap quoted is that recommended by Champion for their specified plugs listed above. If spark plugs of any other type are to be fitted, refer to their manufacturer's recommendations.

Brakes
Brake pad friction material minimum thickness:
  Up to 1988 model year .... 1.0 mm
  From 1988 model year onwards .... 4.0 mm

Tyre pressures
Refer to the end of "Weekly Checks"

Wiper blades
Windscreen and tailgate:
  Saab 900 .... Champion X-4103

Torque wrench settings

	Nm	lbf ft
Spark plugs	28	21
Cylinder block drain plug	55	41
Engine oil drain plug:		
Manual transmission models	50	37
Automatic transmission models	37	27
Cylinder block coolant drain plug	25	19
Manual transmission filler and drain plugs	49	36
Automatic transmission drain plug	7	5
Roadwheel nuts	98	72

Notes

Chapter 2 Part A:
Engine in-car repair procedures

Contents

Camshaft(s) and cam followers - removal, inspection and refitting ...5
Compression test - description and interpretation 2
Crankshaft oil seals - renewal 10
Cylinder head - removal and refitting 6
Cylinder head cover - removal and refitting 4
Engine oil and filter renewal ....See Chapter 1
Engine oil level check ....See "Weekly Checks"
Engine/transmission mountings - inspection and renewal 12
Flywheel - removal, inspection and refitting 11
General information 1
Oil cooler - removal and refitting 8
Oil pressure warning light switch - removal and refitting 9
Oil pump - removal, inspection and refitting 7
Top dead centre (TDC) for No 1 piston - locating 3

Degrees of difficulty

| Easy, suitable for novice with little experience | Fairly easy, suitable for beginner with some experience | Fairly difficult, suitable for competent DIY mechanic | Difficult, suitable for experienced DIY mechanic | Very difficult, suitable for expert DIY or professional |

Specifications

Engine (general)
Designation:
  1985 cc engine with single overhead camshaft (SOHC):
    Pre-October 1981 .... B201* Type B
    October 1981-on .... B201* Type H
  1985 cc engine with twin overhead camshafts (DOHC) B202*
*B indicates petrol engine
Bore .... 90.00 mm
Stroke .... 78.00 mm
Direction of crankshaft rotation .... Anti-clockwise (viewed from front of vehicle)
No 1 cylinder location .... At timing chain (rear) end of engine
Compression ratio:
  Single carburettor (CM, CA) - 1979/1980 .... 9.2 : 1
  Single carburettor (CM, CA) - 1981-on .... 9.5 : 1
  Twin carburettor (TM, TA) - 1979/1980 .... 9.2 : 1
  Twin carburettor (TM, TA) - 1981-on .... 9.5 : 1
  Fuel injection (IM, IA) - 1979/1980 .... 9.2 : 1
  Fuel injection (IM, IA) - 1981-on .... 9.5 : 1
  Turbo without APC (SM, SA) - 1979 to 1982 .... 7.2 : 1
  Turbo with APC (SAM, SAA) - 1982-on .... 8.5 : 1
  Turbo 16 (SLM, SLA) - 1984-on .... 9.0 : 1
Maximum power:
  Single carburettor (CM, CA) .... 73 kW (98 bhp)
  Twin carburettor (TM, TA) .... 79 kW (106 bhp)
  Fuel injection (IM, IA) .... 87 kW (117 bhp)
  Turbo without APC (SM, SA) - 1979 to 1982 .... 107 kW (144 bhp)
  Turbo with APC (SAM, SAA) - 1982-on .... 107 kW (144 bhp)
  Turbo 16 (SLM, SLA) - 1984-on .... 129 kW (173 bhp)
Maximum torque:
  Single carburettor (CM, CA) .... 162 Nm at 3500 rpm
  Twin carburettor (TM, TA) .... 164 Nm at 3300 rpm
  Fuel injection (IM, IA) .... 167 Nm at 3700 rpm
  Turbo without APC (SM, SA) - 1979 to 1982 .... 235 Nm at 3000 rpm
  Turbo with APC (SAM, SAA) - 1982-on .... 235 Nm at 3000 rpm
  Turbo 16 (SLM, SLA) - 1984-on .... 270 Nm at 3000 rpm

2A•2 Engine in-car repair procedures

Maximum torque:
- Single carburettor (CM, CA) .... 162 Nm at 3500 rpm
- Twin carburettor (TM, TA) .... 164 Nm at 3300 rpm
- Fuel injection (IM, IA) .... 167 Nm at 3700 rpm
- Turbo without APC (SM, SA) - 1979 to 1982 .... 235 Nm at 3000 rpm
- Turbo with APC (SAM, SAA) - 1982-on .... 235 Nm at 3000 rpm
- Turbo 16 (SLM, SLA) - 1984-on .... 270 Nm at 3000 rpm

Camshaft(s)
Drive .... Chain from crankshaft
Number of bearings .... 5 each camshaft
Camshaft bearing journal diameter (outside diameter):
- B201 engine .... 28.94 mm
- B202 engine .... 28.922 to 28.935 mm

Endfloat:
- B201 engine .... 0.08 to 0.25 mm
- B202 engine .... 0.08 to 0.35 mm

Lubrication system
Oil pump type:
- B201 engine .... Bi-rotor type, driven from idler shaft
- B202 engine .... Bi-rotor type, driven off the crankshaft
Minimum oil pressure at 80°C .... 3.0 bars at 2000 rpm
Oil pressure warning switch operating pressure .... 0.3 to 0.5 bars
Clearance between pump outer rotor and housing (B201 engine) ..... 0.05 to 0.09 mm
Clearance between pump outer rotor and housing (B202 engine) ..... 0.03 to 0.08 mm

Torque wrench settings

	Nm	lbf ft
Main bearings	110	81
Big-end bearings	55	41
Camshaft bearing caps:		
B201 engine	18	13
B202 engine	15	11
Cylinder head cover (1979 to 1980 models)	2	1.5
Cylinder head cover (B201 1981-on models)	5	3.7
Cylinder head cover (B202 1981-on models)	15	11
Crankshaft pulley:		
Up to 1990	190	140
From 1991	175	129
Rear engine plate (flywheel end):		
To 1980	20	15
From 1981	25	19
Flywheel:		
17.0 mm A/F bolts	60	44
19.0 mm A/F bolts	85	63
Oil pump (1979 to 1980 models)	18	13
Oil pump (1981-on models)	8	6
Chain tensioner:		
B201	12	9
B202	63	47
Chain guide (B201)	12	9
Camshaft sprocket:		
B201	20	15
B202	63	47
Timing cover:		
To 1980	20	15
From 1981	25	19
Oil pressure switch:		
To 1982	10	7
From 1982 (M14)	35	26
Engine oil drain plug:		
Manual transmission models	50	37
Automatic transmission models	37	27
Cylinder block coolant drain plug	25	19

Engine in-car repair procedures 2A•3

Torque wrench settings - continued

	Nm	lbf ft
Cylinder head bolts:		
B201 (1984-on) and B202 with 15 mm hex-head or Torx M12 up to 1987:		
Stage 1	60	44
Stage 2	80	59
Stage 3	Run engine to normal operating temperature and allow to cool for 30 minutes	
Stage 4 (slacken in turn and re-tighten)	80	59
Stage 5	Angle-tighten through 90°	
B201 (1984-on) and B202 with 15 mm hex-head or Torx M12 from 1988:		
Stage 1	60	44
Stage 2	80	59
Stage 3	Angle-tighten through 90°	
Idler shaft plate (B-type engine)	20	15

1 General information

How to use this Chapter

This Part of Chapter 2 describes those repair procedures that can reasonably be carried out on the engine while it remains in the car. If this work is being carried out as part of an engine overhaul as described in Part B, any preliminary dismantling procedures can be ignored.

Part B describes the removal of the engine/transmission from the vehicle, and the full overhaul procedures that can then be carried out.

Engine description

The engine is of in-line four-cylinder type mounted longitudinally at the front of the car with the flywheel/driveplate at the front. Either a single-overhead camshaft (SOHC) 8-valve or double-overhead camshaft (DOHC) 16-valve cylinder head is fitted. Up to October 1981 a type-B SOHC engine was fitted, but this was modified to the lighter type-H SOHC engine from this date. Both engines are similar but the type-B engine incorporates an idler shaft to drive the distributor, water pump, oil pump, and fuel pump. Drive to the front wheels is through the gearbox attached to the bottom of the engine. Drive to the front wheels is through the transmission attached to the bottom of the engine.

The crankshaft runs in five main bearings. Thrustwashers are fitted to the centre main bearing (upper half only) to control crankshaft endfloat.

The connecting rods rotate on horizontally-split bearing shells at their big-ends. The pistons are attached to the connecting rods by fully floating gudgeon pins, which are retained in the pistons by circlips. The aluminium-alloy pistons are fitted with three piston rings - two compression rings and an oil control ring.

The cylinder block is of cast iron and the cylinder bores are an integral part of the cylinder block.

The inlet and exhaust valves are closed by coil springs, and operate in guides pressed into the cylinder head; the valve seat inserts are also pressed into the cylinder head, and can be renewed separately if worn. On SOHC engines there are two valves per cylinder, on DOHC engines there are four valves per cylinder.

Both single and twin camshafts are driven by a single row timing chain. On SOHC engines the camshaft operates the valves via cam followers fitted with separate shims which can be changed in order to adjust the valve clearance. On DOHC engines the camshafts operate the 16 valves via hydraulic cam followers. The hydraulic cam followers maintain a predetermined clearance between the low-point of the cam lobe and the end of the valve stem using hydraulic chambers and a tension spring. The followers are supplied with oil from the main engine lubrication circuit.

On the B201 engine the engine mountings are of rubber and are designed to be progressive in action - the greater the load the higher the resistance. On the B202 engine hydraulic engine mountings were gradually introduced.

Lubrication is by means of a bi-rotor oil pump driven either from the idler shaft on the B201 engine or from the front of the crankshaft on the B202 engine. A relief valve limits the oil pressure during high engine speeds by returning excess oil to the transmission. Oil is drawn from the transmission through a strainer and, after passing through the oil pump, forced through an externally-mounted full-flow filter and oil cooler into galleries in the cylinder block/crankcase. From there, the oil is distributed to the crankshaft (main bearings), camshaft bearings and cam followers. On Turbo models it also lubricates the turbocharger. The big-end bearings are supplied with oil via internal drillings in the crankshaft, while the camshaft lobes and valves are lubricated by splash, as are all other engine components. The oil filter is located on the left-hand side of the cylinder block, either at the front (B-type engines) or at the rear (H-type and B202 engines).

Repair operations possible with the engine in the car

The following work can be carried out with the engine in the car:
a) Compression pressure - testing.
b) Cylinder head cover - removal and refitting.
c) Camshaft(s) - removal, inspection and refitting
d) Cylinder head - removal and refitting.
e) Cylinder head and pistons - decarbonising (refer to Part B of this Chapter)
f) Crankshaft oil seal (flywheel end) - renewal
g) Oil pump - removal, inspection and refitting
h) Flywheel (manual transmission models only) - removal, inspection and refitting
i) Engine/transmission mountings - inspection and renewal

2 Compression test - description and interpretation

1 When engine performance is down, or if misfiring occurs which cannot be attributed to the ignition or fuel systems, a compression test can provide diagnostic clues as to the engine's condition. If the test is performed regularly, it can give warning of trouble before any other symptoms become apparent.

2 The engine must be fully warmed-up to normal operating temperature, the battery must be fully charged, and all the spark plugs must be removed. The aid of an assistant will also be required.

3 Disconnect the coil HT lead from the centre of the distributor cap, and earth it on the cylinder block. Use a jumper lead or similar wire to make a good connection. Alternatively disconnect the low tension wiring plug from the distributor. On models with a catalytic converter the fuel pump must also be disabled by removing the relevant fuse.

4 Fit a compression tester to the No 1 cylinder spark plug hole (No 1 cylinder is at the timing chain end of the engine (ie at the bulkhead end) - the type of tester which screws into the plug thread is to be preferred.

2A•4 Engine in-car repair procedures

2.6 Checking the compression pressures

3.1 TDC mark on the engine plate. Also note advance timing marks on the flywheel

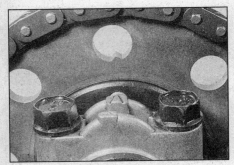

3.4a TDC alignment marks on the camshaft sprocket and bearing cap (H-type engine)

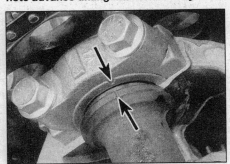

3.4b TDC alignment marks on the camshaft and bearing cap (B202 engine)

5 Have the assistant hold the throttle wide open, and crank the engine on the starter motor; after one or two revolutions, the compression pressure should build up to a maximum figure, and then stabilise. Record the highest reading obtained.
6 Repeat the test on the other cylinders, recording the pressure in each (see illustration).
7 All cylinders should produce very similar pressures; a difference of more than 2 bars between any two cylinders indicates a fault. Note that the compression should build up quickly in a healthy engine; low compression on the first stroke, followed by gradually-increasing pressure on successive strokes, indicates worn piston rings. A low compression reading on the first stroke, which does not build up during successive strokes, indicates leaking valves or a blown head gasket (a cracked head could also be the cause).
8 Although Saab do not specify exact compression pressures, as a guide, any cylinder pressure of below 10 bars can be considered as less than healthy. Refer to a Saab dealer or specialist if in doubt as to whether a given pressure reading is acceptable.
9 If the pressure in any cylinder is low, carry out the following test to isolate the cause. Pour a little clean oil into that cylinder through its spark plug hole, and repeat the test.
10 If the addition of oil temporarily improves the compression pressure, this indicates that bore or piston wear is responsible for the pressure loss. No improvement suggests that leaking or burnt valves, or a blown head gasket, may be to blame.

11 A low reading only from two adjacent cylinders is almost certainly due to the head gasket having blown between them; the presence of coolant in the engine oil will confirm this.
12 If one cylinder is about 20 percent lower than the others and the engine has a slightly rough idle, a worn camshaft lobe could be the cause.
13 On completion of the test, refit the spark plugs and reconnect the ignition system and fuel pump as necessary.

3 Top dead centre (TDC) for No 1 piston - locating

1 Timing marks are provided on the flywheel/driveplate perimeter through a window in the transmission cover (see illustration). Note: *With the timing marks correctly aligned, No 1 piston (at the timing chain end of the engine) will be at top dead centre (TDC).*
2 Using a socket or spanner on the crankshaft pulley turn the engine until the TDC '0' mark on the flywheel/driveplate is aligned with the timing mark on the transmission cover. Number 1 piston (at the timing chain end of the engine) should be at the top of its compression stroke. The compression stroke can be confirmed by removing the No 1 spark plug and checking for compression with a finger as the piston nears the top of its stroke.
3 Remove the cylinder head cover with reference to Section 4.

4 Check that the TDC marks on the sprocket end(s) of the camshaft(s) or in one of the sprocket holes, are aligned with the corresponding TDC marks on the camshaft bearing cap(s) (see illustrations). If necessary, turn the crankshaft to align the marks.

4 Cylinder head cover - removal and refitting

Removal

B201 engine (B-type to 1980)

1 Disconnect the crankcase ventilation hose from the cylinder head cover.
2 Disconnect the ignition HT leads from the spark plugs and from the cylinder head cover. If necessary identify them for position first.
3 Unscrew the retaining screws and lift the cylinder head cover from the top of the cylinder head. Recover the gasket.

B201 engine (H-type from 1981)

4 Turn the engine to the TDC position with reference to Section 3. If this is not done the distributor drive will prevent the removal of the cover because the drive dog will not be vertical.
5 Disconnect the crankcase ventilation hoses (see illustration).
6 Disconnect the HT leads from the spark plugs and coil and disconnect the wiring from the distributor.
7 Using an Allen key, unscrew the bolts and lift the cylinder head cover from the top of the cylinder head (see illustration).

4.5 Crankcase ventilation hoses on the valve cover on the H-type engine

4.7 Unscrewing the cylinder head cover mounting bolts (H-type engine)

8 Prise the gasket from the groove in the cylinder head cover.

B202 engine

9 Unscrew the screws and remove the inspection cover from the centre of the cylinder head cover.
10 Disconnect the ignition HT leads from the spark plugs and remove the distributor cap. Refer to Chapter 5 if necessary.
11 Disconnect the crankcase ventilation hoses and pull out the adapter from the cylinder head cover (see illustrations).
12 Unbolt and remove the cylinder head cover and remove the rubber gaskets and special split rubber plugs (see illustrations). Note on early models the split rubber plugs are separate whereas on later models they are incorporated into the outer gasket. If the cover is stuck, tap it gently with the palm of your hand to free it.

Refitting

13 Clean the contact surfaces of the cylinder head cover and cylinder head. Obtain new gaskets.

B201 engine (B-type to 1980)

14 Check the cylinder head cover for distortion due to over-tightening the retaining screws. If this has occurred the contact pressure between the screw holes will be reduced - restore the contact surface before refitting the cover.
15 Refit the cylinder head cover using a new gasket, and tighten the retaining screws securely to the specified torque.
16 Reconnect the ignition HT leads to the spark plugs and attach them to the fasteners on the cylinder head cover.
17 Reconnect the crankcase ventilation hose.

B201 engine (H-type from 1981)

18 Locate the new gasket in the groove in the cylinder head cover.
19 Refit the cylinder head cover making sure that the distributor drive dog engages the end of the camshaft correctly. Tighten the retaining bolts.
20 Reconnect the HT leads to the spark plugs and coil. Reconnect the wiring to the distributor.
21 Reconnect the crankcase ventilation hoses.

B202 engine

22 On early models, locate the special split rubber plugs on the cylinder head, then apply a 4 mm thick bead of silicone sealant to the corners of the head and over the rubber plugs as shown (see illustration). On later models locate the combined rubber ring in the outer groove in the cylinder head cover. Locate the inner rubber ring in the inner groove.
23 Refit the cylinder head cover and insert the securing bolts. Tighten the bolts progressively to the specified torque starting with the bolts at the distributor end and the central bolt at the timing end.

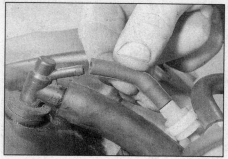

4.11a Disconnect the crankcase ventilation hoses . . .

4.12a Removing the cylinder head cover . . .

24 Reconnect the crankcase ventilation hoses.
25 Refit the distributor cap and reconnect the ignition HT leads to the spark plugs.
26 Refit the inspection cover and tighten the retaining screws.

5 Camshaft(s) and cam followers - removal, inspection and refitting

Note: The following procedure describes removal and refitting of the camshaft(s) and cam followers with the cylinder head in position in the car, however if it is required to remove the cylinder head for other reasons, the work can be carried out on the bench after removing the cylinder head.

Removal

1 Position the engine at TDC compression on No 1 cylinder as described in Section 3. Disconnect the battery negative lead.

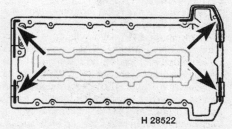

4.22 Areas to apply silicone sealant on the cylinder head cover

4.11b . . . and pull out the adaptor

4.12b . . . and rubber gaskets

2 Remove the cylinder head cover as described in Section 4.

B201 engine (B-type to 1980)

3 Remove one of the camshaft sprocket bolts, then use this bolt to secure the camshaft sprocket to the mounting plate. Tighten the bolt securely. If the sprocket slips downwards, the chain tensioner will extend and the engine will have to be removed to reset it.
4 Unscrew and remove the remaining three sprocket bolts. The sprocket is now held securely on the mounting plate.
5 Unscrew the bolts from the camshaft bearing caps evenly and in diagonal sequence. Note that the bolts extend through the camshaft bearing housing into the cylinder head. Also note that the two bolts at the timing chain end are different to the others.
6 Identify the bearing caps for position, then remove them and lift the camshaft from the head.
7 With the camshaft removed lift out the cam followers keeping them identified for position. If necessary use a valve grinding suction tool or a magnet to remove them.
8 Remove the shims from the tops of the valves, keeping them identified for position.
9 Unscrew the remaining mounting bolts and lift the camshaft bearing housing from the top of the cylinder head. Note the position of the locating dowels.

B201 engine (H-type from 1981)

10 Unbolt the fuel pump from the cylinder head and move the plunger away from the camshaft. Refer to Chapter 4A if necessary.

2A•6 Engine in-car repair procedures

5.16a No 1 camshaft bearing cap (H-type engine)

5.16b No 5 camshaft bearing cap (H-type engine)

5.16c Lifting the camshaft from the cylinder head (H-type engine)

11 Before removing the sprocket from the camshaft the timing chain tensioner must be released and the timing chain wedged so that it remains fully engaged with the crankshaft sprocket.

12 On models up to approximately 1984, obtain a suitable hooked tool or use a length of metal dowel rod bent at right-angles at one end, then insert the hooked end in the slotted hole at the top of the tension spring and pull the tool upwards to release the tension. Locate a suitable wedge (such as a screwdriver handle) between the timing chain and release the tensioner so that the chain is held firmly.

13 On models from approximately 1984-on, a ratchet-type timing chain tensioner is fitted. Before unbolting the camshaft sprocket this tensioner and the timing chain can be locked by inserting a suitable wedge (such as a screwdriver handle) between the fixed and pivoting chain guides. This will hold the timing chain fully engaged with the crankshaft sprocket.

14 With the timing chain tension released, unbolt the sprocket from the end of the camshaft. Position the sprocket and timing chain on the chain guides. Make sure that the sprocket remains fully engaged with the chain in order to maintain the valve timing - use locking wire to keep the chain on the sprocket.

15 Unscrew the bolts from the camshaft bearing caps evenly and in diagonal sequence. Note that the bolts extend through the camshaft bearing housing into the cylinder head. Also note that the two bolts at the timing chain end are different to the others.

16 Identify the bearing caps for position then remove them and lift the camshaft from the head. Number the caps from the timing end of the engine (see illustrations).

17 With the camshaft removed lift out the cam followers keeping them identified for position (see illustration). If necessary use a valve grinding suction tool or a magnet to remove them.

18 Remove the shims from the tops of the valves again keeping them identified for position (see illustration).

19 Unscrew the remaining mounting bolts and lift the camshaft bearing housing from the top of the cylinder head (see illustration). Note the position of the locating dowels.

B202 engine

20 Remove the distributor (see Chapter 5).

21 Unscrew and remove the timing chain tensioner from the left-hand rear of the cylinder head. On later models, first unscrew the centre bolt and remove the spring, then unscrew and remove the tensioner from the cylinder head (see illustrations).

22 While holding each camshaft stationary

5.17 Removing a cam follower (H-type engine)

5.18 Removing a shim from the top of a valve (H-type engine)

5.19 Lifting the camshaft bearing housing from the cylinder head (H-type engine)

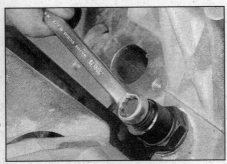

5.21a On later models, unscrew the centre bolt . . .

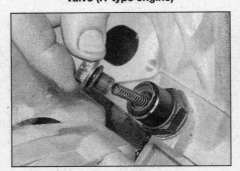

5.21b . . . and remove the bolt . . .

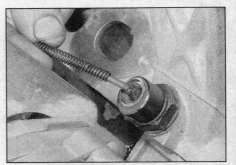

5.21c . . . and spring

Engine in-car repair procedures 2A•7

5.22 Removing the timing chain sprocket retaining bolts

5.25a The camshaft bearing caps are marked for position here

5.25b Markings on another camshaft bearing cap

with a spanner on the special flats at the flywheel/driveplate end of the camshafts, unscrew the bolts then withdraw the sprockets and allow them to rest on the timing chain guides (see illustration). Keep the sprockets fully engaged with the timing chain using locking wire. Note that the sprockets have projections which engage with cut-outs in the ends of the camshafts. The timing cover incorporates a shoulder which prevents the timing chain coming off the crankshaft sprocket teeth.

23 **Do not** rotate the crankshaft or camshafts with the timing chain or sprockets removed, or piston/valve contact may occur.

24 On early engines fitted with oil supply pipes, remove the oil supply pipes from the camshaft bearing caps, keeping them identified for position to ensure correct refitting. Take care not to twist the pipes as they are removed as this may damage the contact surfaces.

25 Check that the camshaft bearing caps and the camshafts are identified for position. The bearing caps are stamped 1 to 5 on the inlet side and 6 to 10 on the exhaust side - do not confuse these marks with the moulded markings on each cap (see illustrations).

26 Progressively unscrew the bearing cap bolts so that the caps are not stressed unduly by the valve springs. Fully remove the bolts and lift off the caps, then lift the camshafts from the cylinder head. Note that the inner bearing cap bolts (except at the timing chain end) have black heads and incorporate drillings for the oil supply to the hydraulic cam followers; always make sure that the correct bolts are fitted. Keep the camshafts identified for location (see illustrations).

27 Obtain sixteen small, clean plastic containers and number them 1I to 8I (inlet) and 1E to 8E (exhaust) or alternatively divide a larger container into sixteen compartments similarly marked for the inlet and exhaust camshafts. Using a rubber sucker or a magnet, withdraw each hydraulic cam follower in turn and place it in its respective container (see illustrations). **Do not** interchange the cam followers. To prevent the oil draining from the hydraulic cam followers, pour fresh oil into the containers until it covers them all.

Inspection

28 Examine the camshaft bearing surfaces and cam lobes for signs of wear ridges and scoring. Renew the camshaft(s) if any of these conditions are apparent. Examine the condition of the bearing surfaces on the camshaft journals, in the camshaft bearing

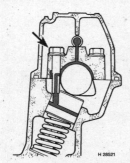

5.26a The camshaft bearing cap inner bolts (arrowed) are hollow for the oil supply to the hydraulic cam followers

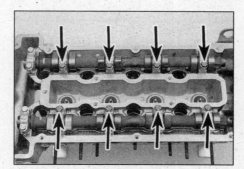

5.26b Locations (arrowed) of the black-headed inner bearing cap bolts incorporating oil drillings

5.26c The two types of camshaft bearing cap bolt - the right-hand bolt is an inner one and is coloured black

5.26d Remove the bolts . . .

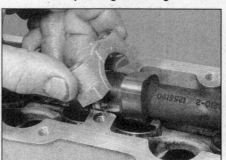

5.26e . . . bearing caps . . .

5.26f . . . and lift the camshafts from the cylinder head

2A•8 Engine in-car repair procedures

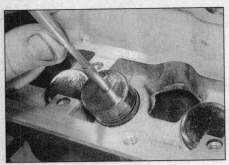

5.27a Using a magnet to remove a hydraulic cam follower

5.27b Hydraulic cam follower removed from the cylinder head. Followers should be stored in an oil bath while removed

caps and in the bearing housing or cylinder head as applicable. On B202 engines, if the head or cap bearing surfaces are worn excessively, the cylinder head will need to be renewed. On B201 engines, the bearing housing and caps can be renewed separately if worn excessively. If the necessary measuring equipment is available, camshaft bearing journal wear can be checked by direct measurement and comparison with the specifications given at the beginning of this Chapter.

29 Camshaft endfloat can be measured by locating the camshaft in the bearing housing or cylinder head as applicable, refitting the sprocket and using feeler blades between the shoulder on the front of the camshaft and the front bearing surface.

30 Check the cam followers where they contact the bores in the bearing housing or cylinder head for wear, scoring and pitting. Occasionally a hydraulic cam follower may be noisy and require renewal, and this will have been noticed when the engine was running. It is not easy to check a hydraulic cam follower for internal damage or wear once it has been removed, but if there is any doubt, the cam followers should be renewed as a set.

31 Clean the internal drillings of the camshaft bearing cap bolts to ensure oil supply to the cam followers.

Refitting

B201 engine (B-type to 1980)

32 Clean the contact faces of the camshaft bearing housing and cylinder head.

33 Locate the camshaft bearing housing on the cylinder head with the feeler blade apertures on the inlet side - incorrect refitting will prevent lubrication oil reaching the camshaft. Make sure the housing locates correctly on the dowels. Refit and tighten the retaining bolts.

34 Fit the shims in their correct positions in the tops of the valves.

35 Oil the camshaft followers and insert them in the bearing housing in their correct locations. Keep the cylinder head upright until the camshaft has been refitted otherwise the shims may be displaced.

36 Oil the bearing surfaces then lower the camshaft into the bearing housing with the lobe peaks for No 1 cylinder pointing upwards.

37 Fit the bearing caps in their correct positions, then insert the bolts and tighten them evenly and in diagonal sequence to the specified torque.

38 With the camshaft sprocket still attached to the mounting plate, insert the two remaining bolts through the sprocket and into the camshaft. Tighten these bolts to the specified torque then remove the bolt from the mounting plate and tighten this one to the specified torque. **Do not** allow the timing chain to slacken.

39 Check that all TDC timing marks are aligned correctly with reference to Section 3.

40 Refit the cylinder head cover as described in Section 4, then reconnect the battery negative lead.

B201 engine (H-type from 1981)

41 Clean the contact faces of the camshaft bearing housing and cylinder head.

42 Locate the camshaft bearing housing on the cylinder head with the feeler blade apertures on the inlet side - incorrect refitting will prevent lubrication oil reaching the camshaft. Make sure the housing locates correctly on the dowels. Refit and tighten the retaining bolts.

43 Fit the shims in their correct positions in the tops of the valves.

44 Oil the camshaft followers and insert them in the bearing housing in their correct locations. Keep the cylinder head upright until the camshaft has been refitted otherwise the shims may be displaced.

45 Oil the bearing surfaces then lower the camshaft into the bearing housing with the lobe peaks for No 1 cylinder pointing upwards.

46 Fit the bearing caps in their correct positions, then insert the bolts and tighten them evenly and in diagonal sequence to the specified torque **(see illustration)**.

47 Keeping the sprocket engaged with the timing chain, locate the sprocket on the camshaft and insert the retaining bolts. Tighten the bolts to the specified torque **(see illustration)**. Remove the locking wire.

48 On pre-1984 models use the special tool to pull up the tensioner and at the same time remove the wedge. Release the tensioner and remove the special tool.

49 On models from 1984-on remove the wedge and allow the tensioner to tension the timing chain.

50 Refit the fuel pump with reference to Chapter 4.

51 Check that all TDC timing marks are aligned correctly with reference to Section 3.

52 Refit the cylinder head cover as described in Section 4, then reconnect the battery negative lead.

B202 engine

53 Lubricate the hydraulic cam followers and their bores in the cylinder head, then insert them in their original positions **(see illustration)**.

54 Lubricate the bearing surfaces of the camshafts in the cylinder head.

5.46 Tightening the camshaft bearing cap bolts (H-type B201 engine)

5.47 Tightening the camshaft sprocket bolts (H-type B201 engine)

5.53 Lubricating the cam followers before inserting them in the cylinder head

5.56 Tightening the camshaft bearing cap bolts with a torque wrench

55 Locate the camshafts in their correct positions in the cylinder head so that the valves of number 1 cylinder (timing chain end) are closed, and the valves of number 4 cylinder are rocking. The timing marks on the sprocket ends of the camshafts should be pointing upwards.
56 Lubricate the bearing surfaces in the bearing caps then locate them in their correct positions and insert the retaining bolts. Make sure the oil supply bolts are in their correct positions. Progressively tighten the bolts to the specified torque **(see illustration)**.
57 On early engines fitted with oil pipes, refit the oil supply pipes to their correct positions on the camshaft bearing caps.
58 Check that each camshaft is at its TDC position - the timing marks are located on the front of the camshafts and must be aligned with the mark on the bearing caps.
59 Check that the TDC '0' mark on the flywheel/driveplate is still aligned with the timing mark on the transmission cover.
60 Locate the sprockets on the camshafts, fitting the exhaust one first followed by the inlet one. Do not fully tighten the bolts at this stage. Check that the timing chain is correctly located on the guides and sprockets.
61 Set the timing chain tensioner as follows. On models manufactured before 1988 push the plunger fully into the tensioner body and turn it to lock it. On later models use a screwdriver to press down the ratchet, then push the plunger fully into the tensioner and release the ratchet. Check the tensioner washer for condition and renew it if necessary.
62 Insert the tensioner in the cylinder head and tighten to the specified torque.
63 On models manufactured up to approximately 1988, trigger the tensioner by pressing the pivoting chain guide against the tensioner, then press the guide against the chain to provide it with its basic tension. When the engine is started, hydraulic pressure will take up any remaining slack.
64 On models manufactured after 1988, insert the spring and plastic guide pin in the tensioner, then fit the plug together with a new O-ring, and tighten it to the specified torque.
Note: *New tensioners are supplied with the tensioner spring held pre-tensioned with a pin.* **Do not** *remove this pin until after the tensioner has been tightened into the cylinder head.* When the engine is started, hydraulic pressure will take up any remaining slack.
65 Turn the engine two complete turns clockwise. Check that the TDC timing marks still align correctly with reference to Section 3.
66 Fully tighten the camshaft sprocket bolts to the specified torque while holding the camshafts with a spanner on the flats.
67 Refit the distributor (see Chapter 5).
68 Refit the cylinder head cover as described in Section 4, then reconnect the battery negative lead.

6 Cylinder head - removal and refitting

Removal
Note: *If the engine has been removed from the car ignore references to removal of wiring and services which will have been carried out during the engine removal procedure*

B201 B-type engine

1 Disconnect the battery negative lead.
2 Drain the cooling system with reference to Chapter 1. The cylinder block should be drained of coolant as well.
3 On carburettor engines disconnect the accelerator and choke cables.
4 On fuel injection engines remove the rubber air duct between the airflow meter and throttle valve housing and disconnect the accelerator cable.
5 Disconnect the wiring from the temperature transmitter.
6 Disconnect the brake servo vacuum hose from the inlet manifold.
7 On carburettor engines disconnect the fuel and vacuum hoses from the carburettor.
8 On the Bosch CI fuel injection system disconnect the fuel lines from the fuel distributor to the injection valves. Identify the lines for position to ensure correct refitting and tape over their ends to prevent entry of dust and dirt. Unbolt the stay from the throttle valve housing.
9 Disconnect the coolant hoses from the thermostat housing, water pump and inlet manifold.
10 Unscrew the nuts/bolts and remove the exhaust downpipe from the exhaust manifold.
11 Remove the ignition cap and HT leads.
12 Remove the cylinder head cover with reference to Section 4.
13 Position the engine at TDC compression on No 1 cylinder as described in Section 3.
14 Remove one of the camshaft sprocket bolts, then use this bolt to secure the camshaft sprocket to the mounting plate. Tighten the bolt securely. If the sprocket slips downwards, the chain tensioner will extend and the engine will have to be removed to reset it. Unscrew and remove the remaining three sprocket bolts. The sprocket is now held securely on the mounting plate.
15 **Do not** try torotate the crankshaft or camshaft with the timing chain or sprockets

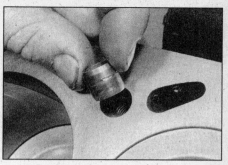

6.19 Removing a cylinder head locating dowel

removed, or piston/valve contact may occur.
16 Remove the two bolts which secure the top of the timing cover to the cylinder head.
17 Slacken the cylinder head bolts in the reverse sequence to that in illustration 6.72. Remove the bolts.
18 Lift the cylinder head from the cylinder block taking care not to disturb the camshaft sprocket and timing chain. If necessary tap the cylinder head free with a wooden mallet. *Do not use a lever otherwise the mating surfaces will be damaged.*
19 Remove the cylinder head gasket from the block noting the two locating dowels. If the location dowels are a loose fit, remove them and store them with the head for safe-keeping **(see illustration)**. Do not discard the gasket - it may be needed for identification purposes.

B201 H-type engine

20 Disconnect the battery negative lead.
21 Drain the cooling system (Chapter 1). The cylinder block should be drained as well.
22 Disconnect and remove the radiator top hose.
23 Remove the cylinder head cover with reference to Section 4.
24 Disconnect the wiring from the temperature transmitter.
25 On carburettor engines disconnect the fuel lines from the fuel pump. Identify them for location.
26 On fuel injection engines remove the warm-up regulator and the auxiliary air valve from the cylinder head.
27 Using a trolley jack and piece of wood support the transmission, then unbolt the stay between the right-hand engine mounting and the cylinder head and move it to one side. Raise the engine slightly and support it on a length of wood located between the crossmember and the transmission case.
28 Unscrew the nuts and remove the exhaust manifold from the cylinder head then support it to one side. Remove the gasket.
29 Unbolt the inlet manifold from the cylinder head and support it to one side. Remove the gasket.
30 Before removing the sprocket from the camshaft the timing chain tensioner must be released and the timing chain wedged so that it remains fully engaged with the crankshaft sprocket.

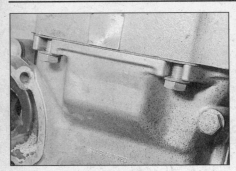

6.34 Timing cover-to-cylinder head bolts (H-type B201 engine)

6.36 Removing the cylinder head (H-type B201 engine)

6.43 Disconnecting the two coolant hoses from the front of the cylinder head

31 On models up to approximately 1984, obtain a suitable hooked tool or use a length of metal dowel rod bent at right-angles at one end, then insert the hooked end in the slotted hole at the top of the tension spring and pull the tool upwards to release the tension. Locate a suitable wedge (such as a screwdriver handle) between the timing chain and release the tensioner so that the chain is held firmly.

32 On models from approximately 1984-on, a ratchet-type timing chain tensioner is fitted. Before unbolting the camshaft sprocket this tensioner and the timing chain can be locked by inserting a suitable wedge (such as a screwdriver handle) between the fixed and pivoting chain guides. This will hold the timing chain fully engaged with the crankshaft sprocket.

33 With the timing chain tension released, unbolt the sprocket from the end of the camshaft. Position the sprocket and timing chain on the chain guides. Make sure that the sprocket remains fully engaged with the chain in order to maintain the valve timing - use locking wire to keep the chain on the sprocket.

34 Remove the two bolts which secure the top of the timing cover to the cylinder head **(see illustration)**.

35 Slacken the cylinder head bolts in the reverse sequence to that given in **illustration 6.72**. Remove the bolts.

36 Lift the cylinder head from the cylinder block taking care not to disturb the camshaft sprocket and timing chain **(see illustration)**. If necessary tap the cylinder head free with a wooden mallet. *Do not use a lever otherwise the mating surfaces will be damaged.*

37 Remove the cylinder head gasket from the block noting the two locating dowels. If the location dowels are a loose fit, remove them and store them with the head for safe-keeping. Do not discard the gasket - it may be needed for identification purposes.

B202 engine

38 Remove the bonnet as described in Chapter 11, and the battery as described in Chapter 5.

39 Drain the cooling system, as described in Chapter 1. The cylinder block should also be drained.

40 Remove the exhaust manifold and where applicable the Turbo unit assembly as described in Chapter 4. Recover the gasket.

41 On models equipped with air conditioning, remove the tensioning pulley and compressor drivebelt.

42 Slacken the power steering pump mounting bolts, remove the drivebelt and push the pump clear of the cylinder head.

43 Disconnect the radiator top hose at the thermostat housing and the two hoses from the front of the cylinder head **(see illustration)**.

44 Release the wiring harness and cable retaining clips on the cylinder head.

45 Remove the fuel pressure regulator.

46 Disconnect the earth leads for the LH-Jetronic fuel injection system.

47 Remove the auxiliary air valve.

48 On models equipped with air conditioning, remove the compressor bracket from the cylinder head.

49 Remove the inlet manifold with fuel injection manifold and fuel injection valves as a complete assembly. Refer to Chapter 4, Part B if necessary.

50 Disconnect the lead at the water temperature transmitter.

51 Remove the two bolts securing the timing cover to the underside of the cylinder head.

52 Remove the right-hand engine mounting bracket bolts and spacer sleeves which are secured to the cylinder head **(see illustration)**.

53 Remove the distributor cap and inspection cover, then disconnect the HT leads from the spark plugs (if necessary remove the distributor as described in Chapter 5). Disconnect the crankcase ventilation hose then unbolt the cylinder head cover and recover the two semi-circular rubber halves. On later models the rubber halves are incorporated in the cover gasket.

54 Turn the engine over until the TDC '0' mark is aligned with the timing mark on the transmission cover and with number 1 piston at TDC firing position. With the engine in this position, the timing notch on each camshaft should be aligned with the mark on the respective bearing caps.

55 Undo and remove the timing chain tensioner assembly from the left-hand rear side of the cylinder head. Take care not to lose the spring and plunger as the tensioner is withdrawn.

56 Using a spanner on the flats of each camshaft to prevent rotation, undo and remove the camshaft sprocket retaining bolt.

57 Withdraw each sprocket from its respective camshaft and disengage it from the timing chain.

58 Using a syringe or old rags, remove as much oil as possible from around the camshafts and in the cylinder head recesses.

59 Raise the engine slightly as necessary so that there is clearance between the cylinder head and the right-hand engine mounting.

60 Using a suitable Torx type socket bit adapter, undo the cylinder head retaining bolts in the reverse order to that shown in **illustration 6.72** and remove them **(see illustration)**.

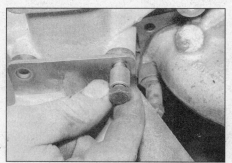

6.52 Removing the right-hand engine mounting bracket bolts and spacer sleeves from the cylinder head

6.60 Removing the cylinder head bolts (B202 engine)

6.61 Removing the cylinder head while an assistant holds the timing chain

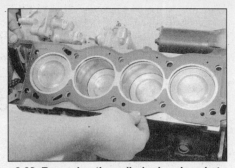

6.63 Removing the cylinder head gasket from the top of the block

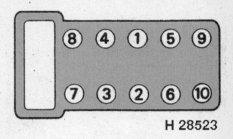

6.72 Cylinder head bolt tightening sequence

61 With all the cylinder head bolts removed, check that the timing chain is positioned so that the pivoting chain guide will not obstruct removal of the head - if necessary enlist the help of an assistant to hold the timing chain while the head is being removed (see illustration). Lift the cylinder head directly from the top of the cylinder block and place it on the workbench. If necessary enlist the help of an assistant since the cylinder head is quite heavy. If the cylinder head is stuck, try rocking it slightly to free it from the gasket - **do not** insert a screwdriver or similar tool between the gasket joint otherwise the contact faces will be damaged.

62 With the cylinder head removed, lay the timing chain over the top of the guide. After removal of the cylinder head do not lay it directly on the bench as this may damage the slightly protruding valves. Use blocks of wood at each end to prevent valve-to-bench contact.

63 Remove the gasket from the top of the block, noting the two locating dowels (see illustration). If the locating dowels are a loose fit, remove them and store them with the head for safe-keeping. Do not discard the gasket - it may be needed for identification purposes. If the cylinder head is to be dismantled for overhaul, remove the camshafts (Section 5).

Preparation for refitting

64 The mating faces of the cylinder head and cylinder block must be perfectly clean before refitting the head. Use a hard plastic or wood scraper to remove all traces of gasket and carbon; also clean the piston crowns but note that the crankshaft must not be turned on B201 engines. Take particular care during the cleaning operations, as the soft aluminium alloy is damaged easily. Also, make sure that the carbon is not allowed to enter the oil and water passages - this is particularly important for the lubrication system, as carbon could block the oil supply to the engine's components. Using adhesive tape and paper, seal the water, oil and bolt holes in the cylinder block. To prevent carbon entering the gap between the pistons and bores, smear a little grease in the gap. After cleaning each piston, use a small brush to remove all traces of grease and carbon from the gap, then wipe away the remainder with a clean rag. Clean all the pistons in the same way.

65 Check the mating surfaces of the cylinder block and the cylinder head for nicks, deep scratches and other damage. If slight, they may be removed carefully with a file, but if excessive, machining may be the only alternative to renewal.

66 If warpage of the cylinder head gasket surface is suspected, use a straight-edge to check it for distortion. Refer to Part B of this Chapter if necessary.

67 Check the condition of the cylinder head bolts, and particularly their threads, whenever they are removed. Wash the bolts in suitable solvent, and wipe them dry. Check each for any sign of visible wear or damage, renewing any bolt if necessary. Measure the length of each bolt, and compare with the length of a new bolt. Although Saab do not actually specify that the bolts must be renewed, it is strongly recommended that the bolts are renewed as a complete set if the engine has completed a high mileage.

Refitting

B201 B-type engine

68 Make sure that the mating faces of the cylinder head and block are perfectly clean and that the locating dowels are fitted. Locate the new gasket on the block making sure that all internal holes are aligned. *Do not use jointing compound.* Ideally two guide pins should be temporarily located in two cylinder head bolt holes to hold the gasket in position while the cylinder head is being refitted.

69 Check that the TDC timing marks are still aligned with reference to Section 3.

70 Lower the cylinder head onto the block, being careful not to dislodge the camshaft sprocket and timing chain.

71 Clean the threads of the cylinder head bolts and lightly oil them. (The 'Torx' type bolt threads are pre-lubricated, and will not need lubricating until they have been used 5 times.)

72 Insert the bolts, where applicable removing the guide pins. Following the correct sequence (see illustration), tighten them to the Stage 1 specified torque, then to the Stage 2 torque.

73 Refit the two bolts which secure the timing cover to the cylinder head.

74 Attach the camshaft sprocket with two bolts. When it is secure, remove the third bolt from the mounting plate and insert it in the sprocket. Do not allow the timing chain to slacken.

75 Check the valve clearances as described in Chapter 1.

76 Refit the cylinder head cover with reference to Section 4.

77 Refit the ignition cap and HT leads.

78 Refit the exhaust downpipe to the exhaust manifold and tighten the retaining bolts/nuts.

79 Reconnect the coolant hoses to the thermostat housing, water pump and inlet manifold.

80 On fuel injection engines, reconnect the fuel lines in their correct positions on the fuel distributor and injection valves. Refit the stay to the throttle valve housing and tighten the bolt.

81 On carburettor engines reconnect the fuel and vacuum hoses to the carburettor.

82 Reconnect the brake servo vacuum hose to the inlet manifold.

83 Reconnect the wiring to the temperature transmitter.

84 On fuel injection engines reconnect the rubber air duct between the airflow meter and throttle valve housing.

85 Reconnect and adjust the accelerator cable.

86 Refit all drain plugs then refill the cooling system with reference to Chapter 1.

87 On carburettor engines reconnect and adjust the choke cable.

88 Reconnect the battery negative lead.

89 Start the engine and allow it to warm up to normal operating temperature. Switch off and allow the engine to cool down for approximately 30 minutes. Remove the cylinder head cover then slacken and re-tighten each cylinder head bolt one at a time in the correct sequence to the Stage 4 specified torque. On 1984-on models with 15 mm hex-head or Torx M12 cylinder head bolts angle-tighten each bolt to the Stage 5 specified angle of 90° in the correct sequence. On completion refit the cylinder head cover.

90 The cylinder head bolts must be re-tightened after 1200 and 6000 miles.

2A•12 Engine in-car repair procedures

6.91 Cylinder head gasket located on the block (H-type B201 engine)

6.95 Tightening the cylinder head bolts (H-type B201 engine)

B201 H-type engine

91 Make sure the mating faces of the head and block are perfectly clean, and that the locating dowels are fitted. Locate the new gasket on the block making sure that all internal holes are aligned **(see illustration)** *Do not use jointing compound.* Ideally two guide pins should be temporarily located in two cylinder head bolt holes to hold the gasket in position while the cylinder head is being refitted.
92 Check that the TDC timing marks are still aligned with reference to Section 3.
93 Lower the cylinder head onto the block, being careful not to dislodge the camshaft sprocket and timing chain.
94 Clean the threads of the cylinder head bolts and lightly oil them. (The 'Torx' type bolt threads are pre-lubricated, and will not need lubricating until they have been used 5 times.)
95 Insert the bolts, where applicable removing the guide pins. Following the correct sequence **(see illustration 6.72)**, tighten them to the Stage 1 specified torque, then to the Stage 2 torque **(see illustration)**. On 1988-on models tighten the bolts in the correct sequence to the Stage 3 torque.
96 Refit the two bolts which secure the timing cover to the cylinder head.
97 Keeping the sprocket engaged with the timing chain, locate the sprocket on the camshaft and insert the retaining bolts. Tighten the bolts to the specified torque. Remove the locking wire.
98 On pre-1984 models use the special tool to pull up the tensioner and at the same time remove the wedge. Release the tensioner and remove the special tool.
99 On models from 1984-on remove the wedge and allow the tensioner to tension the timing chain. As an additional precaution, use a screwdriver to depress the ratchet on the tensioner; this will ensure that the tensioner is positioned correctly.
100 Check the inlet manifold gasket and renew it if necessary. Refit the inlet manifold and gasket and tighten the retaining bolts to the specified torque. **Note:** *Make sure the correct gasket is fitted; there are different gaskets for carburettor and fuel injection engines.* **Do not fit a B 20 inlet manifold gasket to the H-type engine otherwise coolant will enter the cylinder head through the EGR channel.**

101 Check the exhaust manifold gasket and renew it if necessary. Refit the exhaust manifold and gasket and tighten the retaining nuts to the specified torque.
102 Refit the stay between the right-hand engine mounting and cylinder head and tighten the bolts. Lower the engine and remove the trolley jack from under the transmission.
103 On fuel injection engines refit the warm-up regulator and the auxiliary air valve.
104 On carburettor engines reconnect the fuel lines to the fuel pump in their correct positions.
105 Reconnect the wiring to the temperature transmitter.
106 Check the valve clearances (Chapter 1).
107 Refit the cylinder head cover with reference to Section 4.
108 Reconnect the radiator top hose and tighten the clips.
109 Refill the cooling system (see Chapter 1).
110 Reconnect the battery negative lead.
111 On models up to 1987, start the engine and allow it to warm up to normal operating temperature. Switch off and allow the engine to cool down for approximately 30 minutes. Remove the cylinder head cover then slacken and re-tighten each cylinder head bolt one at a time in the correct sequence to the Stage 4 specified torque. Angle-tighten each bolt to the Stage 5 specified angle of 90° in the correct sequence. On completion refit the cylinder head cover.
112 On models up to 1987, the cylinder head bolts must be re-tightened after 1200 and 6000 miles.

B202 engine

113 Where removed, refit the camshafts with reference to Section 5.
114 Wipe clean the mating surfaces of the cylinder head and cylinder block/crankcase. Check that the two locating dowels are in position on the cylinder block.
115 Position a new gasket on the cylinder block surface making sure it is fitted the correct way round **(see illustration)**. **Do not use any jointing compound on the gasket.**
116 Check that each camshaft is at its TDC position - the timing marks are located on the front of the camshaft and must be aligned with the marks on the bearing caps.
117 Check that the TDC '0' mark on the flywheel/driveplate is still aligned with the timing mark on the transmission.
118 Make up a guide pin by sawing the head off an old cylinder head bolt, chamfering its end then fitting it to the right-hand rear end bolt hole.
119 Check that the timing chain is located correctly on the chain guides, then carefully lower the cylinder head over the guide pin then, using this pin as a pivot, twist the cylinder head to clear the pivoting timing chain guide. Now lower the head fully into place on the locating dowels and remove the guide pin.
120 Apply a smear of grease to the threads, and to the underside of the heads, of the cylinder head bolts. Insert the bolts and screw them in finger tight.
121 Working progressively and in the sequence shown **(see illustration 6.72)**, tighten the cylinder head bolts to their stage 1 torque setting, using a torque wrench **(see illustration)**.
122 Using the same sequence, tighten the cylinder head bolts to their stage 2 torque setting. On 1988-on models angle-tighten the bolts in the correct sequence to the Stage 3 angle **(see illustration)**.
123 Insert and tighten the two bolts securing the timing cover to the cylinder head.
124 Locate the sprockets on the camshafts, fitting the exhaust one and timing chain first followed by the inlet one. Do not fully tighten the bolts at this stage. Check that the timing chain is correctly located on the guides and sprockets.

6.115 Locating a new cylinder head gasket on the cylinder block

6.121 Tightening the cylinder head bolts

Engine in-car repair procedures 2A•13

6.122 Angle-tightening the cylinder head bolts

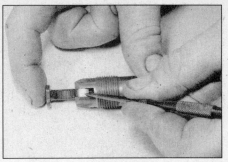
6.125 Set the later timing chain tensioner by pressing down the ratchet and pushing in the plunger

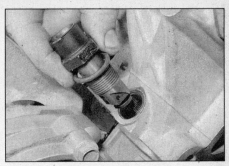
6.126 Inserting the timing chain tensioner in the cylinder head

125 Set the timing chain tensioner as follows. On models manufactured before 1988 push the plunger fully into the tensioner body and turn it to lock it. On later models use a screwdriver to press down the ratchet, then push the plunger fully into the tensioner and release the ratchet **(see illustration)**. Check the tensioner washer for condition and renew it if necessary.

126 Insert the tensioner in the cylinder head and tighten to the specified torque **(see illustration)**.

127 On models manufactured up to approximately 1988, trigger the tensioner by pressing the pivoting chain guide against the tensioner, then press the guide against the chain to provide it with its basic tension. When the engine is started, hydraulic pressure will take up any remaining slack.

128 On models manufactured after 1988, insert the spring and plastic guide pin in the tensioner, then fit the plug together with a new O-ring, and tighten it to the specified torque. **Note:** *New tensioners are supplied with the tensioner spring held pre-tensioned with a pin.* ***Do not*** *remove this pin until after the tensioner has been tightened into the cylinder head.* When the engine is started, hydraulic pressure will take up any remaining slack.

129 Turn the engine two complete turns clockwise. Check that the TDC timing marks still align correctly with reference to Section 3.

130 Fully tighten the camshaft sprocket bolts to the specified torque while holding the camshafts with a spanner on the flats.

131 Remove the engine support used to provide clearance between the cylinder head and the right-hand engine mounting.

132 Clean the contact surfaces of the cylinder head cover and cylinder head. On early models locate the special split rubber plugs on the cylinder head, then apply a 4 mm thick bead of silicone sealant to the corners of the head and over the rubber plugs. On later models the rubber halves are incorporated in the cover gasket. and it is not necessary to apply the bead of silicone sealant. Also locate the inner rubber gasket in the groove.

133 Refit the cylinder head cover and insert the securing bolts. Tighten the bolts progressively to the specified torque starting with the bolts at the flywheel/driveplate end and the central bolt at the timing end.

134 Reconnect the crankcase ventilation hose, reconnect the HT leads, refit the cylinder head cover lid and refit the distributor cap. Where necessary refit the distributor with reference to Chapter 5.

135 Refit the right-hand engine mounting bolts and spacer sleeves and tighten the bolts.

136 Reconnect the wiring at the water temperature transmitter.

137 Refit the inlet manifold with fuel injection manifold and fuel injection valves to the cylinder head together with a new gasket and tighten the mounting bolts. Refer to Chapter 4, Part B if necessary.

138 On models equipped with air conditioning refit the compressor bracket and tighten the mounting bolts.

139 Refit the auxiliary air valve and reconnect the earth leads to the LH-Jetronic fuel injection system.

140 Refit the fuel pressure regulator.

141 Refit the wiring harness and cable retaining clips.

142 Reconnect the radiator top hose and tighten the clips.

143 Refit the power steering pump and drivebelt and tension with reference to Chapter 1.

144 On models equipped with air conditioning, refit the compressor drivebelt and tensioning pulley. Adjust the tension of the drivebelt with reference to Chapter 1.

145 Using a new gasket, refit the exhaust manifold and where applicable the Turbo unit assembly with reference to Chapter 4.

146 Refill the cooling system with reference to Chapter 1.

147 Refit the battery and bonnet.

148 On models up to 1987, start the engine and allow it to warm up to normal operating temperature. Switch off and allow the engine to cool down for approximately 30 minutes. Remove the cylinder head cover then slacken and re-tighten each cylinder head bolt one at a time in the correct sequence to the Stage 4 specified torque. Angle-tighten each bolt to the Stage 5 specified angle of 90° in the correct sequence. On completion refit the cylinder head cover.

149 On models up to 1987, the cylinder head bolts must be re-tightened after 1200 and 6000 miles.

7 Oil pump - removal, inspection and refitting

Removal

B201 B-type engine

1 The oil pump is attached to the oil filter adapter located on the left-hand side of the cylinder block. First position a suitable container beneath the oil pump position.

2 Unscrew and remove the through-bolts securing the oil pump to the adapter. Withdraw the oil pump and recover the O-ring. The driveshaft will remain in the crankcase attached to the distributor drivegear.

B201 H-type and B202 engines

3 The oil pump is located at the rear (timing chain) end of the engine on the timing cover. First clean the timing cover and crankshaft pulley.

4 Remove the auxiliary drivebelt(s) with reference to Chapter 1, then where necessary disconnect the speedometer cable.

5 Jack up the front of the car and support on axle stands (see *"Jacking and Vehicle Support"*).

6 Have an assistant hold the crankshaft stationary by inserting a wide-blade screwdriver or metal bar in the starter ring gear through the ignition timing aperture. Unscrew the crankshaft pulley bolt and remove the pulley using a lever or puller if necessary **(see illustrations)**.

7 Unscrew the bolts and remove the oil pump from the front of the timing cover **(see illustrations)**.

8 Prise the rubber O-ring seal from the groove in the oil pump **(see illustration)**.

Inspection

B201 B-type engine

9 Remove the two screws which secure the cover to the housing **(see illustration)**.

2A•14 Engine in-car repair procedures

7.6a Removing the crankshaft pulley bolt (H-type B201 engine)

7.6b Unscrew the crankshaft pulley bolt (B202 engine) . . .

7.6c . . . and remove the bolt . . .

7.6d . . . spacer . . .

7.6e . . . and pulley

7.7a Oil pump (B201 H-type engine) and mounting bolts

7.7b Removing the oil pump (B201 H-type engine)

7.8 Removing the oil pump rubber sealing ring (B201 H-type engine)

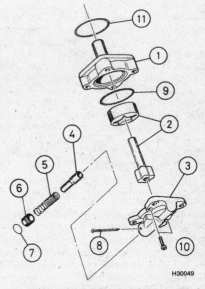

7.9 Exploded view of the oil pump (B201 B-type engine)

1 Pump body
2 Rotors
3 Cover
4 Piston
5 Spring
6 Stop
7 Seal
8 Split pin
9 Seal
10 Screws
11 Seal

10 Remove the cover and extract the rotors and O-ring.
11 The pressure relief valve can be removed if the split pin is first withdrawn.
12 The endfloat of the rotors should be between 0.05 and 0.09 mm. If it exceeds this tolerance, the end face of the oil pump housing can be rubbed down by holding it squarely on a sheet of abrasive placed on a piece of plate glass. Excessive wear between the rotors can only be rectified by renewing the pump.
13 Reassembly is a reversal of dismantling but note that the chamfered edge of the outer rotor is nearest the driveshaft.

B201 H-type and B202 engines

14 Remove the two rotors from the pump housing and check both the rotors and housing for wear. Modifications have been made to the oil pump so it is worthwhile making a thorough inspection. Note also that wear on the timing cover can cause noise from the pressure relief valve by the introduction of air into the oil. An air leak at the pick up tube connection to the timing cover can also cause the same noise (see illustrations).
15 Unscrew the plug and remove the relief valve spring and plunger from the timing cover, noting which way round they are fitted. Recover the plug washer.
16 Clean all components and examine them for wear and damage. Examine the pump rotors and body for signs of wear ridges and scoring. Using a feeler blade check the clearance between the outer rotor and the timing cover with reference to the Specifications (see illustration). If worn excessively, the complete pump assembly must be renewed.
17 Examine the relief valve plunger for signs of wear or damage, and renew if necessary. The condition of the relief valve spring can only be measured by comparing it with a new

Engine in-car repair procedures 2A•15

7.14a Removing the early type inner rotor from the oil pump on the H-type engine

7.14b Removing the late type inner rotor from the oil pump on the H-type engine

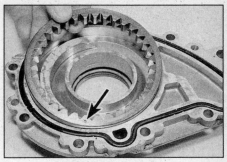
7.14c Removing the outer rotor. The indentation (arrowed) faces away from the oil pump housing

7.14d Dismantled oil pump on the H-type engine

7.14e Showing the two types of crankshaft sprocket necessary for late (left) and early (right) oil pumps on the H-type engine

7.16 Checking the clearance between the pump rotors and housing

7.20 The indentation (arrowed) must face away from the oil pump housing

7.24 Prising the oil seal from the oil pump on the H-type engine

7.28 Tightening the oil pump retaining bolts

one; if there is any doubt about its condition, it should also be renewed.

18 If there are any signs of dirt or sediment in the oil pump, it will be necessary to remove the engine and transmission, separate the transmission and clean the pick-up/strainer.

19 Insert the relief valve plunger and spring, then refit the plug together with a new washer and tighten the plug.

20 Lubricate the rotors with fresh engine oil then insert them in the oil pump body in their original positions. The outer rotor must be positioned with the indentation facing away from the oil pump housing (see illustration).

Refitting

B201 B-type engine

21 Clean the mating faces then fit the oil pump to the adapter on the cylinder block together with a new O-ring. To allow the rotor to engage with the driveshaft it may be necessary to turn the oil pump housing slightly.

22 Insert and tighten the through-bolts to the specified torque.

23 Run the engine at idling speed and check the oil pump for leaks.

B201 H-type and B202 engines

24 Prise the oil seal from the oil pump casing and fit a new seal using a block of wood (see illustration).

25 Locate a new O-ring seal in the groove in the casing.

26 Fit the oil pump to the timing cover and insert the bolts loosely.

27 If centring holes are provided, insert close fitting pins then refit the oil pump to the timing cover. It is most important to centralise the oil pump correctly otherwise damage may occur. Where there are no centring holes locating dowels should be provided to ensure correct centralisation.

28 Insert the retaining bolts and tighten them progressively to the specified torque (see illustration).

29 Prime the oil pump by unbolting the oil filter adapter from the left-hand side of the cylinder block and filling the oilway to the oil pump with oil (see illustration). Refit the adapter and tighten the bolts.

30 Smear a little grease on the seal contact face then locate the pulley on the crankshaft.

31 Insert and tighten the crankshaft pulley bolt while holding the crankshaft stationary (see illustration).

32 Refit and adjust the auxiliary drivebelt(s) with reference to Chapter 1. Where necessary

2A•16 Engine in-car repair procedures

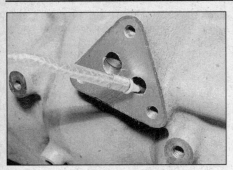

7.29 Priming the oil pump on the H-type engine

7.31 Tightening the crankshaft pulley bolt

reconnect the speedometer cable and tighten the collar.
33 Lower the car to the ground.
34 Run the engine at idling speed and check the oil pump for leaks.

8 Oil cooler - removal and refitting

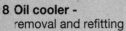

Removal

1 An oil cooler for engine oil is fitted to Turbo models only, being located below the left-hand headlight (see illustration). It is connected to ports on an adapter fitted to the oil filter, and the adapter also incorporates an oil thermostat. Automatic transmission models are fitted with a separate transmission fluid cooler located in the coolant hose between the radiator and water pump (refer to Chapter 7, Part B).

B201 Turbo models up to and including 1985

2 Remove the radiator grille as described in Chapter 11.
3 Remove the left-hand headlight as described in Chapter 12.
4 Refer to Chapter 12 and remove the wiper motor however do not disconnect the wiring. Position the motor to one side.
5 Unscrew the mounting screws securing the oil cooler to the radiator.
6 Position a suitable container beneath the oil cooler, then unscrew the union bolts and disconnect the hydraulic lines.

8.1 Oil cooler located below the left-hand headlight

B201 Turbo models from 1986-on and B202 Turbo models from 1984

7 Remove the left-hand headlight cluster, baffle plate and headlight unit as described in Chapter 12.
8 Remove the spoiler grille for the oil cooler.
9 Unscrew the nut from the bolt in the radiator member, then remove the bolt from under the member.
10 Unscrew the mounting bolts securing the oil cooler and shroud to the body. Lower the shroud from the oil cooler.
11 Using two spanners, unscrew the unions and disconnect the hydraulic lines.

Refitting

B201 Turbo models up to and including 1985

12 Check and if necessary renew the union copper washers, then reconnect the hydraulic lines and tighten the union bolts.
13 Refit the oil cooler to the radiator and tighten the mounting screws.
14 Refit the wiper motor and left-hand headlight with reference to Chapter 12.
15 Refit the radiator grille (see Chapter 11).
16 Top-up the engine with oil with reference to "Weekly Checks". On completion start the engine and run it at a fast idle speed for several minutes to allow the oil to fill the oil cooler. Check and if necessary top-up the engine oil level.

B201 Turbo models from 1986-on and B202 Turbo models from 1984

17 Reconnect the hydraulic lines to the oil cooler and tighten the unions.
18 Locate the oil cooler in the shroud, then refit the shroud leaving the mounting bolts loose at this stage.
19 Insert the bolt in the radiator member from below, then tighten the nut followed by the remaining mounting bolts.
20 Refit the spoiler grille.
21 Pull the hydraulic lines forward so that they are positioned straight up from the oil cooler.
22 Refit the headlight unit, baffle plate and headlight cluster.
23 Top-up the engine with oil with reference to "Weekly Checks". On completion start the engine and run it at a fast idle speed for several minutes to allow the oil to fill the oil cooler. Check and if necessary top-up the engine oil level.

9 Oil pressure warning light switch - removal and refitting

Removal

1 The oil pressure warning switch is screwed into the top of the oil filter adapter on the left-hand rear of the cylinder block.
2 Disconnect the wiring from the switch terminal.
3 Unscrew the switch from the oil filter adapter. Be prepared for slight loss of oil. If the switch is to be left removed for any length of time, plug the hole to prevent entry of dust and dirt.

Refitting

4 Wipe clean the threads of the switch and the location aperture, then insert the switch and tighten securely.
5 Reconnect the wiring to the switch terminal.
6 Start the engine and check for leakage.

10 Crankshaft oil seals - renewal

Flywheel end oil seal

Note: *It is not possible to renew the crankshaft oil seal at the driveplate end on automatic transmission models with the engine in the car; see Part B of this Chapter.*

1 Remove the clutch (Chapter 6) and the flywheel (Section 11 of this Chapter).
2 Make a note of the fitted depth of the seal in the engine plate. Punch or drill two small holes opposite each other in the seal. Screw a self-tapping screw into each, and pull on the screws with pliers to extract the seal. Alternatively, use a screwdriver to prise out the oil seal.
3 Wipe clean the seal location in the engine plate, then lubricate the lips of the new seal with clean engine oil, and carefully locate the seal on the end of the crankshaft making sure the seal lips face inwards.
4 Using a suitable tubular drift, which bears only on the hard outer edge of the seal, drive the seal into position, to the same depth in the housing as the original was prior to removal.
5 Wipe clean the oil seal, then refit the flywheel as described in Section 11 and the clutch as described in Chapter 6.

Timing chain end oil seal

6 Remove the auxiliary drivebelt(s) with reference to Chapter 1.
7 Hold the crankshaft stationary using a wide-blade screwdriver or metal bar inserted in the starter ring gear through the ignition timing aperture. Unscrew the crankshaft pulley bolt and remove the pulley using a lever or puller if necessary.

Engine in-car repair procedures 2A•17

11.3 Removing the flywheel (H-type engine)

11.6 Spigot bearing located in the centre of the flywheel (H-type engine)

8 Make a note of the fitted depth of the seal in the oil pump housing. Use a screwdriver to prise out the oil seal.
9 Wipe clean the seal location in the oil pump housing, then lubricate the lips of the new seal with clean engine oil, and carefully locate the seal on the end of the crankshaft making sure that the seal lips face inwards.
10 The oil seal must now be pressed into the oil pump housing to the previously noted depth. Because there is very little room between the rear of the engine and the bulkhead, it is not possible to use a tubular drift and mallet, however the seal can be pressed in using the crankshaft pulley bolt, a length of metal tube which locates on the outer part of the seal, and a large washer.
11 Wipe clean the oil seal, then refit the crankshaft pulley and tighten the bolt to the specified torque while holding the crankshaft stationary using the method described in paragraph 7. Due to the lack of room Saab technicians use an extension tool on the pulley bolt, but if this method is used the additional length of the extension tool must be taken into account when calculating the torque. The torque wrench must be operated in line with the extension tool.
12 Refit and adjust the auxiliary drivebelt(s) with reference to Chapter 1.

11 Flywheel - removal, inspection and refitting
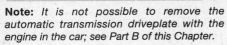

Note: *It is not possible to remove the automatic transmission driveplate with the engine in the car; see Part B of this Chapter.*

Removal
1 Remove the clutch (see Chapter 6).
2 Prevent the flywheel from turning by locking the ring gear teeth with a wide-blade screwdriver or similar tool. Alternatively, bolt a metal link between the flywheel (using the clutch bolt holes) and the cylinder block/crankcase.
3 Unscrew and remove the retaining bolts, remove the locking tool, then remove the flywheel from the crankshaft flange **(see illustration)**. Note that the unit is located by a single dowel pin and cannot be fitted incorrectly.

Inspection
4 If the flywheel's clutch mating surface is deeply scored, cracked or otherwise damaged, the flywheel must be renewed. However, it may be possible to have it surface-ground; seek the advice of a Saab dealer or engine reconditioning specialist.
5 If the ring gear is badly worn or has missing teeth, it may be possible to renew it. This job is best left to a Saab dealer or engine specialist. The temperature to which the new ring gear must be heated for installation is critical and, if not done accurately, the hardness of the teeth will be destroyed.
6 Check the spigot bearing in the flywheel for roughness and if necessary use metal tubing to drive out the old bearing and drive in the new bearing **(see illustration)**. The metal tube should only contact the outer race of the new bearing when fitting it.

Refitting
7 Clean the mating surfaces of the flywheel and crankshaft. Clean the threads of the retaining bolts and the crankshaft holes.

HAYNES HINT *If a suitable tap is not available, cut two slots into the threads of an old flywheel bolt and use the bolt to clean the threads. Examine the bolts and if necessary obtain new ones.*

8 Ensure the dowel is in position, then offer up the flywheel and locate it on the dowel.
9 Apply locking fluid to the bolt threads, then

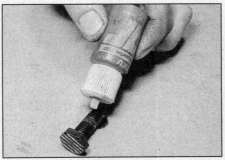

11.9a Applying locking fluid to the flywheel bolts

insert them and tighten to the specified torque while holding the flywheel stationary using one of the methods described in paragraph 2 **(see illustrations)**. Note that new bolts are pre-coated with sealant.
10 Refit the clutch assembly (Chapter 6).

12 Engine/transmission mountings - inspection and renewal

Inspection
1 The engine and transmission is supported on three mountings, one at the front of the engine and the others on each side at the rear of the engine. Vibration damping is progressive depending on the load applied, and is operative for both horizontal and vertical movement.
2 As from chassis number F1033151 on Turbo 16 manual transmission models, the two rear mountings are of hydraulic type. All manual transmission models from 1986 have hydraulic engine mountings at the front and rear. All automatic transmission models from 1986 have hydraulic engine mountings at the rear only. The Hydraulic engine mountings have an inner chamber filled with oil.
3 Check each mounting rubber to see if it is cracked, hardened or separated from the metal at any point. Check that there is clearance between the centre rubber section and the mounting body at the top and bottom of the mounting. Renew the mounting if any damage or deterioration is evident.
4 Check that the mounting bolts and the mounting bracket nuts are securely tightened; use a torque wrench to check if possible.
5 Using a large screwdriver or a crowbar, check for wear in the mounting by carefully levering against it to check for free play. Where this is not possible, enlist the aid of an assistant to move the engine/transmission side to side while you watch the mounting. While some free play is to be expected even from new components, excessive wear should be obvious. If excessive free play is found, check first that the mountings are correctly secured, then renew any worn components as described overleaf.

11.9b Tightening the flywheel bolts (H-type engine)

12.9 The front engine mounting with the engine removed from the engine compartment

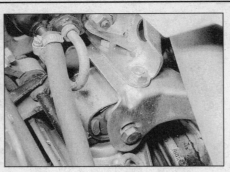

12.14 Right-hand side rear engine mounting on a 900 model

Renewal

Front mounting

6 Loosen the front engine mounting through-bolt but do not remove the nut completely at this stage.
7 Using a trolley jack and piece of wood beneath the front of the transmission, raise the front of the engine slightly so that the through-bolt is free to move in the front engine mounting bracket.
8 Unscrew the bolts and remove the clamp plate securing the mounting to the transmission primary housing.
9 Unscrew the nut then remove the through-bolt and recover the thrustwashers. Withdraw the mounting from the engine compartment **(see illustration)**.
10 Locate the new mounting in the slotted mounting bracket and insert the through-bolt together with the thrustwashers. Screw on the nut loosely.
11 Refit the clamp to the transmission primary housing, but make sure that the clamp is positioned centrally over the mounting before tightening the bolts to their specified torque.
12 Lower the engine/transmission then tighten the through-bolt to the specified torque.

Rear mountings

13 Support the rear of the engine and transmission on a trolley jack and piece of wood.
14 Where applicable, unscrew and remove the through-bolt from the rear engine mounting **(see illustration)**. Note: *On later models the right-hand rear engine mounting is of hydraulic type and is not fitted with a through-bolt.*
15 On later models unscrew and remove the nut from the right-hand rear mounting.
16 Unscrew the nuts/bolts securing the mounting to the body panel.
17 Raise the rear of the engine slightly then remove the mounting from the engine compartment. If required, unbolt the mounting bracket from the rear of the engine.
18 Locate the new mounting on the body panel and tighten the nuts.
19 If removed refit the upper mounting bracket and tighten the bolts.
20 Lower the engine/transmission slightly then insert the through-bolt and tighten to the specified torque.
21 Lower the trolley jack and remove from under the car.

Chapter 2 Part B:
Engine removal and overhaul procedures

Contents

Crankshaft - inspection 17	Engine oil level check .... see "Weekly Checks"
Crankshaft - refitting and main bearing running clearance check ... 21	Engine overhaul - dismantling sequence 5
Crankshaft - removal 14	Engine overhaul - general information 2
Crankshaft oil seal (driveplate end on automatic transmission	Engine overhaul - reassembly sequence 19
models) - renewal 9	Engine removal - methods and precautions 3
Cylinder block/crankcase - cleaning and inspection 15	General information 1
Cylinder head - dismantling 10	Idler shaft (B201 B-type engine) - removal, inspection and refitting .. 7
Cylinder head - reassembly 12	Main and big-end bearings - inspection 18
Cylinder head and valves - cleaning and inspection 11	Piston rings - refitting 20
Driveplate (automatic transmission models) - removal, inspection	Piston/connecting rod assembly - inspection 16
and refitting 8	Piston/connecting rod assembly - refitting and big-end bearing
Engine - initial start-up after overhaul 23	running clearance check 22
Engine and transmission - removal, separation and refitting 4	Piston/connecting rod assembly - removal 13
Engine oil and filter renewal .... see Chapter 1	Timing chain and sprockets - removal, inspection and refitting 6

Degrees of difficulty

Easy, suitable for novice with little experience	**Fairly easy,** suitable for beginner with some experience	**Fairly difficult,** suitable for competent DIY mechanic 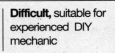	**Difficult,** suitable for experienced DIY mechanic	**Very difficult,** suitable for expert DIY or professional

Specifications

Cylinder head
Height (new):
- B201 engine .... 92.75 ± 0.05 mm
- B202 engine .... 140.5 ± 0.1 mm

Height (minimum):
- B201 engine .... 92.35 ± 0.05 mm
- B202 engine .... 140.1 ± 0.1 mm

Valve guide to valve stem clearance (max) .... 0.50 mm (measured on valve head raised 3.0 mm above seat)

Valves
Valve head diameter:
- B201 engine:
  - Inlet .... 42.0 mm
  - Exhaust .... 34.5 mm
- B202 engine:
  - Inlet .... 32.0 mm
  - Exhaust .... 29.0 mm

Valve head machining angle .... 44.5°
Valve seat cutting angle .... 45°

Valve stem diameter:
- B201 engine:
  - Inlet .... 7.960 to 7.975 mm
  - Exhaust .... 7.955 to 7.970 mm
- B202 engine:
  - Inlet .... 6.960 to 6.975 mm
  - Exhaust .... 6.955 to 6.980 mm

Valve spring free length:
- B201 engine .... 45.7 ± 1.5 mm
- B202 engine .... 45.0 ± 1.5 mm

Depth of valve stem below the camshaft bearing surface (B202 engine):
- Checking .... 19.5 ± 0.05 mm min to 20.5 ± 0.05 mm max
- Setting .... 20.0 mm min to 20.4 mm max (20.2 mm nominal)

Cylinder block
Cylinder bore diameter:
- Standard (A) .... 90.00 to 90.01 mm
- Standard (B) .... 90.010 to 90.020 mm
- B+ oversize .... 90.02 to 90.03 mm
- First oversize .... 90.500 to 90.512 mm
- Second oversize .... 91.000 to 91.012 mm

Idler shaft
End float (B201 Type-B engine) .... 0.05 to 0.13 mm

Pistons
Note: *Piston diameter is measured at right angles to the piston boss, at a distance B from the bottom of the skirt.*

Piston measuring distance B:
- Mahle piston .... 16.00 mm
- Schmidt piston .... 26.00 mm
- Kolben Schmidt piston .... 22.00 mm

Piston diameter:

Turbo (Mahle - 1979 to 1982) - early classification:
- Standard A .... 89.960 to 89.970 mm
- Standard AB .... 89.970 to 89.976 mm
- Standard B .... 89.976 to 89.986 mm
- Standard C .... 89.986 to 90.002 mm
- First oversize (0.5 mm) .... 90.460 to 90.475 mm
- Second oversize (1.0 mm) .... 90.960 to 90.975 mm
- Nominal piston clearance (new) .... 0.020 to 0.050 mm

Non-Turbo (Kolben Schmidt) - early classification:
- Standard A .... 89.972 to 89.980 mm
- Standard AB .... 89.980 to 89.986 mm
- Standard B .... 89.986 to 89.994 mm
- Standard C .... 89.994 to 90.010 mm
- First oversize (0.5 mm) .... 90.472 to 90.487 mm
- Second oversize (1.0 mm) .... 90.972 to 90.987 mm
- Nominal piston clearance (new) .... 0.014 to 0.040 mm

Turbo and carburettor (Mahle) - later classification:
- Standard A .... 89.960 to 89.970 mm
- Standard AB .... 89.970 to 89.978 mm
- Standard B .... 89.978 to 89.986 mm
- Standard C .... 89.986 to 90.002 mm
- First oversize (0.5 mm) .... 90.460 to 90.475 mm
- Second oversize (1.0 mm) .... 90.960 to 90.975 mm
- Nominal piston clearance (new) .... 0.022 to 0.050 mm

Turbo and fuel injection engines (Mahle, Kolben Schmidt):
- Standard A .... 89.978 to 89.988 mm
- Standard AB .... 89.988 to 89.996 mm
- Standard B .... 89.996 to 90.004 mm
- Standard C .... 90.004 to 90.020 mm
- First oversize (0.5 mm) .... 90.482 to 90.497 mm
- Second oversize (1.0 mm) .... 90.982 to 90.977 mm
- Nominal piston clearance (new) .... 0.004 to 0.032 mm

Fuel injection engines (Hepolite):
- Standard A .... 89.977 to 89.985 mm
- Standard AB .... 89.985 to 89.991 mm
- Standard B .... 89.991 to 89.999 mm
- Standard C .... 89.999 to 90.015 mm
- First oversize (0.5 mm) .... 90.477 to 90.492 mm
- Second oversize (1.0 mm) .... 90.977 to 90.992 mm
- Nominal piston clearance (new) .... 0.009 to 0.035 mm

Connecting rods
Maximum weight difference between any two connecting rods:
- To 1980 .... 6.0 g
- From 1981 .... 9.0 g

Crankshaft

Endfloat:
- To 1980 (all engines) .... 0.08 to 0.28 mm
- From 1981 (all engines) .... 0.06 to 0.30 mm

Main bearing journal diameter:
- Standard .... 57.981 to 58.000 mm
- First undersize .... 57.731 to 57.750 mm
- Second undersize .... 57.481 to 57.500 mm
- Third undersize (to 1980 only) .... 57.237 to 57.250 mm
- Fourth undersize (to 1980 only) .... 56.987 to 57.000 mm

Big-end bearing journal diameter:
- Standard .... 51.981 to 52.000 mm
- First undersize .... 51.731 to 51.750 mm
- Second undersize .... 51.481 to 51.500 mm
- Third undersize (to 1980 only) .... 51.237 to 51.250 mm
- Fourth undersize (to 1980 only) .... 50.987 to 51.000 mm

Maximum bearing journal out-of-round (all engines) .... 0.05 mm
Main bearing running clearance .... 0.020 to 0.062 mm
Big-end bearing running clearance .... 0.026 to 0.062 mm

Piston rings

End gaps:
- Top compression ring:
  - To 1980 .... 0.35 to 0.55 mm
  - From 1981 .... 0.35 to 0.48 mm
- Second compression ring:
  - To 1980 .... 0.30 to 0.45 mm
  - From 1981 .... 0.25 to 0.38 mm
- Oil control ring segment .... 0.38 to 1.40 mm

Side clearance in groove:
- Top compression ring:
  - To 1980 .... 0.050 to 0.082 mm
  - From 1982 .... 0.050 to 0.082 mm
- Second compression ring:
  - To 1980 .... 0.040 to 0.072 mm
  - From 1981 .... 0.034 to 0.07 mm
- Oil control ring .... Not applicable

Torque wrench settings

Refer to Chapter 2, Part A Specifications

1 General information

Included in this Part of Chapter 2 are details of removing the engine from the vehicle and general overhaul procedures for the cylinder head, cylinder block/crankcase and all other engine internal components.

The information given ranges from advice concerning preparation for an overhaul and the purchase of replacement parts, to detailed step-by-step procedures covering removal, inspection, renovation and refitting of engine internal components.

For information concerning in-car engine repair, as well as the removal and refitting of those external components necessary for full overhaul, refer to Part A of this Chapter. Ignore any preliminary dismantling operations in Part A that are no longer relevant once the engine has been removed from the vehicle.

Apart from torque wrench settings, which are given at the beginning of Part A, all specifications relating to engine overhaul are at the beginning of this Part of Chapter 2.

2 Engine overhaul - general information

It is not always easy to determine when, or if, an engine should be completely overhauled, as a number of factors must be considered.

High mileage is not necessarily an indication that an overhaul is needed, while low mileage does not preclude the need for an overhaul. Frequency of servicing is probably the most important consideration. An engine which has had regular and frequent oil and filter changes, as well as other required maintenance, should give many thousands of miles of reliable service. Conversely, a neglected engine may require an overhaul very early in its life.

Excessive oil consumption is an indication that piston rings, valve seals and/or valve guides are in need of attention. Make sure that oil leaks are not responsible before deciding that the rings and/or guides are worn. Perform a compression test, as described in Part A of this Chapter, to determine the likely cause of the problem.

Check the oil pressure with a gauge fitted in place of the oil pressure switch, and compare it with that specified. If it is extremely low, the main and big-end bearings, and/or the oil pump, are probably worn out.

Loss of power, rough running, knocking or metallic engine noises, excessive valve gear noise, and high fuel consumption may also point to the need for an overhaul, especially if they are all present at the same time. If a complete service does not remedy the situation, major mechanical work is the only solution.

An engine overhaul involves restoring all internal parts to the specification of a new engine. During an overhaul, the cylinders are rebored (where necessary) and the pistons and the piston rings are renewed. New main and big-end bearings are generally fitted; if necessary, the crankshaft may be renewed or reground, to restore the journals. The valves are also serviced as well, since they are usually in less-than-perfect condition at this point. While the engine is being overhauled, other components, such as the distributor,

starter and alternator, can be overhauled as well. The end result should be an as-new engine that will give many trouble-free miles.
Note: *Critical cooling system components such as the hoses, thermostat and water pump should be renewed when an engine is overhauled. The radiator should be checked carefully, to ensure that it is not clogged or leaking. Also, it is a good idea to renew the oil pump whenever the engine is overhauled.*

Before beginning the engine overhaul, read through the entire procedure, to familiarise yourself with the scope and requirements of the job. Overhauling an engine is not difficult if you follow carefully all of the instructions, have the necessary tools and equipment, and pay close attention to all specifications. It can, however, be time-consuming. Plan on the car being off the road for a minimum of two weeks, especially if parts must be taken to an engineering works for repair or reconditioning. Check on the availability of parts and make sure that any necessary special tools and equipment are obtained in advance. Most work can be done with typical hand tools, although a number of precision measuring tools are required for inspecting parts to determine if they must be renewed. Often the engineering works will handle the inspection of parts and offer advice concerning reconditioning and renewal.

Note: *Always wait until the engine has been completely dismantled, and until all components (especially the cylinder block/crankcase and the crankshaft) have been inspected, before deciding what service and repair operations must be performed by an engineering works. The condition of these components will be the major factor to consider when determining whether to overhaul the original engine, or to buy a reconditioned unit. Do not, therefore, purchase parts or have overhaul work done on other components until they have been thoroughly inspected.* As a general rule, time is the primary cost of an overhaul, so it does not pay to fit worn or sub-standard parts.

As a final note, to ensure maximum life and minimum trouble from a reconditioned engine, everything must be assembled with care, in a spotlessly clean environment.

3 Engine removal - methods and precautions

If you have decided that the engine must be removed for overhaul or major repair work, several preliminary steps should be taken.

Locating a suitable place to work is extremely important. Adequate work space, along with storage space for the vehicle, will be needed. If a workshop or garage is not available, at the very least, a flat, level, clean work surface is required.

Cleaning the engine compartment and engine/transmission before starting will help keep tools clean and organised.

An engine hoist or A-frame will also be necessary. Make sure the equipment is rated in excess of the combined weight of the engine and transmission. Safety is of primary importance, considering the potential hazards involved in lifting the engine/transmission out of the vehicle.

If this is the first time you have removed an engine, an assistant should ideally be available. Advice and aid from someone more experienced would also be helpful. There are many instances when one person cannot simultaneously perform all of the operations required when lifting the engine out of the vehicle.

Plan the operation ahead of time. Before starting work, arrange for the hire of or obtain all of the tools and equipment you will need. Some of the equipment necessary to perform engine/transmission removal and installation safely and with relative ease (in addition to an engine hoist) is as follows: a heavy duty trolley jack, complete sets of spanners and sockets as described in the front of this manual, wooden blocks, and plenty of rags and cleaning solvent for mopping up spilled oil, coolant and fuel. If the hoist must be hired, make sure that you arrange for it in advance, and perform all of the operations possible without it beforehand. This will save you money and time.

Plan for the vehicle to be out of use for quite a while. An engineering works will be required to perform some of the work which the do-it-yourselfer cannot accomplish without special equipment. These places often have a busy schedule, so it would be a good idea to consult them before removing the engine, in order to accurately estimate the amount of time required to rebuild or repair components that may need work.

Always be extremely careful when removing and refitting the engine/transmission. Serious injury can result from careless actions. Plan ahead and take your time, and a job of this nature, although major, can be accomplished successfully.

The engine and transmission is removed by lifting upwards from the engine compartment.

4 Engine and transmission - removal, separation and refitting

Removal

Note: *The engine can be removed from the car only as a complete unit with the transmission; the two are then separated on the bench. The engine/transmission is lifted upwards out of the engine compartment.*

Models up to 1980

1 Remove the battery (see Chapter 5).
2 Drain the cooling system as described in Chapter 1. **Note:** *The engine should be cold before draining the coolant.*
3 Remove the bonnet (see Chapter 11).
4 Loosen the clip and remove the top hose.

4.5 Engine earth strap location showing front engine mounting

5 Unbolt the earth strap from the gearbox/transmission **(see illustration)**.
6 Disconnect the wiring from the starter motor. Identify each lead for location to ensure correct refitting.
7 Disconnect the HT lead from the coil, and the LT lead from the distributor.
8 Disconnect the wiring from the water temperature sender unit. If applicable, detach the wiring harness from the clutch cover.
9 If the clutch is to be removed (ie for engine overhaul) remove it at this stage with reference to Chapter 6. Alternatively, remove the clutch cover then disconnect and plug the hydraulic line from the slave cylinder.
10 Loosen the clip and remove the bottom hose from the radiator.

Carburettor engines

11 Remove the air cleaner and inlet hose with reference to Chapter 4. Also remove the preheater hose and disconnect the crankcase ventilation hose(s).
12 Disconnect the accelerator and choke cables.
13 Disconnect the fuel supply and return (where applicable) hoses from the fuel pump and carburettor.

Fuel injection engines

14 Disconnect the hot air hose and the air inlet hose.
15 Disconnect the accelerator cable.
16 Identify the fuel injection system wiring then disconnect it from the warm-up regulator/boost control, cold start valve, thermo-time switch and auxiliary air valve.
17 Disconnect the fuel supply pipes from the mixture control unit and warm-up regulator.
18 If necessary for additional working room remove the air cleaner and mixture control unit.
19 Where applicable on APC versions disconnect the wiring from the solenoid valve and remove the valve, connector, and knock detector.
20 Where applicable disconnect the wiring from the oxygen sensor on the lambda system and also disconnect the wiring from the throttle valve switch.

All engines

21 Disconnect the wiring from the oil pressure sender unit.
22 Disconnect the hose from the expansion tank and heater - note that the top heater

Engine removal and overhaul procedures 2B•5

4.29 Removing the front panel and radiator on 99 models

4.33 Speedometer cable connection on the gearbox

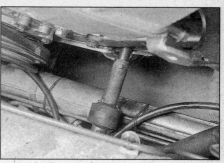

4.36 Gearshift rod connection on manual gearbox models

hose connects to the water pump, the bottom hose to the inlet manifold.
23 Disconnect the vacuum hoses from the brake servo and heater control supply.
24 Disconnect the wiring from the radiator cooling fan motor and thermo-switch, and where applicable the headlights and headlight wipers.
25 On automatic transmission models disconnect the wiring from the switch on the transmission.
26 Identify for location then disconnect the wiring from the alternator.
27 Disconnect the main wiring harness connector and remove the harness from the engine.

99 models

28 Disconnect the headlight wiper linkage.
29 Remove the grille, disconnect the bonnet cable, then remove the retaining screws and withdraw the complete front panel and radiator from the car (see illustration).

All models

30 Where applicable disconnect and plug the hydraulic hoses from the oil cooler.
31 Jack up the front of the car and support it on axle stands (see "Jacking and Vehicle Support"). Chock the rear wheels.
32 Loosen the clips and disconnect the rubber bellows from the inner ends of the driveshafts.
33 Disconnect the speedometer cable from the rear of the gearbox/transmission (see illustration).
34 Remove both front wheels then jack up the suspension on each side in turn and place a metal or hardwood block between the upper control arm and the body. Lower the jack.
35 Unscrew and remove the bolts securing the lower control arms to the lower balljoints then pull out each driveshaft and disconnect them from the inner driveshafts.

Manual gearbox models

36 With the gear lever in neutral unscrew the nut from the gearshift rod, tap out the tapered pin, and separate the two rods (see illustration).

Automatic transmission models

37 Unscrew the screw for the selector cable at the transmission, then withdraw the cable with the selector rod fully forward in position "P". Slide back the sleeve and unhook the cable.

All models

38 Disconnect the exhaust downpipe from the manifold (see illustration).
39 Where applicable remove the power steering pump belt with reference to Chapter 1 then with the mounting bolts/nuts loosened detach the pump and place it to one side of the engine compartment. Do not disconnect the hydraulic hoses.
40 Attach a suitable hoist to the engine and take the weight of the unit.
41 Unscrew and remove the rear engine mounting bolts (see illustration).
42 Loosen only the front engine mounting bolt - do not remove the bolt as the bracket has open ended slots.
43 Lift the engine from the engine compartment taking care not to damage components on the bulkhead and side panels (see illustration).

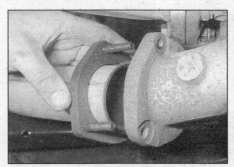

4.38 Disconnecting the exhaust downpipe from the manifold

4.43 Lifting out the engine

Models from 1981

44 Remove the battery with reference to Chapter 5. Also remove the battery tray.
45 Drain the cooling system as described in Chapter 1. Note: *The engine should be cold before draining the coolant*. Drain the cylinder block as well by unscrewing the drain plug.
46 Remove the bonnet (see Chapter 11).
47 At this stage Saab technicians insert a spacer tool between the right-hand front suspension upper wishbone and the inner body in order to relieve the pressure in the suspension when the engine/transmission is removed. A piece of wood approximately 2 inches thick may be used instead of the tool - turn the steering on full right-hand lock and position the wood beneath the upper wishbone and body (see illustration).
48 Disconnect the downpipe from the exhaust manifold with reference to Chapter 4.

4.41 Right-hand side rear engine mounting on a 900 model

4.47 Wooden spacer inserted between the right-hand front suspension upper wishbone and body

2B•6 Engine removal and overhaul procedures

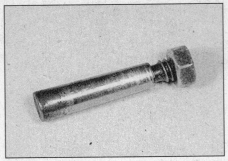

4.51a Transmission selector rod linkage pin viewed from under the car

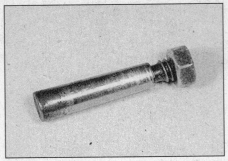

4.51b The selector rod linkage removed from the car

4.53 Disconnecting the speedometer cable from the transmission

4.54a Unbolting the exhaust downpipe bracket from the transmission . . .

4.54b . . . and from the downpipe

4.55 Loosening the driveshaft gaiter clips

49 Loosen the right-hand front wheel bolts then jack up the front of the car and support on axle stands (see "*Jacking and Vehicle Support*"). Remove the wheel.
50 Syphon the fluid from the power steering fluid reservoir.
51 Working under the car on manual transmission models, disconnect the selector rod from the linkage by unscrewing the nut and tapping out the pin **(see illustrations)**.
52 Working under the car on automatic transmission models, remove the screw securing the selector cable to the transmission. Select position 'P' so that the selector rod is fully extended, then slide back the spring-loaded sleeve and unhook the end of the cable.
53 Unscrew the knurled collar and disconnect the speedometer cable from the transmission **(see illustration)**.
54 Unbolt the exhaust downpipe bracket from the bottom of the transmission and from the downpipe itself **(see illustrations)**.
55 Loosen the clips on both inner driveshaft joints and pull the rubber gaiters from the joint housings **(see illustration)**.
56 Disconnect the right-hand lower suspension arm (and anti-roll bar where applicable) from the steering knuckle with reference to Chapter 10 **(see illustration)**.
57 Pull out the steering knuckle and disconnect the inner end of the driveshaft from the cup then move the driveshaft in front of the cup. Support the driveshaft on the lower arm.
58 Position a suitable container beneath the power steering pump then unscrew the unions using two spanners and disconnect the pressure pipe from the pump. Take care not to allow fluid onto the engine mounting **(see illustration)**.
59 On models with air conditioning remove the compressor drivebelt with reference to Chapter 1.
60 On models from 1981 disconnect the oil filler pipe.
61 Identify the hose locations then loosen the clips and disconnect the coolant hoses to the heater at the bulkhead, the coolant hose from the bottom of the expansion tank, and the top hose from the thermostat housing **(see illustration)**.
62 Identify the vacuum hoses for position then disconnect them from the solenoid valve on the engine compartment front crossmember.
63 Loosen the clips and disconnect the Turbo hose between the intercooler and the throttle housing.
64 On models with Bosch CI disconnect and remove the air duct from the top of the airflow meter.
65 Remove the cover from the top of the

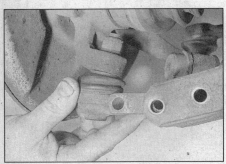

4.56 Disconnecting the right-hand lower suspension arm from the steering knuckle

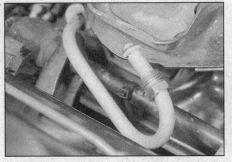

4.58 Disconnecting the power steering pressure pipe

4.61 Disconnecting the coolant hose from the bottom of the expansion tank

Engine removal and overhaul procedures 2B•7

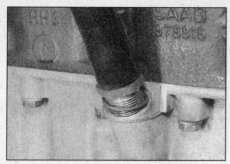
4.69a Unscrew the dipstick tube . . .

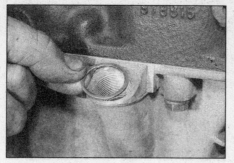
4.69b . . . and recover the copper washer

4.70a Earth points on the front of the cylinder head

4.70b Engine temperature sensor located below the thermostat

4.70c Oil pressure sensor location on the oil filter housing

4.70d Support on right-hand side of engine compartment for the battery positive lead

4.70e Disconnecting the battery positive lead from the junction box

4.70f Support on right-hand corner of engine compartment for starter motor lead

intercooler (and where applicable the baffle plate from the side) then remove the intercooler together with the air duct from the engine compartment. Refer to Chapter 4 if necessary. Cover the Turbocharger inlet with a plastic cap or tape to prevent entry of dust and dirt.

66 Remove the air cleaner and the airflow meter, with reference to Chapter 4. On models with LH-Jetronic also disconnect the fuel hoses from the fuel rail. The following is a summary of the work involved:

Models with Bosch CI
a) Disconnect the air cleaner inlet hose
b) Disconnect the fuel hose from the warm-up regulator
c) Disconnect the hose and unscrew the fuel booster pressure switch securing screw. Move the switch to one side.
d) Disconnect the fuel hose from the cold start valve.
e) Disconnect the hoses from the airflow meter, identifying them for position.
f) Unscrew the unions and disconnect the fuel lines from the injectors.
g) Disconnect the earth wire from the airflow meter, then release the clip and remove the airflow meter together with the air cleaner and hoses.

Models with LH-Jetronic
h) Disconnect the air ducts, disconnect the wiring and remove the airflow meter.
i) Remove the air cleaner together with the air inlet.
i) Loosen the unions and disconnect the fuel hoses from the fuel rail and fuel pressure regulator.

67 Disconnect the accelerator cable with reference to Chapter 4.
68 Disconnect the brake servo and fuel tank breather hoses from the inlet manifold and throttle housing.
69 Unscrew and remove the dipstick tube and recover the copper washer **(see illustrations)**.
70 Note the location of all wires from the engine wiring loom, then disconnect them and position the loom on the engine compartment front crossmember. Refer to the relevant Chapter of this manual if necessary. The wires are connected to the fuel injection components, temperature sensors/switch, earth points, starter motor, oil pressure sensor, air conditioning components and the alternator. Also disconnect the battery positive leads from the junction box on the right-hand side of the engine compartment **(see illustrations)**.

71 Unscrew the union and disconnect the coolant pipe for the Turbocharger from the clip on the timing cover. Also remove the earth lead.
72 Disconnect and plug the hydraulic hose from the clutch slave cylinder with reference to Chapter 6.
73 Loosen the clip and disconnect the bottom hose from the radiator.
74 On models with air conditioning, remove the auxiliary cooling fan.
75 Unclip the battery positive lead from the engine and place it to one side.
76 Refer to Chapter 5 and remove the ignition coil.
77 Disconnect the Lambda sensor wiring above the right-hand wheel arch and place the wiring on the engine.
78 On models with air conditioning unbolt the compressor and position it on the heat exchanger. Do not disconnect the refrigerant hoses.

2B•8 Engine removal and overhaul procedures

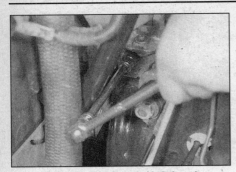

4.80 Unscrewing the left-hand engine mounting bolt

4.81a Removing the front engine mounting clamp plate . . .

4.81b . . . and mounting nut

4.82 Right-hand engine mounting nut

4.84 The return line from the power steering pump

4.85 Disconnecting the oil cooler hoses from the oil filter housing

79 Attach a suitable hoist to the engine and take the weight of the unit.
80 Unscrew and remove the left-hand engine mounting bolt and spacer **(see illustration)**.
81 Unscrew and remove the front engine mounting clamp plate bolts and mounting nut **(see illustrations)**.
82 Unscrew and remove the right-hand engine mounting nut **(see illustration)**.
83 Carefully raise the engine and separate the left-hand driveshaft from its cup.
84 Disconnect the return line from the power steering pump **(see illustration)**.
85 Unscrew and remove the oil pressure switch from the oil filter housing, then unscrew the unions and disconnect the oil cooler hoses from the housing. Identify the hoses for position **(see illustration)**.
86 Lift the engine and transmission out of the engine compartment, taking care not to damage the radiator or the solenoid valve **(see illustration)**.

Separation
87 Clean the engine and transmission and wipe dry.
88 Unscrew the drain plugs and drain the engine oil and transmission oil/fluid.
89 Unbolt the cover from the flywheel/driveplate if still fitted, remove the EGR pipe where applicable, then remove the starter motor as described in Chapter 5 except on automatic transmission models manufactured after 1984.

Manual transmission models
90 Remove the clutch (refer to Chapter 6). Alternatively, pull out the transmission input shaft and unscrew the slave cylinder mounting bolts. Using the latter method the slave cylinder and release bearing is removed as the engine is separated from the transmission.

4.86 Lifting the engine and transmission out of the engine compartment

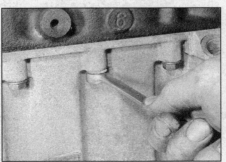

4.91b Unscrewing a left-hand side bolt

91 Unscrew all the bolts and nut securing the transmission to the engine noting the location of any brackets **(see illustrations)**. Also unbolt the oil filler pipe bracket at the throttle control on the inlet manifold.

4.91a Unscrew the nut and stud securing timing end of engine to the transmission

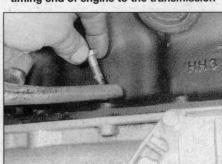

4.91c Removing a bolt securing the engine mounting bracket to the right-hand side of the engine

Engine removal and overhaul procedures 2B•9

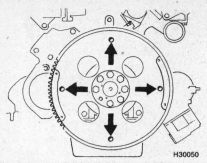

4.95 Torque converter bolt locations on the driveplate - automatic transmission models

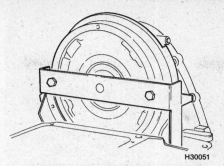

4.97 Torque converter support plate on automatic transmission models

4.99a Locating the gasket on the transmission

4.99b Applying sealing compound to the slots in the gasket

92 Attach a hoist to the engine and carefully lift it from the transmission. Note the location of stays and brackets.
93 Remove the gasket from the contact flange of the transmission.

Automatic transmission models

94 On carburettor models disconnect the downshift cable.
95 Unscrew the four bolts securing the driveplate to the torque converter. To do this turn the driveplate until the bolts are accessible just above the oil pump **(see illustration)**.

96 Unscrew the bolts securing the transmission to the engine. Also where applicable, unscrew the unions and disconnect the hydraulic hoses from the oil cooler.
97 Turn the driveplate so that the angle irons are horizontal then lift the engine from the transmission. Remove the gasket. Retain the torque converter in the transmission by fitting a support plate to it. The special Saab tool consists of a length of metal bolted to the driveplate with angled ends to engage the transfer housing **(see illustration)**.

Reconnection of transmission to engine

All models

98 Check that the two guide sleeves are located in the transmission. On automatic transmission models remove the support plate.

99 Clean the contact surfaces then locate a new gasket on the transmission. On pre-1980 models, Saab recommend that sealing compound is applied to both sides of the gasket at the front and rear flanges - on later models slots are provided at these positions for the sealing compound **(see illustrations)**.
100 On automatic transmission models remove the support plate from the torque converter and make sure that the angle irons are horizontal.
101 Carefully lower the engine onto the transmission then refit the bolts and nut noting that sealing compound must be applied to the threads of the six bolts shown **(see illustration)**. Refit the stays and brackets as noted during removal. Tighten the bolts to the specified torque.

Manual transmission models

102 Refit the oil filler pipe bracket to the inlet manifold and tighten the bolt.
103 Refit the transmission input shaft if removed, or refit the clutch (see Chapter 6).

Automatic transmission models

104 Turn the driveplate to align the bolt holes then insert the bolts and tighten to the specified torque.
105 Where applicable, reconnect the unions to the oil cooler; tighten to the specified torque.
106 On carburettor models reconnect the downshift cable with reference to Chapter 7B.

All models

107 Where applicable refit the starter motor and EGR pipe.
108 Refit the flywheel cover.

Refitting

109 Refitting is a reversal of removal but note the following additional points.
  a) Tighten all nuts and bolts to the specified torque **(see illustration)**.
  b) Pack the driveshaft inner joint housings with the specified grease before lowering the engine into the engine compartment **(see illustration)**.
  c) Renew the copper washers on unions as applicable.
  d) Where applicable, tension the air conditioning compressor drivebelt with reference to Chapter 1.
  e) If applicable, bleed the clutch hydraulic

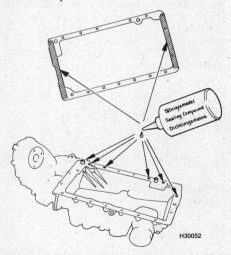

4.101 Apply sealing compound as indicated when fitting the engine to the transmission

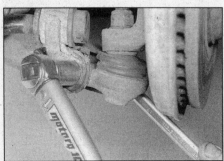

4.109a Tightening the front suspension lower arm to steering knuckle bolts

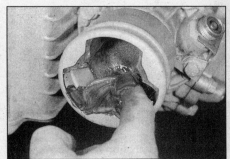

4.109b Packing the driveshaft inner joint housings with grease

2B•10 Engine removal and overhaul procedures

system with reference to Chapter 6.
f) Reconnect and adjust the accelerator cable with reference to Chapter 4.
g) Ensure all wiring has been reconnected and all nuts and bolts tightened.
h) Refill the engine with the correct quantity and grade of oil/fluid (see Chapter 1).
i) Check and if necessary top-up the manual transmission oil level with reference to Chapter 1.
j) Check and if necessary top-up the automatic transmission fluid level with reference to Chapter 1.
k) Refill the cooling system (see Chapter 1).
l) Check and if necessary top-up the power steering fluid (see "Weekly Checks").

5 Engine overhaul - dismantling sequence

1 It is much easier to dismantle and work on the engine if it is mounted on a portable engine stand. These stands can often be hired from a tool hire shop. Before the engine is mounted on a stand, the flywheel/driveplate should be removed, so that the stand bolts can be tightened into the end of the cylinder block/crankcase.
2 If a stand is not available, it is possible to dismantle the engine with it blocked up on a sturdy workbench, or on the floor. Be extra-careful not to tip or drop the engine when working without a stand.
3 If you are going to obtain a reconditioned engine, all the external components must be removed first, to be transferred to the replacement engine (just as they will if you are doing a complete engine overhaul yourself). These components include the following, however check with your supplier first:
a) Alternator mounting bracket (Chapter 5A).
b) The distributor, HT leads and spark plugs (Chapter 1 and Chapter 5).
c) Thermostat and housing (Chapter 3).
d) The dipstick tube.
e) Air conditioning compressor mounting bracket (Chapter 3).
f) The carburettor or fuel injection system, and emission control components (Chapter 4, Parts A, B and C).
g) All electrical switches and sensors and the engine wiring harness.
h) Inlet and exhaust manifolds (Chapter 4).
i) Oil filter (Chapter 1) and where applicable on the H-type engine the oil filter housing (see illustration).
j) Oil pump pick-up tube and O-ring (see illustrations).
k) Water pump (Chapter 3).
l) Engine mounting brackets (Part A of this Chapter).
m) Flywheel (Part A of this Chapter).
n) Driveplate (Section 8 of Chapter 2B).

Note: When removing external components from the engine, pay close attention to details that may be helpful or important during refitting. Note the fitted position of gaskets, seals, spacers, pins, washers, bolts, and other small items.
4 If you are obtaining a 'short' engine (which consists of the engine cylinder block/crankcase, crankshaft, pistons and connecting rods all assembled), then the cylinder head will have to be removed also.
5 If you are planning a complete overhaul, the engine can be dismantled, and the internal components removed, in the order given below, referring to Part A of this Chapter unless otherwise stated.
a) Inlet and exhaust manifolds (Chapter 4A).
b) Timing chain, sprockets and tensioner (Chapter 2B, Section 6).
c) Cylinder head (Chapter 2A, Section 6).
d) Flywheel (Chapter 2A, Section 11) or driveplate (Chapter 2B, Section 8).
e) Piston/connecting rod assemblies (Chapter 2B, Section 10).
f) Crankshaft (Chapter 2B, Section 11).
6 Before beginning the dismantling and overhaul procedures, make sure that you have all of the correct tools necessary. Refer to "Tools and working facilities" for further information.

6 Timing chain and sprockets - removal, inspection and refitting

Removal

B201 B-type engine

1 With the engine removed, position the crankshaft at TDC on compression for No 1 piston (timing chain end of the engine)

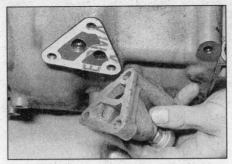
5.3a Removing the oil filter housing on the H-type engine

5.3c Unscrew the bolts . . .

5.3b Oil pump pick-up tube on the H-type engine

5.3d . . . and remove the oil pump pick-up tube and O-ring

with reference to Chapter 2A, Section 3.
2 Jam the starter ring gear and unscrew the crankshaft pulley bolt.
3 Remove the valve cover and detach the sprocket from the camshaft as described in Chapter 2A, Section 6.
4 Remove the crankshaft pulley, using a puller if necessary. Unbolt and remove the timing chain cover.
5 At this point wear in the timing chain can be ascertained using the following method. Using a steel rule measure the distance from the tensioner pad to the upper face of the cylinder head (see illustration). If it is less than 300 mm the chain and/or tensioner are worn excessively.
6 Note how the tensioner is fitted then unbolt and remove it. Remove the guide plate (see illustrations).

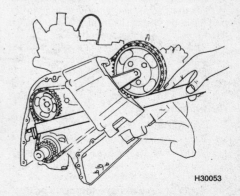
6.5 Using a steel rule to check the timing chain and tensioner

Engine removal and overhaul procedures 2B•11

6.6a Timing chain tensioner on the B-type engine

6.6b Removing the timing chain tensioner guide plate on the B-type engine

6.13a Alternator link location on the timing cover on the H-type engine

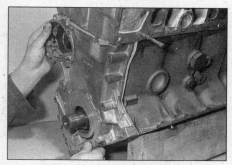

6.13b Removing the timing cover on the H-type engine

6.13c Showing the timing cover gasket locations on the H-type engine

6.14 Removing the timing chain and sprockets on the H-type engine

7 Unbolt the camshaft sprocket mounting plate and guide and withdraw them together with the chain.
8 Using a suitable puller, remove the sprocket from the front of the crankshaft. Remove the shims.
9 Removal of the idler shaft sprocket is described in Section 7 of this Chapter.

B201 H-type engine

10 With the engine removed, position the crankshaft at TDC compression for No 1 piston (timing chain end of the engine) as described in Chapter 2A, Section 3.
11 Remove the water pump (Chapter 3) and the oil pump (Chapter 2A, Section 7).
12 Remove the valve cover and detach the sprocket from the camshaft as described in Chapter 2A, Section 6.
13 Unbolt and remove the timing cover and remove the gasket and where applicable the pick-up tube O-ring. Note the location of the alternator link **(see illustrations)**.
14 Remove the timing chain together with the camshaft and crankshaft sprockets **(see illustration)**.
15 Unbolt the chain guide and tensioner from the front of the cylinder block **(see illustrations)**.

B202 engine

Note: *This procedure describes removal of the timing cover leaving the cylinder head in position, however the alternative method which is less likely to damage the cylinder head gasket is to remove the cylinder head first.*

16 With the engine removed, position the crankshaft at TDC compression for No 1 piston (timing chain end of the engine) as described in Chapter 2A, Section 3.
17 Remove the auxiliary drivebelt(s) with reference to Chapter 1, and the power steering pump with reference to the following illustrations **(see illustrations)**.
18 Have an assistant hold the crankshaft stationary using a wide-bladed screwdriver engaged with the starter ring gear through the ignition timing hole at the top of the transmission.
19 Loosen the crankshaft pulley bolt using a long socket bar. The bolt is tightened to a high torque.
20 Fully unscrew the crankshaft pulley bolt and slide the pulley off the end of the crankshaft.
21 Unscrew and remove the bolts and nut securing the timing cover to the cylinder block, transmission and cylinder head. Note that the bolts are of different lengths. Note the two upper bolts on the cylinder head and the single nut and bolt in the transmission. If the timing cover is being removed with the cylinder head still in position, it will be necessary to unscrew the stud from the transmission in order to allow the cover to slide out. Tighten two nuts on the stud in order to remove it.
22 Taking care not to damage the cylinder head gasket, carefully withdraw the timing cover complete with the oil pump from the nose of the crankshaft **(see illustration)**. Where fitted, remove the gaskets from the cylinder block.
23 Thoroughly clean all traces of gasket or sealant from the contact faces of the timing cover, transmission, cylinder head and block.

6.15a Timing chain guide/tensioner on the H-type engine - note manual release/adjustment point for tensioner (arrowed)

6.15b Removing the timing chain tensioner on the H-type engine

2B•12 Engine removal and overhaul procedures

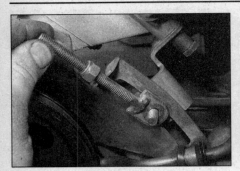
6.17a Unscrew the nut and remove the tensioner...

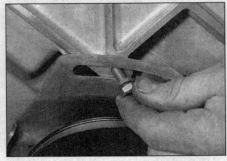
6.17b ...then remove the locking bolt...

6.17c ...unscrew the pivot nut...

6.17d ...and remove the washer...

6.17e ...outer rubber bush, followed by the pump...

6.17f ...inner rubber bush and spacer...

6.17g ...and mounting bracket

6.22 Removing the timing cover (oil pump and cylinder head already removed)

6.27 Removing the sprocket and chain from the end of the inlet camshaft

24 Remove the cylinder head cover as described in Chapter 2A, Section 4.
25 Unscrew and remove the timing chain tensioner from the cylinder head. On later models, first unscrew the centre bolt and remove the spring, then unscrew and remove the tensioner from the cylinder head.
26 While holding each camshaft stationary with a spanner on the flats at the flywheel/driveplate end of the camshaft, loosen but do not remove the camshaft sprocket securing bolts.
27 Unscrew and remove the bolt and withdraw the sprocket from the end of the inlet camshaft **(see illustration)**. Hold the timing chain with one hand and release the sprocket from it with the other hand.
28 Identify each sprocket for position. Note that each sprocket has a projection which engages with a cut-out in the end of the camshaft.

29 Unscrew the bolt and withdraw the sprocket from the end of the exhaust camshaft, then disengage it from the chain **(see illustration)**.
30 Remove the tensioner guide, then disengage the timing chain and remove the

6.29 Removing the sprocket and chain from the end of the exhaust camshaft

sprocket from the end of the crankshaft **(see illustrations)**. Note that the sprocket incorporates the drive dogs for the oil pump. If necessary, remove the Woodruff key from the groove in the crankshaft using a screwdriver.
31 Unscrew the bolts and remove the timing chain fixed guide from the cylinder block **(see illustration)**.

Inspection

32 Examine the teeth on the crankshaft, camshaft and idler shaft (B-type engine) sprockets for wear. If worn the side of each tooth will be slightly concave in shape and in this case the sprockets should be renewed.
33 Examine the links of the chain for side slackness and renew the chain if any slackness is noticeable when compared with a new chain. It is a sensible precaution to renew the chain regardless of its apparent condition if the engine is stripped down for a

Engine removal and overhaul procedures 2B•13

6.30a Remove the tensioner guide ...

6.30b ... and remove the sprocket from the end of the crankshaft

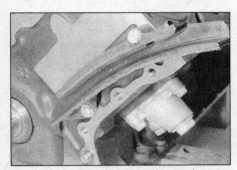

6.31 Timing chain fixed guide on the front of the cylinder block

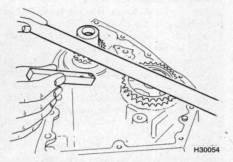

6.35 Checking the sprocket alignment on the B-type engine

major overhaul. The actual rollers on a very badly worn chain may be slightly grooved.
34 Examine the chain tensioner and guide(s) for wear and renew them as necessary.

Refitting

B201 B-type engine

35 Locate the sprocket on the front of the crankshaft together with the shims removed. If the crankshaft or camshaft sprocket has been renewed use a steel rule or straightedge to determine the shim thickness required to align both sprockets **(see illustration)**.
36 Check that the camshaft and flywheel TDC marks are aligned, and turn the idler shaft so that the bulge in the sprocket hole is aligned with the mark on the locking plate **(see illustration)**.
37 Fit the camshaft sprocket to the mounting plate if removed then locate the sprocket in the timing chain.
38 Lower the chain and sprocket through the head aperture and align the sprocket holes with the camshaft with the hole bulge uppermost. Carefully loop the chain over the crankshaft and idler shaft sprockets without misaligning them and insert the sprocket bolts and tighten them.
39 Locate the chain guide then position the mounting plate and insert the bolts. Slightly depress the chain guide to pre-tension the chain, then tighten the mounting bolts.
40 Check that the timing marks are still aligned.
41 Where a Reynolds type chain tensioner is fitted first remove the pad, spring, and ratchet from the housing then fit the spring and ratchet to the pad and use an Allen key to fully compress the spring while turning the ratchet clockwise **(see illustration)**. Fit the pad and spring in the housing with a 0.5 mm spacer.
42 Where a JWIS type chain tensioner is fitted press and turn the pad until it is fully

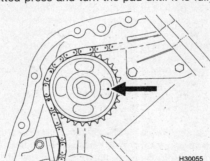

6.36 Align the bulge in the idler sprocket with the TDC mark on the B-type engine

6.44 Checking the timing chain tensioner gap on the B-type engine

entered in the housing. Keep the pad in this position until fitted and the timing chain tensioned.
43 Fit the chain tensioner and guide plate to the block and insert and tighten the bolts.
44 Loosen the bolts then depress the chain guide to take up any slack. Remove the spacer where applicable and adjust the chain guide until the gap between the pad and the housing is 0.5 mm **(see illustration)**. Tighten the bolts.
45 Unscrew the camshaft sprocket centre bolt and refit it to the sprocket.
46 Turn the crankshaft two complete turns making sure that the gap on the chain tensioner does not exceed 1.5 mm or decrease to less than 0.5 mm.
47 Fit the timing chain cover together with a new gasket, insert and tighten the bolts. Trim any excess gasket from the lower face. Do not forget to fit the alternator link.
48 Fit the pulley on the crankshaft then insert and tighten the bolt while holding the crankshaft stationary.

B201 H-type engine

49 Fit the chain guide and tensioner to the front of the cylinder block and insert and tighten the bolts **(see illustration)**. Flat washers should be fitted beneath the bolt heads for the chain guide.
50 Check that the flywheel/driveplate 0° mark is aligned with the TDC timing mark. Check that the camshaft is set for No 1 cylinder firing. If either the crankshaft or camshaft needs to be turned, be careful that piston/valve contact does not occur.

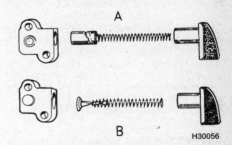

6.41 Timing chain tensioner components on the B-type engine

A Reynolds type   B JWIS type

6.49 Timing chain tensioner on the cylinder block on the H-type engine

2B•14 Engine removal and overhaul procedures

6.57 The bright link (arrowed) must be kept in alignment with the slot in the sprocket

51 Locate the camshaft and crankshaft sprockets in the timing chain with the key slot and TDC pointer uppermost.
52 Slide the sprocket onto the crankshaft over the Woodruff key and place the camshaft sprocket between the tensioner and guide.
53 The camshaft sprocket must now be refitted and the timing chain tensioner released. On early models hook a piece of wire into the release hole and pull upwards before locating the sprocket on the camshaft and inserting the retaining bolts. On later models, set the tensioner to its primed position by depressing the ratchet tooth and pushing the plunger fully inwards - retain it in this position using a plastic cable-tie or the transit lock. Make sure that the camshaft sprocket and bearing cap marks are correctly aligned then tighten the retaining bolts to the specified torque. On later models release the tensioner by cutting and removing the plastic cable or removing the transit lock. To adjust the tensioner, disengage it by pressing down on the outside of the ratchet tooth, then press the chain guide to depress the tensioner.
54 Refit the timing cover, using a new gasket and jointing compound, and where applicable a new O-ring for the oil pump pick-up tube. Refit the alternator link in the previously noted position.
55 Refit the valve cover, the water pump and the oil pump.

B202 engine

56 Locate the Woodruff key in the groove in the crankshaft. Tap it fully into the groove making sure that it is parallel to the crankshaft.

57 Engage the timing chain with the crankshaft sprocket, then locate the crankshaft sprocket on the end of the crankshaft making sure it locates correctly on the Woodruff key. Where the timing chain has bright links, locate the single bright link at the bottom of the sprocket and aligned with the slot in the sprocket (see illustration). Keep the chain tensioned at this stage to prevent it from disengaging from the sprocket.
58 Locate the timing chain in the fixed guide then refit the guide and tighten the bolts. Keep the chain in tension.
59 Refit the sprocket to the end of the exhaust camshaft, insert the bolt and finger-tighten it at this stage. **Do not** apply thread locking fluid to the threads of the bolt.
60 Check that the crankshaft and camshafts are still at their TDC positions.
61 Feed the timing chain up through the cylinder head aperture and locate it on the exhaust camshaft sprocket making sure that it is taut between the two sprockets. Check that it is correctly located on the guides. Where the chain has a bright link make sure that it is aligned with the timing mark.
62 Engage the inlet sprocket with the timing chain so that the engagement cut-out and projection are in alignment, then locate the sprocket on the inlet camshaft and insert the bolt. Finger-tighten the bolt at this stage. **Do not** apply thread locking fluid to the threads of the bolt. Where the chain has a bright link make sure that it is aligned with the timing mark.
63 Set the timing chain tensioner as follows. On models built before approximately 1988, push the plunger fully into the tensioner body and turn it to lock it. On later models use a screwdriver to press down the ratchet, then push the plunger fully into the tensioner and release the ratchet. Check the condition of the tensioner washer, and renew it if necessary.
64 Insert the tensioner in the cylinder head and tighten to the specified torque.
65 On models manufactured up to approximately 1988, trigger the tensioner by pressing the pivoting chain guide against the tensioner, then press the guide against the chain to provide it with its basic tension. When the engine is started, hydraulic pressure will take up any remaining slack.

6.68 Tightening camshaft sprocket bolts, holding the camshafts with a spanner

66 On models manufactured after 1988, insert the spring and plastic guide pin in the tensioner, then fit the plug together with a new O-ring, and tighten it to the specified torque. **Note:** *New tensioners are supplied with the tensioner spring held pre-tensioned with a pin.* **Do not** *remove this pin until after the tensioner has been tightened into the cylinder head. When the engine is started, hydraulic pressure will take up any remaining slack.*
67 Temporarily refit the crankshaft pulley bolt and rotate the engine two complete turns clockwise. Check that the timing marks still align correctly. Remove the pulley bolt. Where the chain has bright links, these will not now be aligned with the timing marks.
68 Tighten the camshaft sprocket bolts to the specified torque while holding the camshafts with a spanner on the flats (see illustration).
69 Refit the cylinder head cover with reference to Chapter 2A, Section 4.
70 Refit the timing cover together with new gaskets or with suitable sealant. Apply suitable sealant to the transmission flange and gasket - see Section 4 of this Chapter. Make sure that the rubber O-ring is correctly located in the bottom of the timing cover and that it engages with the oil pump pick-up tube as the cover is being refitting. Refit and tighten the stud then tighten the nut. Tighten all bolts to the specified torque (see illustrations).
71 Refit the crankshaft pulley. Tighten the pulley bolt to the specified torque while an assistant holds the crankshaft with a screwdriver engaged with the starter ring gear.
72 Refit and adjust the auxiliary drivebelt(s) with reference to Chapter 1.

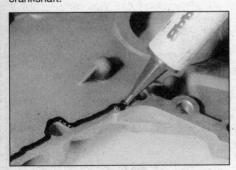

6.70a Applying sealant to the timing cover contact face

6.70b The oil pump pick-up tube sealing ring in the timing cover

6.70c Tightening the timing cover bolts

Engine removal and overhaul procedures 2B•15

7 Idler shaft (B201 B-type engine) - removal, inspection and refitting

Removal

1 Remove the distributor (Chapter 5), water pump (Chapter 3), oil pump (Chapter 2A, Section 7), and fuel pump (Chapter 4).
2 Remove the timing chain (see Section 6).
3 Hold the idler shaft sprocket with a screwdriver inserted through one of the holes then unscrew the bolt and remove the sprocket.
4 Remove the two socket-head screws from the locking plate, extract the plate and withdraw the idler shaft from the cylinder block (see illustrations).

Inspection

5 Examine the bearing surfaces of the idler shaft; if these are scored or worn excessively, renew the shaft.
6 If the fuel pump cam or the gears which drive the water pump, oil pump or distributor are worn then the idler shaft will have to be renewed.

Refitting

7 Oil the bearing surfaces of the idler shaft then insert it into the cylinder block.
8 Fit the locking plate then insert and tighten the two socket-head screws.
9 Locate the sprocket on the idler shaft, insert the bolt together with a new lockwasher, and tighten the bolt while inserting a screwdriver through one of the holes. Bend the lockwasher onto one of the bolt head flats.

7.4a Remove the idler shaft locking plate on the B-type engine ...

7.4b ... and withdraw the idler shaft

10 Refit the timing chain (see Section 6).
11 Refit the distributor (Chapter 5), water pump (Chapter 3), oil pump (Chapter 2A, Section 7), and the fuel pump (Chapter 4).

8 Driveplate (automatic transmission models) - removal, inspection and refitting

Removal

1 Separate the transmission from the engine as described in Section 4.
2 Prevent the driveplate from turning by locking the ring gear teeth with a wide-blade screwdriver or similar tool. Alternatively, bolt a metal link between the flywheel (using the torque converter bolt holes) and the cylinder block/crankcase.
3 Unscrew and remove the retaining bolts, remove the locking tool, then remove the driveplate from the crankshaft flange. Note that the unit is located by a single dowel pin and cannot be fitted incorrectly.

Inspection

4 If the ring gear is badly worn or has missing teeth, it may be possible to renew it. This job is best left to a Saab dealer or engine reconditioning specialist. The temperature to which the new ring gear must be heated for installation is critical and, if not done accurately, the hardness of the teeth will be destroyed.
5 Check the driveplate for any signs of cracking or distortion.

Refitting

6 Clean the mating surfaces of the driveplate and crankshaft. Clean the threads of the retaining bolts and the crankshaft holes. If a suitable tap is not available, cut two slots into the threads of an old driveplate bolt and use the bolt to clean the threads. Examine the bolts and if necessary obtain new ones.
7 Ensure that the locating dowel is in position, then offer up the driveplate and locate it on the dowel.
8 Apply locking fluid to the threads of the retaining bolts, then insert them and tighten to the specified torque while holding the driveplate stationary using one of the methods described in paragraph 2. Note that new bolts may be pre-coated with sealant.
9 Refit the transmission to the engine with reference to Section 4.

9 Crankshaft oil seal (driveplate end on automatic transmission models) - renewal

1 Remove the driveplate (see Section 8).
2 Make a note of the fitted depth of the seal in the engine plate. Punch or drill two small holes opposite each other in the seal. Screw a self-tapping screw into each, and pull on the screws with pliers to extract the seal. Alternatively, use a screwdriver to prise out the oil seal.
3 Wipe clean the seal location in the engine plate, then lubricate the lips of the new seal with clean engine oil, and carefully locate the seal on the end of the crankshaft making sure the seal lips face inwards.
4 Using a suitable tubular drift, which bears only on the hard outer edge of the seal, drive the seal into position, to the same depth in the housing as the original was prior to removal.
5 Wipe clean the oil seal, then refit the driveplate as described in Section 8.

10 Cylinder head - dismantling

Note: *New/reconditioned cylinder heads are obtainable from the manufacturer or from engine overhaul specialists. Be aware that some specialist tools are required for the dismantling and inspection procedures, and new components may not be readily available. It may therefore be more practical and economical for the home mechanic to purchase a reconditioned head, rather than dismantle, inspect and recondition the original head.*

1 Remove the cylinder head as described in Part A then unbolt the external components - these include the engine lifting bracket and thermostat housing (see illustration).

10.1 Removing the thermostat housing on the H-type engine

2 Remove the camshafts and cam followers with reference to Part A, Section 5.
3 Before removing the valves, consider obtaining plastic protectors for the cam follower bores. When using certain valve spring compressors, the bores can easily be damaged should the compressor slip off the end of the valve.

 HAYNES HINT *The protectors can be obtained from a Saab dealer, or alternatively a protector may be made out of plastic cut from a washing-up liquid container or similar.*

2B•16 Engine removal and overhaul procedures

10.4a Using a compressor, compress the valve springs to remove the split collets

10.4b Removing the spring retainer . . .

10.4c . . . valve spring . . .

10.4d . . . and spring seat

4 Position the protector in the cam follower bore, then using a valve spring compressor, compress the valve spring until the split collets can be removed. Release the compressor, and lift off the spring retainer, spring and spring seat. On B202 engines using a pair of pliers, carefully extract the valve stem oil seal from the top of the guide **(see illustrations)**.

5 If, when the valve spring compressor is screwed down, the spring retainer refuses to free and expose the split collets, gently tap the top of the tool, directly over the retainer, with a light hammer. This will free the retainer.

6 Withdraw the valve through the combustion chamber **(see illustration)**.

7 It is essential that each valve is stored together with its collets, retainer, spring, and spring seat. The valves should also be kept in their correct sequence, unless they are so badly worn that they are to be renewed. If they are going to be kept and used again, place each valve assembly in a labelled polythene bag or similar small container **(see illustration)**. Note that No 1 cylinder is nearest to the timing chain end of the engine.

11 Cylinder head and valves - cleaning and inspection

1 Thorough cleaning of the cylinder head and valve components, followed by a detailed inspection, will enable you to decide how much valve service work must be carried out during the engine overhaul. **Note:** *If the engine has been severely overheated, it is best to assume that the cylinder head is warped - check carefully for signs of this.*

Cleaning

2 Scrape away all traces of old gasket material from the cylinder head.

3 Scrape away the carbon from the combustion chambers and ports, then wash the cylinder head thoroughly with paraffin or a suitable solvent.

4 Scrape off any heavy carbon deposits that may have formed on the valves, then use a power-operated wire brush to remove deposits from the valve heads and stems.

Inspection

Note: *Be sure to perform all the following inspection procedures before concluding that the services of a machine shop or engine overhaul specialist are required. Make a list of all items that require attention.*

Cylinder head

5 Inspect the head very carefully for cracks, evidence of coolant leakage, and other damage. If cracks are found, a new cylinder head should be obtained.

6 Use a straight-edge and feeler blade to check that the cylinder head surface is not distorted. If the surface shows any warping in excess of 0.1 mm, it may be possible to have it machined provided that the cylinder head is not reduced to less than the specified minimum height.

7 Examine the valve seats in each of the combustion chambers. If they are severely pitted, cracked, or burned, they will need to be renewed or re-cut by an engine overhaul specialist. If they are only slightly pitted, this can be removed by grinding-in the valve heads and seats with fine valve-grinding compound, as described below.

8 Check the valve guides for wear by inserting the relevant valve, and checking for side-to-side motion of the valve. A very small amount of movement is acceptable. If the movement seems excessive, remove the valve. Measure the valve stem diameter (see below), and renew the valve if it is worn excessively. If the valve stem is not worn, the wear must be in the valve guide, and the guide must be renewed. The renewal of valve guides is best carried out by a Saab dealer or engine overhaul specialist, who will have the necessary tools available.

Valves

9 Examine the head of each valve for pitting, burning, cracks, and general wear. Check the

10.6 Removing a valve

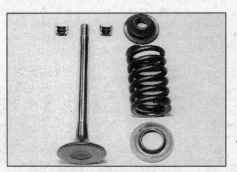

10.7a The valve spring components

10.7b Place each valve and its associated components in a labelled polythene bag

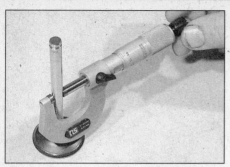

11.10 Measuring a valve stem diameter

11.13 Grinding-in a valve

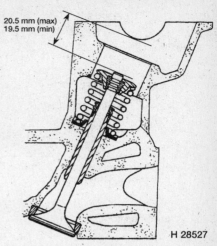

11.16a Check the depth of the valve stems below the camshaft bearing surface

valve stem for scoring and wear ridges. Rotate the valve, and check for any obvious indication that it is bent. Look for pits and excessive wear on the tip of each valve stem. Renew any valve that shows any such signs of wear or damage.

10 If the valve appears satisfactory at this stage, measure the valve stem diameter at several points using a micrometer **(see illustration)**. Any significant difference in the readings obtained indicates wear of the valve stem. Should any of these conditions be apparent, the valve(s) must be renewed.

 Warning: On fuel injection engines, the exhaust valves are sodium-filled, and must not be cut open or melted with other substances. The correct disposal procedure is to drill holes in the valve head and stem, and to throw the valve into a bucket of water from a distance of 3 metres. On contact with the water, a powerful explosive reaction will occur.

11 If the valves are in satisfactory condition, they should be ground (lapped) into their respective seats, to ensure a smooth, gas-tight seal. Note that the exhaust valves have a hardened coating and, although they may be ground-in with paste they must not be machined. If the seat is only lightly pitted, or if it has been re-cut, fine grinding compound should be used to produce the required finish. Coarse valve-grinding compound should *not* be used unless a seat is badly burned or deeply pitted. If this is the case, the cylinder head and valves should be inspected by an expert, to decide whether seat re-cutting, or even the renewal of the valve or seat insert is required.

12 Valve grinding is carried out as follows. Place the cylinder head upside-down on a bench.

13 Smear a trace of (the appropriate grade of) valve-grinding compound on the seat face, and press a suction grinding tool onto the valve head. With a semi-rotary action, grind the valve head to its seat, lifting the valve occasionally to redistribute the grinding compound **(see illustration)**. A light spring placed under the valve head will greatly ease this operation.

14 If coarse grinding compound is being used, work only until a dull, matt even surface is produced on both the valve seat and the valve, then wipe off the used compound, and repeat the process with fine compound. When a smooth unbroken ring of light grey matt finish is produced on both the valve and seat, the grinding operation is complete. *Do not* grind-in the valves any further than absolutely necessary, or the seat will be prematurely sunk into the cylinder head.

15 When all the valves have been ground-in, carefully wash off *all* traces of grinding compound using paraffin or a suitable solvent, before reassembling the cylinder head.

16 To ensure that the hydraulic cam followers operate correctly on the B202 engine, the depth of the valve stems below the camshaft bearing surface must be within certain limits. It may be possible to obtain a Saab checking tool from a dealer, however the check may be made using a steel rule and straight-edge **(see illustrations)**. Check that the dimension is within the limits given in the Specifications by inserting each valve it its guide in turn and measuring the dimension between the end of the valve stem and the camshaft bearing surface.

17 If the dimension is not within the specified limits, adjustment must be made either to the end of the valve stem or to the valve seat height. If lower than the minimum amount the length of the valve stem must be reduced, and if more than the maximum amount the valve seat must be milled.

Valve components

18 Examine the valve springs for signs of damage and discoloration, and for free length **(see illustration)**.

19 Stand each spring on a flat surface, and check it for squareness **(see illustration)**. If any of the springs are less than the minimum free length, damaged, distorted or have lost their tension, obtain a complete new set of springs.

20 On the B202 engine obtain new valve stem oil seals regardless of their apparent condition.

11.16b Using a steel rule and straight-edge to check the depth of the valve stems below the camshaft bearing surface

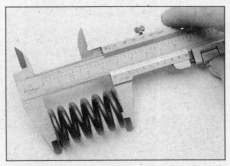

11.18 Checking the valve spring free length

11.19 Checking the valve springs for squareness

2B•18 Engine removal and overhaul procedures

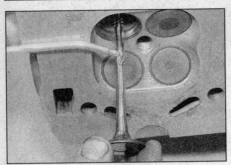

12.1 Lubricating a valve before inserting it in the cylinder head

12.2 Using a socket to fit the valve stem seals

12.4 Inserting the split collets

12 Cylinder head - reassembly

1 Lubricate the stems of the valves, and insert the valves into their original locations. If new valves are being fitted, insert them into the locations to which they have been ground **(see illustration)**.
2 On the B202 engine, working on the first valve dip the new valve stem seal in fresh engine oil. Carefully locate it over the valve and onto the guide. Take care not to damage the seal as it is passed over the valve stem. Use a suitable socket or metal tube to press the seal firmly onto the guide **(see illustration)**.
3 Refit the valve spring seat and spring followed by the spring retainer, then locate the plastic protector in the cam follower bore.
4 Compress the valve spring, and locate the split collets in the recess in the valve stem **(see illustration)**. Release the compressor and remove the protector, then repeat the procedure on the remaining valves.

HAYNES HiNT *Use a little dab of grease to locate the collets on the valve stems and to hold them in place while the spring compressor is released.*

5 With all the valves installed place the cylinder head upright on blocks of wood and, using a hammer and interposed block of wood, tap the end of each valve stem to settle the components.
6 Refit the cam followers and camshaft(s) with reference to Part A, Section 5.
7 Refit the external components removed in Section 10.
8 The cylinder head may now be refitted as described in Part A of this Chapter.

13 Piston/connecting rod assembly - removal

1 Separate the transmission from the engine as described in Section 4.
2 Remove the timing chain (see Section 6).
3 Remove the oil pump pick-up/ strainer/transfer tube. On B201 B-type engines unbolt the pick-up and remove the gasket. On B201 H-type and B202 engines unbolt or remove the oil pump pick-up tube from the timing cover and remove the O-ring **(see illustration)**.
4 If there is a pronounced wear ridge at the top of any bore, it may be necessary to remove it with a scraper or ridge reamer, to avoid piston damage during removal. Such a ridge indicates excess wear of the cylinder bore.
5 Check if the connecting rods and big-end bearing caps are marked for position **(see illustration)**. If not, use a hammer and centre-punch, paint or similar to mark them with their respective cylinder number on the flat machined surface provided; if the engine has been dismantled before, note carefully any identifying marks made previously. Note that No 1 cylinder is at the timing chain end of the engine.
6 Turn the crankshaft to bring pistons 1 and 4 to BDC (bottom dead centre).

13.3 Oil pump pick-up tube on H-type engine

13.7 Removing a big-end bearing cap

7 Unscrew the nuts from No 1 piston big-end bearing cap. Take off the cap, and recover the bottom half bearing shell. If the bearing shells are to be re-used, tape the cap and the shell together **(see illustration)**.
8 To prevent the possibility of damage to the crankshaft bearing journals, tape over the connecting rod stud threads.
9 Using a hammer handle, push the piston up through the bore, and remove it from the top of the cylinder block. Recover the bearing shell, and tape it to the connecting rod for safe-keeping.
10 Loosely refit the big-end cap to the connecting rod, and secure with the nuts - this will help to keep the components in their correct order **(see illustration)**.
11 Remove No 4 piston assembly in the same way.
12 Turn the crankshaft through 180° to bring pistons 2 and 3 to BDC (bottom dead centre), and remove them in the same way.

13.5 Connecting rod and big-end cap markings

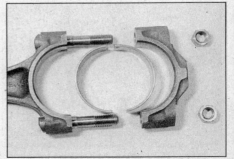

13.10 Big-end bearing components

Engine removal and overhaul procedures 2B•19

14.4 Removing the crankshaft rear oil seal housing (engine plate)

14.5a Main bearing caps are numbered from the timing chain end of the engine

14.5b No 3 main bearing cap showing holes for the oil pump pick-up tube bolts

14 Crankshaft - removal

1 Remove the flywheel (Chapter 2A, Section 11) or driveplate (Chapter 2B, Section 8).
2 Remove the pistons and connecting rods, as described in Section 13. **Note:** *If no work is to be done on the pistons and connecting rods, there is no need to remove the cylinder head, or to push the pistons out of the cylinder bores. The pistons should just be pushed far enough up the bores that they are positioned clear of the crankshaft journals.*
3 Check the crankshaft endfloat (see Section 17), then proceed as follows.
4 Unbolt and remove the crankshaft rear oil seal housing (engine plate) from the end of the cylinder block, noting the correct fitted locations of the locating dowels (see illustration). If the locating dowels are a loose fit, remove them and store them with the housing for safe-keeping. Remove the gasket where fitted (on later models sealant may be used instead of a gasket).
5 Identification numbers should already be cast onto the base of each main bearing cap (see illustrations). If not, number the cap and crankcase using a centre-punch, as was done for the connecting rods and caps.
6 Unscrew and remove the main bearing cap retaining bolts, and withdraw the caps, complete with bearing shells (see illustration). Tap the caps with a wooden or copper mallet if they are stuck.
7 Remove the bearing shells from the caps but keep them with their relevant caps and identified for position to ensure correct refitting.
8 Carefully lift the crankshaft from the crankcase (see illustration).
9 Remove the upper bearing shells from the crankcase, keeping them identified for position. Also remove the thrustwashers at each side of the centre main bearing and store them with the bearing cap (see illustrations).

15 Cylinder block/crankcase - cleaning and inspection

Cleaning

1 Remove all external components and electrical switches/sensors from the block. For complete cleaning, the core plugs should ideally be removed. Drill a small hole in the plugs, then insert a self-tapping screw into the hole. Pull out the plugs by pulling on the screw with a pair of grips, or by using a slide hammer. Also where applicable unbolt the oil filter housing from the block and where fitted the four oil jets from the crankcase (see illustrations).

14.6 Removing the main bearing caps

14.8 Lifting the crankshaft from the crankcase

14.9a Thrustwasher locations on the centre main bearing

14.9b Removing the upper centre main bearing shell

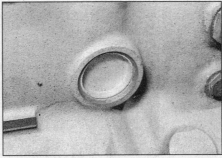
15.1a Core plug in the cylinder block

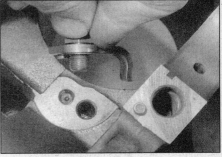
15.1b Removing an oil jet from the crankcase

2B•20 Engine removal and overhaul procedures

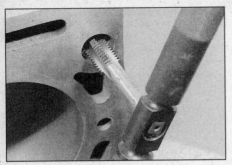

15.7 Cleaning a cylinder head bolt hole in the cylinder block using a tap

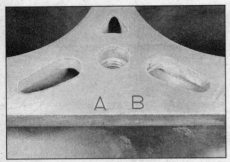

15.12b Cylinder bore classification on the front of the block

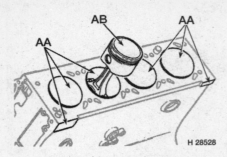

15.12a Piston and cylinder bore classification code locations

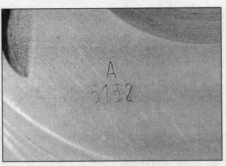

15.12c Piston classification on the piston crown

2 Scrape all traces of sealant from the cylinder block/crankcase, taking care not to damage the gasket/sealing surfaces.

3 If necessary, remove all oil gallery plugs (where fitted). The plugs are usually very tight - they may have to be drilled out, and the holes re-tapped. Use new plugs when the engine is reassembled.

4 If the cylinder block/crankcase is extremely dirty, it should be steam-cleaned.

5 Clean all oil holes and oil galleries and flush all internal passages with warm water until the water runs clear. Dry thoroughly, and apply a light film of oil to all mating surfaces, to prevent rusting. Also oil the cylinder bores. If you have access to compressed air, use it to speed up the drying process, and to blow out all the oil holes and galleries.

 Warning: Wear eye protection when using compressed air!

6 If the cylinder block is not very dirty, you can do an adequate cleaning job with hot (as hot as you can stand!), soapy water and a stiff brush. Take plenty of time, and do a thorough job. Regardless of the cleaning method used, be sure to clean all oil holes and galleries very thoroughly, and to dry all components well. On completion, protect the cylinder bores as described above, to prevent rusting.

7 All threaded holes must be clean, to ensure accurate torque readings during reassembly. To clean the threads, run the correct-size tap into each of the holes to remove rust, corrosion, thread sealant or sludge, and to restore damaged threads **(see illustration)**. If possible, use compressed air to clear the holes of debris produced by this operation.

 HAYNES HiNT *A good alternative is to inject aerosol-applied water dispersant lubricant into each hole, using the long tube usually supplied. Wear eye protection when cleaning out these holes in this way!*

8 Apply suitable sealant to the new oil gallery plugs, and insert them into the holes in the block. Tighten them securely. Where applicable, refit and tighten the oil jets to the bottom of the crankcase.

9 If the engine is not going to be reassembled right away, cover it with a large plastic bag to keep it clean; protect all mating surfaces and the cylinder bores as described above, to prevent rusting.

Inspection

10 Visually check the cylinder block for cracks and corrosion. Look for stripped threads in the threaded holes. If there has been any history of internal water leakage, it may be worthwhile having an engine overhaul specialist check the cylinder block/crankcase with special equipment. If defects are found, have them repaired if possible, or renew the assembly.

11 Check each cylinder bore for scuffing and scoring. Check for signs of a wear ridge at the top of the cylinder, indicating that the bore is excessively worn.

12 The cylinder bores and pistons are matched and classified according to five codes - AB, B, C, 1 (0.5 mm oversize), and 2 (1.0 mm oversize). The code is stamped on the piston crowns and on the left-hand side of the cylinder block **(see illustrations)**. Note that all classifications may occur in the same cylinder block.

13 Wear of the cylinder bores and pistons can be measured by inserting the relevant piston (without piston rings) in its bore and using a feeler blade. Make the check with the piston near the top of its bore. If the clearance is more than the nominal (new) amount given in the Specifications, a rebore should be considered and the opinion of an engine reconditioner sought.

16 Piston/connecting rod assembly - inspection

1 Before the inspection process can begin, the piston/connecting rod assemblies must be cleaned, and the original piston rings removed from the pistons **(see illustration)**.

2 Carefully expand the old rings over the top of the pistons. The use of two or three old feeler blades will be helpful in preventing the rings dropping into empty grooves **(see illustrations)**. Be careful not to scratch the piston with the ends of the ring. The rings are brittle, and will snap if they are spread too far. They're also very sharp - protect your hands and fingers. Note that the third ring incorporates an expander. Always remove the rings from the top of the piston. Keep each set of rings with its piston if the old rings are to be re-used.

3 Scrape away all traces of carbon from the top of the piston. A hand-held wire brush (or a piece of fine emery cloth) can be used, once the majority of the deposits have been scraped away.

4 Remove the carbon from the ring grooves in the piston, using an old ring. Break the ring in half to do this (be careful not to cut your fingers - piston rings are sharp). Be careful to remove only the carbon deposits - do not remove any metal, and do not nick or scratch the sides of the ring grooves.

5 Once the deposits have been removed, clean the piston/connecting rod assembly with paraffin or a suitable solvent, and dry

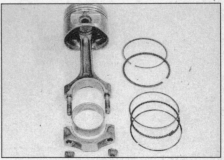

16.1 Piston/connecting rod assembly components

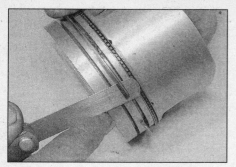

16.2a Removing a piston compression ring with the aid of a feeler blade

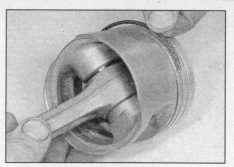

16.2b Removing the oil control ring

thoroughly. Make sure that the oil return holes in the ring grooves are clear.

6 If the pistons and cylinder bores are not damaged or worn excessively, and if the cylinder block does not need to be rebored, the original pistons can be refitted. Normal piston wear shows up as even vertical wear on the piston thrust surfaces, and slight looseness of the top ring in its groove.

7 Carefully inspect each piston for cracks around the skirt, around the gudgeon pin holes, and at the piston ring 'lands' (between the ring grooves).

8 Look for scoring and scuffing on the piston skirt, holes in the piston crown, and burned areas at the edge of the crown. If the skirt is scored or scuffed, the engine may have been suffering from overheating, and/or abnormal combustion which caused excessively high operating temperatures. The cooling and lubrication systems should be checked thoroughly. Scorch marks on the sides of the pistons show that blow-by has occurred. A hole in the piston crown, or burned areas at the edge of the piston crown, indicates that abnormal combustion (pre-ignition, knocking, or detonation) has been occurring. If any of the above problems exist, the causes must be investigated and corrected, or the damage will occur again. The causes may include incorrect ignition timing and/or fuel/air mixture.

9 Corrosion of the piston, in the form of pitting, indicates that coolant has been leaking into the combustion chamber and/or the crankcase. Again, the cause must be corrected, or the problem may persist in the rebuilt engine.

10 Where needed, pistons can be purchased from a Saab dealer.

11 Examine each connecting rod carefully for signs of damage, such as cracks around the big-end and small-end bearings. Check that the rod is not bent or distorted. Damage is highly unlikely, unless the engine has been seized or badly overheated. Detailed checking of the connecting rod assembly can only be carried out by a Saab dealer or engine repair specialist with the necessary equipment.

12 The gudgeon pins are of the floating type, secured in position by two circlips, and the pistons and connecting rods can be separated and reassembled as follows.

13 Using a small flat-bladed screwdriver, prise out the circlips, and push out the gudgeon pin **(see illustrations)**. Hand pressure should be sufficient to remove the pin. Identify the piston, gudgeon pin and rod to ensure correct reassembly.

14 Examine the gudgeon pin and connecting rod small-end bearing for signs of wear or damage. Wear can be cured by renewing both the pin and bush. Bush renewal, however, is a specialist job - press facilities are required, and the new bush must be reamed accurately.

15 The connecting rods themselves should not be in need of renewal, unless seizure or some other major mechanical failure has occurred. Check the alignment of the connecting rods visually, and if the rods are not straight, take them to an engine overhaul specialist for a more detailed check.

16 Examine all components, and obtain any new parts from your Saab dealer. If new pistons are purchased, they will be supplied complete with gudgeon pins and circlips. Circlips can also be purchased individually.

17 Position the piston so that the notch on the edge of the crown faces the timing chain end of the engine and the numbers on the connecting rod and big-end cap face the exhaust side of the cylinder block. With the piston held in your hand and the notch facing the left, the connecting rod numbering should face towards you **(see illustration)**. Apply a smear of clean engine oil to the gudgeon pin. Slide it into the piston and through the connecting rod small-end. Check that the piston pivots freely on the rod, then secure the gudgeon pin in position with the circlips. Ensure that each circlip is correctly located in its groove in the piston.

18 Measure the piston diameters, and check that they are within limits for the corresponding bore diameters. If the piston-to-bore clearance is excessive, the block will have to be rebored, and new pistons and rings fitted.

19 Examine the mating surfaces of the big-end caps and connecting rods, to see if they have ever been filed, in a mistaken attempt to take up bearing wear. This is extremely unlikely, but if evident, the offending connecting rods and caps must be renewed.

17 Crankshaft - inspection

Checking crankshaft endfloat

1 If the crankshaft endfloat is to be checked, this must be done when the crankshaft is still installed in the cylinder block/crankcase, but is free to move (see Section 14).

2 Check the endfloat using a dial gauge in contact with the end of the crankshaft. Push the crankshaft fully one way, and then zero the gauge. Push the crankshaft fully the other way, and check the endfloat **(see illustration)**. The result can be compared with the specified amount, and will give an indication as to whether new thrustwashers are required.

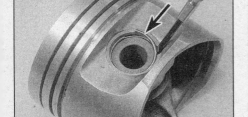

16.13a Prise out the gudgeon pin circlip . . .

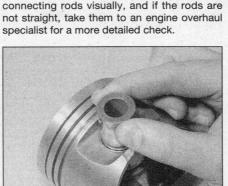

16.13b . . . then withdraw the gudgeon pin, and separate the piston from the connecting rod

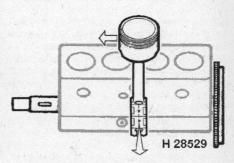

16.17 Relationship of the piston and connecting rod

17.2 Using a dial gauge to check the crankshaft endfloat

17.3 Using feeler blades to check the crankshaft endfloat

17.10 Measuring a crankshaft big-end bearing journal diameter

3 If a dial gauge is not available, feeler blades can be used. First push the crankshaft fully towards the flywheel end of the engine, then use feeler blades to measure the gap between the No 3 crankpin web and the centre main bearing thrustwasher **(see illustration)**.

Inspection

4 Clean the crankshaft using paraffin or a suitable solvent, and dry it, preferably with compressed air if available. Be sure to clean the oil holes with a pipe cleaner or similar probe, to ensure that they are not obstructed. *Warning: Wear eye protection when using compressed air!*

5 Check the main and big-end bearing journals for uneven wear, scoring, pitting and cracking.

6 Big-end bearing wear is accompanied by distinct metallic knocking when the engine is running (particularly noticeable when the engine is pulling from low speed) and some loss of oil pressure.

7 Main bearing wear is accompanied by severe engine vibration and rumble - getting progressively worse as engine speed increases - and again by loss of oil pressure.

8 Check the bearing journal for roughness by running a finger lightly over the bearing surface. Any roughness (which will be accompanied by obvious bearing wear) indicates that the crankshaft requires regrinding (where possible) or renewal. Note that it is permissible to regrind the crankshaft to the first undersize without rehardening, however further regrinding will necessitate rehardening by Tenifer treatment.

9 If the crankshaft has been reground, check for burrs around the crankshaft oil holes (the holes are usually chamfered, so burrs should not be a problem unless regrinding has been carried out carelessly). Remove any burrs with a fine file or scraper, and thoroughly clean the oil holes as described previously.

10 Using a micrometer, measure the diameter of the main and big-end bearing journals, and compare the results with the Specifications **(see illustration)**. By measuring the diameter at a number of points around each journal's circumference, you will be able to determine whether or not the journal is out-of-round. Take the measurement at each end of the journal, near the webs, to determine if the journal is tapered. Compare the results obtained with those given in the Specifications.

11 Check the oil seal contact surfaces at each end of the crankshaft for wear and damage. If the seal has worn a deep groove in the surface of the crankshaft, consult an engine overhaul specialist; repair may be possible, but otherwise a new crankshaft will be required.

18 Main and big-end bearings - inspection

1 Even though the main and big-end bearings are renewed during the engine overhaul, the old bearings should be retained for close examination, as they may reveal valuable information about the condition of the engine. The bearing shells are graded by thickness, the grade of each shell being indicated by the colour code marked on it - they may also have markings on their backing faces **(see illustration)**. Note that the following table only applies up to the second undersize - for the third and fourth undersizes, only one shell thickness is available. Both thin and thick sizes may be combined together on the same journal to obtain the correct clearance.

B201/B202	Thin	Thick
Standard	Red	Blue
First undersize	Yellow	Green
Second undersize	White	Brown

18.1 'STD' marking on the backing of a big-end bearing shell

2 Bearing failure can occur due to lack of lubrication, the presence of dirt or other foreign particles, overloading the engine, or corrosion **(see illustration)**. Regardless of the cause of bearing failure, the cause must be corrected (where applicable) before the engine is reassembled, to prevent it from happening again.

3 When examining the bearing shells, remove them from the cylinder block/crankcase, the main bearing caps, the connecting rods and the connecting rod big-end bearing caps. Lay them out on a clean surface in the same general position as their location in the engine. This will enable you to match any bearing problems with the corresponding crankshaft journal. *Do not* touch any shell's bearing surface with your fingers while checking it, or the delicate surface may be scratched.

4 Dirt and other foreign matter gets into the engine in a variety of ways. It may be left in the engine during assembly, or it may pass through filters or the crankcase ventilation system. It may get into the oil, and from there into the bearings. Metal chips from machining operations and normal engine wear are often present. Abrasives are sometimes left in

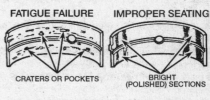

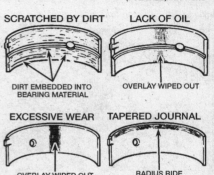

18.2 Typical bearing failures

Engine removal and overhaul procedures 2B•23

engine components after reconditioning, especially when parts are not thoroughly cleaned using the proper cleaning methods. Whatever the source, these foreign objects often end up embedded in the soft bearing material, and are easily recognised. Large particles will not embed in the bearing, and will score or gouge the bearing and journal. The best prevention for this cause of bearing failure is to clean all parts thoroughly, and keep everything spotlessly-clean during engine assembly. Frequent and regular engine oil and filter changes are also recommended.

5 Lack of lubrication (or lubrication breakdown) has a number of interrelated causes. Excessive heat (which thins the oil), overloading (which squeezes the oil from the bearing face) and oil leakage (from excessive bearing clearances, worn oil pump or high engine speeds) all contribute to lubrication breakdown. Blocked oil passages, which may be the result of misaligned oil holes in a bearing shell, will also oil-starve a bearing, and destroy it. When lack of lubrication is the cause of bearing failure, the bearing material is wiped or extruded from the steel backing of the bearing. Temperatures may increase to the point where the steel backing turns blue from overheating.

6 Driving habits can have a definite effect on bearing life. Full-throttle, low-speed operation (labouring the engine) puts very high loads on bearings, tending to squeeze out the oil film. These loads cause the bearings to flex, which produces fine cracks in the bearing face (fatigue failure). Eventually, the bearing material will loosen in pieces, and tear away from the steel backing.

7 Short-distance driving leads to corrosion of bearings, because insufficient engine heat is produced to drive off the condensed water and corrosive gases. These products collect in the engine oil, forming acid and sludge. As the oil is carried to the engine bearings, the acid attacks and corrodes the bearing material.

8 Incorrect bearing installation during engine assembly will lead to bearing failure as well. Tight-fitting bearings leave insufficient bearing running clearance, and will result in oil starvation. Dirt or foreign particles trapped behind a bearing shell result in high spots on the bearing, which lead to failure.

9 *Do not* touch any shell's bearing surface with your fingers during reassembly; there is a risk of scratching the delicate surface, or of depositing particles of dirt on it.

10 As mentioned at the beginning of this Section, the bearing shells should be renewed as a matter of course during engine overhaul; to do otherwise is false economy. Refer to Section 21 for details of bearing shell selection.

19 Engine overhaul - reassembly sequence

1 Before reassembly begins, ensure that all new parts have been obtained, and that all necessary tools are available. Read through the entire procedure to familiarise yourself with the work involved, and to ensure that all items necessary for reassembly of the engine are at hand. In addition to all normal tools and materials, thread-locking compound will be needed. A suitable tube of sealant will also be required for the joint faces that are fitted without gaskets.

2 In order to save time and avoid problems, engine reassembly can be carried out in the following order:
 a) Crankshaft (Section 21).
 b) Piston/connecting rod assemblies (Section 22).
 c) Flywheel (Chapter 2A, Section 11) or driveplate (Chapter 2B, Section 8).
 d) Cylinder head (Chapter 2A, Section 6).
 e) Timing chain, sprockets and tensioner, (Chapter 2B, Section 6).
 f) Inlet and exhaust manifolds (Chapter 4A).
 g) Engine external components (Section 5).

3 At this stage, all engine components should be absolutely clean and dry, with all faults repaired. The components should be laid out (or in individual containers) on a completely clean work surface.

20 Piston rings - refitting

1 Before fitting new piston rings, the ring end gaps must be checked as follows.

2 Lay out the piston/connecting rod assemblies and the new piston ring sets, so that the ring sets will be matched with the same piston and cylinder during the end gap measurement and subsequent engine reassembly.

3 Insert the top ring into the first cylinder, and push it down the bore using the top of the piston **(see illustration)**. This will ensure that the ring remains square with the cylinder walls. Position the ring near the bottom of the cylinder bore, at the lower limit of ring travel. Note that the top and second compression rings are different.

4 Measure the end gap using feeler blades, and compare the measurements with the figures given in the Specifications **(see illustration)**.

5 If the gap is too small (unlikely if genuine Saab parts are used), it must be enlarged, or the ring ends may contact each other during engine operation, causing serious damage. Ideally, new piston rings providing the correct end gap should be fitted. As a last resort, the end gap can be increased by filing the ring ends very carefully with a fine file. Mount the file in a vice equipped with soft jaws, slip the ring over the file with the ends contacting the file face, and slowly move the ring to remove material from the ends. Take care, as piston rings are sharp, and are easily broken.

6 With new piston rings, it is unlikely that the end gap will be too large. If the gaps are too large, check that you have the correct rings for your particular engine.

7 Repeat the checking procedure for each ring in the first cylinder, and then for the rings in the remaining cylinders. Remember to keep rings, pistons and cylinders matched up.

8 Once the ring end gaps have been checked and if necessary corrected, the rings can be fitted to the pistons.

9 Fit the piston rings using the same technique as for removal. Fit the bottom (oil control) ring first, and work up. When fitting the oil control ring, first insert the expander, then fit the lower and upper rings with the ring gaps both on the non-thrust side of the piston at 60° intervals. Ensure that the second compression ring is fitted the correct way up, with the word 'TOP' uppermost. Arrange the gaps of the top and second compression rings on opposite sides of the piston and above the ends of the gudgeon pin **(see illustration)**. *Note: Always follow any*

20.3 Using the top of a piston to push a piston ring into the bore

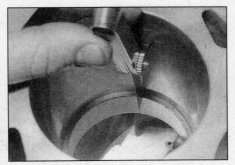

20.4 Measuring a piston ring end gap

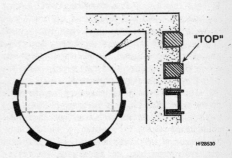

20.9 Piston ring cross-section and gap positioning

2B•24 Engine removal and overhaul procedures

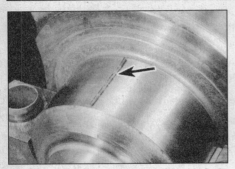

21.8 Plastigage in place on the crankshaft main bearing journal

21.9 Tightening a main bearing cap bolt

21.11 Measuring the width of the deformed Plastigauge using the card gauge

instructions supplied with the new piston ring sets - different manufacturers may specify different procedures. Do not mix up the top and second compression rings, as they have different cross-sections.

21 Crankshaft - refitting and main bearing running clearance check

Selection of new bearing shells

1 The main bearing shells are classified for thickness as described in Section 18. Note that up to the second undersize, it is possible to mix different thicknesses of shell in order to obtain the correct running clearance. Commence the procedure with the two thinnest shells, then if the clearance is too great, fit one thick shell with one thin shell and make the check again. If the clearance is still too great, fit two thick shells.

Main bearing running clearance check

2 Clean the backs of the bearing shells, and the bearing locations in both the cylinder block and the main bearing caps.
3 Press the bearing shells into their locations, ensuring that the tab on each shell engages in the notch in the cylinder block or main bearing cap location. Take care not to touch any shell's bearing surface with your fingers.
4 The running clearance may be checked using one of two methods.
5 One method (which will be difficult to achieve without a range of internal micrometers or internal/external expanding calipers) is to refit the main bearing caps to the cylinder block/crankcase, with the bearing shells in place. With the cap retaining bolts correctly tightened, measure the internal diameter of each assembled pair of bearing shells. If the diameter of each corresponding crankshaft journal is measured and then subtracted from the bearing internal diameter, the result will be the main bearing running clearance.
6 The second (and more accurate) method is to use a product known as 'Plastigauge'. This consists of a fine thread of perfectly-round plastic, which is compressed between the bearing shell and the journal. When the cap and shell are removed, the plastic is deformed, and can be measured with a special card gauge supplied with the kit. The running clearance is determined from this gauge. Plastigauge may be available from your Saab dealer; otherwise, enquiries at one of the larger specialist motor factors should produce the name of a stockist in your area. The procedure for using Plastigauge is as follows.
7 With the main bearing upper shells in place, carefully lay the crankshaft in position. Do not use any lubricant at this stage; the crankshaft journals and bearing shells must be perfectly clean and dry to make the check.
8 Cut several lengths of the appropriate-size Plastigauge (they should be slightly shorter than the width of the main bearings), and place one length on each crankshaft journal axis **(see illustration)**. The length of Plastigauge should be placed about 6.0 mm to one side of the centre-line of the journal.
9 With the main bearing lower shells in position refit the main bearing caps, then insert the bolts and tighten them progressively to the specified torque **(see illustration)**. Take care not to disturb the Plastigauge, and *do not* rotate the crankshaft at any time during this operation.
10 Unscrew the bolts and remove the main bearing caps, again taking great care not to disturb the Plastigauge or rotate the crankshaft.
11 Compare the width of the crushed Plastigauge on each journal to the scale printed on the Plastigauge envelope, to obtain the main bearing running clearance **(see illustration)**. Compare the clearance measured with that given in the Specifications at the start of this Chapter.
12 If the clearance is significantly different from that expected, the bearing shells may be the wrong size (or excessively worn, if the original shells are being re-used). Before deciding that different-size shells are required, make sure that no dirt or oil was trapped between the bearing shells and the main bearing caps or block when the clearance was measured. If the Plastigauge was wider at one end than at the other, the crankshaft journal may be tapered.
13 If the clearance is not as specified, use the reading obtained, along with the shell thicknesses quoted above, to calculate the necessary grade of bearing shells required. When calculating the bearing clearance required, bear in mind that it is always better to have the running clearance towards the lower end of the specified range, to allow for wear in use.
14 Where necessary, obtain the required grades of bearing shell, and repeat the running clearance checking procedure as described above.
15 On completion, carefully scrape away all traces of the Plastigauge material from the crankshaft and bearing shells. Use your fingernail, or a wooden or plastic scraper which is unlikely to score the bearing surfaces.

Final crankshaft refitting

16 Carefully lift the crankshaft out of the cylinder block once more.
17 Using a little grease, stick the upper thrustwashers to each side of the centre main bearing upper location; ensure that the oilway grooves on each thrustwasher face outwards (away from the cylinder block).
18 Place the bearing shells in their locations in the caps as described earlier. If new shells are being fitted, ensure that all traces of protective grease are cleaned off using paraffin. Wipe dry the shells and connecting rods with a lint-free cloth. Liberally lubricate each bearing shell in the cylinder block/crankcase with clean engine oil **(see illustration)**.

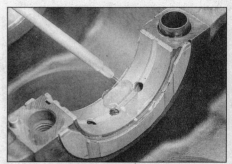

21.18 Lubricating the main bearing shells

Engine removal and overhaul procedures 2B•25

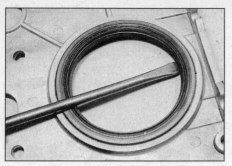
21.25a Prising the oil seal from the crankshaft rear oil seal housing

21.25b Driving the crankshaft rear oil seal into the housing

21.25c Fitting the crankshaft rear oil seal using a block of wood in a vice

19 Lower the crankshaft into position so that Nos 2 and 3 cylinder crankpins are at TDC; Nos 1 and 4 cylinder crankpins will be at BDC ready for fitting No 1 piston. Check the crankshaft endfloat (see Section 17).

20 Lubricate the lower bearing shells in the main bearing caps with clean engine oil. Make sure that the locating lugs on the shells engage with the corresponding recesses in the caps.

21 Fit the main bearing caps to their correct locations, ensuring that they are fitted the correct way round (the bearing shell lug recesses in the block and caps must be on the same side). Insert the bolts loosely.

22 Progressively tighten the main bearing cap bolts to the specified torque wrench setting.

23 Check that the crankshaft rotates freely.

24 Refit the piston/connecting rod assemblies to the crankshaft as described in Section 22.

25 Before refitting the crankshaft rear oil seal housing (engine plate), fit a new rear oil seal in the housing with reference to Chapter 2A. Use a mallet and block of wood to drive it into the housing, or alternatively use the block of wood in a vice **(see illustrations)**.

26 Where a gasket was removed, locate a new gasket on the rear of the cylinder block and retain with a little grease **(see illustration)**. **Do not** use adhesive as this may melt from the heat of the engine and reduce the torque loading on the housing bolts. Where no gasket was removed, apply suitable sealant to the contact faces of the rear oil seal housing.

27 Smear a little oil on the oil seal lips and refit the locating dowels where necessary. Locate the housing on the rear of the cylinder block. To prevent damage to the oil seal as it locates over the crankshaft, make up a guide out of a plastic container or alternatively use adhesive tape. Once the housing is in position remove the guide or tape, insert the bolts and tighten them to the specified torque **(see illustrations)**.

28 Trim the excess gasket from the bottom face of the housing **(see illustration)**.

29 Refit the flywheel/driveplate and pick-up/strainer/transfer tube with reference to the relevant Sections of this Chapter.

22 Piston/connecting rod assembly - refitting and big-end bearing running clearance check

Selection of bearing shells

1 The big-end bearing shells are classified for thickness as described in Section 18. Note that up to the second undersize, it is possible to mix different thicknesses of shell in order to obtain the correct running clearance. Commence the procedure with the two thinnest shells, then if the clearance is too great, fit one thick shell with one thin shell and make the check again. If the clearance is still too great, fit two thick shells.

Big-end bearing running clearance check

2 The clearance can be checked in either of two ways.

3 One method is to refit the big-end bearing cap to the connecting rod before refitting the pistons to the cylinder block, ensuring that they are fitted the correct way round, with the bearing shells in place. With the cap retaining nuts correctly tightened, use an internal micrometer or vernier caliper to measure the internal diameter of each assembled pair of bearing shells. If the diameter of each corresponding crankshaft journal is measured and then subtracted from the bearing internal diameter, the result will be the big-end bearing running clearance.

4 The second, and more accurate, method is to use Plastigauge (see Section 21) after

21.26 Crankshaft rear oil seal housing gasket on the cylinder block

21.27a Adhesive tape over the end of the crankshaft will prevent damage to the oil seal when refitting

21.27b Applying sealant to the rear oil seal housing

21.28 Trimming the excess gasket from the bottom face of the crankshaft rear oil seal housing

22.11a Piston crown showing notch which faces the timing chain end of the engine

22.11b Using a hammer handle to tap the piston down the cylinder bore

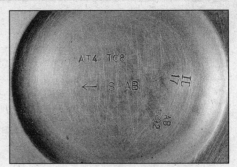
22.11c Arrow and markings on the piston on a B202 engine

refitting the pistons to the cylinder block. The following paragraphs describe the latter method together with the refitting of the pistons to the cylinder block.

5 Position the cylinder block either on its side or on the flywheel/driveplate end.
6 Lay out the assembled pistons and rods in order, with the bearing shells, connecting rod caps and nuts.
7 Clean the backs of the bearing shells, and the bearing locations in both the connecting rod and bearing cap. If new shells are being fitted, ensure that all traces of protective grease are cleaned off, using paraffin.
8 Press the bearing shells into their locations, ensuring that the tab on each shell engages in the notch in the connecting rod and cap. Take care not to touch any shell's bearing surface with your fingers. If the original bearing shells are being used for the check, ensure that they are refitted in their original locations.
9 Lubricate the cylinder bores, pistons and piston rings with clean engine oil then lay out each piston/connecting rod assembly in its respective position. Do not lubricate the bearing shells at this stage.
10 Start with assembly No 1. Make sure that the piston rings are still spaced as described in Section 20, then clamp them in position with a piston ring compressor.
11 Insert the piston/connecting rod assembly into the top of cylinder No 1. Ensure that the notch or arrow on the piston crown is pointing towards the timing chain end of the engine. Using a block of wood or hammer handle against the piston crown, tap the assembly into the cylinder bore until the piston crown is flush with the top of the cylinder. Make sure that the ends of the big-end bearing cap bolts do not scratch the bore walls **(see illustrations)**.
12 With the No 1 crankpin at the bottom of its stroke, guide the connecting rod onto it while tapping the top of the piston with the hammer handle.
13 Place a strand of Plastigauge on the crankpin journal.
14 Refit the big-end bearing cap, using the marks made or noted on removal to ensure that they are fitted the correct way round. Tighten the bearing cap nuts to the specified torque. Take care not to disturb the Plastigauge or rotate the crankshaft during the tightening sequence.
15 Dismantle the assembly, then use the scale printed on the Plastigauge envelope to obtain the big-end bearing running clearance.
16 If the clearance is significantly different from that specified, the bearing shells may be the wrong size (or excessively worn, if the original shells are being re-used). If necessary, select different shells as described in paragraph 1. Make sure that no dirt or oil was trapped between the bearing shells and the cap or connecting rod when the clearance was measured. If the Plastigauge was wider at one end than at the other, the crankpins may be tapered.
17 Push the No 1 piston/connecting rod assembly to the top of the cylinder, then refit the No 4 piston/connecting rod assembly and repeat the bearing running clearance check. With Nos 1 and 4 pistons at the top of their bores, refit Nos 2 and 3 pistons and repeat the bearing running clearance on these.
18 On completion, carefully scrape away all traces of the Plastigauge material from the crankshaft and bearing shells. Use your fingernail, or some other object which is unlikely to score the bearing surfaces.

Final piston/connecting rod refitting

19 Position No 1 crankpin at the bottom of its stroke. Liberally lubricate the crankpin and both bearing shells. Taking care not to mark the cylinder bores, tap the piston/connecting rod assembly down the bore and onto the crankpin. Refit the big-end bearing cap, tightening its retaining nuts finger-tight at first. Note that the faces with the identification marks must match (which means that the bearing shell locating tabs abut each other).
20 Tighten the bearing cap retaining nuts evenly and progressively to the specified torque setting.
21 Rotate the crankshaft. Check that it turns freely; some stiffness is to be expected if new components have been fitted, but there should be no signs of binding or tight spots.
22 Refit the remaining three piston/connecting rod assemblies to their crankpins in the same way.
23 Refit the oil pump pick-up/strainer/transfer tube together with a new gasket or O-ring as applicable. Tighten all bolts to the specified torque.
24 Refit the timing chain (see Section 6).
25 Reconnect the transmission to the engine as described in Section 4.

23 Engine - initial start-up after overhaul

1 With the engine refitted in the vehicle, double-check the engine oil and coolant levels. Make a final check that everything has been reconnected, and that there are no tools or rags left in the engine compartment.
2 Remove the spark plugs. Disable the ignition system by disconnecting the ignition HT coil lead from the distributor cap, and earthing it on the cylinder block. Use a jumper lead or wire to make a good connection.
3 Turn the engine on the starter until the oil pressure warning light goes out. Refit the spark plugs, and reconnect the coil lead to the distributor cap.
4 Start the engine, noting that this may take a little longer than usual, due to the fuel system components having been disturbed.
5 While the engine is idling, check for fuel, water and oil leaks. Don't be alarmed if there are some odd smells and smoke from parts getting hot and burning off oil deposits.
6 Assuming all is well, keep the engine idling until hot water is felt circulating through the top hose then switch off the engine.
7 Check the ignition timing and the idle speed settings (as appropriate), then switch off the engine.
8 After a few minutes, recheck the oil and coolant levels as described in "Weekly Checks", and top-up as necessary.
9 On models up to 1987, the cylinder head bolts must be re-tightened after 1200 and 6000 miles.
10 If new pistons, rings or crankshaft bearings have been fitted, the engine must be treated as new and run-in for the first 500 miles (800 km). *Do not* operate the engine at full-throttle, or allow it to labour at low engine speeds in any gear. It is advisable to change the engine oil and filter at the end of this period.

Chapter 3
Cooling, heating and ventilation systems

Contents

Air conditioning components - removal and refitting 11
Air conditioning system - general information and precautions 10
Air conditioning system refrigerant check ....See Chapter 1
Coolant level check ....See "Weekly Checks"
Coolant renewal ....See Chapter 1
Cooling system electrical switches - testing, removal and refitting ...6
Cooling system hoses - disconnection and renewal 2
Electric cooling fan - testing, removal and refitting 5
General information and precautions 1
Heating/ventilation system - general information 8
Heating/ventilation components - removal and refitting 9
Radiator - removal, inspection and refitting 3
Thermostat - removal, testing and refitting 4
Water pump - removal and refitting 7

Degrees of difficulty

| Easy, suitable for novice with little experience | Fairly easy, suitable for beginner with some experience | Fairly difficult, suitable for competent DIY mechanic | Difficult, suitable for experienced DIY mechanic | Very difficult, suitable for expert DIY or professional |

Specifications

General
Expansion tank cap opening pressure .... 0.9 to 1.2 bars

Thermostat
Opening temperature:
  Standard .... 89° ± 2°C
  Winter thermostat (Nordic countries only) .... 92° ± 2°C

Electric cooling fan thermostatic switch
Cut-in temperature .... 90° to 95°C
Cut-out temperature .... 85° to 90°C

LH-Jetronic temperature sensor
Resistance:
  Bosch:
    At 0°C .... 5800 ohms
    At 20°C .... 2600 ohms
    At 80°C .... 320 ohms
  Lucas:
    -10°C .... 7000 to 11600 ohms
    20°C .... 2100 to 2900 ohms
    80°C .... 270 to 390 ohms

Torque wrench settings	Nm	lbf ft
Thermostat housing (B201)	22	16
Thermostat housing cover	22	16

1 General information and precautions

General information

The cooling system is of pressurised type, comprising a water pump (driven by gear on the B201 B-type engine and driven by an auxiliary drivebelt on all other engines), a crossflow radiator, electric cooling fan, a thermostat, heater matrix, and all associated hoses. The expansion tank is located on the left-hand side of the engine compartment. On models equipped with air conditioning, an auxiliary radiator and electric fan are fitted in addition to the normal radiator.

The system functions as follows. With the engine cold the thermostat is shut and circulation is from the water pump through the cylinder block, cylinder head, inlet manifold and returning to the pump. With the heater valve open, the water returning to the pump passes through the interior heater matrix. With the engine at normal operating temperature circulation is also through the radiator, however at high engine temperature, when the thermostat is fully open, the circulation through the inlet manifold is closed by the bottom of the thermostat and all coolant passes through the radiator. On 99 models the return from the radiator is via the expansion tank, however on 900 models a separate return hose is fitted.

When the engine is at normal operating temperature, the coolant expands, and some of it is displaced into the expansion tank. Coolant collects in the tank, and is returned to the radiator when the system cools.

The electric cooling fan mounted on the rear of the radiator is controlled by a thermostatic switch. At a predetermined

3•2 Cooling, heating and ventilation systems

coolant temperature, the switch actuates the fan.

Automatic transmission models have a transmission fluid cooler located in the hose between the radiator and the water pump.

Precautions

Do not attempt to remove the expansion tank filler cap, or to disturb any part of the cooling system, while the engine is hot, as there is a high risk of scalding. If the expansion tank filler cap must be removed before the engine and radiator have fully cooled (even though this is not recommended), the pressure in the cooling system must first be relieved. Cover the cap with a thick layer of cloth, to avoid scalding, and slowly unscrew the filler cap until a hissing sound is heard. When the hissing has stopped, indicating that the pressure has reduced, slowly unscrew the filler cap until it can be removed; if more hissing sounds are heard, wait until they have stopped before unscrewing the cap completely. At all times, keep well away from the filler cap opening, and protect your hands.

Do not allow antifreeze to come into contact with your skin, or with the painted surfaces of the vehicle. Rinse off spills immediately, with plenty of water. Never leave antifreeze lying around in an open container, or in a puddle in the driveway or on the garage floor. Children and pets are attracted by its sweet smell, but antifreeze can be fatal if ingested.

If the engine is hot, the electric cooling fan may start even if the engine is not running. Be careful to keep your hands, hair, and any loose clothing well clear when working in the engine compartment.

Refer to Section 10 for precautions to be observed when working on models equipped with air conditioning.

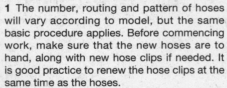

2 Cooling system hoses - disconnection and renewal

1 The number, routing and pattern of hoses will vary according to model, but the same basic procedure applies. Before commencing work, make sure that the new hoses are to hand, along with new hose clips if needed. It is good practice to renew the hose clips at the same time as the hoses.
2 Drain the cooling system, as described in Chapter 1, saving the coolant if it is fit for re-use. Squirt a little penetrating oil onto the hose clips if they are corroded.
3 Unscrew the clips and release the hose clips from the hose concerned.
4 Unclip any wires, cables or other hoses which may be attached to the hose being removed. Make notes for reference when reassembling if necessary.
5 Release the hose from its stubs with a twisting motion. Be careful not to damage the stubs on delicate components such as the radiator. If the hose is stuck fast, try carefully prising the end of the hose with a screwdriver or similar, taking care not to use excessive force. The best course is often to cut off a stubborn hose using a sharp knife, but again be careful not to damage the stubs.
6 Before fitting the new hose, smear the stubs with washing-up liquid or a suitable rubber lubricant to aid fitting. Do not use oil or grease, which may attack the rubber.
7 Fit the hose clips over the ends of the hose, then fit the hose over its stubs. Work the hose into position. When satisfied, locate and tighten the hose clips.
8 Refill the cooling system as described in Chapter 1. Run the engine, and check that there are no leaks.
9 Top-up the coolant level if necessary.

3 Radiator - removal, inspection and refitting

Note: If the reason for removing the radiator is to cure a leak, it is worth trying the effect of a radiator sealing compound first - this is added to the coolant, and will often cure minor leaks with the radiator in situ.

Removal

1 The radiator is removed upwards from the engine compartment complete with the electric cooling fan. First disconnect the battery negative lead.
2 Drain the cooling system (see Chapter 1).
3 Remove the distributor cap (Chapter 5), then loosen the clips and disconnect the top hose from the radiator and thermostat housing and the bottom hose from the radiator (see illustrations).
4 Loosen the clip and disconnect the expansion tank hose from the radiator.
5 Remove the ignition coil (see Chapter 5).
6 Disconnect the wiring from the electric cooling fan and thermo switch.

99 models (except Turbo)

7 Disconnect the headlamp lens wiper pushrod from the wiper motor crank in front of the radiator.
8 Disconnect the wiring from the headlamp wiper motor and temperature sensing switch, and unclip the harness from the electric cooling fan cowling.
9 Unscrew the radiator securing screws and lift the radiator from the engine compartment. The fan motor and the wiper motor can be released from the radiator if required, by removing the attachment bolts.

99 models (Turbo)

10 Disconnect the wiring from the headlamp wiper motors and temperature sensing switch, and unclip the harness from the electric cooling fan cowling.
11 Remove the screws securing the front panel section then lift away the panel and radiator assembly.
12 The fan motor can be removed after undoing the bolts which secure the fan cover to the radiator frame. If the fan blades are being removed from the motor shaft note that the securing nut has a left-hand thread.

900 models (except Turbo APC)

13 Loosen the clip and disconnect the air inlet duct to the air cleaner with reference to Chapter 4.
14 Unscrew and remove the two upper mounting bolts, move the top of the radiator slightly rearwards, and lift the radiator from the engine compartment. For better access

3.3a Disconnecting the top hose from the radiator

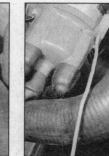

3.3b Top hose connection to the thermostat housing

3.3c Radiator bottom hose connection

Cooling, heating and ventilation systems 3•3

3.14a Radiator upper mounting screw (900 model H-type engine)

3.14b Removing the radiator (900 model H-type engine)

3.14c Radiator upper mounting screw (late 900 model)

the electric cooling fan can be removed first **(see illustrations)**.

900 models (Turbo APC)
15 On 1984 and earlier models loosen the clips and remove the air inlet duct.
16 Remove the APC valve and switch with reference to Chapter 4.
17 Disconnect and remove the turbocharger pressure pipe with reference to Chapter 4.
18 Remove the electric cooling fan as described in Section 5. On models with air conditioning remove both electric fans.
19 Unscrew the mounting bolts then move the top of the radiator slightly rearwards and lift it from the engine compartment.
20 Remove the rubber bush and the fan lower mounting bolt.

Inspection
21 If the radiator has been removed due to suspected blockage, reverse-flush it as described in Chapter 1. Clean dirt and debris from the radiator fins, using an air line (in which case, wear eye protection) or a soft brush. Be careful, as the fins are sharp, and easily damaged.
22 If necessary, a radiator specialist can perform a 'flow test' on the radiator, to establish whether an internal blockage exists.
23 A leaking radiator must be referred to a specialist for permanent repair.
24 If the radiator is to be sent for repair, or is to be renewed, remove the cooling fan thermostatic switch (see Section 6).
25 Inspect the condition of the radiator mounting rubbers and hoses, and renew them if necessary.

Refitting
26 Refitting is a reversal of removal, bearing in mind the following points:

a) Ensure that the lower lugs on the radiator are correctly engaged with the mounting rubbers in the body crosspanel.
b) Refit the thermostat switch (Section 6).
c) Reconnect the hoses (see Section 2).
d) On completion, refill the cooling system as described in Chapter 1.

4 Thermostat - removal, testing and refitting

Removal
1 The thermostat is located on the front of the cylinder head. First partially drain the cooling system with reference to Chapter 1 so that the coolant level is below the thermostat housing.
2 Unscrew the bolts from the thermostat cover and remove the cover from the housing - there is no need to disconnect the top hose although this will make access to the cover mounting bolts easier **(see illustrations)**.
3 Note its fitted position, then extract the thermostat from the housing and ease off the sealing ring **(see illustration)**.

Testing
4 A rough test of the thermostat may be made by suspending it with a piece of string in a container full of water. Heat the water to bring it to the boil - the thermostat must open by the time the water boils. If not, renew it.
5 If a thermometer is available, the precise opening temperature of the thermostat may be determined; compare with the figures given in the Specifications. The opening temperature is also marked on the thermostat.
6 A thermostat which fails to close completely as the water cools must also be renewed.

Refitting
7 Refitting is a reversal of removal, bearing in mind the following points.
a) Examine the sealing ring for signs of damage or deterioration, and if necessary, renew.
b) Ensure that the thermostat is fitted the correct way round, as noted before removal. The arrow must point towards the outlet on the cover.
c) Refill and bleed the cooling system as described in Chapter 1.

4.2a Removing the thermostat cover on an early 900 model

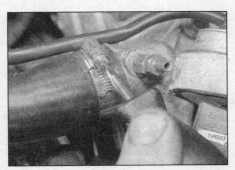

4.2b Disconnecting the top hose from the thermostat housing

4.2c Removing the thermostat cover on a late 900 model

4.3 Removing the thermostat from the housing

5.8a Electric cooling fan location on the radiator (engine removed from car)

5.8b Removing the electric cooling fan assembly

5.8c Electric cooling fan upper mounting (late B202 engine)

5 Electric cooling fan - testing, removal and refitting

Testing

1 Current supply to the cooling fan is direct from the battery (see Chapter 5), a fuse (see Chapter 12) and a cooling fan relay unit located in the main fuse and relay box. The circuit is completed by the cooling fan thermostatic switch, which is mounted in the right-hand radiator tank. On models with air conditioning, the cooling fan is also controlled by the air conditioning control unit - see Section 6.

2 If the fan does not appear to work, run the engine until normal operating temperature is reached, then allow it to idle. The fan should cut in within a few minutes (just before the temperature gauge needle enters the red section). On air conditioning models, switching on the air conditioning will also operate the fan, since the same fan is used to draw air though the air conditioning condenser. If the fan does not operate, switch off the ignition and disconnect the wiring plug from the cooling fan switch on the radiator. Bridge the two contacts in the wiring plug using a length of spare wire. If the fan now operates, the switch is probably faulty, and should be renewed.

3 If the fan still fails to operate, check that battery voltage is available at the feed wire to the switch; if not, then there is a fault in the feed wire (possibly due to a blown fuse). If there is no problem with the feed, check that there is continuity between the switch earth terminal and a good earth point on the body; if not, then the earth connection is faulty, and must be re-made.

4 If the switch and the wiring are in good condition, the fault must lie in the motor itself. The motor can be checked by disconnecting it from the wiring loom, and connecting a 12-volt supply directly to it.

Removal

5 Disconnect the battery negative lead.
6 Disconnect the wiring for the electric cooling fan.
7 Disconnect and remove the HT lead from the distributor cap to the ignition coil.
8 Unscrew the mounting screws and remove the cooling fan assembly from the radiator and withdraw it from the engine compartment (see illustrations).
9 With the unit on the bench, unscrew the motor mounting nuts/screws and remove the motor from the cowling. Unscrew the central nut or prise off the clip and withdraw the fan from the spindle.

Refitting

10 Refitting is a reversal of removal.

6 Cooling system electrical switches - testing, removal and refitting

Electric cooling fan thermostatic switch

Testing

1 The electric cooling fan thermostatic switch is located on the left-hand side of the radiator beneath the expansion tank hose connection (see illustration). On models with air conditioning an additional switch is located in the radiator top hose.

2 To test the switch pull back the cover to expose the two terminals, then, with the ignition switched on, use a screwdriver to bridge the two terminals. The cooling fan should operate proving that the circuit is in order. Switch off the ignition.

3 Further testing can be made after removing

6.1 Electric cooling fan thermostatic switch location (H-type engine on 900 model)

the switch as described in the following paragraphs. Suspend the switch with a piece of string in a container of water and connect a test lamp and leads with 12 volt supply to the two terminals. Gradually heat the water and check that the test lamp lights up at the cut-in temperature given in the Specifications. Allow the water to cool and check that the light goes out at the cut-out temperature given in the Specifications. Renew the switch if it is faulty.

Removal

4 The engine and radiator should be cold before removing the switch. First drain the cooling system as described in Chapter 1. Alternatively, have ready a suitable bung to plug the switch aperture when the switch is removed. If this method is used, take great care not to damage the switch aperture, and do not use anything which will allow foreign matter to enter the cooling system.

5 Disconnect the wiring plug from the switch.
6 Carefully unscrew the switch and recover the sealing washer. If the system has not been drained, immediately plug the switch aperture to prevent further coolant loss.

Refitting

7 Clean the switch threads thoroughly, then refit the switch together with a new sealing washer using a reversal of the removal procedure. Make sure the switch is tightened securely, then refill (or top-up) the cooling system as described in Chapter 1.
8 On completion, start the engine and run it until it reaches normal operating temperature. Continue to run the engine, and check that the cooling fan cuts in and out correctly.

Coolant temperature gauge transmitter

Testing

9 The temperature gauge transmitter is located on the thermostat housing (see illustration).
10 The temperature gauge is fed with voltage via the ignition switch and a fuse. The gauge earth is controlled by the transmitter. The transmitter contains a thermistor - an electronic component whose electrical resistance decreases at a predetermined rate as its temperature rises. When the coolant is cold, the transmitter resistance is high,

Cooling, heating and ventilation systems

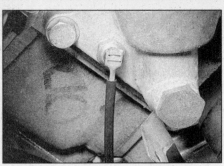

6.9 Temperature gauge transmitter location on the thermostat housing (H-type engine on 900 model)

current flow through the gauge is reduced, and the gauge needle points towards the 'cold' (low) end of the scale. As the coolant temperature rises and the transmitter resistance falls, current flow increases, and the gauge needle moves towards the 'hot' (high) end of the scale. If the transmitter is faulty, it must be renewed.

11 If the gauge develops a fault, first check the other instruments; if they do not work at all, check the instrument panel electrical feed. If the readings are erratic, there may be a fault in the voltage stabiliser, which will mean renewal of the stabiliser (the stabiliser is integral with the instrument panel printed circuit board - see Chapter 12). If the fault lies in the temperature gauge alone, check it as follows.

12 If the gauge needle remains at the 'cold' (low) end of the scale when the engine is hot, disconnect the transmitter wiring plug, and earth the relevant wire to the cylinder head. If the needle then deflects when the ignition is switched on, the transmitter unit is proved faulty, and should be renewed. If the needle still does not move, remove the instrument panel (Chapter 12) and check the continuity of the wire between the transmitter and the gauge, and the feed to the gauge unit. If continuity is shown, and the fault still exists, then the gauge is faulty, and the gauge unit should be renewed.

13 If the gauge needle stays at the 'hot' end of the scale when the engine is cold, detach the transmitter wire. If the needle then returns to the 'cold' end of the scale, the transmitter is proved faulty, and should be renewed. If the needle still does not move, check the rest of the circuit as described previously.

Removal and refitting

14 The procedure is similar to that described previously in this Section for the electric cooling fan thermostatic switch.

7 Water pump - removal and refitting

Removal

B201 B-type engine

1 Disconnect the battery negative lead.
2 Drain the cooling system (see Chapter 1).
3 Remove the inlet manifold and gasket as described in Chapter 4 and use masking tape to cover the inlet ports and water passage in the cylinder head.
4 Remove the alternator and bracket (see Chapter 5). On some models it will be necessary to unbolt both rear engine mountings and jack up the rear of the engine/transmission in order to remove the alternator bracket bolt from the transmission cover. Also pivot the bracket away from the engine without removing the lower bracket bolt.
5 Unscrew the remaining bolts and remove the outer pump cover and gasket from the cylinder block **(see illustration)**.
6 In order to remove the pump it may be necessary to use a SAAB extractor tool. An alternative method depends on the type of pump; if the impeller is retained by a nut or bolt, remove the nut/bolt (left-hand thread) and refit it over a large washer. Apply leverage at opposite points on the washer to prise the pump free. If the impeller is a press fit (ie no nut), the removal procedure remains the same, but it will be necessary to obtain a nut that fits (also left-hand thread). Shock loads, as applied from a slide hammer, should be avoided to prevent damage to the bearing.
7 The water pump may be overhauled as follows if necessary. First support the underside of the impeller and drive out the shaft using a soft metal drift.
8 Note the location of the seals then remove them from the driveshaft.
9 Extract the circlip then support the bearing

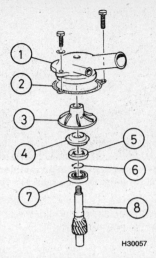

7.5 Exploded view of the water pump fitted to the B201 B-type engine

1 Cover
2 Gasket
3 Impeller
4 Pump seal
5 Bearing seal
6 Circlip
7 Ball bearing
8 Driveshaft

with the drivegear downwards and drive out the shaft using a soft metal drift.
10 Clean all the components in paraffin and wipe dry. Spin the bearing by hand and check it for roughness and excessive play. Examine the driveshaft and gear teeth for wear and damage. Check the impeller for corrosion. Renew the components as necessary and obtain a set of seals.

B201 H-type and B202 engines

11 The water pump is driven by the auxiliary drivebelt, and is located on the rear left-hand side of the engine. First disconnect the battery negative lead.
12 Drain the cooling system (see Chapter 1).
13 Remove the auxiliary drivebelt(s) as described in Chapter 1.
14 Unscrew the bolts and remove the pulley from the water pump flange **(see illustration)**.
15 On 99 models unbolt the cover from the heater unit.
16 Unscrew the bolts and remove the water pump from the timing cover. Remove the gasket **(see illustrations)**.

7.14 Removing the pulley from the water pump flange

7.16a Water pump showing mounting bolts on the timing cover

7.16b Removing the water pump from the timing cover

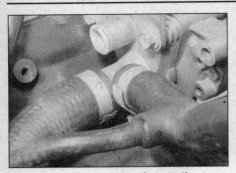

7.17a Hose connections to the water pump cover

7.17b Disconnecting the hoses from the water pump cover (B202 engine)

17 If necessary, loosen the clips and disconnect the hoses from the water pump cover **(see illustrations)**. Identify the hoses for location to ensure correct refitting. Unbolt the cover from the timing cover and remove the gasket. Note that the rear engine hanger is located on the upper bolts.
18 No repairs are possible, and if the water pump is faulty it should be renewed.
19 Clean the mating faces of the timing cover, water pump and cover as applicable.

Refitting

B201 B-type engine

20 Commence reassembly by driving the bearing fully onto the shaft using a length of metal tubing on the inner race.
21 Fit the bearing retaining circlip to the shaft.
22 Clean the cylinder block seating then insert the water pump shaft and engage the gear with the idler shaft.
23 Using a length of metal tubing on the bearing outer race, drive the bearing and shaft fully into the cylinder block.
24 Locate the bearing seal over the shaft and use metal tubing to position the seal in the block.
25 Locate the pump seal over the shaft and use metal tubing to position it in the block.
26 Clean the end of the shaft and the impeller bore then locate the impeller on the shaft.
27 Using tool 83 92 649 or a home made equivalent carefully press the impeller onto the shaft taking care not to push the shaft through the bearing inner race. Press on the impeller a little at a time and make sure that the shaft turns freely. Check that the impeller is fully home by temporarily fitting the cover and checking that there is clearance. If the shaft jams, it has probably moved through the bearing against the cylinder block in which case the pump must be removed and the bearing repositioned.
28 Clean the mating faces then fit the water pump cover to the cylinder block together with a new gasket.
29 Fit the alternator bracket then insert the bolts in the cover and tighten them evenly. Where applicable, lower the engine/transmission and reconnect the rear engine mountings.

30 Refit the alternator (Chapter 5) and tension the drivebelt (Chapter 1).
31 Remove the masking tape then refit the inlet manifold together with a new gasket as described in Chapter 4.
32 Reconnect the battery negative lead.
33 Fill the cooling system (see Chapter 1).

B201 H-type and B202 engines

34 Where removed, refit the cover to the timing cover together with a new gasket and tighten the bolts. Make sure that the rear engine hanger is located on the upper bolts.
35 Reconnect the hoses to the water pump cover and tighten the clips.
36 Ensure that the contact surfaces are clean. Locate a new gasket on the timing cover.
37 Locate the water pump on the gasket then insert the mounting bolts and tighten them securely in a progressive manner.
38 On 99 models refit the cover to the heater unit and tighten the bolts.
39 Refit the pulley to the water pump flange and tighten the bolts.
40 Refit and tension the auxiliary drivebelt(s) as described in Chapter 1.
41 Reconnect the battery negative lead.
42 Fill the cooling system (see Chapter 1).

8 Heating/ventilation system - general information

The heating/ventilation system consists of a three-speed blower (90/99: two-speed) motor located on the bulkhead in the engine compartment, face level vents in the centre and at each end of the facia, air ducts to the front and rear footwells, and demister vents on top of the facia.

The control unit is located centrally in the facia, and the controls operate flap valves to deflect and mix the air flowing through the various parts of the heating/ventilation system. The flap valves are contained in the air distribution housing, which acts as a central distribution unit, passing air to the various ducts and vents. The control valves are operated by vacuum from the inlet manifold, and cables on the 90/99 models **(see illustrations)**.

Cold air enters the system through the grille on the right-hand rear of the bonnet and then passes through a pollen filter (not 90/99 models). If required, the airflow is boosted by the blower, and then flows through the various ducts, according to the settings of the controls. Stale air is expelled through ducts at the rear of the vehicle. If warm air is required, the cold air is passed over the heater matrix, which is heated by the engine coolant. A thermostatically controlled water valve regulates the flow of coolant through the heater according to the manual setting on the control unit.

On 1984-on models fitted with air conditioning, a recirculation switch enables the outside air supply to be closed off, while the air inside the vehicle is recirculated. This can be useful to prevent unpleasant odours entering from outside the vehicle, but should only be used briefly, as the recirculated air inside the vehicle will soon become stale. With the recirculation switch off, air entering the passenger compartment is cooled while passing over the air conditioning evaporator.

9 Heating/ventilation components - removal and refitting

Heater blower motor

Removal

99 models

1 Disconnect the battery negative lead.
2 Remove the wiper motor (see Chapter 12).
3 Disconnect the fan motor wiring.
4 Unscrew the motor retaining screws and withdraw it slightly, then separate the motor

8.2a Heater control vacuum reservoir inside the right-hand front wing panel

8.2b Heater vacuum control unit located beneath the facia panel

Cooling, heating and ventilation systems 3•7

9.9 Heater fan motor and wiring

from the fan and withdraw the motor followed by the fan.
5 If necessary, remove the fan bearing and plate.

900 models
6 Disconnect the battery negative lead.
7 Remove the switch panel.
8 Remove the upper section of the facia panel with reference to Chapter 11.
9 Disconnect the fan motor wiring **(see illustration)**.
10 Remove the right-hand side defroster valve housing screws.
11 Remove the retaining screws and withdraw the fan motor.

Refitting
12 Refitting is a reversal of removal.

Heater matrix
Removal
99 models
13 Drain the cooling system (see Chapter 1).
14 Remove the alternator (see Chapter 5).
15 With the battery negative lead disconnected remove the radiator fan relay from the bulkhead.
16 Remove the screws and withdraw the fan casing panel, then remove the screws and withdraw the matrix retaining plate.
17 Remove the water valve cap and the control cable, then remove the valve screw.
18 Loosen the heater hose clips and disconnect the hoses.
19 Disconnect the thermostat coil from the matrix. Remove the water valve and the coil.
20 Withdraw the matrix from the thermostat casing.

900 models
21 Disconnect the battery negative lead.
22 Working inside the car, remove the steering column lower shroud (Chapter 11). Where necessary also remove the centre console.
23 Unscrew the screw and nuts from the bulkhead and withdraw the lower facia panel **(see illustrations)**.
24 Remove the screws and withdraw the air duct **(see illustrations)**.
25 Prise out the left-hand side defroster/speaker grille, then slide the water valve control rod forwards so that it is released from the control knob. Then pull it from the water valve. Remove the heater control switches.
26 Remove the screws and withdraw the matrix lower cover **(see illustrations)**.
27 Drain the cooling system (see Chapter 1).
28 Loosen the clips and disconnect the heater hoses from the matrix tubes on the engine compartment side of the bulkhead **(see illustrations)**. Plug the outlets.

9.23a Remove screw behind the ashtray ...

9.23b ... and bolts from the bulkhead ...

9.23c ... and remove lower facia panel

9.24a Removing air duct outer screws ...

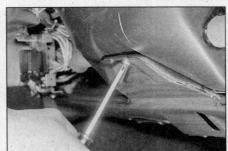
9.24b ... and central screw

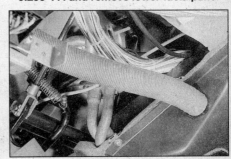
9.26a Matrix lower cover and air tube

9.26b Removing the matrix lower cover

9.28a Loosen the clips ...

9.28b ... and disconnect the hoses from the matrix tubes

3•8 Cooling, heating and ventilation systems

9.30a Water valve connection to the heater matrix

9.30b Matrix located in its housing

9.35 Pollen filter location on an early model

29 Unhook the brake pedal return spring and hold the pedal down on left-hand drive models.
30 Separate the matrix and water valve from the heater housing and withdraw it from inside the car **(see illustrations)**.
31 Where applicable remove the capillary tube from the matrix taking care not to damage it.
32 Unbolt the water valve and remove the gasket.

Refitting
33 Refitting is a reversal of removal, but always fit a new gasket to the water valve. After reconnecting the hoses and before refitting the remaining components, refill the cooling system with reference to Chapter 1 and check for leaks.

Pollen filter
Removal
34 The inlet pollen filter is located on the right-hand side of the bulkhead in the engine compartment.
35 Remove the four screws and slide out the filter **(see illustration)**. Note that from 1983 the cover is incorporated in the filter casing and retained with plastic clips; use a pair of pliers to turn the clips 90° in either direction to release them. To prevent skin irritation while handling the glass fibre filter, it is recommended that protective gloves are worn.

Refitting
36 Refitting is a reversal of removal.

Heater control panel - 900 models
Removal
37 Disconnect the battery negative lead.
38 Remove the steering wheel as described in Chapter 10.
39 Unscrew the mounting screws on the lower edge of the panel. Note that the screws are all of different lengths and are marked with 1 to 4 lines for identification.
40 Tilt the panel, then disconnect the wiring while noting its location.
41 Disconnect the hoses from the vacuum distributor, then withdraw the panel from the facia.
42 If necessary remove the vacuum distributor by unscrewing the retaining screws.

Refitting
43 Refitting is a reversal of removal.

Heater controls - 90/99 models
Removal
44 Remove the safety padding on both sides, and the tidy box from the instrument panel.
45 Disconnect the cables from the heater controls. Release the cable clamps by pulling them up and twisting them off at the same time.
46 Undo the four retaining screws from the heater control assembly.
47 Lift the control assembly up until it has detached itself from the fresh air channel. The control assembly can now be removed.

Refitting
48 Refitting is a reversal of removal.

10 Air conditioning system - general information and precautions

General information
Air conditioning is fitted as standard on certain models. It enables the temperature of air inside the car to be lowered, and also dehumidifies the air, which makes for rapid demisting and increased comfort **(see illustration)**.

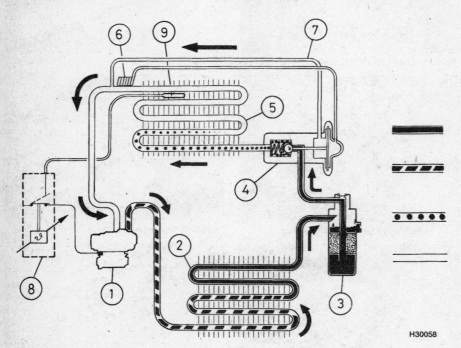

10.1 Air conditioning system circulation diagram

1. Compressor
2. Condenser
3. Receiver shell
4. Expansion valve
5. Evaporator
6. Temperature-sensitive expansion valve body
7. Compensating hose
8. Anti-freezing thermostat
9. Temperature-sensitive anti-frost thermostat body

Cooling, heating and ventilation systems

The cooling side of the system works in the same way as a domestic refrigerator. Refrigerant gas is drawn into a belt-driven compressor, and passes into a condenser mounted in front of the radiator, where it loses heat and becomes liquid. The liquid passes through a receiver and expansion valve to an evaporator, where it changes from being liquid under high pressure to gas under low pressure. This change is accompanied by a drop in temperature, which cools the evaporator. The refrigerant returns to the compressor, and the cycle begins again.

Air drawn through the evaporator passes to the air distribution unit. The air conditioning system has two control buttons located on the heater control panel, one button switches the air conditioning on and the other operates the interior air only circulation system.

The heating side of the system works in the same way as on models without air conditioning (see Section 8).

The operation of the system is controlled by an electromagnetic clutch on the compressor drive pulley. Any problems with the system should be referred to a Saab dealer.

Precautions

When working on the air conditioning system, it is necessary to observe special precautions. If for any reason the system must be disconnected, entrust this task to your Saab dealer or a refrigeration engineer.

Warning: The refrigeration circuit contains a liquid refrigerant, and it is dangerous to disconnect any part of the system without specialised knowledge and equipment. The refrigerant is potentially dangerous, and should only be handled by qualified persons. If it is splashed onto the skin, it can cause frostbite. It is not itself poisonous, but in the presence of a naked flame (including a cigarette) it forms a poisonous gas. Uncontrolled discharging of the refrigerant is dangerous, and potentially damaging to the environment. Do not operate the air conditioning system if it is known to be short of refrigerant, as this may damage the compressor.

11 Air conditioning components - removal and refitting

1 The only operation which can be carried out easily without discharging the refrigerant is renewal of the compressor drivebelt. This is described in Chapter 1, Section 24. All other operations must be referred to a Saab dealer or an air conditioning specialist.

2 If necessary for access to other components, the compressor can be unbolted and moved aside, without disconnecting its flexible hoses, after removing the drivebelt.

3 Access to the condenser is gained by unscrewing the front spoiler lower mounting screws and lifting the spoiler.

Warning: Do not attempt to open the refrigerant circuit. Refer to the precautions given in Section 10.

Notes

Chapter 4 Part A:
Fuel/exhaust systems - carburettor models

Contents

Accelerator cable - removal, refitting and adjustment 4
Accelerator pedal - removal and refitting 5
Air cleaner assembly - removal and refitting 2
Air cleaner automatic air temperature control system -
 general information and component renewal 3
Air cleaner filter element renewal . . . . See Chapter 1
Carburettor - overhaul 12
Carburettor(s) - removal and refitting 11
Choke cable - removal, refitting and adjustment 10
Exhaust manifold - removal and refitting .14

Exhaust system - general information and component renewal 15
Fuel filter renewal . . . . .See Chapter 1
Fuel gauge sender unit - removal and refitting 8
Fuel pump - testing, removal and refitting .7
Fuel tank - removal and refitting 9
General fuel system checks . . . . .See Chapter 1
General information and precautions 1
Idle speed and mixture adjustment . . . . .See Chapter 1
Inlet manifold - removal and refitting 13
Unleaded petrol - general information and usage 6

Degrees of difficulty

| Easy, suitable for novice with little experience | | Fairly easy, suitable for beginner with some experience | | Fairly difficult, suitable for competent DIY mechanic | | Difficult, suitable for experienced DIY mechanic | | Very difficult, suitable for expert DIY or professional |

Specifications

Air cleaner type	Automatic air temperature control with renewable paper element
Temperature rating:	
B-type engine	8° to 18°C
H-type engine	23° to 37°C

Fuel pump

Type	Mechanical, diaphragm, operated from idler shaft on B-type engine or camshaft on H-type engine
Fuel pressure at starter speed	0.17 to 0.25 bar

Carburettor

Single carburettor engines (to 1987)

Type:	
To 1984	Zenith/Stromberg 175 CD
From 1985	Zenith/Stromberg 175 CDSEVX
Diameter	1.75 in
Metering needle (CM, CA to 1979)	B1 DS
Float setting	16.0 to 17.0 mm
Needle valve	2.0 mm
Damper oil level	Level with or a minimum of 10.0 mm below the upper edge of the air valve inner cylinder
Float idle cam clearance (B-type engine)	1.0 mm
Fast idling speed:	
CM, CA (1981 to 1982)	1100 ± 50 rpm
CM (1983)	1350 ± 50 rpm
CA (1984)	1350 ± 50 rpm
Jet position	2.5 ± 0.1 mm from seating
Initial metering needle position	Shoulder level with bottom of piston
Temperature compensator opening at 20°C	0.1 to 0.3 mm
Piston return spring colour coding	Red
Damper lubrication type/specification	Type F ATF to Ford specification M2C 33G, or Dexron type ATF

4A•2 Fuel/exhaust systems - carburettor models

Single carburettor engine (from 1987)
Type .... Pierburg 175 CDUS
Diameter .... 1.75 in
Metering needle .... YC
Fast idling speed .... 1350 ± 50 rpm
Float level .... 8.0 ± 1.0 mm
Damper oil level .... Level with or a minimum of 10.0 mm below the upper edge of the air valve inner cylinder
Fuel jet position under jet face mounting boss .... 2.5 mm

Twin carburettor engines
Type .... Zenith/Stromberg 150 CD 2
Diameter .... 1.5 in
Metering needle:
  TM, TA 1979 to 1980 .... B5 EJ
  TM, TA 1981-on .... B5 EQ
Float setting .... 16.0 to 17.0 mm
Needle valve .... 1.5 mm
Damper oil level .... Level with or a minimum of 10.0 mm below the upper edge of the air valve inner cylinder
Fast idle cam clearance (B-type engine) .... 1.0 mm
Fast idle speed (TM, TA, 1981 to 1984) .... 1100 ± 50 rpm
Jet position:
  B-type engine .... 2.5 ± 0.1 mm from seating
  H-type engine .... 2.3 ± 0.1 mm from seating
Initial metering needle position .... Shoulder level with bottom of piston
Temperature compensator opening at 20°C .... 0.1 to 0.3 mm
Piston return spring colour coding .... Blue
Damper lubrication type/specification .... Type F ATF to Ford specification M2C 33G, or Dexron type ATF

Recommended fuel
All models to 1984 .... 98 RON leaded only
All models from 1985 .... 95 RON unleaded

Torque wrench settings   **Nm**   **lbf ft**
Inlet manifold 18   13
Exhaust manifold 25   18

---

1 General information and precautions

The fuel system consists of a fuel tank mounted under the rear of the car, a mechanical fuel pump, and either single or twin side-draught carburettors. The fuel pump is operated by an eccentric on the idler shaft (B201 B-type engine) or camshaft (B201 H-type engine). The air cleaner contains a disposable paper filter element. It incorporates an automatic flap valve air temperature control system, which allows cold air from the outside of the car and warm air from the exhaust manifold to enter the air cleaner in the correct proportions.

The fuel pump lifts fuel from the fuel tank, and incorporates an internal gauze filter. On early models the filter can be removed for cleaning, however on later models it is not possible to dismantle the fuel pump.

Single carburettor models are fitted with either a Zenith/Stromberg 175 or Pierburg 175 CDU carburettor. Twin carburettor models are fitted with a Zenith/Stromberg 150 carburettor. On all carburettors, mixture enrichment for cold starting is by a cable-operated choke control.

From 1985 all carburettors were modified by fitting a disc valve to the carburettor throttle butterfly, its purpose being to assist in regulating the fuel/air mixture on the overrun. This valve replaces the deceleration valve (see Chapter 1) fitted to some earlier models.

 Warning: Many of the procedures in this Chapter require the removal of fuel lines and connections, which may result in some fuel spillage. Before carrying out any operation on the fuel system, refer to the precautions given in "Safety first!" at the beginning of this manual, and follow them implicitly. Petrol is a highly-dangerous and volatile liquid, and the precautions necessary when handling it cannot be overstressed.

2 Air cleaner assembly - removal and refitting

Removal

1 The air cleaner is located on the left-hand side of the engine compartment and it incorporates an automatic air temperature control.

2 To remove the element either release the clips or remove the screws from the air cleaner cover, move the cover to one side, and withdraw the element. Where fitted remove the insert from the bottom of the body.

3 If cleaning the element between service intervals, tap it to release the accumulated dust and if available use an air line. Do not attempt to wash the element.

4 Wipe clean the inside surfaces of the air cleaner.

5 To remove the air cleaner loosen the clips and withdraw the air ducts and crankcase ventilation hose, then release the unit from the mounting bracket.

Refitting

6 Refitting is a reversal of the removal procedure, but check all associated hoses and air ducts for condition and security.

3 Air cleaner automatic air temperature control system - testing, removal and refitting

Testing

1 The system is controlled by a heat-sensitive bi-metal vacuum switch mounted on the air

Fuel/exhaust systems - carburettor models 4A•3

inlet duct. When the engine is started from cold, the switch is open, and allows inlet manifold depression to act on the air temperature control valve diaphragm in the inlet duct. This vacuum causes the diaphragm to move a flap valve across the cold-air inlet, thus allowing only (warmed) air from the exhaust manifold to enter the air cleaner.

2 As the temperature of the exhaust-warmed air in the air cleaner-to-carburettor duct rises, the bi-metal strip in the vacuum switch deforms and closes the vacuum supply to the air temperature control valve assembly. As the vacuum supply is cut, the flap is gradually lowered across the hot-air inlet until, when the engine is fully warmed up to normal operating temperature, only cold air from the front of the car is entering the air cleaner.

3 To make a basic check of the system, allow the engine to cool down completely, then loosen the clip and disconnect the control valve assembly from the inlet duct at the front of the engine compartment. Start the engine; the flap should immediately close off the cold-air inlet, and should then open steadily as the engine warms up, until it eventually allows only cold air to enter the air cleaner.

Removal

4 To remove the components and to make a more accurate test, loosen the clips and remove the control flap unit from the air cleaner body. Also remove the thermostatic vacuum switch from the inlet duct leading to the carburettor - the switch may be left on the removed duct if preferred as a new clip may be required to refit it.

5 Check the operation of the unit by starting the engine then immersing the thermostatic vacuum switch in water and checking the operating temperatures with reference to the Specifications. At the lower temperature or under, the flap must admit only heated air, and at the higher temperature or over, only cold air.

6 Renew any faulty components.

Refitting

7 Refitting is a reversal of removal, but before reconnecting the control valve assembly to the inlet duct, make sure that the flap is over the cold air aperture at room temperature. If not, loosen the plastic nut on the cable and turn the cable through 180° as necessary. Tighten the nut afterwards and reconnect the control valve assembly.

4 Accelerator cable - removal, refitting and adjustment

Removal

1 Turn the throttle lever at the carburettor and disconnect the inner cable at the engine end.
2 Loosen the nuts and disconnect the outer cable from the bracket.
3 Remove the lower facia panel.
4 Disconnect the cable from the accelerator pedal or arm and withdraw it through the bulkhead. Where applicable unscrew the grommet in the bulkhead. Note: *On right-hand drive models the accelerator pedal extends into the left-hand passenger footwell area.*

Refitting and adjustment

5 Refitting is a reversal of removal, but adjust the ferrule on the outer cable so that there is a minimal amount of slack in the cable with the accelerator pedal fully released. Tighten the adjustment nuts on completion. Check that the throttle is fully open with the accelerator pedal pressed to the floor.

5 Accelerator pedal - removal and refitting

Removal

1 Turn the throttle lever at the carburettor and disconnect the inner cable at the engine end.
2 Inside the car remove the insulation above the pedals, and on left-hand drive models remove the panel from the left-hand side of the centre console. On right-hand drive models remove the insulation on the left-hand side of the passenger. Note: *On right-hand drive models the accelerator pedal extends into the left-hand passenger footwell area.*
3 Lift the pedal and disconnect the inner cable after releasing the clip.
4 Unscrew the bolts on the pedal bracket and remove the pedal.

Refitting

5 Refitting is a reversal of removal but tighten the mounting bolts securely and adjust the accelerator cable to eliminate slackness.

6 Unleaded petrol - general information and usage

Note: *The information given in this Chapter is correct at the time of writing, and applies only to petrols currently available in the UK. If updated information is thought to be required, check with a Saab dealer. If travelling abroad, consult one of the motoring organisations (or a similar authority) for advice on the petrols available and their suitability for your vehicle.*

1 The fuel recommended by Saab is given in the Specifications at the start of this Chapter.
2 The fuel ratings are in AKI units which are an average of RON and MON ratings. RON and MON are different testing standards; RON stands for Research Octane Number (also written as RM), while MON stands for Motor Octane Number (also written as MM).
3 Under no circumstances should leaded fuel be used in models fitted with a catalytic converter, as this will permanently damage the catalytic converter.

7 Fuel pump - testing, removal and refitting

Note: *Refer to the warning note at the end of Section 1 before proceeding.*

Testing

1 The fuel pump is located on the left-hand side of the cylinder block (B201 B-type engine) or cylinder head (B201 H-type engine). To test its operation disconnect the outlet hose and hold a wad of rag by the outlet. Disconnect the ignition HT lead from the distributor cap and connect it to a suitable earthing point using a length of wire. Have an assistant spin the engine on the starter and check that well defined spurts of fuel are ejected. *Keep your hands away from the electric cooling fan.*
2 If a pressure gauge is available check that the fuel pump delivers the specified pressure at starter speed.
3 The pump can also be tested by removing it. With the pump outlet pipe disconnected but the inlet pipe still connected, hold the wad of rag by the outlet. Operate the pump lever by hand; if the pump is in a satisfactory condition, the lever should move and return smoothly, and a strong jet of fuel should be ejected.
4 If the fuel pump is not functioning correctly on early models, clean the filter gauze with reference to Chapter 1 then check the pump again.

Removal

5 To remove the fuel pump, first identify the positions of the inlet and outlet hoses then disconnect them.
6 Unscrew the mounting bolts and withdraw the pump from the cylinder block (B201 B-type engine) **(see illustration)** or cylinder head (B201 H-type engine). Remove the gasket.
7 On H type engines withdraw the pushrod from the cylinder head, and also unbolt the pump from the adapter and remove the O-ring **(see illustration)**.
8 Clean away all traces of gasket from the mating surfaces. If the fuel pump is faulty it must be renewed as it is not possible to dismantle it.

7.6 Removing the fuel pump from the cylinder block (B201 B-type engine)

4A•4 Fuel/exhaust systems - carburettor models

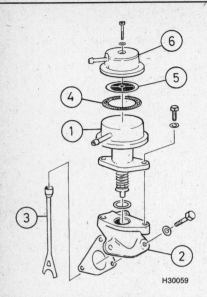

7.7 Fuel pump fitted to the B201 H-type engine

1 Body   4 Seal
2 Adaptor   5 Filter
3 Pushrod   6 Cover

Refitting

9 Refitting is a reversal of removal, but always fit a new gasket, and on H-type engines a new O-ring seal between the adapter and pump. The manufacturers recommend that suitable sealant is applied to the cylinder head flange before a new gasket is fitted. On B-type engines locate the pump operating lever on the idler shaft cam. On H-type engines make sure that the pushrod is correctly located in the groove on the camshaft and hold it in this position with a small screwdriver while fitting the pump.

8 Fuel gauge sender unit - removal and refitting

Note: *Refer to the warning note at the end of Section 1 before proceeding.*

Removal

1 Disconnect the battery negative lead.
2 Remove the floor panel (or carpet on some 99 models) from the luggage compartment, then carefully prise out the sender unit rubber cover or remove the screws and cover plate as applicable **(see illustration)**.
3 Disconnect the wiring (and return hose where applicable) **(see illustration)**. Tape the hose to the floor to prevent it disappearing out of reach.
4 Remove the screws or unscrew the plastic cap as applicable, then carefully withdraw the sender unit from the tank **(see illustrations)**. Saab technicians use a special tool to engage the cap, however a large pair of grips may be used instead. On pre-1981 models the unit incorporates a float and lever, whereas on later models the float is sealed in a plastic tube. Also note that sender units for 1981 and 1982 models cannot be fitted to 1983-on models.
5 Remove the rubber seal from the top of the sender unit **(see illustration)**.

Refitting

6 Refitting is a reversal of removal, but always fit a new rubber seal and make sure that the plastic cap (where fitted) is tightened securely.

9 Fuel tank - removal and refitting

Note: *Refer to the warning note at the end of Section 1 before proceeding.*

Removal

1 Before removing the fuel tank, all the fuel must be drained from the tank. Since a fuel tank drain plug is not provided, it is therefore preferable to carry out the removal operation when the tank is nearly empty. Before proceeding, disconnect the battery negative lead, then remove the filler cap and syphon or hand-pump the remaining fuel from the tank.
2 Chock the front wheels, then jack up the rear of the car and support on axle stands (see "*Jacking and Vehicle Support*").
3 Remove the floor panel (or carpet on some 99 models) from the luggage compartment, then prise out the rubber cover or remove the screws and cover as applicable.
4 Disconnect the wires from the fuel gauge sender unit, noting their location for refitting.
5 Disconnect the filler and ventilation hoses, and the fuel supply and return hoses **(see illustration)**. Remove the fuel line clips where applicable.

8.2 Fuel gauge sender unit beneath the rear floor

8.3 Disconnecting the wiring

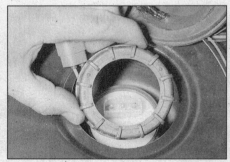

8.4a Unscrewing the plastic cap from the fuel gauge sender unit

8.4b Withdrawing the fuel gauge sender unit from the tank

8.5 Removing the rubber seal from the top of the sender unit

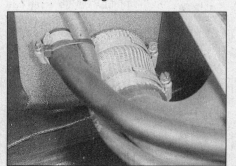

9.5 Fuel tank filler hose

Fuel/exhaust systems - carburettor models 4A•5

9.6 Fuel tank securing strap

6 Support the tank then release the straps by unscrewing the nuts **(see illustration)**.
7 Lower the fuel tank and withdraw it from under the car.
8 Remove the fuel gauge sender unit with reference to Section 8.
9 If the tank is contaminated with sediment or water, swill it out with clean fuel. If the tank leaks or is damaged, it should be repaired by specialists or alternatively renewed. Do not, under any circumstances, attempt to repair a fuel tank yourself.

Refitting

10 Refitting is a reversal of removal, but make sure that all hoses are securely fitted and correctly routed. On completion, refill the tank with fuel and check for signs of leakage prior to taking the car out on the road.

10 Choke cable - removal, refitting and adjustment

Removal

1 Note the fitted position of the choke cable for correct refitting.
2 Disconnect the inner cable from the carburettor lever(s), and unclip the outer cable from the bracket(s).
3 Release the cable from the clips in the engine compartment.

Pre-1982 900 models

4 Remove the gear lever housing cover screws and lift it sufficiently to disconnect the inner and outer cables. Where applicable disconnect the cable from the intermediate lever.
5 Release the grommet from the bulkhead and withdraw the cable.

1982-on 900 models

6 Unscrew the choke control knob then remove the gear lever housing cover screws and lift it sufficiently to disconnect the cable and wiring.
7 Release the grommet from the bulkhead and withdraw the cable.

99 models

8 Remove the lower facia panel then unscrew the control knob and the plastic washer on the warning light.
9 Disconnect the warning light wiring and unscrew the cable nut.

10 Release the grommet from the bulkhead and withdraw the cable.

Refitting

All models

11 Refitting is a reversal of removal, but adjust the cable to remove slackness.

11 Carburettor(s) - removal and refitting

Note: *Refer to the warning note at the end of Section 1 before proceeding.*

Removal

Single Zenith/Solex carburettor

1 Disconnect the inlet duct.
2 Disconnect the fuel supply hose, and choke and throttle cables from the carburettor.
3 Disconnect the vacuum advance and retard hose. Also disconnect the wire from the idle shut-off valve.
4 On the Pierburg carburettor disconnect the warm start valve, float chamber vent shut-off valve and the idle speed adjustment valve (as applicable).
5 Remove the upper mounting screw from the dipstick tube.
6 Unscrew the nuts and withdraw the carburettor from the inlet manifold. Remove the gasket.

Twin Zenith/Solex carburettors

7 On pre-1983 models disconnect the inlet hose from the air cleaner then unclip and remove the inlet air box.
8 On 1983-on models disconnect the inlet hose from the air cleaner. Unscrew the dipstick tube mounting bolts then remove the screws and withdraw the inlet air box.
9 Disconnect the choke and throttle cables.
10 Remove the clips and disconnect the choke linkage from the operating rod.
11 Remove the screws and withdraw the air box together with the throttle cable bracket, choke lever and gasket.
12 Disconnect the fuel pipes, and also detach the ignition advance and retard vacuum hoses. Detach the idle shut-off valve wires.
13 Unscrew the nuts and withdraw both carburettors together, then separate them. Remove the gaskets and insulators.

Pierburg 175 CDUS carburettor

14 Disconnect the following items.
 a) Vacuum advance/retard hose and the EGR hose (if fitted)
 b) Warm start valve, float chamber vent shut-off valve and idle speed adjustment valve (as applicable)
 c) The accelerator and choke cables
 d) The air inlet hose from the carburettor flange
15 Detach and remove the inlet hose flange and gasket from the carburettor.
16 Detach the fuel hose from the carburettor.

Move the hose, complete with the vapour trap and clamp, out of the way.
17 Undo the retaining nuts and withdraw the carburettor from the inlet manifold.

Refitting

Zenith/Solex carburettor(s)

18 Refitting is a reversal of removal, but fit a new gasket (single carburettor) or insulator side gaskets (twin carburettors) and adjust the choke and throttle cables as described in Sections 4 and 10. If necessary top-up the oil level in the damper level with, or a minimum of 10.0 mm below the top of the damper inner cylinder. On twin carburettor engines make sure that the intermediate spring is correctly located, and also check that both choke controls touch their stops at the same time when the choke is operated. If not, adjust the spindle linkages.
19 On B-type engines, check that the distance between the adjusting screw on the throttle lever and the choke cam is 1.0 mm with the choke knob fully inserted **(see illustration)**. On twin carburettor engines make this check on the front carburettor only, and repeat the check after completing the slow running adjustments.
20 Check and adjust the idling speed and mixture with reference to Chapter 1.
21 On H-type engines check the fast idling speed with the engine warm by inserting an 8.0 mm twist drill behind the choke cam. Check that with the engine idling the fast idling speed is as given in the Specifications. If not, loosen the locknut and adjust the screw as necessary. Tighten the locknut and remove the drill.

Pierburg 175 CDUS carburettor

22 Refitting is a reversal of the removal procedure. Fit new gaskets where applicable. Ensure that the hose and wiring connections are secure. When fitted, top-up the damper cylinder with oil to the depth indicated in the accompanying illustration **(see illustration)** and also check that the ventilation hole in the damper piston is clear. On restarting the engine run it up to its normal operating temperature then check and adjust the carburettor as necessary.

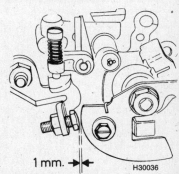

11.19 Basic choke setting clearance on the B201 B-type engine

4A•6 Fuel/exhaust systems - carburettor models

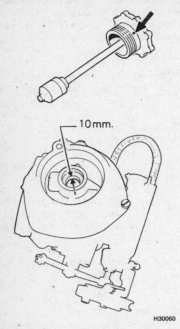

11.22 Damper cylinder oil level must be 10 mm at point indicated. Check ventilation hole (arrowed) in damper piston is clear - Pierburg carburettor

12 Carburettor - overhaul

Zenith/Solex

1 Clean the external surfaces of the carburettor.
2 Refer to the accompanying illustration **(see illustration)**. Unscrew the damper and cap.
3 Mark the vacuum chamber cover in relation to the body then remove the screws and withdraw the cover and spring.
4 Lift out the piston and diaphragm.
5 If necessary remove the metering needle from the piston by unscrewing the screw then using Saab tool 83 93 035 and turning the spindle anti-clockwise.
6 Remove the screws from the top of the piston and withdraw the two washers followed by the diaphragm. Only one washer is fitted to the twin carburettors.
7 Remove the screws and withdraw the float chamber and gasket.
8 Unclip and remove the float and spindle.
9 Unscrew and remove the needle valve and washer.
10 Remove the countersunk screws and withdraw the choke mechanism.
11 Remove the screws and withdraw the temperature compensator. Recover the two O-ring seals.
12 Clean all the components and examine them for wear and damage. The diaphragm should be checked for distortion and cracks. Check that the temperature compensator

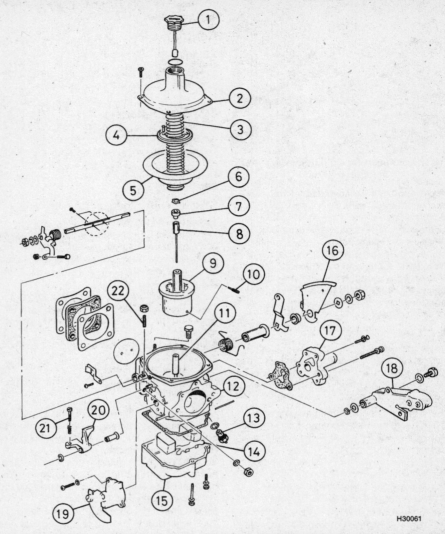

12.2 Exploded view of the front carburettor (twin carburettor engine) or single carburettor arrangement - typical pre-1984 arrangement

1 Cover with damper piston
2 Vacuum chamber cover
3 Spring
4 Washer
5 Diaphragm
6 Circlip
7 Adjusting screw
8 Fuel needle
9 Vacuum piston
10 Lock screw with spring-loaded plunger
11 Jet
12 Carburettor body
13 Float valve
14 Float
15 Float chamber
16 Plate
17 Deceleration valve
18 Temperature compensator
19 Cold start device with cam disc
20 Arm, float chamber ventilation
21 Adjusting screw, idling (pre-1984 front twin carburettor and single carburettor) OR adjustment screw, float chamber ventilation (rear twin carburettor)
22 Adjustment screw, float chamber ventilation (pre-1984 front twin carburettor and single carburettor) - not fitted to rear twin carburettor

valve moves freely. Clean the body using an air line or tyre pump.
13 Commence assembly by locating the diaphragm on the piston with the tab in the cut-out.
14 Fit the washer(s) then insert and tighten the screws taking care not to distort the diaphragm.
15 Insert the metering needle housing in the piston and use tool 83 93 035 to secure the needle. Fit and tighten the lock screw into the groove in the housing and use the tool to adjust the needle so that its shoulder is level with the bottom of the piston.
16 Fit the piston and diaphragm to the body making sure that the diaphragm rim locates in the body groove.
17 Fit the spring and vacuum chamber cover

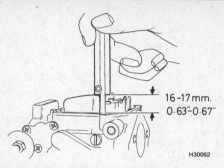

12.19 Using vernier calipers to check the float level setting

in its previously noted position, then insert and tighten the screws.
18 Clip the float and spindle in position with the flat side of the float away from the body.
19 Invert the carburettor and use vernier calipers to check that the float setting is as given in the Specifications. The measurement is taken from the top of the float to the float chamber mating surface **(see illustration)**. If necessary bend the metal tab which contacts the needle valve.
20 Insert and tighten the needle valve and washer.
21 Fit the float chamber together with a new gasket and slide on the chamber until the O-ring makes contact. Insert all the screws loosely then press down the chamber and tighten them all evenly.
22 Fit the choke unit and tighten the screws.
23 Fit the temperature compensator together with two new O-ring seals and tighten the screws evenly.
24 Top-up the oil level in the damper level with, or a minimum of 10.0 mm below the top of the damper inner cylinder.
25 With the carburettor fully assembled, check and if necessary adjust the vent valve as follows. The vent valve facilitates good starting when the engine is hot by preventing vaporized fuel entering the inlet manifold. On twin carburettor engines it also prevents 'run-on' after the ignition has been switched off.
26 To check the valve, connect a hose to the atmosphere air aperture **(see illustration)**. If the carburettor is removed from the engine and the fuel inlet pipe is off, plug the fuel inlet.
27 Fully close the throttle and check that it is not possible to blow through the hose (ie the float chamber is open to the atmosphere).
28 Open the throttle lever 0.5 to 1.0 mm and check that it is now possible to blow through the hose proving that the valve is half open.
29 Further opening of the throttle should prevent blowing through the hose as the valve will only allow venting to the air cleaner.
30 If necessary adjust the valve by loosening the locknut and turning the screw. Tighten the locknut afterwards. If adjustment has been made, recheck the idle adjustment (Chapter 1).

Idle shut-off valve - single carburettor (Zenith/Solex)

31 The idle shut-off valve is located on the underside of the carburettor and is screwed into the throttle body. The purpose of the valve is to prevent the engine from running on when the engine is switched off. It achieves this by closing off the idle fuel mixture through the throttle bypass port, the valve being electrically activated by the ignition switch.
32 To check the idle shut-off valve, connect up a tachometer and run the engine at its normal idle speed. Detach the lead from the shut-off valve and check that the engine speed drops by a minimum of 200 rpm. Reconnect the lead to the valve. If the valve is defective it must be renewed.
33 Running-on can be caused by any one or more of the following.
 a) Incorrect ignition timing (too far advanced)
 b) Incorrect fuel grade used (higher octane rating required)
 c) Carburettor idle speed set too high, or mixture setting too weak
 d) Heavy carbon build-up in combustion chambers caused by use of excessive choke and engine running too cool

Idle shut-off valve - twin carburettors (Zenith/Solex)

34 The idle shut-off valve on twin carburettor engine variants is located on a bracket beneath the air cleaner unit.
35 The shut-off valve is electrically operated and regulates the communication between the area of the float chamber above the fuel level and the constant pressure chamber. As the engine is switched off, a timed relay shuts a circuit, the solenoid valve opens a connection to the float chamber, and the depression above the fuel rises. This does away with the difference in pressure required to draw fuel through the carburettor needle valve and the engine stops. The relay is de-energised after a period of 6 seconds.
36 To check the idle shut-off valve, detach

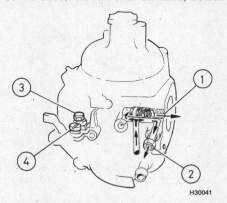

12.26 Carburettor float chamber ventilation
1 To air cleaner
2 To atmosphere
3 Vent valve adjustment (pre-1984 front carburettor)
4 Idle speed adjustment (pre-1984 front carburettor) OR vent valve adjustment (rear carburettor) OR vent valve adjustment (1984 front carburettor)

the communication hoses from the valve, then blow through the connections to the float chamber vent valves. Initially blow through within six seconds of turning off the ignition. In this instance, the float chamber to suction chamber passage should be open.
37 Make a similar check, but this time with the ignition switched on, or 6 seconds or more after the ignition has been turned off. The float chamber vent valve should be open, allowing the entry of air.
38 If the idle shut-off valve is proved to be defective it must be renewed.
39 Other possible causes of running-on are listed in paragraph 33.

Pierburg

40 The dismantling, inspection and reassembly procedures of this carburettor closely follow those described in the previous paragraphs for the Zenith/Solex carburettor **(see illustration)**.
41 Check that the axial play of the damper piston is 10.5 to 1.5 mm. Ensure that the ventilation port in the damper piston is not blocked.
42 When fitting the fuel metering needle into position, align the flat section of the needle shoulder with the locking screw and set the needle so that the base of its shoulder is level with the base of the piston.
43 When fitting the float, ensure that its adjustment tongue is located under the locking needle of the float valve, fit the plastic retaining bracket, then check the float level.
44 To check the float level, press the plastic bracket down so that the float is correctly fitted, then tilt the carburettor to the point where the float arm is just in contact with the needle valve ball, but with no pressure applied to the ball. In this position measure the distance between the float and the gasket face. If required, adjust the float height by bending the tongue at the needle valve to suit.
45 The jet spring, jet and float chamber cover can now be fitted and the jet adjusted for basic setting as follows.
46 Turn the adjuster screw in the float chamber cover to set the jet 2.5 mm under the face of the jet mounting boss.
47 When fitting the piston and diaphragm into the carburettor, ensure that the diaphragm guide locates in its recess in the carburettor body. The carburettor cover alignment marks must correspond as it is fitted.
48 The remainder of the assembly procedures are a reversal of the dismantling procedures.

13 Inlet manifold - removal and refitting

Note: *Refer to the warning note at the end of Section 1 before proceeding.*

Removal

1 Remove the carburettor(s) (see Section 11).
2 Drain the cooling system (see Chapter 1).

4A•8 Fuel/exhaust systems - carburettor models

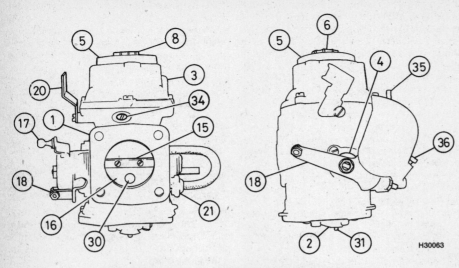

12.40 General external views of the Pierburg carburettor

1 Carburettor body
2 Float chamber cover
3 Carburettor cover
4 Choke disc
5 Cover
6 Oil filler plug
8 Damper piston cap
15 Throttle spindle
16 Throttle
17 Throttle lever
18 Choke lever
20 Choke cable holder
30 Deceleration valve
31 CO adjuster screw
34 Idle speed adjuster valve plug (air conditioning models)
35 Vacuum connection (ignition advance/retard)
36 Idle speed adjustment screw

3 Disconnect the coolant hose.
4 Disconnect the vacuum hose.
5 Disconnect the crankcase ventilation hose from the top of the inlet manifold.
6 Unscrew the mounting bolts and withdraw the inlet manifold from the cylinder head. Note the location of the lifting eye bracket on the front two bolts.
7 Remove the gasket.

Refitting

8 Refitting is the reverse of the removal procedure, noting the following points:
 a) Ensure that the manifold and cylinder head mating surfaces are clean and dry, and always fit a new gasket.
 b) Ensure that all relevant hoses are reconnected to their original positions, and are securely held (where necessary) by their retaining clips.
 c) Refit the carburettor(s) as described in Section 11.
 d) On completion, refill the cooling system as described in Chapter 1.

14 Exhaust manifold - removal and refitting

Removal

1 Disconnect the hot-air inlet hose from the manifold shroud.
2 Where necessary remove the power steering pump and bracket.
3 Unbolt the gearbox oil dipstick tube from the exhaust manifold.
4 Unscrew the nuts securing the exhaust downpipe to the exhaust manifold. Position the downpipe to one side out of the work area, and remove the gasket from the manifold.
5 Unscrew and remove the mounting nuts and remove the exhaust manifold from the studs in the cylinder head.
6 Remove the heatshield followed by the gaskets.

Refitting

7 Refitting is the reverse of the removal procedure, noting the following points:
 a) Examine all the exhaust manifold studs for signs of damage and corrosion; remove all traces of corrosion, and repair or renew any damaged studs.
 b) Ensure that the manifold and cylinder head sealing faces are clean and flat, and fit the new manifold gaskets. Check the heatshield and renew it if necessary.
 c) Renew the exhaust downpipe to manifold gasket and tighten the nuts securely.

15 Exhaust system - general information and component renewal

General information

1 The exhaust system is of three section type and varies slightly in design according to date of production and model (see illustration).
2 The centre and rear mountings are of flexible rubber type (see illustrations).
3 Examination of the exhaust pipe and silencers at regular intervals is worthwhile as small defects may be repairable with proprietry products. If small holes or other defects are left, without attention, the whole system, or at least part of it, will probably need to be completely renewed at a later stage. Also, any leaks, apart from the noise factor, may cause poisonous exhaust gases to find their way into the car which could be unpleasant, to say the least, even in mild concentrations. Prolonged inhalation could cause sickness and giddiness or even death.

Component renewal

4 As the sleeve connections and clamps are usually very difficult to separate, it is better to remove the complete system from the car then separate the sections when full access is

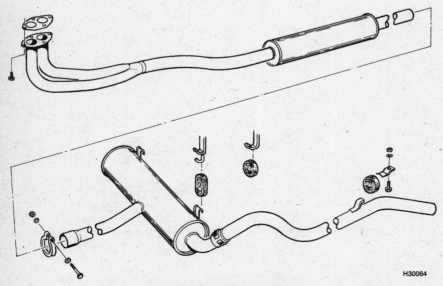

15.1 Typical exhaust system

15.2a Exhaust centre mounting rubber

15.2b Exhaust rear mounting rubber

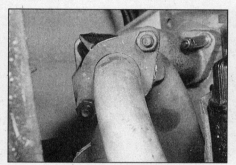
15.8a Exhaust downpipe flange

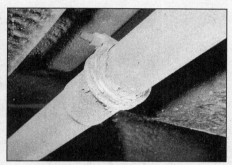
15.8b Exhaust section joint

available around the connections. However, one of the sleeve connections must be separated first in order to remove the system.
5 To remove the complete system first raise the front and rear of the car and support on axle stands (see *'Jacking and Vehicle Support'*).
6 Disconnect the centre and rear mounting rubbers.
7 Just in front of the rear axle loosen the clamp securing the two sections of the exhaust together. Carefully tap around the sleeve joint in order to release the two sections, then withdraw the rear section(s) from the rear of the car.
8 Unscrew the nuts and disconnect the downpipe from the exhaust manifold flange, remove the gasket, then withdraw the front section (and also the intermediate section where applicable) from under the car. Loosen the clamp and carefully separate the remaining sections from each other **(see illustrations)**.
9 When separating a damaged section, cut away the damaged part from the adjoining good section rather than risk damaging the latter.
10 If small repairs are being carried out it is best, if possible, not to try and pull the sections apart.
11 Check the mounting rubbers for condition and renew them if necessary.
12 When refitting the exhaust sections, de-burr and grease the sockets and make sure that the clamps are in good condition and not distorted before reconnecting them. Do not fully tighten the clamps at this stage.
13 Connect the system to the manifold together with a new gasket, and reconnect the rubber mountings. Now adjust the exhaust sections in relation to each other so that the mounting rubbers are not strained unduly and each joint is fully engaged.
14 Tighten the clamp bolts and flange nuts. Check that the exhaust system will not knock against any part of the vehicle when deflected slightly in a sideways or upwards direction.

Notes

Chapter 4 Part B:
Fuel/exhaust systems - fuel injection models

Contents

Accelerator cable - removal, refitting and adjustment 3
Accelerator pedal - removal and refitting 4
Air cleaner assembly - removal and refitting 2
Cruise control system - description and component renewal 19
Exhaust manifold - removal and refitting 21
Exhaust system - general information and component removal 22
Fuel gauge sender unit - removal and refitting 10
Fuel injection system - general information 6
Fuel injection system - precautions and depressurisation 7
Fuel injection system - testing, checking and adjustment 12
Fuel injection system components (Bosch CIS/K-Jetronic) - removal and refitting 13
Fuel injection system components (Bosch LH-Jetronic) - removal and refitting 14
Fuel injection system components (Lucas CU14) - removal and refitting 15
Fuel pump - testing, removal and refitting 8
Fuel pump relay - testing and renewal 9
Fuel tank - removal, repair and refitting 11
General information and precautions 1
Inlet manifold - removal and refitting 20
Intercooler - removal and refitting 18
Turbocharger - description and precautions 16
Turbocharger and control components - removal, refitting and adjustment 17
Unleaded petrol - general information and usage 5

Degrees of difficulty

Easy, suitable for novice with little experience	Fairly easy, suitable for beginner with some experience	Fairly difficult, suitable for competent DIY mechanic	Difficult, suitable for experienced DIY mechanic	Very difficult, suitable for expert DIY or professional

Specifications

System type
1981-1985 models*	Bosch CI (Continuous Injection) K-Jetronic fuel injection system
1985-on models*	Bosch LH-Jetronic fuel injection system
1990-on models*	Lucas CU14 fuel injection system

*Approximate dates according to model

Air cleaner
	Non-Turbo models	Turbo models
Automatic temperature control (pre-1985 models only)		
Pre-heated air only	23°C	-5°C
Cold air only	37°C	+5°C

Fuel system data (Bosch CIS/K-Jetronic fuel injection system)

B201 engine

Fuel pump capacity (return pipe discharge in 30 seconds)	900 cc
Control pressure (warm engine/all models)	3.4 to 3.8 bars
Full load control pressure (Turbo models)	2.4 to 2.8 bars
Line pressure:	
Non-Turbo models	4.7 to 4.9 bars
Turbo models	5.3 to 5.5 bars
Minimum leakage pressure after 20 minutes	1.5 bars
Injector opening pressure:	
1979 models to date code 828	2.5 to 3.6 bars
1979 models from date code 829	2.7 to 3.8 bars
1980-on models	3.0 to 4.1 bars

4B•2 Fuel/exhaust systems - fuel injection models

Fuel system data (LH-Jetronic fuel injection system)

B202 engine

Fuel pump capacity (return pipe discharge in 30 seconds)	900 cc
System pressure:	
Turbo models up to 1986	2.5 bars higher than the pressure in the inlet manifold
Turbo models from 1987-on	2.8 bars higher than the pressure in the inlet manifold
B202i models from 1986-on	3.0 bars higher than the pressure in the inlet manifold
Residual pressure (engine switched off)	2.3 bars
Temperature transmitter resistance:	
Bosch:	
0°C	5800 ohms
20°C	2600 ohms
80°C	320 ohms
Lucas:	
-10° ± 1°C	7000 to 11600 ohms
20° ± 1°C	2100 to 2900 ohms
80° ± 1°C	270 to 390 ohms
Auxiliary air valve resistance	40 to 60 ohms
Idle air control valve resistance at 20°C:	
LH 2.2	20 ± 5 ohms
LH 2.4	7 ± 5 ohms
LH 4.2	12 ± 3 ohms
Full load enrichment system:	
Throttle position switch valve angle when contacts close (LH 2.2, and 2.4)	Approximately 72°
Throttle position sensor voltage (LH 2.4.2):	
Idling position	0.25 volts
Wide open throttle	4.0 volts

Fuel system data (Lucas CU14 fuel injection system)

B202 engine

Fuel pump capacity (return pipe discharge in 30 seconds)	900 cc
System pressure	3.0 bars higher than the pressure in the inlet manifold
Residual pressure (engine switched)	2.8 bars
Temperature transmitter resistance:	
0°C	5700 to 5900 ohms
20°C	2400 to 2600 ohms
40°C	1100 to 1300 ohms
60°C	500 to 700 ohms
80°C	300 to 400 ohms
Road speed sensor voltage	0.5 to 11.0 volts
Lambda sensor preheater resistance	Less than 10.0 ohms
Lambda sensor signal voltage	0 to 1.0 volts
Automatic idle control valve (AIC) resistance	40 to 60 ohms
Injector resistance at 20°C	2.0 to 2.8 ohms
Ballast resistor resistance	2.0 to 3.0 ohms
Throttle potentiometer resistance	4.0 to 6.0 ohms
Airflow meter resistance	331 o 341 ohms

Turbocharger

Type	Garrett Airesearch or Mitsubishi
Maximum charging pressure:	
B201:	
1981 to 1985	0.7 ± 0.05 bars
1986	0.75 ± 0.05 bars
1981 cat.	0.50 ± 0.05 bars
1982 cat.	0.60 ± 0.05 bars
1983 to 1986 cat.	0.65 ± 0.05 bars
1987-on cat.	0.67 ± 0.05 bars
B202:	
1984-on non-cat.	0.85 ± 0.05 bars
1984-on cat.	0.75 ± 0.05 bars
B202 LTT:	
1990-on cat.	0.45 +0.0/-0.03 bars
Pressure regulator spring length (approx)	18.0 mm

Turbocharger - continued
Pressure switch actuating pressure:
   B201 .... 0.95 ± 0.05 bars
   B202 .... 1.10 ± 0.05 bars
Turbo shaft bearing clearance:
   Endfloat:
      Garrett (to 1980) .... 0.025 to 0.100 mm
      Garrett (from 1981) .... 0.013 to 0.081 mm
      Mitsubishi .... 0.057 to 0.103 mm
   Radial play:
      Models to 1980 .... 0.075 to 0.180 mm
      Models from 1981 .... 0.076 to 0.145 mm

Turbo fuel boost (1979 models)
Type .... Full-load enrichment, speed and throttle controlled
Throttle valve switch (valve opening with circuit closed) .... 62° (except Turbo APC)
Speed transmitter (closing speed) .... 80 ± 3 mph
Regulator reduced control pressure .... 2.5 to 2.9 bars
CO value with throttle valve switch depressed,
   and idling CO value 1.0 to 2.0% .... 4.0 to 6.0%

Turbo fuel boost (1980-on models)
Type .... Charging pressure controlled full-load enrichment
Warm up regulator/boost control simulated charging
pressure when control pressure reduced:
   Standard .... 0.33 to 0.40 bar
   APC .... 0.15 to 0.32 bar
Reduced pressure with charging pressure over 0.4 bar .... 2.5 to 2.9 bar

Basic boost pressure APC
B201:
   To 1985 .... 0.30 ± 0.03 bar
   1986-on .... 0.32 ± 0.03 bar
B202:
   To 1985 .... 0.40 ± 0.03 bar
   1985-on cat. .... 0.35 ± 0.03 bar

Recommended fuel
Non-Turbo models without a catalytic converter:
   Models to 1984 .... 98 RON leaded only
   Models from 1985 .... 95 RON unleaded
Turbo models without a catalytic converter:
   1981-1983 models without APC .... 98 RON leaded only
   1984-on models without APC .... 95 RON unleaded (but 98 RON unleaded recommended)
   1982-1986 models with APC .... 95 RON unleaded (but 98 RON unleaded recommended)
Models with a catalytic converter .... 95 RON unleaded only

Torque wrench settings

	Nm	lbf ft
Airflow sensor plate retaining bolt (Bosch CIS/K-Jetronic system)	5.3	4
Fuel flow meter (Bosch CIS/K-Jetronic system)	3.5	3
Line pressure regulator plug (Bosch CIS/K-Jetronic system)	14	10
Fuel pump screw top cover (Bosch LH-Jetronic system)	55	41
Knock detector:		
Up to 1982	8	6
1983 to 1986	14	10
1986-on (from engine number G122319)	20	15

1 General information and precautions

The fuel system consists of a fuel tank mounted under the rear of the car with an electric fuel pump immersed in it, a fuel filter, and the fuel feed and return lines. The fuel pump supplies fuel to the fuel distributor (CIS fuel injection) or fuel rail (LH-Jetronic and Lucas CU14 fuel injection), and it is then fed to the injectors which inject the fuel into the inlet tracts near the inlet valves. A fuel filter is incorporated in the feed line from the pump to the engine compartment, to ensure that the fuel supplied to the injectors is clean.

Refer to Section 6 for further information on the operation of the relevant fuel injection system, and to Section 22 for information on the exhaust system.

A cruise control system is fitted as standard equipment on certain models. This system allows the driver to preselect the speed of the car, and then remove his foot from the accelerator pedal. The system is disengaged automatically when the clutch or brake pedals are depressed or when the system is switched off.

A Turbocharger is fitted to certain models. Refer to Section 16 for more information.

4B•4 Fuel/exhaust systems - fuel injection models

⚠️ **Warning:** *Many of the procedures in this Chapter require the removal of fuel lines and connections, which may result in some fuel spillage. Before carrying out any operation on the fuel system, refer to the precautions given in 'Safety first!' at the beginning of this manual, and follow them implicitly. Petrol is a highly-dangerous and volatile liquid, and the precautions necessary when handling it cannot be overstressed.*

2.10a Loosen the clamp bolt . . .

2.10b . . . and remove the clamp completely

2 Air cleaner assembly - removal and refitting

Note: *Some pre-1985 models are fitted with an automatic air temperature control in the air cleaner similar to the type fitted to carburettor engines described in Part A, however with different operating temperatures.*

Removal

CIS/K-Jetronic

1 The air cleaner assembly is located in the left-hand front corner of the engine compartment. First loosen the clips and disconnect the inlet duct from between the airflow meter and the throttle valve housing.
2 Remove the screws securing the lower section of the air cleaner to the airflow meter.
3 Lift the airflow meter slightly taking care not to damage the fuel line, and withdraw the air cleaner element from the body.
4 Remove the holder from the bottom of the body.
5 Loosen the clip and disconnect the inlet duct from the body, then loosen the clamp and remove the assembly from the engine compartment.

LH-Jetronic/Lucas CU

6 Disconnect the wiring from the airflow meter.
7 Release the clip and lift the airflow meter from the top of the air cleaner cover. Recover the O-ring seal.
8 Release the toggle clips and lift the cover from the top of the air cleaner body.

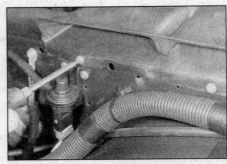

2.11a Unscrew the screws . . .

9 Withdraw the air cleaner element from the body.
10 Loosen the clamp bolt securing the air cleaner body to the inner wing panel. Once loosened the clamp may be removed completely **(see illustrations)**.
11 Using a Torx key unscrew the screws securing the upper inlet duct to the inner wing panel, then lift the complete body from the engine compartment **(see illustrations)**. Loosen the clip and separate the two sections of the assembly.

Refitting

All models

12 Refitting is a reversal of the removal procedure, but wipe clean the interior of the air cleaner body and tap the element to release any accumulated dust and dirt. Refer to Chapter 1 for the refitting procedure for the air cleaner element if necessary.

2.11b . . . and remove the air cleaner assembly from the engine compartment

3 Accelerator cable - removal, refitting and adjustment

Removal

1 Turn the throttle lever at the throttle housing and detach the inner cable at the engine end. On later models, the stop on the end of the inner cable is held by a clamp; prise the clamp open with a screwdriver **(see illustration)**.
2 Loosen the nuts and disconnect the outer cable from the bracket **(see illustration)**.
3 Remove the lower facia panel.
4 Disconnect the cable from the accelerator pedal or arm and withdraw it through the bulkhead. Where applicable unscrew the adapter screws in the bulkhead **(see illustration)**. **Note:** *On right-hand drive models the accelerator pedal extends into the left-hand passenger footwell area.*

3.1 Accelerator cable inner cable connection to the throttle lever

3.2 Accelerator cable and bracket

3.4 Accelerator cable adapter on the left-hand side of the bulkhead

Fuel/exhaust systems - fuel injection models 4B•5

3.5a Securing the stop on the end of the accelerator cable using a pair of pliers

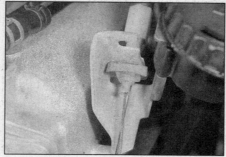

3.5b Accelerator cable and clip on the inlet manifold bracket

Refitting

5 Refitting is a reversal of removal, but adjust the outer cable so that there is a minimal amount of slack in the cable with the accelerator pedal fully released. Tighten the adjustment nuts/clip on completion. Check that the throttle fully opens with the accelerator pedal pressed to the floor. On later models secure the stop in the clamp by squeezing the clamp with a pair of pliers **(see illustrations)**.

4 Accelerator pedal - removal and refitting

Removal

1 Turn the throttle lever at the throttle housing and disconnect the inner cable at the engine end.
2 Inside the car remove the insulation above the pedals, and on left-hand drive models remove the panel from the left-hand side of the centre console. On right-hand drive models remove the insulation on the left-hand side of the passenger. **Note:** *On right-hand drive models the accelerator pedal extends into the left-hand passenger footwell area.*
3 Lift the pedal and disconnect the inner cable after releasing the clip **(see illustration)**.
4 Unscrew the bolts on the pedal bracket and remove the pedal.

Refitting

5 Refitting is a reversal of removal but tighten the mounting bolts securely and adjust the accelerator cable to eliminate slackness.

5 Unleaded petrol - general information and usage

Note: *The information given in this Chapter is correct at the time of writing, and applies only to petrols currently available in the UK. If updated information is thought to be required, check with a Saab dealer. If travelling abroad, consult one of the motoring organisations (or a similar authority) for advice on the petrols available and their suitability for your vehicle.*

The fuel recommended by Saab is given in the Specifications at the start of this Chapter.

The fuel ratings are in AKI units which are an average of RON and MON ratings. RON and MON are different testing standards; RON stands for Research Octane Number (also written as RM), while MON stands for Motor Octane Number (also written as MM).

Under no circumstances should leaded fuel be used in models fitted with a catalytic converter, as this will permanently damage the catalytic converter.

6 Fuel injection system - general information

Bosch CIS (K-Jetronic) fuel injection system

The CI (Continuous Injection) fuel injection system is fitted to models fitted with the B201 engine **(see illustration)**. It is often referred to as a 'mechanical' injection system as it does not make use of an electronic control unit, although simple electrics are used. Each cylinder has one injector and the injectors are continuously injecting fuel all the time that the engine is running. Fuel is supplied from the fuel pump located in the fuel tank through a filter to a distribution metering head which supplies fuel continuously to the injectors but at differing rates of flow according to the needs of the engine. The fuel supply is controlled by an airflow meter incorporating a sensor plate which rises or falls according to the quantity of air entering the engine; the air-to-fuel mixture is therefore regulated automatically. For cold starting, additional enrichment is provided by an independent cold start injector. The mixture is further regulated during the warm-up stage by a warm-up regulator which senses the heat of the engine. Turbo models incorporate a fuel boosting

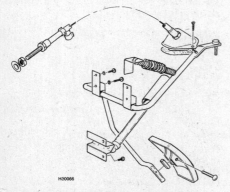

4.3 Accelerator pedal components (left-hand drive models shown)

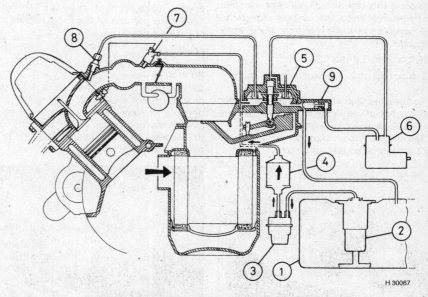

6.1 Diagram of the CIS (K-Jetronic) fuel injection system

1 Fuel tank	4 Fuel filter	7 Cold start valve
2 Fuel pump	5 Fuel distributor	8 Injectors
3 Fuel accumulator	6 Warm-up regulator	9 Line pressure regulator

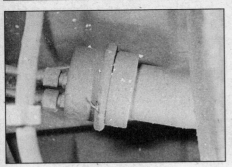

6.2a Fuel accumulator

6.2b Airflow meter and air cleaner unit

system to provide extra fuel for rapid acceleration and to assist internal cooling of the engine during sustained periods of load. On models fitted with a catalytic converter an ECU, throttle position switch, Lambda sensor, and modulating valve are included in the system.

The main components of the system are as follows:

a) *Fuel pump*: the fuel pump is housed in the fuel tank.
b) *Fuel accumulator*: the fuel accumulator is located in the fuel line leading from the fuel tank **(see illustration)**. Its purpose is to maintain pressure in the fuel lines when the engine is switched off. It includes a spring tensioned diaphragm which is displaced when the fuel pump is working. When the engine is switched off, the injectors close and a non-return valve in the fuel pump prevents loss of pressure into the tank. It also prevents fuel vaporizing in the fuel lines and delays system pressure rise at cold starting whcih would result in too much fuel being injected into the engine cylinders.
c) *Fuel filter*: the fuel filter is located beneath the rear of the car, in the fuel line leading from the fuel tank.
d) *Airflow meter*: the airflow meter comprises an air venturi tube in which an air flow sensor plate moves **(see illustration)**. The plate is connected through a lever to the fuel distributor and automatically controls the quantity of fuel-injected according to the engine speed and load.
e) *Warm-up regulator*: the warm-up regulator enriches the mixture during the warm-up stage by reducing the control pressure of the fuel entering the airflow meter. This has the effect of allowing the control plunger to rise so that more fuel is injected into the engine.
f) *Cold start valve*: the cold start valve is actuated via the thermostatic time switch and supplies additional fuel when the starter motor is operating and also for a short period after.
g) *Thermostatic time switch*: the thermostatic time switch operates the cold start valve when the engine temperature is below 45°C independently of the starter motor. It is sensitive to engine temperture in order to vary the operational period of the cold start valve.
h) *Injectors*: each injector consists of a spring loaded disc and needle valve. The injectors will inject fuel continuously when the fuel pressure from the airflow meter exceeds apporoximately 3.3 bars.
i) *Auxiliary air valve*: the auxiliary air valve by-passes the throttle valve and supplies additional air in order to maintain the idling speed. It includes a bi-metal strip and a heating coil.
j) *ECU*: the ECU is fitted to models with a catalytic converter. It processes data from the Lambda sensor and then activates the modulating valve accordingly.
k) *Lambda sensor*: the Lambda sensor is only fitted to models with a catalytic converter. It provides the ECU with constant feedback on the oxygen content of the exhaust gases.
l) *Modulating valve*: the modulating valve is only fitted to models with a catalytic converter. It provides fine tuning of the fuel/air mixture by changing the fuel line pressure.

LH-Jetronic fuel injection system

The LH-Jetronic fuel injection system is a microprocessor-controlled fuel management system which continuously monitors the engine using various sensors, and provides the correct amount of fuel necessary for complete combustion under all engine conditions **(see illustration)**. Data from the sensors is processed in the fuel system electronic control unit (ECU) in order to determine the opening period of the injectors for the exact amount of fuel to be injected into the inlet manifold. The system is of the simultaneous type, which means that all injectors open and close at the same time. They open once for each revolution of the crankshaft, except during cold starting when they open twice for each revolution. Where a catalytic converter is fitted (later models), a Lambda sensor is incorporated in the LH-Jetronic system. On later models the ECU includes an emergency 'Limp-home' mode which is actuated in the event of the airflow meter or temperature sensor failing. Should this occur a 'CHECK ENGINE' warning lamp lights up on the instrument panel.

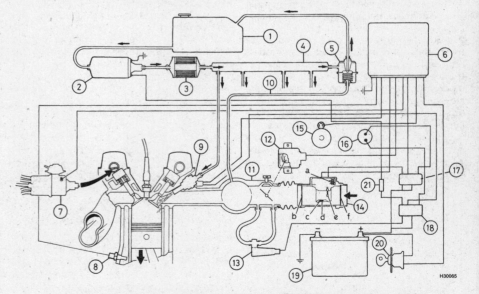

6.3 Bosch LH-Jetronic fuel injection system (as fitted to 1984 models)

1 Fuel tank
2 Fuel pump
3 Fuel filter
4 Fuel injection manifold
5 Fuel pressure regulator
6 Electronic control unit
7 Distributor
8 Temperature transmitter
9 Fuel injector
10 Vacuum line
11 Inlet manifold
12 Throttle position switch
13 Auxiliary air valve
14 Airflow meter
 a) CO adjusting screw
 b) Fine mesh filter
 c) Detector resistor
 d) Filament
 e) Compensation resistor
 f) Fine mesh filter
15 Starter motor
16 Pressure switch
17 System relay
18 Fuel pump relay
19 Battery
20 Ignition switch
21 Line fuse

Fuel/exhaust systems - fuel injection models

The main components of the system are as follows.

a) **ECU:** the electronic control unit controls the entire operation of the fuel injection system. LH2.4 ECU's have an expanded memory in relation to the LH2.2 version.
b) **Injectors:** each fuel injector consists of a solenoid operated needle valve which opens under the commands from the ECU. Fuel from the fuel rail is then delivered through the injector nozzle into the inlet manifold.
c) **Airflow meter:** the airflow meter measures the amount of air entering the engine by means of a hot wire.
d) **Temperature sensor:** the coolant temperature sensor monitors the engine temperature.
e) **Auxiliary air valve:** the auxiliary air valve provides additional air when the engine is cold. It includes a bi-metallic strip.
f) **Idle air control valve:** the idle air control (IAC or AIC) valve controls the volume of air bypassing the throttle butterfly. The valve maintains the idle speed under certain conditions such as when the power steering pump or air conditioning compressor is working or when the alternator is under load. The valve controls the flow of air by-passing the throttle valve and is operated by the ECU. It incorporates a two-stage stepper motor.
g) **Throttle position switch/sensor:** the throttle position switch informs the ECU whether the throttle butterfly is in its open or closed position. As from 1991 a throttle position sensor is fitted instead of the switch; this continuously informs the ECU of the throttle position.
h) **Fuel pump:** the fuel pump is housed in the fuel tank. On later models the pump housing incorporates a separate feed pump which supplies the main fuel pump with pressurised fuel, free of air bubbles.
i) **Fuel pressure regulator:** the fuel pressure regulator is located next to (or attached to) the fuel supply rail. Its purpose is to control the pressure of the fuel in the fuel supply rail in relation to the air pressure in the inlet manifold so that the mixture can be varied by varying the duration of the injector opening. It incorporates a diaphragm and spring which moves according to the air pressure in the inlet manifold and returns excess fuel to the fuel tank.
j) **Fuel filter:** on early models the fuel filter is located in the engine compartment, however on later models it is located beneath the rear of the car next to the fuel tank.
k) **Lambda sensor:** the Lambda sensor provides the ECU with constant feedback on the oxygen content of the exhaust gases. The Lambda sensor is only fitted to models with a catalytic converter.

Lucas CU14 fuel injection system

The Lucas CU14 fuel injection system is fitted to some models manufactured from 1990 onwards. Like the LH-Jetronic system it is a microprocessor-controlled fuel management system and it functions in the same manner.

7 Fuel injection system - precautions and depressurisation

Note: *Refer to the warning notes at the end of Section 1 before proceeding.*

 Warning: The following procedure will merely relieve the pressure in the fuel system - remember that fuel will still be present in the system components, and to take precautions accordingly before disconnecting any of them.

1 The fuel system referred to in this Section is defined as the tank-mounted fuel pump, the fuel filter, the fuel injectors, the fuel rail and the pressure regulator, and the metal pipes and flexible hoses of the fuel lines between these components. All these contain fuel, which will be under pressure while the engine is running and for some time after the engine has been switched off. The pressure must be relieved before any of these components are disturbed for servicing work.
2 Disconnect the battery negative lead.

Bosch CIS/K-Jetronic system

3 Place a suitable container beneath the relevant connection/union to be disconnected, and have a large rag ready to soak up any escaping fuel not being caught by the container.
4 Slowly loosen the connection or union nut (as applicable) to avoid a sudden release of pressure, and position the rag around the connection to catch any fuel spray which may be expelled. Once the pressure is released, disconnect the fuel line. Plug or cap the fuel line/union, to minimise fuel loss and to prevent the entry of dirt into the fuel system.

LH-Jetronic/Lucas CU14 system

5 Position some cloth rags beneath the front or fuel pressure regulator end of the fuel distribution rail, then slowly unscrew the banjo and allow the fuel pressure to escape **(see illustration)**. Tighten the banjo on completion.

8 Fuel pump - testing, removal and refitting

Note: *Refer to the warning note at the end of Section 1 before proceeding.*

Testing

1 To test the fuel pump without removing it, obtain a graduated container with a capacity

7.5 Loosening the banjo on the front of the fuel distribution rail

of at least 900 cc. The pump is located in the fuel tank. Make sure that the ignition is switched off.
2 In the engine compartment, disconnect the return line at the fuel distributor (Bosch CIS/K-Jetronic) or fuel pressure regulator (Bosch LH-Jetronic/Lucas CU14) and connect a hose to the container.
3 On models with the CI (K-Jetronic) fuel injection system, remove the fuel pump relay then connect a bridging wire between terminals 30 and 87 to supply power to the pump. On models with the LH-Jetronic or Lucas CU fuel injection system, remove fuse 30 and connect the bridging wire between the input of fuse 30 and fuse 27, 28 or 29 to supply power to the pump.
4 Run the pump for 30 seconds exactly then stop it. Check that the pump delivers a minimum of 900 cc of fuel.
5 A more comprehensive test may be carried out to ascertain the actual pressure of the fuel. This test should be carried out by your Saab dealer as it involves the use of a pressure gauge. A further test may be carried out by carrying out the procedure given in paragraph 3, then accessing the fuel pump as described later in this Section and lifting the edge of the fuel pump rubber collar to check that fuel is discharging through the safety valve on top of the receiver. If this method is used, an assistant should connect the bridging wire and care must be taken to prevent fuel entering the luggage compartment.

Removal

6 Disconnect the battery negative terminal, and depressurise the fuel injection system as described in Section 7.

Pre-1989 models

7 Remove the cover plate from the luggage compartment and disconnect the wiring from the pump, and on later models the feed pump and fuel flow transmitter **(see illustration)**.
8 Hold the pump still, then disconnect the delivery pipe by unscrewing the banjo union.
9 On early models turn the top plate anti-clockwise and withdraw the pump and O-ring.
10 On late models insert a jointed screwdriver through the special hole and release the clamp. Withdraw the pump.

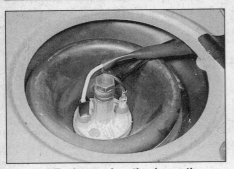

8.7 Fuel pump location beneath the rear floor

8.12a Removing the upper cover . . .

8.12b . . . and lower cover from over the fuel pump

11 On early models, release the clips and remove the splash guard and mounting from the pump. On later models, disconnect the fuel return hose from the pump and release the feed pump wiring from the gland in the tank. Withdraw the pump assembly and remove it from the car.

1989-on models

12 Remove the luggage compartment floor panel and the covers over the fuel pump (see illustrations).
13 Remove the locking cover and disconnect the wiring and fuel lines from the fuel pump (see illustrations). Secure them to one side using a plastic cable-tie.
14 The screw cap must now be loosened. Saab technicians use an elaborate tool to hold the pump stationary while the cap is loosened, however careful use of a large pair of water pump pliers may be used by expanding them onto the serrations of the cap. Make sure that the pump remains stationary and if necessary make up a retaining tool using long bolts on a length of metal bar to hold the inlet and outlet extensions.
15 Unscrew the cap, On non-Turbo models, carefully remove the rubber seal (see illustration).
16 Carefully lift the pump from the tank while tilting it as necessary. Before actually removing it let the fuel drain from it into the tank. On Turbo and Lucas CU models remove the O-ring seal from the top of the tank (see illustrations).

Refitting

Pre-1989 models

17 Refitting is a reversal of removal but note the following points. On early models when reassembling the splash guard to the pump and the pump to the mounting, make sure that the parts take up the attitude shown (see illustration).
18 Make sure that the height of the splash guard is as shown, and as from 1985 models position the fuel pump in the mounting as shown (see illustrations).
19 Install the pump using a new seal (early models) and reconnect the leads and fuel line to it. On later models the suction strainer inlet must be positioned facing to the rear and to the right at an angle of 45°.

1989-on models

Non-Turbo models

20 When inserting the pump into the tank make sure that the pressure and return lines are in alignment with the longitudinal axis of the car, and locate the bottom of the pump in the special ribs in the tank. The mark on top of

8.13a Remove the screw . . .

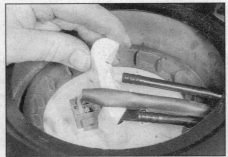

8.13b . . . and lift off the locking cover . . .

8.13c . . . then disconnect the wiring . . .

8.13d . . . the return fuel line . . .

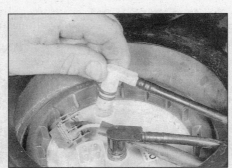

8.13e . . . and the feed fuel line

8.15 Removing the fuel pump retaining screw cap

Fuel/exhaust systems - fuel injection models 4B•9

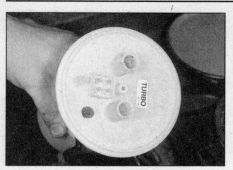

8.16a Removing the fuel pump from the tank

8.16b On Turbo and CU models, remove the O-ring from the top of the tank

the pump must line up with the mark on the top of the tank.
21 Apply vaseline to the new rubber seal and locate it on the top of the pump with the chamfered edge uppermost.
22 Locate the screw cap so that the alignment mark is in line with the other marks. Press down on the cap and turn it half a turn clockwise.

Turbo and Lucas CU models
23 Locate a new O-ring seal in the top of the tank then insert the pump making sure that the alignment mark lines up with the mark on the tank.
24 Refit the screw cap and tighten by hand at this stage.

All models
25 Using the retaining tool described in paragraph 15 hold the pump stationary, then press the cap into contact with the tank flange and tighten it fully. **Do not** allow the pump to move otherwise the ejector pump will be damaged. When fully tightened the alignment marks must not deviate by more than 30°.
26 Reconnect the fuel lines using new O-ring seals. The return line with the non-return valve must be facing the front of the car.
27 Reconnect the wiring and fit the locking cover.
28 Refit the cover and floor panel.

9 Fuel pump relay - testing and renewal

Testing

Bosch CI fuel injection system
1 Remove the cover plate from the luggage compartment for access to the fuel pump.
2 Connect a voltmeter across the positive and negative terminals of the fuel pump.
3 Remove the fuel pump relay (see Chapter 12) and connect a bridging wire between terminals 30 and 87.
4 Check that battery voltage or at least 11.5 volts is reaching the pump. If not check all the relevant wiring between the battery, relay and pump. If not, check the voltage supplied to the relay. Renew the relay if necessary.

LH-Jetronic and Lucas CU fuel injection system
5 Disconnect the wiring from the ECU, airflow meter and auxiliary air valve.
6 Connect a bridging wire between pins 17 and 25 of the ECU.
7 Remove fuse 30 from the fusebox and connect a bridging wire between fuse 30 and fuse 27, 28 or 29 so that the fuel pump is supplied with power.
8 Using a multimeter check that voltage is present between the 'Limp-Home' connector's grey/red lead and an earthing point.
9 If no voltage is present pull back the carpet and remove the relay (see Chapter 12), then check for voltage between the relay pin 30 (red) and an earthing point, and between the relay pin 86 (brown/white) and an earthing point.
10 If there is still no voltage check the red lead between pin 30 and the positive terminal. Also check the brown/white lead between pin 86 and main relay pin 87.
11 If voltage is now present, connect a bridging wire between pin 85 (violet) and an earthing point. Using the multimeter check for voltage on the relay pins 87 (blue/red) and 87b (grey/red).
12 If there is no voltage, renew the relay. If voltage is present, check the grey/red lead from the 'Limp-Home' connector to the relay pin 87b.

Renewal
13 Remove the trim panel from the outside of the right-hand footwell beneath the facia. The fuel pump relay is located just in front of the ECU - there are two relays and the fuel pump relay is the rear one, the front one is the main relay for the fuel injection system.
14 To remove the relay disconnect the wiring and remove the relay from the bracket.
15 Fit the new relay using a reversal of the removal procedure.

10 Fuel gauge sender unit - removal and refitting

Refer to Part A of this Chapter, Section 8.

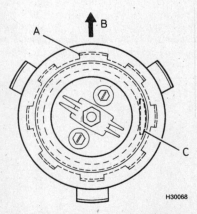

8.17 Correct fitted position of the fuel pump on pre-1989 models

A Wide tongue
B Front of car
C Splash guard indent

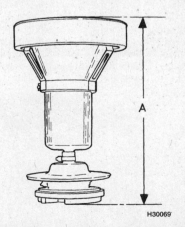

8.18a Splash guard fitting dimension (A) on pre-1989 models

Pre-1980 models = 218 mm
1980-on models = 236 mm

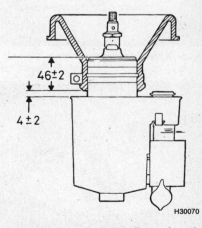

8.18b Fuel pump fitting dimensions for 1985 to 1989 models

Dimensions in mm

4B•10 Fuel/exhaust systems - fuel injection models

11 Fuel tank - removal, repair and refitting

Note: *Refer to the warning note at the end of Section 1 before proceeding.*

Removal

1 Chock the front wheels, then jack up the rear of the car and support on axle stands (see "*Jacking and Vehicle Support*").
2 Before removing the fuel tank, all the fuel must be drained from the tank. Since a fuel tank drain plug is not provided, it is therefore preferable to carry out the removal operation when the tank is nearly empty. Before proceeding, remove the filler cap and syphon or hand-pump the remaining fuel from the tank. Alternatively the fuel tank outlet hose can be disconnected and the pump located in the tank can be used to remove the fuel for access to the fuel gauge sender unit and fuel pump.
3 Disconnect the battery negative lead.
4 Remove the floor panel from the luggage compartment; then prise out the rubber cover or remove the screws and cover as applicable.
5 Disconnect the wires from the fuel gauge sender unit and pump, noting their location for correct refitting.
6 Disconnect the filler and ventilation hoses, and the fuel supply and return hoses. Remove the fuel line clips where applicable.
7 Support the tank then release the straps by unscrewing the nuts **(see illustrations)**.
8 Lower the fuel tank and withdraw it from under the car.
9 Remove the fuel gauge sender unit and fuel pump with reference to Sections 8 and 10.
10 Models manufactured from 1984-on are fitted with a roll-over valve located in the right-hand side of the luggage compartment. The valve prevents fuel escaping from the tank in the event of an accident. To remove the valve, first remove the trim panel then disconnect the hose and unscrew the mounting screws.

Repair

11 If the tank is contaminated with sediment or water, swill it out with clean fuel. If the tank leaks or is damaged, it should be repaired by

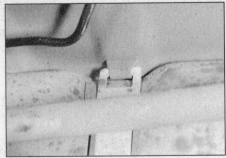

11.7a Upper strap securing the fuel tank to the floor of the car

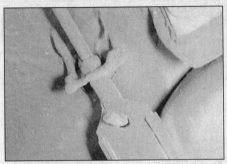

11.7b Lower strap and threaded hook securing the fuel tank to the bracket

specialists or alternatively renewed. *Do not, under any circumstances, attempt to repair a fuel tank yourself.*

Refitting

12 Refitting is a reversal of removal, but make sure that all hoses are securely fitted and correctly routed. On completion, refill the tank with fuel and check for signs of leakage prior to taking the car out on the road.

12 Fuel injection system - testing, checking and adjustment

Note: *Refer to the warning note at the end of Section 1 before proceeding.*

Bosch CIS (K-Jetronic) fuel injection system

Testing and checking

Auxiliary air valve
1 Make sure that the engine is cold, and on B type engines, that the safety circuit wiring plug at the airflow sensor is disconnected.
2 If the cold engine control pressure is also to be checked disconnect the wiring from the warm up regulator.
3 On H type engine models remove the fuel pump relay and connect terminals 30 and 87 with a wire link. Switch on the ignition.
4 Using a torch and mirror, check that there is an opening (oval in shape) within the valve. Switch on the ignition and observe that the opening closes after a period of about five minutes **(see illustration)**.

Warm up regulator
5 Disconnect the lead from the warm-up regulator and bridge the contacts in the lead plug with a voltmeter **(see illustration)**.
6 On B type engines disconnect the safety circuit wiring plug at the airflow sensor, and switch on the ignition.
7 On H type engines remove the fuel pump relay and connect terminals 30 and 87 with a wire link. Switch on the ignition.
8 Check that the recorded voltage is at least 11.5 volts.

Fuel injectors
9 Remove the flexible bellows from the airflow sensor.
10 Unscrew the retaining plate screws and remove the injectors from the inlet manifold (leaving the fuel lines connected) and place them in a clean container **(see illustration)**.
11 On B type engines disconnect the safety circuit wiring plug at the airflow sensor, and switch on the ignition.
12 On H type engines remove the fuel pump relay and connect terminals 30 and 87 with a wire link. Switch on the ignition.
13 Raise the lever in the airflow sensor and observe the fuel spray from the injection nozzles. If the spray is particularly restricted or poorly defined, the valves should be removed for cleaning by your dealer or renewed.
14 Switch off the ignition and wipe the ends of the injector nozzles dry. Lift the airflow sensor lever and check the nozzles for leakage of fuel. If a drop of fuel forms in under fifteen seconds, then the injectors must be cleaned or renewed.

12.4 Auxiliary air valve location on an early model

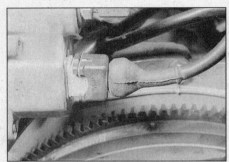

12.5 Warm-up regulator lead connector

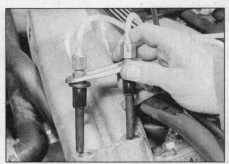

12.10 Removing the fuel injectors

Fuel/exhaust systems - fuel injection models 4B•11

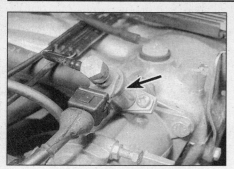

12.15 Cold start valve location

12.21 Cold start valve thermo time switch location

Cold start valve
15 Disconnect the lead from the cold start valve and unscrew it from the throttle valve housing (see illustration). Do not disconnect the fuel line.
16 Connect two leads between the cold start valve terminals and a main beam terminal of one of the headlamps and earth.
17 On B type engines disconnect the safety circuit wiring plug at the airflow sensor, and switch on the ignition.
18 On H type engines remove the fuel pump relay and connect terminals 30 and 87 with a wire link. Switch on the ignition.
19 Place the cold start valve in a container and have an assistant switch the headlamps to main beam for a period not exceeding thirty seconds. During this time, fuel should spray out of the cold start valve.
20 With the headlamps switched off but the ignition still switched on (fuel pump operating) dry the cold start valve nozzle and check that no fuel leaks from the valve during a period of 60 seconds. If it does, renew the valve.

Thermo-time switch
21 When the engine temperature is below 45°C current flows while the starter motor is actuated. To check that the switch closes when the starter is actuated, connect a test-lamp in series across the contacts of the connector plug of the cold start valve (see illustration).

Airflow sensor
22 Check that there is a steady resistance as the lever is raised, but no resistance as it is lowered.
23 The plate should be central within the venturi and positioned as shown when at rest (see illustration). Adjustment is made by bending the wire clip beneath the plate after removing the unit.

Adjustments
Throttle valve
24 Remove the air hose and check that the throttle valve is central in its bore (see illustration).
25 Loosen the locknut and unscrew the throttle stop screw so that the valve is completely closed.
26 Turn the screw until it just touches the lever, then turn it a further third of a turn and tighten the locknut. The clearance between the valve and bore should now be approximately 0.05 mm.

Idle speed and mixture
27 The idle speed and mixture adjustments are described in Chapter 1.

Bosch LH-Jetronic and Lucas CU14 fuel injection system

Testing, checking and adjustment
28 Full testing of the fuel injection system should be carried out by a Saab dealer. A diagnostic connector is incorporated in the vehicle wiring circuit, into which a Saab special electronic diagnostic tester can be plugged. The tester will locate the fault quickly and simply, alleviating the need to test all the system components individually.
29 The following adjustments are only possible on certain models fitted with the LH-Jetronic fuel injection system. The idle speed may be adjusted on engines not fitted with

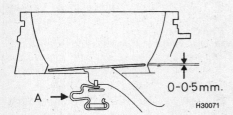

12.23 Correct 'rest' position for the airflow sensor plate

A = wire clip

12.24 Air hose located on the airflow sensor

idle control, and the idle mixture may be adjusted on engines not fitted with a catalytic converter. On models with the Lucas CU14 system, the idle speed and mixture are automatically controlled by the ECU. The *checking* of idle speed and mixture is possible on all models by using a tachometer and exhaust gas analyser.
30 Before checking and adjusting the idle speed or mixture settings, always check the following first:
 a) Check the ignition timing (models without a catalytic converter) (Chapter 1).
 b) Check that the spark plugs are in good condition and correctly gapped (Chapter 1).
 c) Check that the accelerator cable is correctly adjusted (Chapter 4A or 4B).
 d) Check that the crankcase breather hoses are secure, with no leaks or kinks (Chapter 1).
 e) Check that the air cleaner filter element is clean (Chapter 1).
 f) Check that the exhaust system is in good condition (Chapter 4A or 4B).
 g) If the engine is running very roughly, check the compression pressures (Chapter 2A).

31 Take the car on a journey of sufficient length to warm the engine to normal operating temperature.
Note: *Checking/adjustment should be completed as soon as possible so that the engine is still at its normal operating temperature. If the radiator electric cooling fan operates, first wait for the cooling fan to stop. Clear any excess fuel from the inlet manifold by racing the engine two or three times to between 2000 and 3000 rpm, then allow it to idle again.*
32 Ensure that all electrical loads are switched off, then stop the engine and connect a tachometer to it following the manufacturer's instructions. Where a tachometer is fitted to the instrument panel, this may be used instead. If the idle mixture is to be checked, connect an exhaust gas analyser in accordance with the manufacturer's instructions.

Models with LH-Jetronic fuel injection (except models with idle air control and catalytic converter)
33 The idle speed adjustment screw is located on the throttle housing. Start the engine and allow it to idle, then check that the idle speed is as given in the Specifications. If adjustment is necessary, loosen the locknut and adjust the screw in to reduce the speed or out to increase the speed. Tighten the locknut after making the adjustment.
34 The idle mixture adjustment screw is located on the airflow meter and may be hidden under a tamperproof cap. First remove the cap. With the engine idling at the correct idling speed, check that the CO level is as given in the Specifications. If adjustment is necessary, use an Allen key or screwdriver to turn the mixture adjustment screw in or out (in very small increments) until the CO level is as

given in the Specifications. Turning the screw in (clockwise) richens the mixture and increases the CO level, turning it out will weaken the mixture and reduce the CO level.
35 If necessary readjust the idling speed.
36 Temporarily increase the engine speed, then allow it to idle and recheck the settings.
37 When adjustments are complete, stop the engine and disconnect the test equipment.

Other models with LH-Jetronic and Lucas CU14 systems

38 The idle speed is controlled by an idle air control valve in conjunction with the system ECU. With the engine at normal operating temperature, check that the idle speed is as given in the Specifications. No adjustment is possible.
39 The idle mixture is controlled by the Lambda sensor in conjunction with the system ECU. With the engine at normal operating temperature, check that the idle mixture is as given in the Specifications. No adjustment is possible.

13 Fuel injection system components (Bosch CIS/K-Jetronic) - removal and refitting

Note: *Refer to the warning note at the end of Section 1 before proceeding.*

Fuel accumulator

Removal

1 The fuel accumulator is located on the side of the fuel tank. First jack up the rear of the car and support on axle stands (see *"Jacking and Vehicle Support"*).
2 Disconnect the fuel lines and pull the accumulator from its bracket.

Refitting

3 Refitting is a reversal of removal, but make sure that when installing the new accumulator the fuel lines are correctly connected (fuel pump line nearest the edge of the accumulator).

Airflow meter (mixture control unit)

Removal

4 The airflow meter (or mixture control unit) is mounted on the air cleaner and consists of a fuel distributor and airflow sensor. To remove the unit, disconnect the fuel lines from the fuel distributor and disconnect the lines to the injectors then the control pressure line **(see illustration)**.
5 Remove the flexible bellows which run between the airflow sensor and the throttle valve housing.
6 Remove the retaining bolts and lift the mixture control unit from the air cleaner. If the fuel distributor is to be separated from the airflow sensor, take care that the control plunger does not fall out. Do not handle the

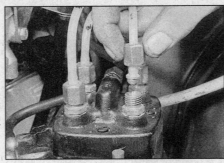

13.4 Disconnecting the fuel lines from the fuel distributor

plunger, however if this is unavoidable, clean it before installation with fuel.

Refitting

7 Refitting is a reversal of removal. When refitting the fuel distributor (which is a sealed unit and must be renewed complete if faulty), check that the O-ring seal is in position and tighten its three bolts no tighter than 4 Nm.

Pressure regulator

Removal

8 The line pressure regulator is screwed into the fuel distributor and should not be removed unless absolutely essential. Pressure adjustment is carried out by varying the thickness of the shims on the end of the spring. To remove the regulator, unscrew and remove the plug then remove the shims, spring and piston. Note that the plug includes a shut-off valve for the return fuel from the control pressure circuit. O-rings are fitted to both the plug and the piston.

Refitting

9 Refitting is a reversal of the removal procedure, but renew the O-rings and tighten the plug to the specified torque.

Airflow sensor

Removal

10 The airflow sensor can be dismantled if the complete mixture control unit is first removed from the engine and the fuel distributor detached from it as described earlier in this Section.
11 Remove the lower plastic section from the airflow sensor and extract the two stop bracket screws. Remove the bracket, spring, insulation and connectors, as applicable.
12 Extract the retaining screws and remove the sensor plate.
13 Extract the circlips from the lever seating and remove the shims, seals, spring and balls.
14 Remove the counter-weight screw and press out the pivot.
15 Withdraw the lever, counter-weight and adjustment arm.

Refitting

16 Reassemble the stop bracket in the reverse order to dismantling and tighten the screws to 6 Nm.
17 Fit the counter-weight to the lever tightening the screw only fingertight. Place the adjustment arm in the lever so that the socket headed screw on the arm is visible.
18 Apply Silicone grease to both bearings and install the lever/arm assembly in the airflow sensor housing and insert the pivot.
19 Apply grease to the balls and fit them together with the spring, seals, shims and circlips. The spring goes on the side which has the longer bearing seat and the circlips should have their sharp edges facing outwards.
20 Centre the sensor plate in the air venturi and then centre the lever so that the threaded holes in both components are in alignment with each other.
21 Tighten the counter-weight screw to a torque of 5 Nm and then fit the sensor plate screw and tighten it to 5.3 Nm. Check that the lever can be moved without binding.
22 Adjust the rest position of the sensor plate by bending the wire loop on the stop bracket underneath the airflow sensor.
23 Now set the position of the adjustment arm. To do this, use a depth gauge and measure the distance between the face with which the fuel distributor mates and the needle bearing. This should be between 18 and 19 mm, if not turn the mixture control screw using an Allen key.
24 With the airflow meter (consisting of the fuel distributor and airflow sensor) installed to the engine, the mixture should be checked as described in Chapter 1.

Injectors

Removal

25 Clean away any surrounding dirt, then disconnect the fuel line using two spanners to prevent the injector turning
26 Unscrew the retaining plate screw and lift the plate from the injector.
27 Remove the injector from the inlet manifold and remove the O-ring seal.

Refitting

28 Refitting is a reversal of removal, but renew the O-ring seal and tighten the screw and fuel line union securely.

Warm-up regulator

Removal

29 The warm up regulator is located by the thermostat housing. First clean the area around the regulator.
30 Disconnect the fuel lines and wiring noting their location, then unbolt and remove the regulator using an Allen key **(see illustration)**.

Refitting

31 Refitting is a reversal of removal, but tighten the mounting and unions securely.

Auxiliary air valve

Removal

32 The auxiliary air valve is located by the

Fuel/exhaust systems - fuel injection models 4B•13

13.30 Fuel lines on the warm-up regulator

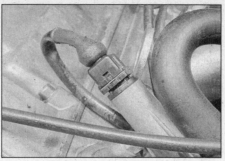

13.32 Wiring connector on the auxiliary air valve

distributor on B type engines and by the warm up regulator on H type engines. Disconnect the hoses and wiring and remove the mounting screws **(see illustration)**.

Refitting

33 Refitting is a reversal of removal.

14 Fuel injection system components (Bosch LH-Jetronic) - removal and refitting

Electronic control unit

Removal

1 Disconnect the battery negative lead.

Pre-1981 models

2 With the bonnet open, remove the cover behind the bulkhead panel on the left-hand side of the engine compartment.
3 Unscrew the mounting screws and withdraw the electronic control unit from the bulkhead.
4 Release the clip and disconnect the multiplug connector from the ECU.

1981-on models

5 Open the right-hand front door and remove the sill scuff plates. Lift the door weatherstrip from the bottom front of the door aperture.
6 Remove the plate securing the carpet on the wheel arch and pull back the carpet.
7 Unscrew the mounting screws and remove the ECU from the wheel arch.
8 Disconnect the wiring connector from the ECU by releasing the plastic clip.

Refitting

9 Refitting is a reversal of the removal procedure.

Temperature sensor

Removal

10 Unclip the crankcase ventilation hose from the cylinder head cover.
11 Remove the fuel pressure regulator as described later in this Section.
12 Disconnect the wiring from the temperature sensor **(see illustration)**.
13 Drain the cooling system as described in Chapter 1. Alternatively, have ready a suitable bung to plug the aperture when the sensor is removed. If this method is used, take great care not to damage the sensor aperture, and do not use anything which will allow foreign matter to enter the cooling system.
14 Unscrew the temperature sensor from the inlet manifold flange.

Refitting

15 Refitting is a reversal of removal but renew the copper washer if necessary, clean the threads of the sensor and inlet manifold, and tighten the sensor to the specified torque.

Throttle position switch/sensor

Removal

16 Disconnect the wiring plug from the throttle position switch/sensor on the throttle housing.
17 On pre-1991 models, mark the position of the switch on the throttle housing.
18 Unscrew the mounting screws and remove the switch/sensor.

Refitting and adjustment

1991-on models

19 Refit the sensor and tighten the mounting screws, then reconnect the wiring plug. On these models the sensor continuously monitors the throttle position, and is adaptive in that it 'learns' the idling and full throttle positions of the throttle.

Pre-1991 models

20 Refit the switch to the throttle housing and insert the mounting screws finger-tight.
21 If the original switch is being refitted, align the previously made marks and tighten the screws.
22 If adjustment is required, disconnect the ECU wiring connector and remove the cover. Connect an ohmmeter to pins 3 (grey) and 11 (black) on the connector and check that the circuit is closed (ie there is continuity). If not, check the grey lead from pin 3 to the throttle position switch, the black lead from pin 11 to earth, and the black/white lead from the throttle position switch to an earthing point.
23 Slowly open the throttle and check that the circuit is broken immediately the throttle moves from its idling position. if not, loosen the mounting screws and adjust the switch as necessary. Tighten the screws after making the adjustment.
24 Connect the ohmmeter between pins 11 (black) and 12 (green/red). Open the throttle by 72° so that it is nearly fully open, and check that the circuit closes at this position. If this is not the case, the switch is probably faulty. To prove this check the green/red lead between pin 12 of the connector and pin 3 of the throttle position switch, the black lead between pin 11 of the ECU and earth, and the black lead between pin 2 of the throttle position switch and an earthing point. If these wires appear to be in good order, renew the switch.

Throttle housing

Removal

25 Refer to Chapter 1 and drain off approximately 2 litres of coolant, so that the coolant level is below the throttle housing. Alternatively use hose clamps to clamp the coolant hoses leading to the throttle housing.
26 Loosen the clip and disconnect the air inlet hose from the throttle housing **(see illustration)**.
27 Disconnect the accelerator cable and remove it from the throttle housing with reference to Section 3.
28 Disconnect the wiring plug from the throttle position switch/sensor.
29 Loosen the clips and disconnect the coolant hoses. If necessary, plug the hoses to prevent loss of coolant.
30 Disconnect the air hoses from the auxiliary air valve.

14.12 Disconnecting the wiring from the temperature sensor on the inlet manifold flange

14.26 Disconnecting the air inlet hose from the throttle housing

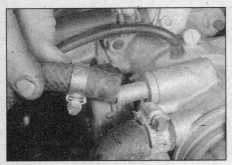  

14.32 Disconnecting the vacuum hoses from the throttle housing

14.41a Disconnecting the wiring from the injectors

14.41b Cable-tie securing wiring to the fuel supply rail

31 Where necessary, remove the Turbo pressure (discharge) pipe.
32 Disconnect the vacuum hoses from the throttle housing **(see illustration)**.
33 Unscrew the three mounting nuts and remove the throttle housing from the inlet manifold.

Refitting

34 Refitting is a reversal of the removal procedure. On completion check and if necessary adjust the throttle position switch and the accelerator cable. Top-up the cooling system with reference to "Weekly Checks".

Airflow meter

Removal

35 Loosen the clip and disconnect the throttle housing hose from the airflow meter.
36 Disconnect the wiring multiplug from the airflow meter.
37 Release the toggle spring clips then lift the airflow meter assembly from the air cleaner lower body.

Refitting

38 Refitting is a reversal of the removal procedure, but align the slot with the air cleaner body.

Fuel supply rail and injectors

Note: *Refer to the warning notes at the end of Section 1 before proceeding.*

Removal

Note: *If a faulty injector is suspected, before condemning the injector, it is worth trying the effect of one of the proprietary injector-cleaning treatments.*

39 Thoroughly clean the fuel supply rail and injectors externally and dry with compressed air or lint-free rags.
40 Depressurise the system (see Section 7).
41 Disconnect the crankcase ventilation hose from the valve cover and the electrical connections at each injector. Release the cable-ties securing the wiring to the fuel supply rail **(see illustrations)**.
42 Release the cable harness cable-tie at the point where the fuel supply rail is attached to the inlet manifold.
43 Disconnect the fuel hose banjo unions at each end of the fuel supply rail. Wipe up any spilled fuel immediately **(see illustration)**.
44 Undo the bolts securing the fuel supply rail to the inlet manifold and lift off the supply rail complete with injectors.
45 Slide off the retaining clips and remove the injectors using a twisting action.

Refitting

46 Refitting is a reversal of the removal procedure, however before locating new rubber O-rings in the inlet manifold, apply a little petroleum jelly to them to facilitate entry of the injectors.

Fuel pressure regulator

Note: *Refer to the warning notes at the end of Section 1 before proceeding.*

Removal

47 Clean the area surrounding the fuel pressure regulator to prevent entry of dust and dirt into the fuel system.
48 Unscrew the union nut and disconnect the fuel supply hose from the regulator. Wipe up any spilled fuel immediately.
49 Disconnect the vacuum hose from the regulator.
50 Unbolt and remove the regulator, complete with bracket, from the cylinder head and disconnect the fuel return hose. Note the location of the LH-Jetronic system earthing leads **(see illustrations)**.
51 Remove the fuel pressure regulator from the bracket.

Refitting

52 Refitting is a reversal of removal.

Auxiliary air valve

Removal

53 Disconnect the wiring plug from the auxiliary air valve.
54 Loosen the clips and disconnect the two hoses from the valve.
55 Unscrew the mounting bolts and remove the auxiliary air valve.

Refitting

56 Refitting is a reversal of removal.

AIC (automatic idle control) valve

Note: *This valve is also referred to as an IAC (idle air control) valve.*

Removal

57 Disconnect the wiring plug from the AIC valve **(see illustration)**.
58 Loosen the clips and disconnect the hoses.

14.43 Disconnecting the fuel hose banjo union at the front of the fuel supply rail

14.50a Disconnecting the fuel return hose from the fuel pressure regulator

14.50b Removing the fuel pressure regulator

Fuel/exhaust systems - fuel injection models 4B•15

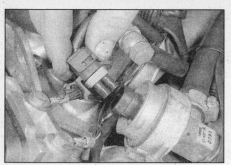

14.57 Disconnecting the wiring from the AIC valve

59 Unscrew the mounting bolts and remove the AIC valve **(see illustration)**.

Refitting
60 Refitting is a reversal of removal.

15 Fuel injection system components (Lucas CU14) - removal and refitting

Electronic control unit
1 Refer to Section 14 **(see illustration)**.

Temperature sensor
2 Refer to Section 14.

Throttle position sensor
Removal
3 Disconnect the AIC valve hose from the bottom of the throttle housing.

14.59 Removing the automatic idle control valve

4 Loosen the clip and disconnect the by-pass valve hose from the throttle housing.
5 Disconnect the wiring plug **(see illustration)**.
6 Unscrew the mounting screws and remove the sensor.

Refitting
7 Refitting is a reversal of removal.

Throttle housing
8 Refer to Section 14.

Airflow meter
Removal
9 Loosen the clip and disconnect the rubber elbow from the airflow meter.
10 Disconnect the wiring plug **(see illustration)**.
11 Release the spring clip then lift the airflow meter from the air cleaner cover **(see illustrations)**.

12 Remove the O-ring seal from the air cleaner cover **(see illustration)**.

Refitting
13 Refitting is a reversal of removal but check that the O-ring seal is correctly located in the groove. Locate the seal on the airflow meter before refitting the unit to the air cleaner cover.

Fuel supply rail and injectors
14 Refer to Section 14.

Fuel pressure regulator
15 Refer to Section 14.

AIC (automatic idle control) valve
16 Refer to Section 14.

16 Turbocharger - description and precautions

Description
The Turbocharger fitted to certain models increases engine efficiency by raising the pressure in the inlet manifold above atmospheric pressure. Instead of the air simply being drawn into the engine, it is forced in.

Energy for the operation of the turbocharger comes from the exhaust gas. The gas flows through a specially-shaped housing (the turbine housing) and in so doing, spins the turbine wheel. The turbine wheel is attached to a shaft, at the end of which is

15.1 Lucas fuel injection ECU location

15.5 Disconnecting the wiring plug from the throttle position sensor

15.10 Disconnecting the wiring plug from the airflow meter

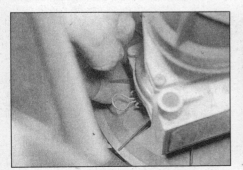

15.11a Release the spring clip . . .

15.11b . . . and lift the airflow meter from the top of the air cleaner cover

15.12 Removing the O-ring seal from the air cleaner cover

4B•16 Fuel/exhaust systems - fuel injection models

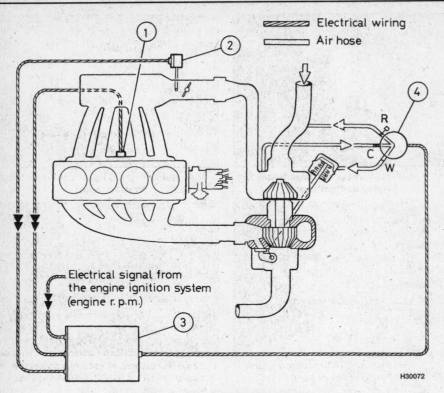

16.6 Diagram of the APC system

1 Knock detector
2 Pressure transducer
3 Control unit
4 Solenoid valve

another vaned wheel known as the compressor wheel. The compressor wheel spins in its own housing, and compresses the inlet air on the way to the inlet manifold.

On some models the compressed air passes through an intercooler located between the turbocharger and the inlet manifold. This is an air-to-air heat exchanger, mounted on the left-hand side of the engine compartment. The purpose of the intercooler is to remove from the inlet air some of the heat gained in being compressed. Because cooler air is denser, removal of this heat further increases engine efficiency.

Boost pressure (the pressure in the inlet manifold) is limited by a wastegate, which diverts the exhaust gas away from the turbine wheel in response to a pressure-sensitive control valve. An overpressure switch, mounted on a bracket and connected to the inlet manifold by a hose, breaks the fuel pump circuit in the event of the charging pressure exceeding the pre-set limit. Over-revving of the engine is prevented by a rotor having a built-in centrifugal cut-out device which breaks the ignition circuit on B type engines. On other engines the fuel pump relay cuts-out at engine speeds above 6000 rpm. A gauge indicating the charge pressure is mounted on the top of the instrument panel.

Prior to October 1987 models the turbocharger is not cooled except by the lubricating oil, however from this date on it is water-cooled in order to reduce the temperature of the bearings.

Later models are fitted with an Automatic Performance Control system **(see illustration)**. Its basic purpose is to adjust the ignition timing and turbocharger operation so that the engine runs at peak efficiency and economy whatever quality of fuel is being used. This is necessary because fuels of the same grade can sometimes give different engine characteristics.

The APC system incorporates its own separate electronic control unit (ECU) located under the rear seat on models up to and including 1985, and on the left-hand front side of the engine compartment on 1986-on models.

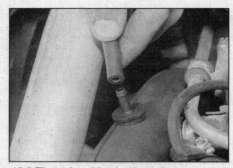

16.8 The solenoid valve hose is connected to the turbocharger output air duct

The APC system monitors knocking in the engine by means of a knock detector, and included in the system is a pressure transducer (which monitors pressure in the inlet manifold) and a solenoid valve (which controls the turbocharger wastegate). The solenoid valve senses the turbocharger output pressure by means of a hose connected to the output air duct **(see illustration)**.

The turbo shaft is pressure-lubricated by an oil feed pipe from the main oil gallery. The shaft 'floats' on a cushion of oil. A drain pipe returns the oil to the sump.

Accurate testing of the system should be carried out by a Saab dealer as special instrumentation is required.

Precautions

The turbocharger operates at extremely high speeds and temperatures. Certain precautions must be observed, to avoid premature failure of the turbo, or injury to the operator.

Do not operate the turbo with any of its parts exposed, or with any of its hoses removed. Foreign objects falling onto the rotating vanes could cause excessive damage, and (if ejected, personal injury).

Do not race the engine immediately after start-up, especially if it is cold. Give the oil a few seconds to circulate.

Always allow the engine to return to idle speed before switching it off - do not blip the throttle and switch off, as this will leave the turbo spinning without lubrication.

Allow the engine to idle for *several* minutes before switching off after a high-speed run.

Observe the recommended intervals for oil and filter changing, and use a reputable oil of the specified quality. Neglect of oil changing, or use of inferior oil, can cause carbon formation on the turbo shaft, leading to subsequent failure.

17 Turbocharger and control components - removal, refitting and adjustment

Turbocharger unit

Note: *On 16-valve engines the turbocharger unit can be removed together with the exhaust manifold, then separated on the bench (refer to Section 21).*

Removal

1 Remove the battery as described in Chapter 5. On post October 1987 models drain the cooling system as described in Chapter 1.
2 On pre-1981 models remove the charge pressure regulator as described later in this Section, and disconnect the hose between the compressor and throttle housing.
3 On 1981-on models remove the suction and pressure connections from the turbocharger/compressor and loosen the pre-heating hose **(see illustrations)**.

Fuel/exhaust systems - fuel injection models 4B•17

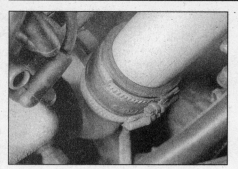
17.3a Loosen the clip attaching the pressure hose/tube to the turbocharger...

17.3b ... and remove the tube from the engine compartment

17.4a Removing the heatshield from the turbocharger

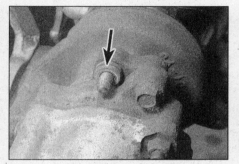
17.4b Note the location of the spacer on the turbocharger

17.5a Removing the Turbocharger unit (B201 H-type engine)

17.5b Turbo unit oil return pipe on the cylinder block (B201 H-type engine)

4 Unscrew the bolts attaching the elbow to the exhaust downpipe and disconnect the downpipe. Also unscrew the nut and bolt and remove the heatshield - note the location of the spacer (see illustrations).

5 On post October 1987 models disconnect the coolant hoses. On all models unbolt the oil supply and return/drain pipes then unbolt the turbo unit from the exhaust manifold. Recover the gaskets and seals (see illustrations).

6 The turbo unit is not reparable; if it is defective, an exchange unit will have to be fitted.

Refitting

7 Refitting is the reverse of the removal

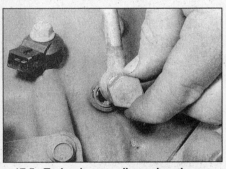
17.5c Turbocharger oil supply union on left-hand side of the block (B202 engine)

17.5d Disconnecting the turbocharger coolant hoses (post October 1987 models)

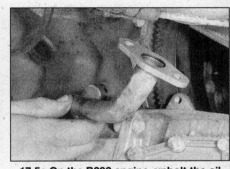
17.5e On the B202 engine, unbolt the oil return/drain pipe from the turbocharger...

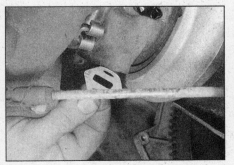
17.5f ... and recover the gasket

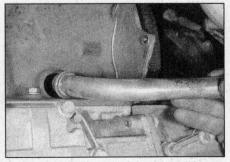
17.5g Disconnect the oil return/drain pipe from the cylinder block ...

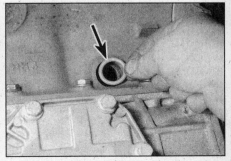
17.5h ... and recover the O-ring seal (arrowed)

4B•18 Fuel/exhaust systems - fuel injection models

17.7a Priming the turbo unit oil inlet

17.7b Fitting the oil supply pipe to the turbo unit

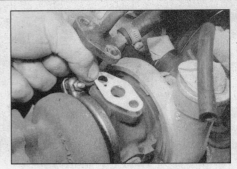

17.7c ALWAYS use new gaskets

17.9 Turbo charge pressure regulator and connecting pipes

17.10a Removing the charge pressure regulator

17.10b Turbo unit with charge pressure regulator removed

procedure. Fill the oil inlet port with engine oil and always use new gaskets (see illustrations). Before starting the engine disable the ignition coil by connecting a lead between the coil HT lead and a suitable earthing point, then turn the engine over on the starter motor for about 30 seconds to prime the turbo unit lubricating system.

Charge pressure regulator

Removal

Pre-1981 models

8 Disconnect the battery leads, then remove the battery, heat shield and battery tray as described in Chapter 5.
9 Disconnect the air cooling pipe and the exhaust pressure pipe from the regulator (see illustration). Remove the bolts securing the exhaust manifold flange to the regulator and collect the sealing ring.
10 Release the tabs of the locking plate, undo the securing bolts and remove the bellows pipe. Remove the turbo securing bolts, after releasing the lockplate tabs, and lift away the charge pressure regulator (see illustrations).

1981-on models

11 The charge pressure regulator is mounted on the turbocharger by a bracket, and the operating arm from the diaphragm unit is fitted with a tamperproof seal.
12 Removal is by disconnecting the bracket and housing, and link pipe.

Dismantling (pre-1981 models)

Note: *The pressure regulator on 1981-on models is a sealed unit and cannot be dismantled.*

13 Undo the securing nuts and bolts and remove the diaphragm housing cover. Measure and record the length of the compressed spring (see illustrations).
14 Mark the position of the valve and outer spring seat to ensure that they can be refitted in their original positions at reassembly.
15 Hold the valve seat with a pair of pipe grip pliers, then using a ring spanner undo the securing nut and remove the spring outer seat, the spring and the spring inner seat (see illustrations).
16 Insert a screwdriver or similar tool in the groove in the valve disc to prevent it from turning, then undo the diaphragm nut. Remove the outer diaphragm washer, plain washer, diaphragm housing, gasket, heat shield, gasket, bearing housing and gasket (see illustrations).

17.13a Removing the diaphragm housing cover from the charge pressure regulator

17 Inspect all the parts for wear and damage. Renew the diaphragm if it shows any signs of deterioration. If the valve and valve seat need grinding this is a job for your Saab garage as a special valve spindle guide is required, also a valve seat cutter and a valve grinding machine.

Reassembly (pre-1981 models)

18 Reassembly is the reverse of the dismantling procedure, but ensure that the diaphragm inner ridge locates in the groove in the diaphragm washer and that the marks made at dismantling are aligned. The outer ridge on the diaphragm locates in the groove in the diaphragm housing. The spring is fitted with its closed coils towards the diaphragm and the end of the coil nearest the housing cover at the six o'clock position.

Refitting and adjusting

19 Refitting is the reverse of the removal

17.13b Measuring the length of the compressed spring

17.15a Removing the securing nut, spring outer seat ...

17.15b ... spring ...

17.15c ... inner seat and gasket ...

17.16a ... nut., and washers ...

17.16b ... diaphragm ...

17.16c ... inner washer ...

17.16d ... diaphragm housing and heat shield ...

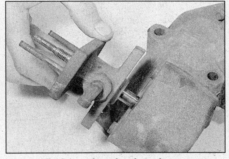

17.16e ... bearing housing ...

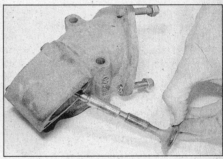

17.16f ... and valve

procedure. Always use a new gasket between the turbo unit and the charge pressure regulator and lock the nuts with the lockplates.

20 Refit the charge pressure regulator on the engine then check, and if necessary adjust, the charging pressure as described in the following paragraphs.

21 On 1981-on models check the basic setting as follows. Hold the control arm in the 'closed' position with the operating arm disconnected, then adjust the length of the operating arm until it aligns with the control arm. Screw the end fitting onto the operating arm six turns then tighten the locknut. Connect the operating arm to the control arm and fit the circlip.

22 A pressure gauge and length of hose are required for checking the charge pressure. Connect the gauge to the connector on the inlet manifold by means of the hose and position the gauge beside the charge pressure gauge on the top of the instrument panel.

23 Drive the car to warm it up to normal operating temperature and then check the charging pressure as follows.

24 Drive the car in 3rd gear (manual gearbox) or position 1 (automatic transmission) at an engine speed of less than 1500 rpm.

25 Accelerate to full throttle, pedal right down to the floor, then as the engine speed approaches 3000 rpm apply the brakes (throttle pedal still pressed down) to put the car under full load at 3000 rpm and note the maximum pressure recorded on the pressure gauge.

26 If the pressure is not as given in the Specifications, adjust the regulator as follows.

Pre-1981 models

27 Remove the heat shield and disconnect the exhaust pressure line from the diaphragm housing cover. Remove the diaphragm housing cover.

28 Hold the spring seat with a pair of pipe grip pliers and slacken the locknut, then adjust the tension of the spring by turning the seat clockwise or anti-clockwise as required.

29 After adjustment, tighten the locknut.

30 Refit the diaphragm cover, use a new gasket, exhaust pressure line and heat shield.

31 Recheck the charging pressure as described previously.

1981-on models

32 Disconnect the operating arm and lengthen it to reduce the pressure, shorten it to increase the pressure. Note that two turns are equivalent to 0.04 bar.

Charge pressure regulator (pre-1981 models)

Cleaning (pre-1981 models only)

33 The diaphragm housing of the charge pressure regulator should be cleaned every 30 000 miles (48 000 km).

4B•20 Fuel/exhaust systems - fuel injection models

17.38 Turbo pressure switch location

34 Remove the heat shield. Disconnect the exhaust pressure line and remove the diaphragm housing cover.
35 Using a suitable brush, thoroughly clean the diaphragm housing.
36 Refit the cover, exhaust line and heat shield.

Pressure switch and gauge

Checking

37 With the engine at idle speed, disconnect the inlet manifold to pressure gauge hose at the inlet manifold connector, then connect a pressure gauge and a suitable pump to the hose.
38 Use the pump to apply pressure to the switch and check that the engine cuts out at the specified pressure, see the Specifications **(see illustration)**.
39 The pressure gauge on the top of the instrument panel can be checked at the same time as the pressure switch. The needle should be within the wide orange zone at maximum charging pressure. At cut-out pressure the needle should be in front of the limit between the orange and red zones.

Fuel boost device (pre-1982 models)

Description and checking

1979 models

40 The fuel boost device is fitted in the fuel supply on turbocharged engines to supply extra fuel under heavy engine load conditions **(see illustration)**. It also improves engine cooling at continual high speed. The device consists of a solenoid valve and a pressure regulator which are connected in parallel with the control pressure regulator in the control pressure system.
41 The pressure regulator is preset approximately 1 kgf/cm2 below the pressure of the control pressure regulator therefore the control pressure will drop from 3.7 kgf/cm2 to 2.7 kgf/cm2 when the solenoid valve opens. This raises the position of the control plunger and thereby boosts the fuel supply to the engine. The solenoid is actuated either by a switch at the throttle valve or by means of the speed transmitter which is connected to the speedometer cable and which closes the circuit when the speed exceeds 80 mph (130 km/h).

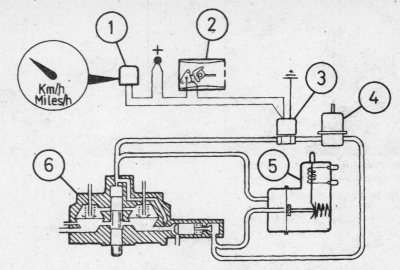

17.40 Diagram of the fuel boosting system on 1979 models

1 Speed transmitter 3 Solenoid (valve) 5 Control pressure regulator
2 Throttle valve switch 4 Pressure regulator 6 Fuel distributor

42 The fuel boost device is checked when measuring the CO emission as follows.
a) Warm up the engine to the normal operating temperature, connect a CO meter to the exhaust and check that the CO content is within the specification.
b) Depress the actuating arm on the throttle valve switch and check that the CO value increases to the figure in the Specifications.
c) Release the actuating arm and check that the CO value is to specification.

1980/81 models

43 The function of the system is similar to that for 1979 models, but the components are different **(see illustration)**.
44 To check the high load (partially open throttle) first run the engine to normal operating temperature and connect a CO meter. Remove the pressure hose from the butterfly housing and connect a pressure gauge and air pump to the hose. Plug the butterfly housing connection. With the engine

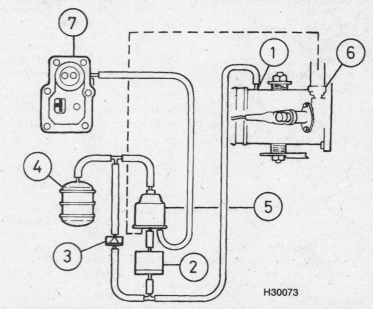

17.43 Diagram of the fuel boosting system on 1980/81 models

1 Pressure outlet in throttle housing (before the butterfly)
2 Delay valve
3 Non-return valve
4 Pressure tank
5 Electrical control valve
6 Throttle valve switch (62° throttle opening)
7 Warm up regulator/boost control

Fuel/exhaust systems - fuel injection models

17.72 Unscrew the screws securing the APC control unit to the inner wing panel

idling increase the pressure to 0.8 bar and check that the CO is between 2% and 6%.

45 To check the wide-open throttle setting first remove the air pump and check that the CO reading returns to normal. Then depress the throttle valve switch and check that the CO reading is between 4% and 6%.

46 To check the residual pressure connect a pressure gauge between the non-return valve and the pressure tank. Using an air pump increase the pressure to 0.8 bar and check that the pressure does not drop below 0.6 bar after 2 minutes.

APC system

General description

47 The automatic performance control (APC) system provides continual adjustment of the Turbo system by means of a knock detector which together with other components controls the charge pressure. Differences in fuel octane rating can therefore be compensated for automatically to provide optimum performance.

48 Accurate testing of the system should be carried out by a Saab dealer as special instrumentation is required.

Charge pressure regulator - basic setting

49 This is identical to that described previously in this Section except that the operating arm end fitting should be screwed onto the operating arm 3.5 turns instead of 6.0 turns, and also the APC solenoid wires should be disconnected while checking the pressure on the road.

APC system - checking

50 Remove the pressure transducer from the inlet manifold and plug the aperture.
51 Connect a pressure gauge in the hose to the pressure transducer (1982 models) or pressure switch (1983-on models).
52 Connect an air pump to the hose then start the engine and run it at 2000 rpm.
53 Increase the pressure with the air pump to 0.5 bar and check that the solenoid valve makes a 'chattering' sound. This proves the system is functioning correctly.

Pressure transducer - checking

54 Up to and including 1982 models disconnect the transducer hose at the inlet manifold and connect a pressure gauge. As from 1983 remove the bellows between the front and rear centre consoles, remove the padding and disconnect the transducer hose then connect the pressure gauge. As from 1986 models remove the cover from the wiring connector and measure the resistance between terminal 23 (black/white) and terminal 10 (green/red).
55 Connect an air pump to the transducer. Disconnect the wiring.
56 Connect an ohmmeter to the pressure switch terminals and check that the resistance is 10.0 ohms at atmospheric pressure.
57 Increase the pressure to 1.0 bar then reduce it to 0.6 bar while tapping the transducer.
58 Now measure the resistance which should be 88.0 ohms.

Solenoid valve - checking

59 Disconnect the wiring from the valve and the hose from the turbo compressor inlet to the valve R connection.
60 Connect a 12 volt supply to the solenoid terminals and check that it is possible to blow through the unit. With the supply disconnected it should not be possible to blow through the unit.
61 Check that the orifice in the solenoid C outlet to the turbo charger is unobstructed.

Knock detector - removal and refitting

62 Disconnect the wiring and unscrew the unit from the cylinder block. As from 1983 models the unit is retained with a centre bolt.
63 Refitting is a reversal of removal, but first clean and oil the threads of the unit or centre bolt. Note the different tightening torques according to year. Where the unit is retained with a centre bolt, the wiring connector must be offset towards the front of the car by 20°.

Pressure transducer (up to and including 1982 models) - removal and refitting

64 Disconnect the wiring and hose, then remove the screws and withdraw the unit.
65 Refitting is a reversal of removal.

Pressure transducer (1983-on models) - removal and refitting

66 Remove the bellows between the front and rear centre consoles, remove the padding, and disconnect the hose and wiring.
67 Remove the bracket and unbolt the unit.
68 Refitting is a reversal of removal.

Control unit - removal and refitting

Up to and including 1985 models

69 Lift the rear seat cushion and remove the multi-connector from the unit.
70 Remove the cross-head screws and withdraw the unit from the floor.
71 Refitting is a reversal of removal.

1986-on models

72 With the bonnet open unscrew the mounting screws securing the APC control unit to the left-hand inner wing panel (see illustration).
73 Disconnect the multiplug connector from the control unit, then remove the control unit (see illustrations).
74 Refitting is a reversal of removal.

Solenoid valve - removal and refitting

75 Disconnect the wiring and hoses after identifying them for position, remove the cross-head screws, and withdraw the unit. (see illustration)
76 Refitting is a reversal of removal.

Fuel boost device - checking

77 Run the engine to normal operating temperature and connect a CO meter.
78 Disconnect the hose from the throttle housing and connect a pressure gauge and air pump. Plug the throttle housing connection.
79 Increase the pressure to 0.6 bar with the engine idling and check that the CO reading is between 2% and 6%.

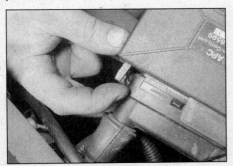

17.73a Release the tab . . .

17.73b . . . and disconnect the multiplug connector from the APC control unit

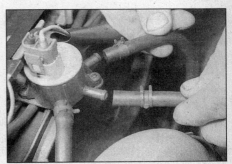

17.75 Disconnecting the hoses from the solenoid valve

4B•22 Fuel/exhaust systems - fuel injection models

18.4 Removing the plastic cover from the top of the intercooler

18.5 Hose and tube connecting the turbocharger to the intercooler

18 Intercooler - removal and refitting

Removal

1 An intercooler is fitted between the turbo compressor and the inlet manifold on later models (B201 from 1986, B202 from 1984). Its purpose is to reduce the temperature of the compressed air, and so to improve engine efficiency and reduce detonation.

Early models

2 To remove the intercooler, first remove the plate which secures it to the radiator crossmember. Also as from 1986 models remove the plastic shroud.
3 Unbolt the intercooler, disconnect the pipes from it and remove it.

Later models

4 Remove the air cleaner assembly as described in Section 2. Also remove the plastic cover from the top of the intercooler **(see illustration)**.
5 Loosen the clips and disconnect the pressure hoses and intermediate tubes from the intercooler **(see illustration)**.
6 Unscrew the upper mounting nut, then lift the intercooler upwards from the upper mounting stud and lower mounting rubbers **(see illustrations)**. Take care not to damage the matrix of the intercooler.
7 If necessary the mounting bracket may be unbolted from the front valance **(see illustration)**.

All models

8 If the intercooler is contaminated with oil, flush it out after removing the plugs provided.

Refitting

9 Refitting is a reversal of removal.

19 Cruise control system - description and component renewal

Description

1 The cruise control system allows the driver to preselect the speed of the car, then remove his foot from the accelerator pedal. The system is deactivated when the clutch or brake pedals are depressed or when the main cruise control switch is switched off.
2 The main components of the system are as follows.
a) *Speed transducer: the speed transducer is located on the rear of the instrument panel. It monitors the speed of the car and sends signals to the vacuum pump to raise or reduce the vacuum in order to adjust the speed of the car.*
b) *Electronic control unit: the electronic control unit (ECU) is located on a bracket beneath the left-hand side of the facia. The ECU stores the speed transmitted by the speed transducer when the 'Set speed' button is depressed. The system is not operative at speeds below 21 mph.*
c) *Relay: power to the cruise control system is via a relay located near the ECU.*
d) *Vacuum pump and valve: the vacuum pump is located in the left-hand side of the engine compartment on the inner wing panel. When the cruise control system is activated the pump runs but the actual vacuum produced is regulated by the vacuum valve. In certain circumstances the vacuum pump will stop in order to reduce the vacuum in the vacuum regulator.*
e) *Vacuum regulator: the vacuum regulator is located on a bracket beneath the left-hand side of the facia. It incorporates a ball-link chain which is clipped to the accelerator cable at the top of the accelerator pedal arm. On right-hand drive models the accelerator pedal is extended into the left-hand footwell, and the vacuum regulator is located under the facia behind the glovebox position.*
f) *Clutch and brake pedal switches: switches are provided on the brake and clutch pedals in order to disengage the system when either pedal is depressed. Each switch has a double function on early models - electrical and vacuum. If either pedal is depressed, the electrical connection will be broken and the vacuum pump will stop. At the same time the system vacuum will be dissipated by the internal port of the switch. On later models the vacuum is dissipated by an unit located on the left-hand side of the engine compartment.*
g) *Selector switch: the selector switch is located on the left-hand side of the steering wheel and is incorporated in the direction indicator switch. It incorporates functions for SET SPEED, RESUME, OFF and ON.*

3 In the event of a fault in the cruise control system, first check all relevant wiring for security. Further testing is best left to a Saab dealer who will have the necessary diagnostic equipment to find the fault quickly.

Component renewal

Speed transducer

4 Remove the instrument panel as described in Chapter 12, then remove the upper section of the panel for access to the transducer.
5 Refitting is a reversal of removal.

18.6a Intercooler upper mounting nut

18.6b Lifting the intercooler from the lower mounting rubbers

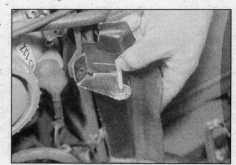

18.7 Removing the intercooler mounting bracket from the front valance

Fuel/exhaust systems - fuel injection models 4B•23

19.6 Removing the cruise control electronic control unit

19.8 Cruise control relay and control unit location

Vacuum release unit

13 Disconnect the wiring from the vacuum release unit located on the left-hand side of the engine compartment (see illustration).
14 Unscrew the screws and disconnect the hose either at the unit or at the connector beneath the vacuum pump (see illustrations).
15 Refitting is a reversal of removal.

Clutch and brake pedal switches

16 The switches are located on the same bracket as the stop light switch. Access is gained by removing the lower facia panel. When refitting the switches, adjust them so that there is a clearance of 1.0 mm between the threaded part of the switch and the actuator tip. Make sure that the pedals are fully released before adjusting the switches.

Selector switch

17 The switch is integral with the direction indicator switch. Refer to Chapter 12 for the removal and refitting procedure.

20 Inlet manifold - removal and refitting

Removal

1 Disconnect the battery negative lead.
2 Disconnect the vacuum hoses and crankcase ventilation hoses from the inlet manifold (see illustrations).
3 Drain the cooling system as described in Chapter 1, then disconnect the coolant hoses from the inlet manifold and, on B202 engines, from the coolant pipe located below the inlet

19.10 Removing the cruise control vacuum pump

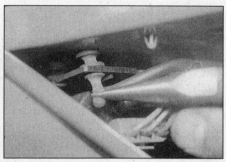

19.12 Make sure the rubber mountings are pulled right through the mounting plate

Electronic Control Unit

6 Remove the lower panel from under the facia panel, then disconnect the wiring, unscrew the mounting screws and remove the unit (see illustration). On early models the ECU is mounted on the same bracket as the APC system.
7 Refitting is a reversal of removal.

Relay

8 Disconnect the wiring from the relay, then unscrew the mounting screw and remove the relay from under the facia (see illustration).
9 Refitting is a reversal of removal.

Vacuum pump

10 The pump is located on the left-hand side of the bulkhead in the engine compartment. Disconnect the wiring and vacuum hoses, then release the pump from the rubber mountings (see illustration).
11 Refitting is a reversal of removal, but make sure that the rubber mountings are pulled through the holes in the mounting plate so that the shoulders are located either side of the plate (see illustration).

Vacuum regulator

12 The vacuum regulator is located on a bracket near the left-hand end of the accelerator pedal (on RHD models). Access is gained by removing the glovebox (right-hand drive models) or lower facia panel (left-hand drive models). Take care not to damage the rubber gaiter when removing and refitting the regulator - the regulator can be removed from the bracket or the bracket can be unbolted

from the bulkhead (see illustration). The ball-link chain should be fitted to the accelerator cable clip near the top of the accelerator pedal so that all slack is taken up without moving the cable.

19.13 Removing the cruise control vacuum regulator and bracket

19.15a Removing the vacuum release unit from the inner wing panel

19.14 Disconnecting the wiring from the vacuum release unit

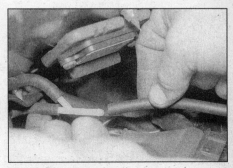

19.15b Disconnecting the vacuum hoses at the connector

4B

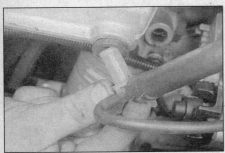
20.2a Vacuum and crankcase ventilation hoses from the rear of the inlet manifold

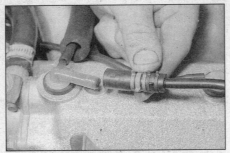
20.2b Disconnecting the hoses from the top of the inlet manifold

20.3a Coolant hose connection to the inlet manifold

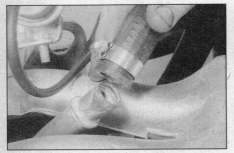
20.3b Disconnecting the coolant hose from the coolant pipe below the inlet manifold

20.4 Unbolting the engine oil dipstick tube support bracket from the inlet manifold

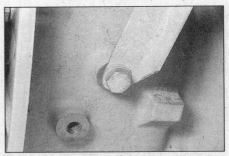
20.7a Inlet manifold bracket mounting on the cylinder block

manifold **(see illustrations)**. Alternatively fit a hose clamp to the coolant hose connected to the inlet manifold.

4 Loosen the clip and disconnect the air duct from the throttle housing, then unbolt the engine oil dipstick tube support bracket from the inlet manifold **(see illustration)**.

5 Remove the throttle housing from the inlet manifold as described in Sections 13, 14 or 15 (as applicable). If preferred, the hoses and accelerator cable may remain attached and the housing positioned to one side.

6 Remove the fuel supply rail from the inlet manifold (Sections 13, 14 or 15, as applicable).

7 Unscrew the mounting bolts securing the inlet manifold to the cylinder head - also unbolt the support bracket from the cylinder block after removing the inspection cover for access to the bolts. Note that the front engine mounting bracket is located is located on the manifold front bolts **(see illustrations)**.

8 Withdraw the inlet manifold from the cylinder head. Recover the gasket **(see illustrations)**.

20.7b Front engine lifting bracket on the front inlet manifold bolts on an early model

20.7c Front engine lifting bracket on the front inlet manifold bolts on a later model

20.8a Removing the inlet manifold . . .

20.8b . . . and gasket (B201 type-H engine)

20.8c Removing the inlet manifold . . .

20.8d . . . and gasket (B202 engine)

Fuel/exhaust systems - fuel injection models 4B•25

20.9 Tightening the inlet manifold mounting bolts

Refitting

9 Refitting is a reversal of removal but fit a new gasket and tighten the mounting bolts to the specified torque **(see illustration)**.

21 Exhaust manifold - removal and refitting

Removal

1 Remove the battery (see Chapter 5).
2 On carburettor models remove the warm air hose from the shroud on the exhaust manifold **(see illustration)**. If necessary unbolt and remove the shroud.
3 On Turbo models, remove the turbocharger as described in Section 17. Alternatively the unit can be separated from the manifold on the bench later. If the latter course of action is taken, it will be necessary to disconnect the Turbo oil pipes at this stage.
4 Remove the distributor as described in Chapter 5.
5 Remove the power steering pump and bracket (where applicable) as described in Chapter 10 **(see illustration)**.
6 On early models unbolt and remove the manual transmission oil dipstick tube from the exhaust manifold **(see illustration)**.
7 Unscrew the nuts and detach the exhaust downpipe from the exhaust manifold flange.
8 Unscrew the mounting nuts and remove the spacers, then ease the manifold off its mounting studs and remove it from the cylinder head **(see illustrations)**. Where the manifold is in two sections on the B202 engine the centre section should be removed first, noting that on the remaining section sleeves are fitted beneath all the nuts.
9 Remove the heat shields and gaskets from the cylinder head studs **(see illustrations)**.

Refitting

10 Refitting is a reversal of removal, but renew the gaskets or single gasket and clean

21.2 Warm air hose connection to the exhaust manifold

21.5 Power steering pump bracket location

21.6 Manual transmission oil dipstick tube attached to the exhaust manifold

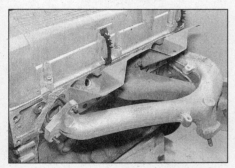

21.8a Removing the exhaust manifold (B201 H-type engine)

21.8b Exhaust manifold mounting nut and spacer (B202 engine)

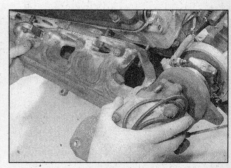

21.8c Removing the exhaust manifold complete with turbocharger (B202 engine)

21.9a Heat shield located on the cylinder head studs

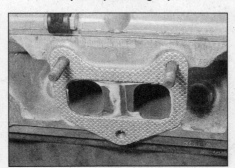

21.9b Centre exhaust manifold gasket

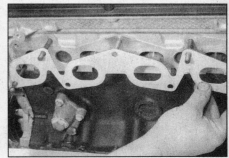

21.9c Removing the exhaust manifold gasket (B202 engine)

21.10 Tightening the exhaust manifold mounting nuts

22.9 Clamp connecting the two rear sections of the exhaust together

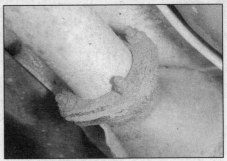

22.12 Joint connecting the catalytic converter to the intermediate exhaust pipe

the contact surfaces of the cylinder head and exhaust manifold. Tighten all nuts and bolts to the specified torque **(see illustration)**. Where removed on Turbo models, refit the turbocharger with reference to Section 17.

22 Exhaust system - general information and component removal

General information

1 The exhaust system is of three or five section type; non-catalytic converter models have a three-section exhaust and catalytic converter models have a five-section exhaust.
2 The centre and rear mountings are of flexible rubber type.
3 Examination of the exhaust pipe and silencers at regular intervals is worthwhile as small defects may be repairable with a proprietry product. If small holes or other defects are left, without attention, the whole system, or at least part of it, will probably need to be completely renewed at a later stage. Also, any leaks, apart from the noise factor, may cause poisonous exhaust gases to find their way into the car which could be unpleasant, to say the least, even in mild concentrations. Prolonged inhalation could cause sickness and giddiness or even death.

Component renewal

4 As the sleeve connections and clamps at each end of the intermediate section are usually very difficult to separate, it is better to remove the complete system from the car then separate the sections with full access to the connections. However, the sleeve connection at the rear of the intermediate section must be separated first in order to remove the system.
5 On models fitted with a catalytic converter, it is not necessary to remove the complete system if the downpipe or catalytic converter is being removed. Both of these items can be removed leaving the intermediate pipe, intermediate silencer and rear sections connected together.
6 To remove the complete system first raise the front and rear of the car and support on axle stands (see "Jacking and Vehicle Support"). Alternatively, position the car over an inspection pit, or on car ramps.
7 Disconnect the centre and rear mounting rubbers.
8 On catalytic converter models, disconnect the wiring from the Lambda sensor located in the front downpipe. To prevent damage to the sensor, remove it completely as described in Part 4C of this Chapter.
9 Just in front of the rear axle loosen the clamp securing the two sections of the exhaust together. Carefully tap around the sleeve joint in order to release the two sections, then withdraw the rear section from the rear of the car **(see illustration)**.
10 Unscrew the nuts and disconnect the downpipe from the exhaust manifold flange or turbocharger flange, remove the gasket, then withdraw the downpipe, catalytic converter and intermediate pipe (where fitted) and intermediate silencer from under the car.
11 To separate the intermediate pipe and silencer, loosen the clamp and separate the two sections from each other.
12 To separate the catalytic converter from the downpipe and intermediate pipe, unscrew the bolts/nuts, separate the joints and recover the rear gasket and front sealing ring **(see illustration)**. Note: *do not drop the catalytic converter as it contains a fragile ceramic element.*
13 When separating a damaged section, cut away the damaged part from the adjoining good section rather than risk damaging the latter.
14 If small repairs are being carried out it is best, if possible, not to try and pull the sections apart.
15 Check the mounting rubbers for condition and renew them if necessary.
16 When refitting the exhaust sections, de-burr and grease the sockets and make sure that the clamps are in good condition and not distorted before reconnecting them. Do not fully tighten the clamps at this stage. Assemble the catalytic converter together with a new gasket and sealing ring and tighten the nuts/bolts.
17 Connect the system to the manifold/turbocharger flange together with a new gasket, and reconnect the rubber mountings. Now adjust the exhaust sections in relation to each other so that the mounting rubbers are not strained unduly and each joint is fully engaged.
18 Tighten the clamp bolts and flange nuts. Check that the exhaust system will not knock against any part of the vehicle when deflected slightly in a sideways or upwards direction.

Chapter 4 Part C:
Emission control systems

Contents

Catalytic converter - general information and precautions 3
Emissions control systems - testing and component renewal 2
General information 1

Degrees of difficulty

Easy, suitable for novice with little experience		Fairly easy, suitable for beginner with some experience		Fairly difficult, suitable for competent DIY mechanic		Difficult, suitable for experienced DIY mechanic		Very difficult, suitable for expert DIY or professional	

Specifications

EGR system (certain markets only)
EGR valve:
  Starts to open . . . . Approximately 1900 rpm
  Fully open . . . . Approximately 2500 rpm

Lambda sensor
Preheater resistance:
  Bosch . . . . 4.0 ± 2.0 ohms
  Lucas . . . . Less than 10.0 ohms

Torque wrench settings	Nm	lbf ft
Lambda sensor	40	29
EGR valve	8	6

1 General information

All models from 1985-on can use unleaded petrol (see Chapter 4 Specifications). Various other features are built into the fuel system to help minimise harmful emissions. All models have a crankcase emissions control system, and all later models are equipped with a three-way catalytic converter. Other systems fitted include exhaust gas recirculation, evaporative emissions control, deceleration device, dashpot, delay valve and pulse air, however the actual components fitted will depend on the model, year of manufacture, and territory in which it is operated.

The emissions systems function as follows.

Crankcase emissions control

To reduce emissions of unburned hydrocarbons from the crankcase into the atmosphere, the engine is sealed. The blow-by gases are drawn from the crankcase up through the block and cylinder head and then through a hose to the air cleaner, from where they are drawn into the engine and burnt in the combustion chambers. A small bore hose also feeds the gases direct to the inlet manifold on all but Turbo models (see illustration).

When manifold depression is high (idling, deceleration) the gases will be sucked positively out of the crankcase. Under conditions of low manifold depression (acceleration, full-throttle running) the gases are forced out of the crankcase by the (relatively) higher crankcase pressure; if the engine is worn, the raised crankcase pressure (due to increased blow-by) will cause some of the flow to return under all manifold conditions.

On carburettor engines a flame guard is fitted to the hose at the air cleaner.

Exhaust emissions control

To minimise the amount of pollutants which escape into the atmosphere, later models with the Bosch CI, Bosch LH-Jetronic and Lucas CU14 fuel injection systems are fitted with a catalytic converter in the exhaust system. The catalytic converter system is of the closed-loop type, in which a Lambda sensor in the exhaust manifold provides the fuel-injection system ECU with constant feedback on the oxygen content of the exhaust gases (see illustration). This enables the ECU to adjust the mixture to provide the best possible conditions for the converter to operate.

1.3 Crankcase ventilation hoses on the valve cover

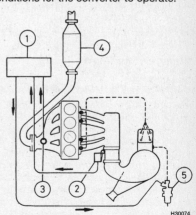

1.6 Lambda sensor system on models with the CIS fuel injection system

1 ECU (control unit)
2 Throttle valve switch, full-load enrichment
3 Lambda sensor
4 Catalytic converter
5 Modulating valve

4C•2 Emission control systems

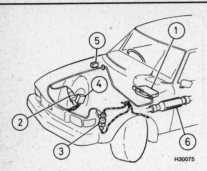

1.7 Lambda sensor components on CIS fuel injection models

1. ECU
2. Lambda sensor
3. Modulating valve
4. Throttle position switch
5. Relay
6. Catalytic converter

Because the Bosch CI fuel injection system is a mechanical system, the Lambda system on these models incorporates a separate ECU together with a throttle position switch and modulating valve. The modulating valve varies the fuel control pressure in response to signals from the ECU, and regulates the mixture accordingly **(see illustration)**.

The Lambda sensor has a built-in heating element, controlled by the ECU through the Lambda sensor relay, to quickly bring the sensor's tip to an efficient operating temperature. The sensor's tip is sensitive to oxygen, and sends the ECU a varying voltage depending on the amount of oxygen in the exhaust gases; if the inlet air/fuel mixture is too rich, the sensor sends a high-voltage signal. The voltage falls as the mixture weakens. Peak conversion efficiency of all major pollutants occurs if the inlet air/fuel mixture is maintained at the chemically-correct ratio for the complete combustion of petrol - 14.7 parts (by weight) of air to 1 part of fuel (the 'stoichiometric' ratio). The sensor output voltage alters in a large step at this point, the ECU using the signal change as a reference point, and correcting the inlet air/fuel mixture accordingly.

Exhaust gas recirculation system

This system circulates a proportion of the exhaust gas back into the combustion chambers during certain operating conditions, in order to reduce harmful exhaust gas emissions. This is achieved by effectively reducing the peak temperatures reached in the combustion chambers by the introduction of the inert exhaust gas. Two systems were fitted - a mechanical two-port system, and an electronic system controlled by the LH-Jetronic ECU. The mechanical two-port system is operated by vacuum and the system operates in conjunction with engine temperature **(see illustration)**.

On later models with the electronic system, if a fault occurs in the EGR system, the 'CHECK ENGINE' warning light will come on

the instrument panel and the corresponding fault code will be stored in the LH-Jetronic ECU.

Evaporative emissions control system

To minimise the escape into the atmosphere of unburned hydrocarbons, an evaporative emissions control system is fitted to certain models. The system is referred to as the evaporative-loss control device (ELCD). The fuel tank filler cap is sealed, and a charcoal canister is mounted on the front left-hand side of the car beneath the left-hand wing to collect the petrol vapours generated in the tank when the car is parked. The vapours are stored until they can be cleared from the canister (under the control of the fuel system ECU via the purge valve, into the inlet tract, to be burned by the engine during normal combustion.

To ensure that the engine runs correctly when it is cold and/or idling, and to protect the catalytic converter from the effects of an over-rich mixture, the purge control valve is not opened by the ECU until the engine has warmed up, and the engine is under load; the valve solenoid is then modulated on and off to allow the stored vapour to pass into the inlet tract.

Deceleration device

The device assists combustion during engine overrun in order to prevent the emission of unburned hydrocarbons.

The electric type incorporates an engine speed transmitter and a solenoid operated idling stop. During engine overrun the idling speed is increased to around 1550 rpm if the speed of the car exceeds 18 mph (30 kph).

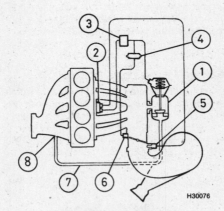

1.9 EGR two-port system

1. EGR valve
2. PVS valve
3. Holding valve (white end towards PVS valve)
4. Release valve
5. Two-port vacuum connection
6. Release vacuum connection
7. EGR crosspipe
8. Restriction

1.16 Throttle dashpot

The vacuum type incorporates a spring-loaded valve actuated by inlet manifold vacuum.

Dashpot system

The dashpot system is a deceleration device to prevent the throttle shutting too quickly which would otherwise result in unburnt hydrocarbons **(see illustration)**.

Delay valve system

This unit is located in the distributor vacuum advance pipe and its purpose is to delay ignition advance during acceleration so reducing nitrous oxide emissions. Note that the white end of the valve must be towards the distributor.

Pulse air system

Some models are equipped with a pulse air system which supplies air to the downstream side of the exhaust valves in order to complete oxidation of any unburned hydrocarbons present in the exhaust gases. The system uses the pressure pulses in the exhaust manifold to draw the air from the air cleaner through two check valves. There are two check valves, one to numbers 1 and 4 cylinders, and the other to numbers 2 and 3 cylinders.

2 Emissions control systems - testing and component renewal

Crankcase emissions control

1 The components of this system require no attention, other than to check at regular intervals that the hoses, and where applicable the flame guard, are clear and undamaged.

Catalytic converter

Testing

2 The performance of the catalytic converter can be checked only by measuring the exhaust gases using a good-quality, accurately calibrated exhaust gas analyser as described in Chapter 1. On some models a plug is provided in the exhaust downpipe for checking the CO content upstream of the catalytic converter.

Emission control systems 4C•3

3 If the CO level at the tailpipe is too high, the vehicle should be taken to a Saab dealer so that the fuel injection and ignition systems, including the Lambda sensor, can be thoroughly checked using special diagnostic equipment. Once these have been checked and are known to be free from faults, the fault must be in the catalytic converter, which must be renewed (refer to Part B of this Chapter).

Removal and refitting

4 Removal and refitting of the catalytic converter is described in Chapter 4, Part B.

Lambda sensor

Testing

5 The Lambda sensor may be tested with a multimeter by disconnecting the two-pin wiring at the connector and connecting an ohmmeter across the two white terminals. **Do not** connect the ohmmeter to the ECU wiring. The resistance should be approximately 4.0 ohms with a cold engine.

6 Disconnect the ohmmeter and connect a voltmeter (2 volt DC scale) between the black lead of the sensor and a good earthing point. Connect an exhaust gas analyser to the end of the exhaust pipe, then start the engine and allow it to idle. As the temperature of the sensor rises, the sensor will produce a voltage of between 100 and 900 mV but this will vary according to the fuel/air mixture. A lean mixture giving a Lambda value greater than 1.0 will produce a voltage of approximately 100 mV. A rich mixture giving a Lambda value greater than 1.0 will produce a voltage of approximately 900 mV. **Note:** *With the wiring disconnected, the system ECU will revert to 'Limp-home' mode.*

7 Disconnect the multimeter and reconnect the wiring after making the test. Further testing of the Lambda sensor system should be carried out by a Saab dealer.

Removal

Note: *The Lambda sensor is DELICATE. It will not work if it is dropped or knocked, if its power supply is disrupted, or if any cleaning materials are used on it. The sensor should be renewed every 25 000 miles/40 000 km.*

8 Disconnect the Lambda sensor wiring at the connector.

9 Unscrew the sensor from the exhaust manifold and remove it **(see illustration)**. The sensor may be tight - it will help if it is turned back and forth on its threads as it is being removed. It is possible to obtain a special slotted socket which locates on the sensor without causing any damage to the wiring.

Refitting

10 Refitting is a reverse of the removal procedure, but tighten the sensor to the specified torque. Prior to installing the sensor, apply a smear of high-temperature grease to the sensor threads. Also, the wiring must be correctly routed, and in no danger of contacting the exhaust system.

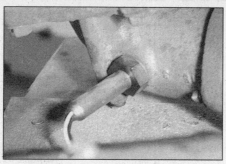

2.9 Lambda sensor location on the exhaust manifold

Modulating valve (CI fuel injection with Lambda sensor system)

Removal

11 To remove the modulating valve disconnect the wiring, the small bore line to the valve and the return lines, and remove the valve.

Refitting

12 Always use new seals, and note that the rubber valve retainer is made of special rubber which must not be exposed to fuel.

ECU (CI fuel injection with Lambda sensor system)

Removal

13 To remove the control unit on early models slide back the passenger seat, remove the unit cover, disconnect the wiring and remove the screws. On later models tilt the rear seat forward to expose the unit.

Refitting

14 Refitting is a reversal of removal.

Throttle valve switch (CI fuel injection with Lambda sensor system)

Removal

15 To remove the throttle valve switch, disconnect the wiring and remove the mounting screws.

Refitting

16 Refitting is a reversal of removal, but position the switch so that with the throttle valve fully open there is a clearance of 0.2 to 0.5 mm at the switch lever.

EGR system

Testing

Mechanical two-port system

17 Remove the throttle valve housing, the EGR pipe, and the EGR valve.

18 Clean the pipe with a suitable solvent or piece of wire, and clean the inlet and outlet of the EGR valve with a wire brush taking care not to damage the valve stem. Blow through the valve with compressed air.

19 Clean the calibrated hole in the exhaust manifold using a 5.0 mm drill (manual transmission models) or 10.0 mm drill (automatic transmission models).

20 Clean the calibrated hole in the inlet manifold using a 10.0 mm drill.

21 Refit the unit using new gaskets and where applicable reset the warning counter unit.

22 To check the operation of the valve disconnect the PVS hose at the switch, suck the end of the hose and then release. The valve should be heard to close. Also suck on the hose with the engine idling and check that this causes the idling speed to drop.

23 To check the PVS valve disconnect the hoses and attempt to blow through the valve. With the engine cold the valve should be shut, but with the engine warm it should be open.

24 To check the cut-in speed run the engine to normal operating temperature and connect a tachometer. Increase the engine speed and check that the valve opens at between 2600 and 3200 rpm on manual transmission models or between 2300 and 2900 rpm on automatic transmission models.

25 To check the holding valve rev the engine to between 3000 and 3500 rpm then pinch the hose between the release valve and three-way nipple. Reduce the engine speed to idling and check that the EGR valve remains open for 6 more seconds.

26 To check the release valve, rev the engine so that the EGR valve is open then reduce the speed to idling and check that the EGR valve closes immediately.

Electronic system

27 If a fault occurs in the EGR system, a fault code is stored in the system ECU and a warning light is lit on the instrument panel. Check all wiring and hoses on the EGR valve for security. If this does not rectify the fault, the advice of a Saab dealer should be sought.

Removal and refitting

28 Removal and refitting of the EGR valve has been described already. Removal and refitting of the other components is simple, but it will be necessary to drain the cooling system before removing the PVS valve.

Evaporative emission canister and purge valve

Testing

29 If the system is thought to be faulty, first disconnect the hoses from the charcoal canister and purge control valve, and check that they are clear by blowing through them. If the purge control valve or charcoal canister are thought to be faulty, the system may be checked using a vacuum pump however this work should be carried out by a Saab dealer.

Removal

30 Remove the left-hand indicator lamp (Chapter 12), APC control unit (Chapter 4B), the cruise control vacuum unit and vacuum release unit (Chapter 4B), and the ignition system amplifier unit (Chapter 5).

31 Unscrew the earth lead screw, then

4C•4 Emission control systems

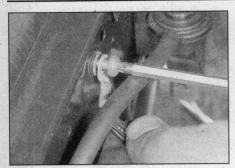

2.31a Unscrew the earth lead screw...

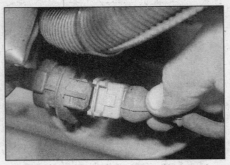

2.31b ...disconnect the wiring, and disconnect the two pipes

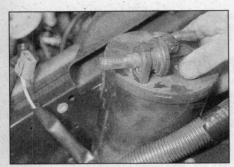

2.32 Removing the evaporative emission canister

2.33 Arrow on the top of the canister must point to the hose leading to the engine

disconnect the wiring from the purge valve and detach the two pipes **(see illustrations)**.
32 Remove the evaporative emission canister from the inner wing panel **(see illustration)**.

Refitting
33 Refitting is a reversal of removal, but make sure that the hoses and wiring are reconnected to their original positions and that the purge valve arrow points in the correct direction **(see illustration)**.

Deceleration device
Testing
34 To check the electric type connect a tachometer and run the engine to normal operating temperature. Disconnect the positive solenoid cable and connect battery voltage. Rev the engine then release the throttle and check that the idling speed is maintained at 1550 rpm. If necessary adjust the solenoid with the screw. Connect a test lamp between the positive solenoid wire and earth, then drive the car at about 25 mph (40 kph) and declutch. Brake the car and check that the test lamp goes out at a speed of 18.6 mph (30 kph).
35 To check the vacuum controlled type run the engine to normal operating temperature and connect a tachometer. Adjust the idling speed to 875 rpm then rev up the engine to 3000 rpm, release the throttle, and check that the time for the engine to return to idling speed is 4 to 6 seconds. Unscrew the unit screw until the valve closes then adjust the idling speed of the engine in the normal way. Now screw in the unit screw until the engine speed is 1600 rpm and back off the screw two complete turns. Finally adjust the idling speed again. Note that the check should be completed without the radiator fan cutting in.

Removal
36 To remove the electric speed transmitter remove the lower facia panel and unplug the unit. The solenoid can be unscrewed after disconnecting the wiring.
37 To remove the vacuum valve disconnect the hose and unscrew the unit from the inlet manifold.

Refitting
38 Refitting is a reversal of removal.

Dashpot system
Testing
39 Run the engine to operating temperature, then adjust the idling speed to 875 rpm. Rev the engine to 3000 rpm, release the throttle, and check that the time for the engine to return to idling speed is 3 to 6 seconds. If not, loosen the locknut and reposition the unit.

Removal
40 Loosen the locknut and remove the unit from the bracket.

Refitting
41 Refitting is a reversal of removal.

Delay valve system
Testing
42 Connect a tachometer and timing light and run the engine to normal operating temperature. Have an assistant open the throttle quickly so that the engine runs at 3000 rpm. Using the timing light check that the time from when the throttle was opened to when the ignition advances is between 4 and 8 seconds. If not, renew the delay valve.

Removal
43 Disconnect the hoses and remove the valve.

Refitting
44 Refitting is a reversal of removal, but make sure that the white end of the valve is towards the distributor.

Pulse air system
Testing
45 Disconnect the hoses from the check valves and check that with the engine running at idling speed suction can be felt with the thumbs placed over the openings.

Removal
46 Unscrew the pipe unions and the mounting bracket and disconnect the air hoses, then unscrew the bracket bolt.

Refitting
47 Refitting is a reversal of removal. When refitting the inlet pipes to the exhaust manifold align them as shown **(see illustration)**.

3 Catalytic converter - general information and precautions

1 DO NOT use leaded petrol - the lead content will "poison" and destroy the catalyst.
2 Always keep the ignition and fuel systems well-maintained in accordance with the manufacturer's schedule (see Chapter 1).
3 If the engine develops a misfire, do not drive the car at all (or at least as little as possible) until the fault is cured.
4 DO NOT push-start the car - this will soak the catalytic converter in unburned fuel.
5 DO NOT use fuel or engine oil additives.
6 DO NOT continue to use the car if the engine is burning oil (blue exhaust smoke).
7 The converter operates at very high temperatures. DO NOT, therefore, park the car in dry undergrowth, over long grass, or over piles of dead leaves, after a long run.
8 The catalytic converter is FRAGILE - do not strike it with tools during servicing work.

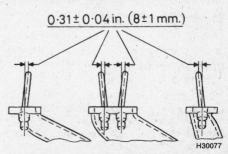

2.47 Showing correct alignment of the pulse air inlet pipes

Chapter 5
Engine electrical systems

Contents

Alternator - removal and refitting 7
Alternator brushes and regulator - removal, inspection and refitting ..8
Alternator drivebelt - removal, refitting and tensioning 6
Battery - removal and refitting 4
Battery - testing and charging 3
Battery terminal check ....See Chapter 1
Charging system - testing 5
Condenser (conventional ignition) - testing, removal and refitting ...17
Distributor - dismantling and reassembly 16
Distributor - removal and refitting 15
Electrical fault-finding - general information 2
Electrical system check ....See "Weekly Checks"
General information and precautions 1
Ignition HT coil - removal, testing and refitting 14
Ignition system - testing 13
Ignition system amplifier unit (electronic ignition) - removal and refitting 18
Ignition system check ....See Chapter 1
Ignition/starter lock cylinder - removal and refitting 12
Ignition/starter switch - removal and refitting 11
Spark plug renewal ....See Chapter 1
Starter motor - removal and refitting 10
Starting system - testing 9

Degrees of difficulty

Easy, suitable for novice with little experience	Fairly easy, suitable for beginner with some experience	Fairly difficult, suitable for competent DIY mechanic	Difficult, suitable for experienced DIY mechanic	Very difficult, suitable for expert DIY or professional

Specifications

Engine electrical system type .... 12-volt, negative earth

Battery
Type .... Lead-acid, 'standard', 'low-maintenance' or 'maintenance-free' (sealed for life)
Battery capacity .... 60/62 amp/hr
Charge condition:
  Poor .... 12.5 volts
  Normal .... 12.6 volts
  Good .... 12.7 volts

Alternator
Type:
  99 models .... Bosch 55 amp, 65 amp, 70 amp, or SEV/Marchal 55 amp
  900 models .... Bosch 55 amp, 65 amp, 70 amp, 80 amp, or Motorola 70 amp
Minimum brush protrusion from holder:
  Bosch and Motorola .... 5.0 mm
  SEV/Marchal .... 4.0 mm

Starter motor
Type .... Bosch pre-engaged

Ignition system
Type .... Conventional with coil and contact breaker points, or electronic with coil, impulse generator, and control unit

HT leads
Resistances:
  Leads to Nos 1 and 2 cylinders .... 3250 ohms (max)
  Leads to Nos 3 and 4 cylinders .... 3000 ohms (max)
  Lead from coil to distributor cap .... 1000 ohms (max)

5•2 Engine electrical systems

Ignition coil

Conventional system
Primary winding resistance (at 20°C)	2.6 to 3.1 ohms
Primary winding current (at 1000 distributor rpm)	1.9 amp
Secondary winding resistance	8000 to 11 000 ohms

Electronic system - Inductive transmitter
Primary winding resistance (at 20°C)	1.0 to 1.4 ohm
Secondary winding resistance	5500 to 8500 ohms
Primary winding current (at 1000 distributor rpm)	3.2 amp

Electromagnetic - Hall transmitter
Primary winding resistance (at 20°C)	0.52 to 0.76 ohm
Secondary winding resistance	7200 to 8200 ohms

Distributor

Rotation (viewed from cap end)	Anti-clockwise

Conventional system
Contact points gap	0.4 mm
Dwell angle	50° ± 3°
Shaft endfloat	0.10 to 0.30 mm
Rotor arm resistance	5000 ohms

Electronic system - Inductive transmitter
Rotor arm ignition cut-out speed	5900 to 6200 rpm (engine)
Pulse transmitter coil resistance	895 to 1285 ohms
Minimum gap between rotor and stator	0.25 mm
Rotor arm resistance	5000 ohms
Compensating (ballast) resistance	0.6 ohm (starter connected), 1.0 ohm (engine running)

Electronic system - Hall transmitter
Rotor arm ignition cut-out speed	6000 rpm (engine)
Rotor arm resistance	1000 ohms

Firing order: 1-3-4-2 (No 1 cylinder at timing chain end)

Torque wrench settings

	Nm	lbf ft
Alternator pulley nut	26	35
Spark plugs	27	20

1 General information and precautions

General information

Because of their engine-related functions, components of the starting, charging and ignition systems are covered separately from the body electrical devices - eg lights, instruments, etc (which are covered in Chapter 12).

The electrical system is of the 12-volt negative earth type.

The battery is of standard, low-maintenance or 'maintenance-free' (sealed for life) type, charged by the alternator, which is belt-driven from the crankshaft pulley.

The starter motor is of the pre-engaged type, incorporating an integral solenoid. On starting, the solenoid moves the drive pinion into engagement with the flywheel/driveplate ring gear before the starter motor is energised. Once the engine has started, a one-way clutch prevents the motor armature being driven by the engine until the pinion disengages from the ring gear. On pre-1985 models the starter motor is mounted on the right-hand front of the engine however on later models it is mounted on the left-hand front of the engine.

The ignition system functions by generating an electrical spark to ignite the fuel/air mixture in the combustion chamber at exactly the right moment in relation to engine speed and load. The ignition system is based on feeding low tension (LT) voltage from the battery to the coil where it is converted to high tension (HT) voltage. The high tension voltage is powerful enough to jump the spark plug gap in the cylinders many times a second under high compression pressures, providing that the system is in good condition.

The ignition system may be of conventional contact breaker type, electronic type with inductive pulse generator, electronic type with Hall generator, or electronic type with knock detector. All systems use a coil, but on the conventional system the primary circuit is switched by contact points whereas on the electronic system the circuit is switched electronically.

Early models with electronic ignition incorporate an inductive pulse generator in the distributor consisting of a magnetic pick-up coil and a reluctor with four arms. This system uses an amplifier unit to amplify the signals from the pulse generator and control the dwell angle. The dwell angle is increased as the engine speed increases. The compensating (ballast) resistor, which is located on the right-hand side of the engine compartment (see illustration), effectively increases the coil HT voltage during starting.

1.7 The compensating (ballast) resistor located on the right-hand side of the engine compartment

Turbo models from 1982 and all models from 1984 use the Hall effect system to trigger the low tension circuit by means of a Hall effect generator and slotted rotor inside the distributor. This system also uses an control unit to amplify the signals from the Hall generator and control the dwell angle. As from 1986 the system includes a pulse amplifier located on the relay box in addition to the control unit.

On more recent models fitted with the EZK electronic ignition system a knock detector in communication with the LH-Jetronic fuel injection system is used to monitor adverse combustion characteristics. With this system the ignition timing is adjusted in relation to the characteristics of the fuel being used.

Advance and retard are adjusted automatically. Except on models with the EZK ignition system, centrifugal weights inside the distributor control the ignition timing according to the speed of the engine, and a vacuum capsule connected to the distributor baseplate retards the ignition timing according to engine load. On models with EZK ignition, ignition timing is controlled automatically by the electronic control unit using a throttle position sensor and knock detector to monitor engine load and combustion.

The ignition system functions in the following manner. Low tension voltage is changed in the coil to high tension voltage by the alternate switching on and off the primary circuit by the contact points (conventional system) or impulse generator, control unit and amplifier (electronic system). The high tension voltage is fed to the relevant spark plug via the distributor cap and rotor arm. The ignition is advanced and retarded automatically to ensure that the spark occurs at the correct instant in relation to the engine speed and load.

When working on electronic ignition systems remember that the high tension voltage can be considerably higher than on a conventional system and in certain circumstances could prove fatal. Depending on the position of the distributor components it is also possible for a single high tension spark to be generated simply by knocking the distributor with the ignition switched on. It is therefore important to keep the ignition system clean and dry at all times, and to make sure that the ignition switch is off when working on the engine.

Precautions

While some repair procedures may be possible on certain electrical components, the usual course of action is to renew the component concerned. The owner whose interest extends beyond mere component renewal should obtain a copy of the 'Automobile Electrical & Electronic Systems Manual', available from the publishers of this manual.

It is necessary to take extra care when working on the electrical system, to avoid damage to semi-conductor devices (diodes and transistors), and to avoid the risk of personal injury. In addition to the precautions given in 'Safety first!' at the beginning of this manual, observe the following when working on the system:

 Always remove rings, watches, etc before working on the electrical system. Even with the battery disconnected, capacitive discharge could occur if a component's live terminal is earthed through a metal object. This could cause a shock or nasty burn.

Do not reverse the battery connections. Components such as the alternator, electronic control units, or any other components having semi-conductor circuitry could be irreparably damaged.

If the engine is being started using jump leads and a slave battery, connect the batteries positive-to-positive and negative-to-negative (see 'Booster battery (jump) starting'). This also applies when connecting a battery charger.

Never disconnect the battery terminals, the alternator, any electrical wiring or any test instruments when the engine is running.

Do not allow the engine to turn the alternator when the alternator is not connected.

Never 'test' for alternator output by 'flashing' the output lead to earth.

Never use an ohmmeter of the type incorporating a hand-cranked generator for circuit or continuity testing.

Always ensure that the battery negative lead is disconnected when working on the electrical system.

Before using electric-arc welding equipment on the car, disconnect the battery, alternator and components such as the fuel injection/ignition electronic control unit, to protect them from the risk of damage.

The radio/cassette unit fitted as standard equipment on later models has a built-in security code, to deter thieves. If the power source to the unit is cut, the anti-theft system will activate. Even if the power source is immediately reconnected, the radio/cassette unit will not function until the correct security code has been entered. Therefore, if you do not know the correct security code for the radio/cassette unit, do not disconnect the battery negative terminal of the battery, or remove the radio/cassette unit from the vehicle. Refer to 'Radio/cassette unit anti-theft system - precaution' at the beginning of this manual for further information.

2 Electrical fault-finding - general information

Refer to Chapter 12.

3 Battery - testing and charging

Standard and low-maintenance battery - testing

1 If the vehicle covers a small annual mileage, it is worthwhile checking the specific gravity of the electrolyte every three months, to determine the state of charge of the battery. Use a hydrometer to make the check, and compare the results with the following table. Note that the specific gravity readings assume an electrolyte temperature of 15°C (60°F); for every 10°C (18°F) below 15°C (60°F), subtract 0.007. For every 10°C (18°F) above 15°C (60°F), add 0.007. However, for convenience, the temperatures quoted in the following table are ambient (outdoor air) temperatures, above or below 25°C (77°F):

	Above 25°C (77°F)	Below 25°C (77°F)
Fully-charged	1.210 to 1.230	1.270 to 1.290
70% charged	1.170 to 1.190	1.230 to 1.250
Discharged	1.050 to 1.070	1.110 to 1.130

2 If the battery condition is suspect, first check the specific gravity of electrolyte in each cell. A variation of 0.040 or more between any cells indicates loss of electrolyte or deterioration of the internal plates.

3 If the specific gravity variation is 0.040 or more, the battery should be renewed. If the cell variation is satisfactory but the battery is discharged, it should be charged as described later in this Section.

Maintenance-free battery - testing

4 In cases where a 'sealed for life' maintenance-free battery is fitted, topping-up and testing of the electrolyte in each cell is not possible. The condition of the battery can therefore only be tested using a battery condition indicator or a voltmeter.

5 A battery with a built-in charge condition indicator may be fitted. The indicator is located in the top of the battery casing, and indicates the condition of the battery from its colour. If the indicator shows green, then the battery is in a good state of charge. If the indicator turns darker, eventually to black, then the battery requires charging, as described later in this Section. If the indicator shows clear/yellow, then the electrolyte level in the battery is too low to allow further use, and the battery should be renewed. **Do not attempt to charge, load or jump-start a battery when the indicator shows clear/yellow.**

5•4 Engine electrical systems

6 If testing the battery using a voltmeter, connect the voltmeter across the battery, and compare the result with those given in the Specifications under 'charge condition'. The test is only accurate if the battery has not been subjected to any kind of charge for the previous six hours, including charging by the alternator. If this is not the case, switch on the headlights for 30 seconds, then wait four to five minutes after switching off the headlights before testing the battery. All other electrical circuits must be switched off, so check (for instance) that the doors and tailgate or bootlid are fully shut when making the test.

7 If the voltage reading is less than 12.2 volts, then the battery is discharged. A reading of 12.2 to 12.4 volts indicates a partially-discharged condition.

8 If the battery is to be charged, remove it from the vehicle (Section 4) and charge it as described in the following paragraphs.

Standard and low-maintenance battery - charging

Note: The following is intended as a guide only. Always refer to the manufacturer's recommendations (often printed on a label attached to the battery) before charging a battery.

9 Charge the battery at a rate of 3.5 to 4 amps, and continue to charge the battery at this rate until no further rise in specific gravity is noted over a four-hour period.

10 Alternatively, a trickle charger used at the rate of 1.5 amps can safely be used overnight.

11 Specially rapid 'boost' charges which are claimed to restore the power of the battery in 1 to 2 hours are not recommended, as they can cause serious damage to the battery plates through overheating.

12 While charging the battery, note that the temperature of the electrolyte should never exceed 37.8°C (100°F).

Maintenance-free battery - charging

Note: The following is intended as a guide only. Always refer to the maker's recommendations (often printed on a label attached to the battery) before charging a battery.

13 This battery type requires a longer period to fully recharge than the standard type, the time taken being dependent on the extent of discharge, but it can take up to three days.

14 A constant-voltage type charger is required, to be set, where possible, to 13.9 to 14.9 volts with a charger current below 25 amps. Using this method, the battery should be usable within three hours, giving a voltage reading of 12.5 volts, but this is for a partially-discharged battery and, as mentioned, full charging can take considerably longer.

15 Use of a normal trickle charger should not be detrimental to the battery, provided excessive gassing is not allowed to occur and the battery is not allowed to become hot.

4 Battery - removal and refitting

Removal

1 The battery is located at the front on the right-hand side of the engine compartment **(see illustration)**.

2 Loosen the clamp nut and disconnect the lead at the negative (earth) terminal.

3 Loosen the clamp nut and disconnect the lead at the positive terminal in the same way.

4 On Turbo models disconnect the pressure pipe from the exhaust manifold to the boost control, and on APC models disconnect the relevant hoses.

5 Unscrew the nuts and remove the clamp bar/heat shield **(see illustration)**, then lift the battery from the platform taking care not to spill any electrolyte on the bodywork.

Refitting

6 Refitting is a reversal of removal, however do not over-tighten the clamp nuts. Smear petroleum jelly on the terminals when reconnecting the leads, and always reconnect the positive lead first, and the negative lead last.

5 Charging system - testing

Note: Refer to the warnings given in 'Safety first!' and in Section 1 of this Chapter before starting work.

1 If the ignition/no-charge warning light does not come on when the ignition is switched on, first check the alternator wiring connections for security. If satisfactory, check that the warning light bulb has not blown, and that the bulbholder is secure in its location in the instrument panel (see Chapter 12). If the light still fails to illuminate, check the continuity of the warning light feed wire from the alternator to the bulbholder. If all is satisfactory, the alternator is at fault, and should be renewed or taken to an auto-electrician for testing and repair.

2 If the ignition warning light comes on when the engine is running, stop the engine and check that the drivebelt is intact and correctly tensioned (see Chapter 1) and that the alternator connections are secure. If all is satisfactory, check the alternator brushes and slip rings as described in Section 8. If the fault persists, the alternator should be renewed, or taken to an auto-electrician for testing and repair.

3 If the alternator output is suspect even though the warning light functions correctly, the regulated voltage may be checked as follows.

4 Connect a voltmeter across the battery terminals, and start the engine.

5 Increase the engine speed until the voltmeter reading remains steady; the reading should be approximately 12 to 13 volts, and no more than 14 volts.

6 Switch on as many electrical accessories (eg, the headlights, heated rear window and heater blower) as possible, and check that the alternator maintains the regulated voltage at around 13 to 14 volts.

7 If the regulated voltage is not as stated, the fault may be due to worn brushes, weak brush springs, a faulty voltage regulator, a faulty diode, a severed phase winding, or worn or damaged slip rings. The brushes and slip rings may be checked (see Section 8), but if the fault persists, the alternator should be renewed or taken to an auto-electrician for testing and repair.

6 Alternator drivebelt - removal, refitting and tensioning

Refer to the procedure given for the auxiliary drivebelt in Chapter 1.

7 Alternator - removal and refitting

Removal

1 On models equipped with power steering, remove the power steering pump drivebelt with reference to Chapter 1.

2 Disconnect the battery negative lead.

3 Disconnect the wiring from the alternator **(see illustration)**.

4 Loosen the mounting and adjustment bolts/nuts, swivel the alternator towards the

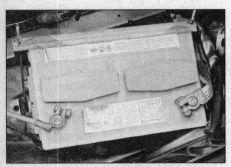

4.1 Battery showing terminals and retaining clamp position

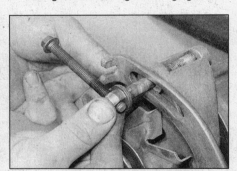

4.5 Removing the battery clamp bar/heat shield

Engine electrical systems 5•5

7.3 Alternator wiring (Bosch)

7.4 Auxiliary drivebelt tension adjustment nut on the alternator

7.5 Removing the tension adjustment bolts from the alternator

7.8a Unscrew the through-bolt . . .

7.8b . . . and remove the alternator from its mounting bracket

7.8c Removing the adjustment link from the timing cover

engine, and remove the drivebelt from the pulleys. **(see illustration)**.
5 Unscrew the nut and remove the tension adjustment bolts from the alternator **(see illustration)**.

Pre-1983 models
6 Remove the mounting and adjustment bolts and withdraw the alternator from the engine.

1983-on models
7 As from 1983 a new mounting was introduced, and the through-bolt cannot be removed with the alternator in position on the engine. First loosen the through-bolt, then unscrew the mounting bracket bolts with an Allen key and remove the alternator complete with the mounting bracket. With the alternator removed, unscrew the through-bolt and remove the alternator from the bracket.
8 If the alternator is being removed with the engine removed from the car, unscrew the through-bolt and remove the alternator from the mounting bracket. If necessary the adjustment link may be removed by unbolting it from the timing cover **(see illustrations)**.

Refitting
9 Refitting is a reversal of removal, but before fully tightening the mounting and adjustment bolts tension the drivebelt so that it can be depressed under firm thumb pressure 10.0 mm midway between the alternator and crankshaft pulleys on type B engines, or 5.0 mm midway between the alternator and water pump pulleys on type H and later engines. Lever the alternator on the drive end bracket and then tighten the bolts. Tension the power steering pump drivebelt with reference to Chapter 1.

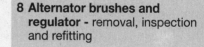

8 Alternator brushes and regulator - removal, inspection and refitting

Removal and inspection
1 Disconnect the battery negative lead. On most models there is no need to remove the alternator as there is sufficient working room on the left-hand side of the engine compartment.

Bosch type
2 Remove the screws and withdraw the regulator and brush box from the rear of the alternator **(see illustrations)**.
3 If the protruding length of either brush is less than the minimum given in the Specifications, unsolder the wiring and renew the

8.2a Remove the screws . . .

8.2b . . . and remove the regulator and brush box from the alternator

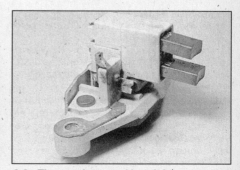
8.2c The regulator and brush box removed from the alternator

5•6 Engine electrical systems

8.10 The alternator slip rings with the regulator/brush box removed

brushes, then resolder the new wires in position. **Note:** *Take care not to overheat the regulator.*

Motorola type
4 Pull the multi-plug from the rear of the alternator.
5 Remove the screws and withdraw the regulator. Note the location of the wires then disconnect them and remove the regulator.
6 Remove the screws and carefully withdraw the brush box.
7 If the protruding length of either brush is less than the minimum given in the Specifications, obtain a new brush box.

SEV/Marchal type
8 Remove the screws and withdraw the regulator and brush box from the rear of the alternator. Detach the wire from the terminal.
9 If the protruding length of either brush is less than the minimum given in the Specifications, obtain a new regulator and brush box.

All types
10 Wipe clean the slip rings with a fuel-moistened cloth - if they are very dirty use fine glasspaper then wipe with the cloth **(see illustration)**.

Refitting
11 Refitting is a reversal of removal, but make sure that the brushes move freely in their holders.

9 Starting system - testing

Note: *Refer to the precautions given in 'Safety first!' and in Section 1 of this Chapter before starting work.*
1 If the starter motor fails to operate when the ignition key is turned to the appropriate position, the following may be the reason:
  a) The battery is faulty.
  b) The electrical connections between the switch, solenoid, battery and starter motor are somewhere failing to pass the necessary current from the battery through the starter to earth.
  c) The solenoid is faulty.
  d) The starter motor is mechanically or electrically defective.

2 To check the battery, switch on the headlights. If they dim after a few seconds, this indicates that the battery is discharged - recharge (see Section 3) or renew the battery. If the headlights glow brightly, operate the starter on the ignition switch and observe the lights. If they dim, this indicates that current is reaching the starter motor - therefore, the fault must lie in the starter motor. If the lights continue to glow brightly (and no clicking sound can be heard from the starter motor solenoid), this indicates that there is a fault in the circuit or solenoid - see the following paragraphs. If the starter motor turns slowly when operated, but the battery is in good condition, then this indicates that either the starter motor is faulty, or there is considerable resistance somewhere in the circuit.
3 If a fault in the circuit is suspected, disconnect the battery leads, the starter/solenoid wiring and the engine/transmission earth strap. Thoroughly clean the connections, and reconnect the leads and wiring, then use a voltmeter or test lamp to check that full battery voltage is available at the battery positive lead connection on the solenoid, and that the earth is sound. Smear petroleum jelly around the battery terminals to prevent corrosion - corroded connections are amongst the most frequent causes of electrical system faults.
4 If the battery and all connections are in good condition, check the circuit by disconnecting the wire from the solenoid terminal. Connect a voltmeter or test light between the wire end and a good earth (such as the battery negative terminal), and check that the wire is live when the ignition switch is turned to the 'start' position. If it is, then the circuit is sound - if not, the circuit wiring can be checked as described in Chapter 12.
5 The solenoid contacts can be checked by connecting a voltmeter or test light between the terminal on the starter side of the solenoid, and earth. When the ignition switch is turned to the 'start' position, there should be a reading or lighted bulb, as applicable. If there is no reading or lighted bulb, the solenoid or contacts are faulty and the solenoid should be renewed.
6 If the circuit and solenoid are proved sound, the fault must lie in the starter motor and it should be taken to an auto electrical specialist for checking. It may be possible to have the starter motor overhauled, but check on the cost of spares before proceeding, as it may prove more economical to obtain a new or exchange motor.

10 Starter motor - removal and refitting

Note: *On pre-1985 models the starter motor is mounted on the right-hand front of the engine however on later models it is mounted on the left-hand front of the engine.*

Removal
1 Disconnect the battery negative lead.
2 Note their locations then disconnect the wires from the starter motor **(see illustrations)**.

Pre-1985 models
3 Remove the plastic flywheel cover. On manual gearbox models, remove the gearbox oil dipstick.

Turbo models
4 Disconnect the preheater hose.
5 Remove the turbocharger suction pipe and the stay from the gearbox.
6 Loosen the oil return pipe sufficiently so that it can be bent to one side, out of the work area.

All models
7 Unbolt and remove the starter motor heat shield and rear mounting bracket **(see illustration)**.

10.2a Starter motor wiring (pre-1985 models)

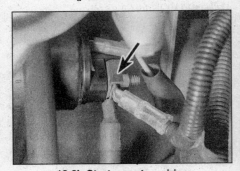

10.2b Starter motor wiring (1985-on models)

10.7 Starter motor rear mounting bracket (pre-1985 models)

Engine electrical systems 5•7

10.8 Starter motor front mounting bolts (pre-1985 models)

10.9a Starter motor mounting bolts (1985-on models)

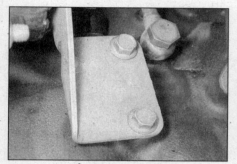

10.9b Starter motor bracket and mounting bolts (1985-on models)

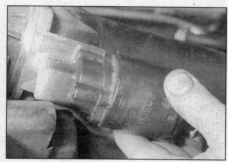

10.9c Removing the starter motor (1985-on models)

8 Remove the front mounting bolts and withdraw the starter from the engine (see illustration). On Turbo models tilt the starter motor downwards then lift out while moving the oil return pipe to one side.

1985-on models
9 Unscrew the mounting bolts securing the starter motor to the engine plate and block mounting bracket, then move the starter motor back and remove it from the engine (see illustrations). If necessary unbolt the mounting bracket from the starter motor.

Refitting
Pre-1985 models
10 Refitting is a reversal of the removal procedure, but on Turbo models fit a new gasket to the turbocharger oil return pipe flange.

1985-on models
11 Refitting is a reversal of removal.

11 Ignition/starter switch - removal and refitting

Removal
1 Disconnect the battery negative lead.
2 Remove the gear lever housing or selector housing as described in Chapter 7A (manual gearbox) or Chapter 7B (automatic transmission).
3 Remove the barrel cover, then disconnect the wiring noting the location of each wire (see illustrations).

4 Unscrew the two retaining screws and remove the ignition/starter switch.

Refitting
5 When refitting the ignition/starter switch, the following procedures must be followed.

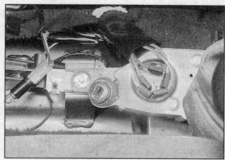

11.3a Ignition switch and wiring (with gear lever housing removed)

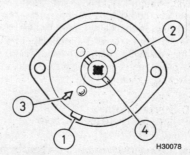

11.6 Ignition/starter switch showing the location stud (1), index mark (2), arrow (3) and cogwheel locating slot (4)

6 Rotate the contact so that the index mark on its end face is opposite the arrow marking (see illustration). Turn the contact by inserting a suitable screwdriver in the slot.
7 With the ignition key in the 'L' (locked) position, locate the switch into the housing and ensure that the location stud is engaged in the groove.
8 Fit the retaining screws to secure the switch, then operate the ignition key to ensure that it moves freely without sticking.
9 Reconnect the wiring in their previously noted positions.
10 Refit the gear lever housing or selector housing.
11 Reconnect the battery negative lead.

12 Ignition/starter lock cylinder - removal and refitting

Removal
1 Disconnect the battery negative lead.
2 Remove the gear lever housing or selector housing - refer to Chapter 7A (manual gearbox) or Chapter 7B (automatic transmission).
3 Turn the ignition key and locate it midway between the 'L' (locking) and 'G' (garage) position (see illustration). Leave the key in this position, then using a suitable wire rod bent to suit, depress the lock cylinder dowel, through the access hole in the base of the lever housing, and remove the cylinder.

Refitting
4 To refit the cylinder, check that the key is between the 'L' and 'G' positions, compress

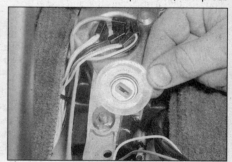

11.3b Removing the barrel cover from the ignition/starter switch

12.3 Ignition key set between the 'L' and 'G' positions

the lock cylinder dowel in, ensure that the toothed segment driver pins are positioned correctly, then press the cylinder into position. It should engage with the toothed segment.

5 Refit the gear lever housing or selector housing and check for satisfactory operation of the ignition/starter.

6 If the key is missing or the lock is jammed, the lock cylinder can be removed after drilling out the dowel cover plug and driving the dowel in with a pin punch. This will release the lock cylinder, but it will also destroy it.

13 Ignition system - testing

Warning: *Voltages produced by an electronic ignition system are considerably higher than those produced by conventional ignition systems. Extreme care must be taken when working on the electronic ignition system with the ignition switched on. Persons with surgically-implanted cardiac pacemaker devices should keep well clear of the ignition circuits, components and test equipment. Refer to the precautions in Section 1 before starting work. Always switch off the ignition before disconnecting or connecting any component, and when using a multi-meter to check resistances.*

Engine will not start

1 If the engine fails to start and the car was running normally when it was last used, first check there is fuel in the fuel tank. If the engine turns over normally on the starter motor and the battery is evidently well charged, then the fault may be in either the high or low tension circuits. First check the HT circuit. **Note:** *If the battery is known to be fully charged: the ignition light comes on and the starter motor fails to turn the engine, check the tightness of the leads on the battery terminals and also the secureness of the earth lead to its connection to the body. It is quite common for the leads to have worked loose, even if they look and feel secure. If one of the battery terminal posts gets very hot when trying to work the starter motor this is a sure indication of a faulty connection to that terminal.*

2 One of the commonest reasons for bad starting is wet or damp spark plug leads and distributor. Remove the distributor cap. If condensation is visible internally, dry the cap with a rag and also wipe over the leads, then refit the cap. It may be helpful to use a proprietry moisture dispersan.

3 If the engine still fails to start, disconnect an HT lead from any spark plug and connect it to a known good plug which is itself connected to a good earth point on the engine by means of a cable or jump lead. While an assistant spins the engine on the starter motor, check that a regular blue spark occurs. If so, remove, clean, and re-gap the spark plugs as described in Chapter 1 as the blue spark indicates that the ignition system is functioning correctly. **Note:** *On models with a catalytic converter* **do not** *spin the engine for more than a few seconds.*

4 If no spark occurs, disconnect the main feed HT lead from the distributor cap and connect it to the test spark plug. Check for a spark as in paragraph 3. If sparks now occur, check the distributor cap, rotor arm, and HT leads as described in Chapter 1 and renew them as necessary. Also check that all wiring and connectors are secure.

Conventional ignition system

5 Use a 12v voltmeter or a 12v bulb and two lengths of wire. With the ignition switched on and the points open test between the low tension wire to the coil (it is marked 15 or +) and earth. No reading indicates a break in the supply from the ignition switch. Check the connections at the switch to see if any are loose. Refit them and the engine should run. If voltage is present, a faulty coil or condenser, or broken lead between the coil and the distributor is indicated.

6 Take a reading between the low tension coil terminal marked 1 or - and earth. No reading indicates a faulty condenser (check as described in Section 17) or faulty coil. A reading shows a broken lead between the coil and distributor. For these tests it is sufficient to separate the points with a piece of dry paper while testing with the points open.

Electronic ignition system (inductive transmitter)

7 Disconnect the wires from the compensating resistor (located by the fusebox), and use an ohmmeter to check that its resistance is 0.6 ohm. If not, renew it.

8 Check that with the ignition on, current is available at the single terminal on the compensating resistor. If not, check the starter relay and connecting wire, and the ignition switch wiring.

9 Check that the wire from the control unit terminal 31 is well earthed to the body. Check that battery voltage is available at terminal 15. Check that the wires on terminals 16 are well insulated.

10 Connect a voltmeter across terminals 7 and 3 at the control unit, Spin the engine on the starter and check that a minimum reading of 1 volt (AC) is obtained. If there is no reading check that the transmitter (located in distributor) air gaps between the rotor and stator posts are 0.25 mm.

11 Using an ohmmeter, check that the transmitter resistance is between 895 and 1285 ohms. If not, renew the induction coil.

Electronic ignition system (Hall transmitter)

12 Locate the control unit on the left hand side of the engine compartment (**not** the LH-Jetronic ECU). Remove the mounting bolt and turn the unit around the earthed bolt. Pull back the rubber cover and connect a voltmeter between terminals 4 and 2. If battery voltage is not available with the ignition switched on, check the ignition switch wiring.

13 Disconnect the wiring from the distributor and check that battery voltage is available with the ignition switched on. If not, check the wiring.

14 Connect a voltmeter between terminals 6 and 3 on the control unit, leaving the fuse connected. Remove the distributor cap and dust cover, then turn the engine until one of the gaps in the slotted rotor is aligned with the Hall transmitter. The voltage reading should be 0.4 volt or less. Now turn the engine so that the rotor covers the transmitter. The voltage reading should be 1.0 volt or more. If the voltage is incorrect renew the transmitter unit.

15 Disconnect the distributor wiring and connect a voltmeter across coil terminals 15 and 1. When the ignition is turned on the reading should be 6 volt dropping to zero volt within 1 to 2 seconds. If not, renew the control unit.

Engine misfires

16 If the engine misfires regularly, the misfiring plug may be identified as follows. First disconnect each HT lead in turn from the spark plugs then obtain four short lengths of fuse wire and reconnect the HT leads with the fuse wire trapped between the lead and the terminal on the plug. Run the engine at a fast idling speed then use an insulated screwdriver to 'short' each plug to earth in turn while listening to the note of the engine. Hold the screwdriver with a dry cloth or rubber glove as additional protection against a shock from the HT supply. No difference in engine running will be noticed when the defective plug is earthed. 'Shorting' the lead from one of the good cylinders will accentuate the misfire.

17 Check the HT supply as previously described. If in order, check the spark plug. The plug may be loose, the insulation may be cracked, or the points may have burnt away giving too wide a gap for the spark to jump. Worse still, one of the points may have broken off. Either renew the plug, or clean it, reset the gap, and then test it. If there is no spark at the end of the plug lead, or if it is weak and intermittent, check the HT lead from the distributor cap to the plug using an ohmmeter. If the insulation is cracked or perished, renew the lead. Check the connections at the distributor cap.

18 If there is still no spark, examine the distributor cap carefully for tracking. This can be recognised by a very thin black line running between two or more electrodes, or between an electrode and some other part of the distributor. These lines are paths which now conduct electricity across the cap thus letting it run to earth. Renew the distributor cap if necessary.

19 Finally check the coil for tracking across the HT tower, and renew it if necessary.

Engine electrical systems 5•9

14.1 Ignition coil and wiring

14.3 Low tension leads and terminals on the ignition coil

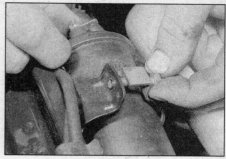

14.4 Removing the ignition coil mounting clamp

14 Ignition HT coil - removal, testing and refitting

Note: *The ignition coil fitted to the conventional ignition system is different from the coil fitted to the electronic ignition system.*

Removal

1 The coil is located on the radiator fan housing (see illustration). First make sure that the ignition is switched off.
2 Roll back the rubber dust cover where fitted and disconnect the HT lead.
3 Identify the low tension leads for position, then disconnect them from the terminals on the coil (see illustration).
4 Unscrew the mounting clamp screws and remove the coil (see illustration).

Testing

Note: *The resistance of the coil windings will vary slightly according to the coil temperature.*
5 Connect an ohmmeter between terminals 1 and 15 (ie the primary windings) and check that the resistance is as given in the Specifications.
6 Connect the ohmmeter between terminal 1 and the HT terminal (ie the secondary windings), and check that the resistance is as given in the Specifications.

Refitting

7 Refitting is a reversal of removal but make sure that the wiring connectors are fitted correctly. Always keep the coil clean and dry.

15 Distributor - removal and refitting

Removal

B201 B-type engines

1 Prise back the spring clips, remove the distributor cap and place it to one side.
2 Disconnect the coil to distributor wiring.
3 Disconnect the vacuum hose.
4 Turn the engine with a spanner on the crankshaft pulley bolt until the rotor arm is approaching the No 1 HT lead position of the distributor cap (No 1 cylinder is at the timing chain end of engine). Alternatively remove No 1 spark plug (timing chain end of the engine), place a finger over the plug hole, and turn the engine until pressure is felt in No 1 cylinder (this indicates that the piston is commencing its compression stroke). Continue turning until the '0' (TDC) ignition timing mark on the flywheel/driveplate is aligned with the mark on the transmission housing cover timing hole. The rotor arm should now point towards the groove in the distributor rim (remove the plastic cover if applicable).
5 Mark the distributor body in relation to the cylinder block then unscrew the mounting bolt and lift out the distributor. As the distributor is being removed, the rotor arm will turn approximately 50° clockwise - make a pencil mark on the rim for this position (see illustration).

B201 H-type and B202 engines

6 Prise back the spring clips, remove the distributor cap and place it to one side.

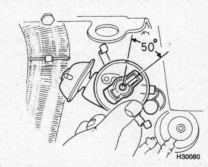

15.5 Showing rotor arm movement when removing distributor on the B-type engine

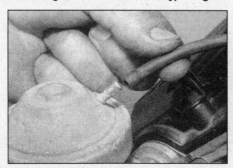

15.7b Disconnecting the vacuum hose from the distributor

7 Disconnect the LT wiring/amplifier lead and the vacuum hose (see illustrations).
8 Mark the distributor body in relation to the cylinder head cover, then unscrew the mounting bolt(s) and withdraw the distributor (see illustrations). On the B202 engine examine the rubber O-ring for condition and renew it if necessary.

Refitting

B201 B-type engines

9 To refit the distributor check that the '0' (TDC) ignition timing mark on the flywheel/driveplate is still aligned with the mark on the transmission housing cover timing hole. Turn the rotor arm so that it is approximately 50° clockwise of the groove in the distributor rim.
10 Position the distributor over the hole in the block with the groove in the distributor rim facing the cylinder head, then insert it fully

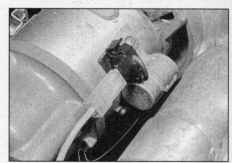

15.7a LT lead connection on the H-type engine distributor

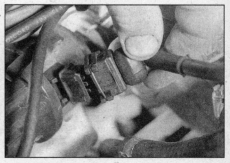

15.7c Disconnecting the amplifier lead on an engine with Hall transmitter ignition

5•10 Engine electrical systems

15.8a Distributor mounting bolts on the B201 H-type engine

15.8b Removing the distributor on the B201 H-type engine

15.8c Removing the distributor from the cylinder head on the B202 engine

into the block so that the gears are engaged. If may be necessary to turn the engine slightly so that the oil pump shaft locates correctly. The rotor arm should now point towards the groove in the distributor rim.
11 Check that the previously made marks on the distributor body and block are aligned, then insert the mounting bolt and tighten lightly.
12 Reconnect the vacuum hose and LT lead, and refit the distributor cap. Where removed refit No 1 spark plug.
13 Check and adjust the ignition timing as described in Chapter 1.

B201 H-type and B202 engines

14 To refit the distributor hold it over the mounting hole with the previously made marks aligned. If a new distributor is being fitted, the vacuum capsule should face the exhaust manifold side of the engine. On the B202 engine make sure that the rubber O-ring is correctly located it its groove on the distributor body.
15 Turn the distributor shaft to align the dogs with the offset groove in the camshaft, then fit the distributor to the cylinder head cover.
16 Align the previously made marks then insert the mounting bolts and tighten them lightly.
17 Reconnect the vacuum hose and LT lead, and refit the distributor cap.
18 Check and adjust the ignition timing as described in Chapter 1.

16 Distributor - dismantling and reassembly

Note: *Complete dismantling of the distributor is not recommended, however the following work is within the scope of the home mechanic and may be undertaken with the distributor still in position on the engine.*

Conventional ignition vacuum control unit

Dismantling

1 Prise back the spring clips, remove the distributor cap and place it to one side. Pull off the rotor arm.
2 Where applicable remove the plastic dust cover then mark the bearing plate for position, loosen the screws and withdraw the plate.
3 Disconnect the vacuum hose.
4 Remove the vacuum control unit retaining screws from the distributor body. Note on B-type engines one of the screws secures the distributor cap clip.
5 Extract the circlip from inside the distributor, disengage the control arm and withdraw the vacuum control unit.

Reassembly

6 Reassembly is a reversal of dismantling.

Electronic ignition vacuum control unit (not fitted to EZK ignition)

Dismantling

7 Remove the distributor cap, rotor arm and plastic dust cover.
8 Disconnect the vacuum hose.
9 Remove the vacuum control unit retaining screws from the distributor body.
10 Disconnect the operating arm, then remove the vacuum control unit.

Reassembly

11 Turn the pulse generator or Hall sensor (as applicable) clockwise as far as possible, then reconnect the operating arm.
12 The remaining procedure is a reversal of dismantling.

Electronic ignition (inductive) induction coil

Dismantling

13 Remove the distributor (see Section 15).
14 Remove the rotor arm and plastic dust cover **(see illustration)**.
15 Remove the clip, unscrew the screw and withdraw the cable terminal.
16 Remove the screws, unhook the control arm and withdraw the vacuum control unit and pad noting the location of the distributor cap clip.

16.14 Exploded view of the distributor on the B201 H-type engine with inductive electronic ignition

1 Distributor cap
2 Rotor arm
3 Dust cover
4 Bearing plate
5 Spring clip
6 Rotor
7 Induction coil
8 Stator
9 Base plate
10 Shaft and centrifugal mechanism
11 Body
12 Vacuum unit
13 Delay valve

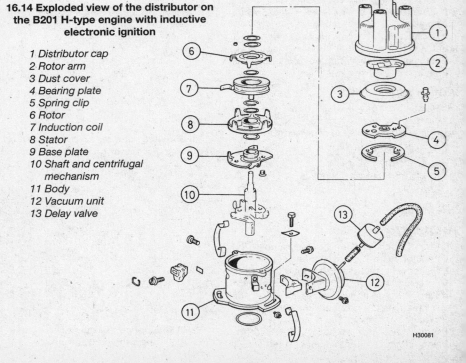

17 Remove the screws securing the transmitter plate and distributor cap clip and remove the clip.
18 Extract the circlip from the shaft, remove the washer, and prise off the rotor. Recover the lock pin.
19 Extract the circlip and withdraw the impulse transmitter.
20 Remove the screws and separate the induction coil from the transmitter plate.

Reassembly
21 Reassembly is a reversal of dismantling.

Electronic ignition (Hall effect) sensor and slotted vane

22 It is not possible to renew the Hall sensor or the slotted vane and if either of these components is found to be faulty, the complete distributor should be renewed.

17 Condenser (conventional ignition) - testing, removal and refitting

Testing
1 The condenser is fitted in parallel with the contact points. Its purpose is to reduce arcing between the points, and also to accelerate the collapse of the coil low tension negative field. A faulty condenser which has shorted internally can cause the complete failure of the ignition system, as the points will be prevented from interrupting the low tension circuit.
2 To test the condenser, remove the distributor cap and on H-type engines the rotor arm and plastic dust cover. Turn the engine until the points are closed then switch on the ignition and separate the points with a screwdriver. If this is accompanied by a

17.10 Condenser location on the B201 H-type engine distributor

strong blue flash, the condenser is faulty (a weak spark is normal).
3 A further test can be made for short circuiting by disconnecting the LT leads from the connector block and using a test lamp and leads connected to the terminal and body. If the test lamp lights when the points are open, the condenser is faulty.
4 If the condenser is suspect, renew it and check whether the fault persists.

Removal
5 To remove the condenser first check that the ignition is switched off.

B-type engines
6 Prise back the spring clips, remove the distributor cap and place it to one side.
7 Mark the distributor body in relation to the cylinder block then loosen the mounting bolt and turn the distributor so that the condenser screw is visible.
8 Disconnect the LT leads then remove the screw and withdraw the condenser.

H-type engines
9 Disconnect the LT lead from the distributor terminal.

18.1 Ignition system amplifier unit on the left-hand side of the engine compartment

10 Remove the mounting screw, withdraw the condenser and disconnect the LT lead (see illustration).

Refitting
11 Refitting is a reversal of removal, but on B-type engines check and if necessary adjust the ignition timing as described in Chapter 1.

18 Ignition system amplifier unit (electronic ignition) - removal and refitting

Removal
1 The ignition system amplifier unit (also known as the control unit) is located on the left-hand side of the engine compartment (see illustration).
2 With the ignition switched off, disconnect the wiring plug from the amplifier unit.
3 Unscrew the retaining screws and remove the amplifier unit.

Refitting
4 Refitting is a reversal of removal.

Notes

Chapter 6
Clutch

Contents

Clutch assembly - removal, inspection and refitting 7
Clutch linings - checking for wear (in situ) 6
Clutch pedal - removal and refitting 2
General information 1
Hydraulic system - bleeding 5
Master cylinder - removal and refitting 3
Operation check .... see Chapter 1
Slave cylinder and release bearing - removal, inspection and refitting 4

Degrees of difficulty

| Easy, suitable for novice with little experience | Fairly easy, suitable for beginner with some experience | Fairly difficult, suitable for competent DIY mechanic | Difficult, suitable for experienced DIY mechanic | Very difficult, suitable for expert DIY or professional |

Specifications

General
Make	Borg and Beck (4 speed); Fichtel & Sachs (5 speed)
Type	Single dry plate with damper, diaphragm spring with spring loaded hub
Operation	Hydraulic, via slave and master cylinders.
Diameter	204 mm (normally-aspirated models), 216 mm (Turbo models)
Pressure plate inner taper (max)	0.03 mm

Torque wrench setting
	Nm	lbft
Slave cylinder-to-transmission screws	10	7

1 General information

The clutch is of single dry plate type with a diaphragm spring pressure plate. The clutch plate (or disc) incorporates a spring cushioned hub to absorb transmission shocks and ensure a smooth take up of drive. On five-speed gearbox models a pre-damper is also fitted to the hub. The clutch assembly is dowelled and bolted to the flywheel.

The clutch plate is splined to the clutch shaft and is held in position between the flywheel and pressure plate by the pressure of the diaphragm spring. Friction linings are riveted to both sides of the clutch plate, separated by a corrugated steel plate damper.

The clutch is hydraulically operated. When the pedal is depressed, the master cylinder piston moves forwards and fluid is forced through the hydraulic pipe to the slave cylinder. The slave cylinder is of annular type and is fitted over the clutch shaft - the release bearing fits on the end of the piston. Hydraulic pressure forces the release bearing against the inner edges of the diaphragm spring fingers, which act as levers, pivoting the pressure plate away from the clutch plate linings. The clutch plate slides clear of the flywheel and no drive is transmitted through the clutch.

When the clutch pedal is released, the clutch plate is gripped between the flywheel and pressure plate and drive is again transmitted through the clutch.

The slave cylinder contains no internal return spring mechanism. When the clutch is released, the pressure plate diaphragm spring pushes the annular piston back inside the cylinder. The friction of the fluid and dust seals prevents any further retraction of the piston. The result of this arrangement is that the wear of the clutch plate linings is compensated for automatically, by friction of the slave cylinder seal and a plastic sleeve and circlip on the slave cylinder piston.

2 Clutch pedal - removal and refitting

Removal

1 Working inside the car, remove the centre console (where applicable) and the lower facia panel; refer to Chapter 11 for greater detail.
2 Unhook and remove the return spring from the clutch pedal **(see illustration)**.

90/99 models

3 Extract the split pin, remove the washer and pivot pin securing the master cylinder clevis to the pedal. Unscrew the locknut, remove the pivot bolt, and remove the clutch pedal.

Left-hand drive 900 models

4 Remove the retaining screws and lower the air duct away from underneath the left-hand side of the facia.
5 Remove the split pin/circlip and washer (as applicable), then withdraw the clip and clevis pin securing the master cylinder push rod to the pedal.
6 Unscrew the locknut, remove the pivot bolt and withdraw the clutch pedal.
7 Remove the pivot centre tube, then check it and the bush bearings for signs of wear or damage. If necessary, drive out the old bushes with a soft metal drift and fit new ones.

Right-hand drive 900 models

8 Remove the retaining screws and lower the air duct away from underneath the left-hand side of the facia.
9 The master cylinder is located on the left hand side of the vehicle. Pedal movement is

2.2 Unhook and remove the return spring from the clutch pedal (arrowed)

6•2 Clutch

2.9a Clutch pedal extension shaft (arrowed, right hand drive models)

2.9b At the left hand end of the shaft, remove the split pin and washer/locking clip (arrowed) . . .

2.9c . . . then withdraw clevis pin securing the master cylinder push rod to the lug on the end of the extension shaft (arrowed)

communicated to it across the bulkhead via an extension shaft. At the left hand end of the shaft, remove the split pin/circlip and washer (as applicable), then withdraw the clevis pin and clip securing the master cylinder push rod to the lug on the end of the extension shaft **(see illustrations)**.
10 Prise the circlip from the left-hand end of the pedal shaft and remove the washer.
11 Remove the three retaining nuts from the pedal assembly mounting bracket, which are accessible from within the engine bay. Working in the right hand footwell, lift off the bracket, then withdraw the clutch pedal and shaft. Temporarily refit the bracket to hold the brake pedal in place.

Refitting

12 Refit the clutch pedal by following the removal procedure in reverse. Lubricate the pivot bush bearings with a molybdenum disulphide-based grease.
Caution: Where applicable, take care to ensure that the cruise control switches are not disturbed when refitting the pedal.

3 Master cylinder - removal, overhaul and refitting

Warning: Hydraulic fluid is poisonous; thoroughly wash off spills from bare skin without delay. Seek immediate medical advice if any fluid is swallowed or gets into the eyes. Certain types of hydraulic fluid are inflammable and may ignite when brought into contact with hot components; when servicing any hydraulic system, it is safest to assume that the fluid is inflammable and to take precautions against the risk of fire as though it were petrol that was being handled. Hydraulic fluid is an effective paint stripper and will also attack many plastics. If spillage occurs onto painted bodywork or fittings, it should be washed off immediately, using copious quantities of fresh water. It is also hygroscopic i.e. it can absorb moisture from the air, which then renders it useless. Old fluid may have suffered contamination and should never be re-used. When topping-up or renewing the fluid, always use the recommended grade, and ensure that it comes from a new sealed container.

Removal

1 Remove the clutch pedal (Section 2).
2 In the engine bay, place some absorbent rag around the master cylinder, to absorb any fluid that may be spilt.
3 On vehicles fitted with anti-lock braking, refer to Chapter 5 and remove the alternator and drivebelt; this will improve access to the master cylinder mounting bolts.
4 On all models except right hand drive 99 models, fit a hose clamp to the fluid reservoir supply hose. Release the clip and disconnect the end of the hose from the master cylinder - be prepared for a small amount of hydraulic fluid loss. Tie the open end of the hose to the bodywork, pointing upwards, to minimise further fluid loss.
5 Unscrew the union at the point where the slave cylinder delivery hydraulic pipe connects to the master cylinder. Use a small container or a wad of absorbent rag to catch any spilt fluid. Plug the open end of the pipe to prevent fluid loss and the ingress of dirt **(see illustration)**.
6 Release the delivery pipe from the clip, where applicable. Position it to one side, to keep it out of the way - tie it back if necessary.
7 Unscrew the retaining nuts from within the footwell, then withdraw the master cylinder from the engine bay side of the bulkhead. Take care not to spill hydraulic fluid on the bodywork **(see illustrations)**.

Overhaul

8 Pour the hydraulic fluid from the cylinder into a sealable container and discard it. On right hand drive 99 models, remove the filler cap from the integrated reservoir to allow it to be drained.
9 Clean the exterior of the cylinder with paraffin or methylated spirit, then dry it thoroughly.
10 Prise off the rubber cap then extract the circlip from the mouth of the cylinder using circlip pliers.
11 Remove the pushrod and spacer.
12 Remove the piston together with the

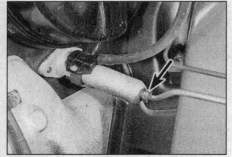
3.5 Unscrew union (arrowed) where slave cylinder delivery hydraulic pipe connects to the master cylinder (900 model shown)

3.7a Unscrew the clutch master cylinder retaining nuts (arrowed) from within the footwell

3.7b Removing the clutch master cylinder on right hand drive 99 modesls

Clutch 6•3

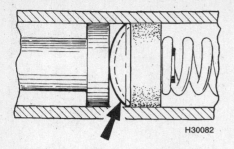

3.18 Fit the washer (arrowed) with its concave side against the seal

washer, seal and spring by tapping the cylinder on a block of wood. Note the position of the components.
13 Prise the seal from the piston.
14 Clean all the components with methylated spirit then examine them for wear and damage. In particular check the cylinder bore for scoring or corrosion. If such deterioration is evident, renew the complete master cylinder. If the bore is in good condition, obtain a master cylinder repair kit, which will contain the new piston seals and dust covers required to complete the overhaul.
15 Fit the new seal to the piston, with the sealing lip towards the centre of the piston. Before fitting, dip the seal in fresh hydraulic fluid, then manipulate the seal into position, but do not use tools or implements that may damage it.
16 Locate the seat on the small diameter end of the spring then dip the new seal in clean, unused hydraulic fluid and locate it on the seat.
17 Insert the spring (large diameter end first), seat and seal in the cylinder until the spring touches the end of the bore.
18 Fit the washer with its concave side against the seal, then insert the pushrod and spacer together with a new dust cap (see illustration).
19 Depress the pushrod and insert the circlip in the groove in the mouth of the cylinder.
20 Locate the lip of the dust cap in the groove at the edge of the cylinder.

Refitting

21 Refit the clutch master cylinder by following the removal procedure in reverse, noting the following points:
(a) Vehicles with ABS: refer to Chapter 5 and refit the alternator and drivebelt, then working from Chapter 1, adjust the drivebelt tension to specification.
(b) On completion, bleed the hydraulic system, as described in Section 5 of this Chapter.

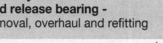

4 Slave cylinder and release bearing - removal, overhaul and refitting

Removal

1 The slave cylinder and release bearing must be removed as part of the clutch assembly, as described in Section 7. It can then be then withdrawn from the pressure plate.

Overhaul

2 Disconnect and plug the hydraulic bleed pipe. Pour the fluid from the slave cylinder and discard it.
3 Expand the circlip and lift the release bearing off the end of the piston (see illustration). If the bearing is tight, support it across the jaws of a bench vice and tap the piston through it, using a soft-faced mallet and a metal tube of an appropriate diameter. Spin the release bearing by hand and check it for roughness. Hold the outer race and attempt to move it laterally against the inner race. If any excessive movement or roughness is evident, renew the bearing. Do not attempt to clean the bearing with solvents, as it factory lubricated and hence non-serviceable.
4 Remove the circlip and lift the plastic dust seal off (see illustration). On earlier models, remove the circlip from the piston and recover the o-ring seal from the release bearig end of the piston.
5 Support the cylinder body, then push out the inner sleeve. Prise out the O-ring from the flange of the sleeve (see illustrations). On earlier models, recover the lip seal from the sleeve body.
6 Push the piston from the cylinder body. On later models, remove both the inner and outer O-ring seals (see illustrations).
7 Clean all the components with methylated spirit then examine them for wear and damage. In particular, check the internal surfaces of the cylinder bore for scoring and/or corrosion. If deterioration is evident, renew the entire slave cylinder. However if the surfaces are in good condition, obtain a repair kit, which will contain new seals and O-rings (see illustration).
8 The reassembly sequences differ slightly from this point onwards, depending on the age of the vehicle and hence the version of slave cylinder fitted.

Earlier models

9 Dip the sleeve O-ring in hydraulic fluid and carefully manipulate it by hand into the groove in the flange of the sleeve.
Caution: Do not use tools or implements that may damage the seal.
10 Smear a small quantity of rubber grease

4.3 Removing the release bearing from the slave cylinder piston

4.4a Remove the circlip . . .

4.4b . . . and lift the plastic dust seal off

4.5a Support the cylinder body, then push out the inner sleeve

4.5b Prise out the O-ring from the flange of the sleeve

6•4 Clutch

4.6a Push the piston from the cylinder body

4.6b On later models, remove both the inner . . .

4.6c . . . and outer O-ring seals

onto the piston lip seal, then slide it onto the sleeve.
Caution: Mineral-based oils will attack the lip seal and will cause it to fail prematurely; do not allow it to come into contact with such lubricants.

11 Press the sleeve into the cylinder, such that the lip seal is pushed about halfway along the cylinder bore.

12 Push the piston into the space between the sleeve and the cylinder bore, such that the end of it just touches the lip seal. Fit the return circlip, plastic dust seal with its retaining circlip, and O-ring seal to the end of the piston.

13 Ensure that the sleeve is pressed fully into the cylinder, then fit the release bearing onto the end of the piston, with the friction surface facing away from the cylinder body.

Later models

14 Dip the O-ring in hydraulic fluid and carefully manipulate it by hand into the groove in the flange of the sleeve -
Caution: Do not use tools or implements that may damage the seal.

15 Dip the piston seals in hydraulic fluid then fit them to the piston; do not use implements that may damage the seals. On earlier models,

16 Insert the sleeve in the cylinder body and press in fully, then fit the piston such that it slides between the sleeve and the cylinder bore.

17 Locate the plastic dust seal and circlip on the end of the piston. Then, expand the circlip and fit release bearing, with the friction surface facing away from the cylinder body.

Refitting

18 Refit the slave cylinder together with the remainder of the clutch components as described in Section 7 of this Chapter. On completion, bleed the hydraulic system as described in Section 5 of this Chapter.

5 Hydraulic system - bleeding

General information

1 Whenever the clutch hydraulic lines are disconnected for service or repair, an amount of air will enter the system. The presence of air in any hydraulic system will introduce a degree of elasticity and in the clutch system, this will translate into poor pedal feel and reduced clutch travel, leading to inefficient gear changes and even clutch system failure. For this reason, the hydraulic lines must be sealed using hose clamps before any work is carried out and then on completion, the system must be topped up and bled to remove any air bubbles.

2 To seal off the hydraulic supply to the clutch slave cylinder, trace the rigid pipe from its point of entry at the clutch housing cover back to the point where it connects to the flexible hydraulic hose. Fit a proprietary brake hose clamp to the flexible hose and tighten it securely.

3 Unlike the braking system, the clutch hydraulic system cannot be bled by simply pumping the clutch pedal and catching the ejected fluid in a receptacle, connected to the bleed pipe. The system must be pressurised externally; the most effective way of achieving this is to use a pressure brake bleeding kit. These are readily availble in motor accessories shops and are extremely effective; their construction ensures that a large reservoir of hydraulic fluid is always available during the bleeding process, eliminating the danger of accidentally drawing air into the system. The following sub-section describes bleeding the clutch system using such a kit.

Bleeding

4 Remove the inspection cover from the clutch housing to expose the bleed pipe and nipple **(see illustration)**.

5 Fit a ring spanner over the bleed nipple head, but do not slacken it at this point. Connect a length of clear plastic hose over nipple and insert the other end into a clean container. Pour hydraulic fluid into the container, such that the end of the hose is covered.

6 Following the manufacturers instructions, pour hydraulic fluid into the bleeding kit vessel.

7 Unscrew the vehicles fluid reservoir cap, then connect the bleeding kit fluid supply hose to the reservoir.

8 Connect the pressure hose to a supply of compressed air - a spare tyre is convenient source.
Caution: Check that the pressure in the tyre does not exceed the maximum quoted by the pressure kit manufacturer; let some air escape to reduce the pressure, if necessary. Gently open the air valve and allow the air and fluid pressures to equalise. Check that there are no leaks before proceeding.

9 Using the spanner, slacken the bleed pipe nipple until fluid and air bubbles can be seen to flow through the tube, into the container. Maintain a steady flow until the emerging fluid is free of air bubbles; keep a watchful eye on the level of fluid in the bleeding kit vessel and the vehicles fluid reservoir - if either is allowed

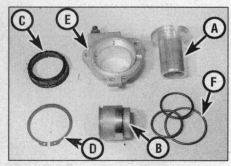

4.7 Slave cylinder dismantled

A Sleeve
B Piston
C Plastic dust seal
D Dust seal circlip
E Slave cylinder body
F Seals

5.4 Clutch hydraulic system bleed nipple (arrowed) - clutch housing cover removed for clarity

6.1 Prising out the rectangular inspection plate from the clutch housing cover

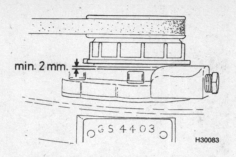

6.5 Check that the distance between the plastic sleeve and the turned surface of the slave cylinder is not less than 2.0 mm

to drop too low, air may be forced into the system, defeating the object of the exercise. To refill during the bleeding process, turn off the compressed air supply, remove the lid and pour in an appropriate quantity of clean fluid from a new container - do not re-use the fluid collected in the receiving container. Repeat as necessary, until the ejected fluid is bubble-free, then tighten the bleed pipe nipple using the spanner.

10 On completion, pump the clutch pedal several times to assess its feel and travel. If firm, constant pedal resistance is not felt throughout the pedal stroke, it is likely that air is still present - repeat the bleeding procedure until pedal feel is restored.

11 Depressurise the bleeding kit and remove it from the vehicle. At this point, the fluid reservoir may be over-full; if this is the case, the excess should be removed using a *clean* pipette (or an old poultry baster), to reduce the level to the "MAX" mark.

12 Remove the receiving container from the bleed nipple, then refit the clutch housing cover.

13 Finally, road test the vehicle and check the operation of the clutch system whilst changing up and down through the gears, whilst pulling away from a standstill, and from a hillstart.

6 Clutch linings - checking for wear (in situ)

1 Wear of the clutch plate linings can be checked without removing the clutch components. First, prise the rectangular inspection plate from the clutch housing cover **(see illustration)**.
2 Fully depress and release the clutch pedal several times.
3 Look through the inspection hole, directly over the slave cylinder and release bearing - on early models a circlip is located on the piston, but on later models a plastic sleeve will be visible.
4 On early models, check that the circlip is in contact with the end of the slave cylinder body. Then check that the distance between the circlip on the piston and the release bearing circlip is not less than 1.0 mm.
5 On later models, check that the plastic sleeve is contacting the release bearing. Then check that the distance between the plastic sleeve and the turned surface of the slave cylinder is not less than 2.0 mm **(see illustration)**.
6 If the measured distances are less than specified, the clutch plate linings are worn excessively and the clutch should be dismantled for further investigation. Otherwise, refit the inspection cover.

7 Clutch assembly - removal, inspection and refitting

Warning: Dust created by clutch wear and deposited on the clutch components may contain asbestos, which is a health hazard. DO NOT blow it out with compressed air or inhale any of it. DO NOT use petrol or petroleum-based solvents to clean off the dust. Brake system cleaner or methylated spirit should be used to flush the dust into a suitable receptacle. After the clutch components are wiped clean with clean rags, dispose of the contaminated rags and cleaner in a sealed, marked container.

Note: *Although some friction materials are now asbestos-free, if their type is unknown it is safest to assume that they do contain asbestos and to take precautions accordingly.*

Removal

1 Disconnect the battery negative cable and position it away from the terminal.
2 To improve access to the clutch, refer to Chapter 11 and remove the bonnet. On 99 models, refer to Chapter 3 and remove the radiator.

Fuel-injected vehicles

3 Slacken the worm drive clips and detach both the air mass meter and the air inlet hose from the throttle body and air cleaner. Blank off the throttle body aperture, to prevent the accidental ingress of debris.
4 Release the battery positive cable from the clip at the top of the primary chain casing.

Turbocharged vehicles

5 Slacken the worm drive clips at the throttle body and intercooler outlet, then disconnect and remove the air delivery pipe. Blank off the throttle body aperture, to prevent the ingress of debris.
6 Slacken the worm drive clips at the turbcharger outlet and intercooler inlet, then disconnect and remove the air pipe. Blank off the turbocharger opening to prevent the ingress of debris.
7 Disconnect the APC solenoid valve hose from the tapping point on the turbocharger inlet pipe. Also unplug the wiring from the preheater unit in the inlet pipe, at the connector. Slacken the worm drive clips and remove the inlet pipe.
8 Referring to Chapter 4 Part A, disconnect the air by-pass valve and hose from the turbocharger inlet pipe; blank off the opening to prevent the ingress of dirt.
9 Where applicable, unplug the LT and HT leads from the ignition coil, labelling them to aid refitting. Remove the retaining screws and lift off the coil.

Vehicles with air-conditioning

10 Unplug the wiring for the left hand cooling fan at the connector. Slacken and withdraw the two fan unit upper mounting bolts. Slacken the lower mounting bolt slighlty, then lift the entire fan assembly away from the radiator.

All models

11 Unbolt and remove the clutch housing cover **(see illustration)**.
12 The clutch diaphragm spring must now be held in the depressed position, until after the clutch has been removed. To do this, have an assistant keep the clutch pedal fully depressed and fit a spacer ring between the diaphragm spring and the pressure plate cover, then have the assistant release the pedal. Saab tool number 83 90 023 should be used if possible, but if this is not available a home made spacer can be made out of length of metal brake pipe or similar, coiled into a ring **(see Tool Tip overleaf)**. If for some reason the clutch hydraulics are inoperative, then a forked lever must be fabricated or Saab tool number 83 93 175 obtained, to enable the diaphragm spring to be depressed mechanically, before spacer ring can be inserted.

7.11 Unbolt and remove the clutch housing cover

6•6 Clutch

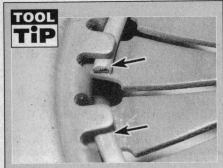

The clutch diaphragm spring must be held depressed until the clutch is removed - a home-made spacer (arrowed) can be made out of length of metal brake pipe or similar, coiled into a ring

7.13a Remove the spring clip . . .

7.13b . . . and withdraw the cap and seal from the front of the primary gear casing

13 Remove the spring clip and withdraw the cap and seal from the front of the primary gear casing **(see illustrations)**.
14 Unscrew the plastic oil thrower propeller from the front of the clutch shaft **(see illustration)**.
15 The clutch shaft must now be partially drawn out through the primary chain housing. To do this, screw an 8 mm bolt to into the end of the clutch shaft. Then lever against the underside of the bolt head using a stout metal bar, using the front of the primary chain casing as a pivot, until the shaft is released from the support bearing, driven plate and primary gear. Alternatively, a puller may be improvised, using a long 8mm bolt, a socket and a drilled plate The shaft does not need to be removed completely **(see illustrations)**.
Caution: Take care not to damage the radiator on 900 models..
16 Unscrew the bolts securing the slave cylinder to the primary gear casing using an Allen key, where applicable **(see illustration)**.
17 Hold the flywheel stationary with a lever, braced against the starter ring gear and starter motor, then begin to unscrew the bolts that secure the pressure plate to the flywheel. Slacken the bolts progressively, in a diagonal sequence, until they can be removed by hand **(see illustration)**.
18 Release the pressure plate from the dowels then withdraw it, together with the driven plate and slave cylinder, from the flywheel. **(see illustration)**. Take care not to damage the slave cylinder components as they pass the diaphragm spring fingers. Note: *Unless the slave cylinder is to be removed completely for overhaul, there is no need to disconnect the hydraulic delivery pipe from it.*
19 If necessary, place the pressure plate assembly on the bench, depress the diaphragm spring fingers with a lever, and remove the spacer or plastic tube.

Inspection

Note: *Due to the amount of work required to remove and refit clutch components, it is usually considered good practice to renew the clutch friction plate, pressure plate assembly and release bearing as a matched set, even if only one of these components is actually worn enough to require renewal*
20 When cleaning clutch components, observe the warning at the beginning of this

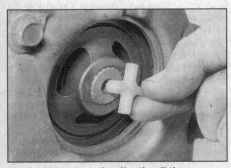

7.14 Unscrew the plastic oil thrower propeller from the front of the clutch shaft

7.15a To release the clutch shaft, use an 8 mm bolt as described in the text

7.15b . . . or use a home-made puller

7.16 Unscrew the slave cylinder-to-primary gear casing bolts using an Allen key

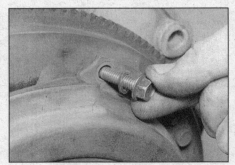

7.17 Slacken the pressure plate bolts progressively, in a diagonal sequence, until they can be removed by hand

7.18 Release the pressure plate from the dowels then withdraw it, with the driven plate and slave cylinder, from the flywheel

7.24 Place a straight edge across the friction face and use feeler blades at the inner edges to check for taper

7.30 Orientation marks on the driven plate casing

7.36 Tighten the pressure plate bolts progressively in diagonal sequence

Section regarding the hazards of handling the friction materials contained in clutch components; remove dust using a clean, dry cloth and working in a well-ventilated atmosphere.

21 Check the driven plate facings for signs of wear, damage or oil contamination. If the friction material is cracked, burnt, scored or damaged, or if it is contaminated with oil or grease (shown by shiny black patches), the friction plate must be renewed.

22 If the friction material is still serviceable, check that the centre boss splines are unworn, that the torsion springs are in good condition and securely fastened and that all the rivets are tightly fastened. If any wear or damage is found, the friction plate must be renewed.

23 If the friction material is fouled with oil, this is probably due to an oil leak from a crankshaft oil seal or possibly from the sump to cylinder block joint. Ensure that the fault is identified and rectified, before the new friction plate is fitted.

24 Check the pressure plate assembly for obvious signs of wear or damage; shake it to check for loose rivets or worn or damaged fulcrum rings and check that the drive straps securing the pressure plate to the cover do not show signs of overheating, indicated by deep yellow or blue discoloration. If the diaphragm spring is worn or damaged, or if its pressure is in any way suspect, the pressure plate assembly should be renewed. Place a straight edge across the friction face and use feeler blades at the inner edges to check for taper **(see illustration)**. If the taper at any point exceeds the maximum amount given in the Specifications or if the pressure plate is otherwise damaged, it should be renewed.

25 Examine the machined bearing surfaces of the pressure plate and of the flywheel; they should be clean, completely flat and free from scratches or scoring. If either is discoloured from excessive heat or shows signs of cracking, it should be renewed, although minor surface imperfections can be polished away, using emery paper.

26 Check that the release bearing contact surface rotates smoothly and easily, with no sign of noise or roughness and that the surface itself is smooth and unworn, with no signs of cracks, pitting or scoring. If there is any doubt about its condition, the bearing must be renewed; refer to Section 4 for guidance.

27 Before refitting the clutch, check the condition of the clutch shaft seal in the primary chain casing. In addition, check the condition of the support bearing race in the flywheel.

Refitting

28 Ensure that the spacer tube is fitted between the diaphragm spring and the pressure plate cover, to keep the clutch in the depressed position.

29 Locate the driven plate against the pressure plate, with the projecting side towards the slave cylinder (ie away from the flywheel). Hold the slave cylinder with the release bearing against the diaphragm spring fingers.

30 Lower the three components into position and locate the pressure plate cover on the flywheel dowels - note the orientation marks on the driven plate casing **(see illustration)**. Insert two bolts and hand tighten them in the flywheel, to retain the assembly.

31 Smear a little molybdenum disulphide grease on the clutch shaft splines, then insert the shaft through the primary gear, into engagement with the clutch plate splines, and into the spigot bearing located in the flywheel. Use a soft-face mallet to tap the shaft fully into position so that the primary gear circlip engages the groove, where applicable.

32 Coat the threads of the slave cylinder bolts with a liquid thread locking compound, locate the cylinder on the primary gear casing then insert and tighten the bolts to the specified torque.

33 Insert all the pressure plate bolts and turn them until finger tight.

34 Insert and tighten the plastic propeller to the front of the clutch shaft.

35 Fit the cap and seal to the front of the primary gear casing and retain it with the spring clip.

36 Hold the flywheel stationary with a lever, inserted in the starter ring gear, then tighten the pressure plate bolts progressively in diagonal sequence **(see illustration)**.

37 Have an assistant depress the clutch pedal just enough to allow the spacer or plastic tube to be removed from the pressure plate assembly.

Caution: Do not fully depress the clutch pedal at this stage, otherwise the piston could be pushed right out of the slave cylinder, damaging the seal and losing hydraulic fluid.

38 While the assistant holds the clutch pedal depressed, use two screwdrivers to slide the return movement circlip against the slave cylinder body (early models), or the plastic sleeve and circlip against the release bearing (later models).

39 Release the clutch pedal. The circlip (early models) or plastic sleeve (later models) will now take up its initial position. On early models, the distance between the circlip on the piston and the release bearing circlip should be approximately 6.0 mm. On later models the distance between the plastic sleeve and the turned surface of the slave cylinder should be approximately 9.0 mm.

40 Refit the clutch housing cover and tighten the bolts.

41 Refit the remainder of the components by following paragraphs 1 to 10 of the removal procedure in reverse order.

Notes

Chapter 7 Part A:
Manual transmission

Contents

Gear lever lock and ignition switch - removal and refitting 7
Gearchange linkage - removal, refitting and adjustment 6
Gearchange lever and housing - removal and refitting 5
General information 1
Manual transmission oil- draining and refilling 8
Manual transmission - removal and refitting 9
Manual transmission oil level check ....See Chapter 1
Manual transmission overhaul - general information 10
Oil seals - renewal 2
Reversing light switch - testing, removal and refitting 4
Speedometer drive - removal and refitting 3

Degrees of difficulty

| Easy, suitable for novice with little experience | | Fairly easy, suitable for beginner with some experience | | Fairly difficult, suitable for competent DIY mechanic | | Difficult, suitable for experienced DIY mechanic | | Very difficult, suitable for expert DIY or professional | |

Specifications

Type .... Longitudinally mounted beneath engine, with integral transaxle differential/final drive. 4/5 forward speeds - all synchromeshed; 1 reverse speed, fitted with braking mechanism. Drive transmitted from clutch to input shaft by triplex chain drive

Lubricant capacity (total):
    4 speed .... 2.5 litres approx.
    5 speed .... 3.0 litres approx.
Recommended lubricant .... Engine oil, SAE 10W30 or 10W40

Designation:
    Pre 84 model year .... G M *nn* x y z 12345
        M = Manual (A = Automatic)
        zz = Version number
        x = Number of forward gears
        y = Highest gear ratio code (see gear ratio chart)
        z = Variant
        3nn nnn = Serial number
    84 model year onwards .... G *nn* x y z 12345
        G = Manual
        nn = Version number
        x = Number of forward gears
        y = Highest gear ratio code (see gear ratio chart)
        z = Variant
        12345 = Serial number

Gear lever latch stud protrusion:
    4 speed .... 25 mm
    5 speed:
        Up to 1988 model year .... 27 mm
        1989 model year onwards .... 22 mm

7A•2 Manual transmission

Typical gear ratios

GM 55704:
Final drive 3.89
Overall ratios:
  1st 13.84
  2nd 8.05
  3rd 5.27
  4th 3.81
  5th 3.04
  Reverse 15.22

G 34402:
Final drive 3.89
Overall ratios:
  1st 12.94
  2nd 7.80
  3rd 5.23
  4th 3.76
  Reverse 14.26

Torque wrench settings

	Nm	lbf ft
Gearcase drain plug	50	37
Transmission-to-engine bolts	35	26

1 General information

The transmission is longitudinally mounted in the engine bay and is bolted to the underside of the engine block. All components are housed in an alloy casing, including the integral transaxle differential/final drive.

The gearbox incorporates four or five forward speeds, according to model, and one reverse speed. Synchromeshes are fitted to all forward speeds. Reverse gear engagement is assisted by a braking device.

Drive is taken from the clutch shaft and is communicated to the transmission input shaft via a triplex chain drive, housed in a casing directly in front of the clutch assembly.

The transmission gearwheels are mounted on two shafts, the output shaft and the layshaft, and are in constant mesh with each other. The gearwheels on the output shaft are free to rotate independently of the shaft, until a speed is selected. The difference in diameter and number of teeth between the gearwheels on the output and layshafts provides the necessary speed reduction and torque multiplication. When fifth gear is engaged on 5 speed transmissions, the nut and output shafts are meshed together to provide direct drive, bypassing the layshaft.

Drive from the output shaft is then transmitted to the final drive gears/differential through the universal joint drivers, which are mounted on the outside of the transmission casing.

All forward gears are fitted with syncromeshes. When a speed is selected, the movement of the cabin floor mounted gear lever is communicated to the gearbox by a selector rod. This in turn actuates a series of selector forks inside the gearbox which are slotted onto the synchromesh sleeves. The sleeves, which are locked to the gearbox shafts but can slide axially by means of splined hubs, press baulk rings into contact with their respective gearwheels. The coned surfaces between the baulk rings and the gearwheels act as a friction clutch, that progressively matches the speed of the synchromesh sleeve (and hence the gearbox shaft) with that of the gearwheel. The dog teeth on the outside of the baulk ring prevent the synchromesh sleeve ring from meshing with the gearwheel until their speeds are exactly matched; this allows gear changes to be carried out smoothly and greatly reduces the noise and mechanical wear caused by rapid gear changes.

When reverse gear is engaged, a pinion that is in constant mesh with the first gear layshaft gearwheel is brought into mesh with the reverse gearwheel, mounted on the output shaft. This arrangement introduces the necessary speed reduction and also causes the output shaft to rotate in the opposite direction, allowing the vehicle to be driven in reverse. To allow smooth engagement of reverse gear, a braking device is fitted. The device consists of a spring on the selector shaft, which applies pressure to the 1st/2nd gear selector fork. The 1st gear synchromesh mechanism then slows down the rotation of the output shaft and hence the reverse gearwheel, allowing the reverse gear pinion to brought into mesh with it smoothly and quietly.

2 Oil seals - renewal

Clutch shaft seal
(on primary chain casing)

1 Remove the clutch assembly, release mechanism and clutch shaft (see Chapter 6).
2 From the clutch side of the primary chain casing, prise out the existing seal with a flat-bladed screwdriver; take care to avoid scoring the internal bore of the seal housing.
3 Lubricate the new seal with clean engine oil, then using a socket as a drift and mallet, drive the seal into the bearing housing until it is flush with the machined surface of the seal housing. Select the socket such that it bears only on the hard outer edge of the seal, rather than the inner sealing lips, which are more easily damaged.
4 Refer to Chapter 6 and refit the clutch shaft, release mechanism and clutch assembly. Ensure that the inner sealing surface of the seal is adequately lubricated before attempting to refit the clutch shaft.

Speedometer drive seals

5 Refer to Section 4.

3 Speedometer drive - removal and refitting

Removal

1 The speedometer drive gear is located on the transmission, at the housing for the left hand inner driveshaft bearing.
2 Unscrew the speedometer cable from the drive gear.
3 Slacken the screw and pivot the retaining bracket away from the drive gear.
4 Extract the drive gear from the bearing housing; doscard the O-ring and shaft seals.
5 Clean the assembly thoroughly and examine the gear teeth closely; renew the assembly if it shows signs of wear.
6 Lubricate the assembly with clean engine oil, then fit new O-ring and shaft seals.

Refitting

7 Refit the drive gear by following the removal procedure in reverse.

Manual transmission 7A•3

4.3 Reversing light switch location on gear lever housing (arrowed)

5.3 Remove three screws (arrowed) to release dished lever cover plate

(a) Remove left hand seat from the cabin.
(b) Remove the front section of the centre console.(On earlier models, remove the heater control buttons then unbolt the gear lever cover and disconnect the warning lamp wiring.)
(c) Remove the left hand door aperture trim panels and fold back the carpet, to expose the gear change lever housing.

10 Remove the retaining screws and lift the ventilation duct away from the underside of the facia. Disconnect the rear seat heating ducting at the joint below the seat mounting member **(see illustrations)**.
11 Unplug the wiring from the reversing light switch and ignition switch at the connectors; label the cables to aid refitting later **(see illustrations)**.
12 Using Saab tool No. 87 90 370 (or the improvised tool - see the note at the beginning of this sub-Section) unscrew the lever housing retaining nuts and withdraw the bolts **(see illustration)**.
13 Lift the housing away from the floorpan slightly, then turn it over and remove the screws and lower the plastic cover panel from its underside **(see illustrations)**.
14 Disengage the base of the gear lever from the selector rod boss and remove the housing from the vehicle **(see illustration)**.
15 To remove the gear lever and ignition lock, refer to Section 7.

4 Reversing light switch - removal and refitting

Removal

1 Disconnect the battery negative cable and position it away from the terminal.
2 With reference to Chapter 11, remove the front section of the centre console.
3 The reversing light switch is located on the gear lever housing **(see illustration)**. Unplug the wiring from it, labelling the connectors to aid refitting later.
4 Use an open ended spanner to unscrew the switch from the housing.

Refitting

5 Refit the switch by following the removal procedure in reverse.

5 Gearchange lever and housing - removal and refitting

Gear change lever

Removal

1 Park the vehicle and apply the handbrake. Move the gear change lever to the neutral position.
2 Unclip the lever gaiter from the surface of the console and fold it up over the lever knob.
3 Remove the three retaining screws to release the dished lever cover plate, then withdraw the lever from its housing **(see illustration)**.

Adjustment

4 Measure the length of latch stud protruding from the base of the gear change lever and compare it with that given in *Specifications*.
5 To alter the amount of protrusion, drive out the tubular dowel from the plastic bearing using a pin punch, then turn the latch stud until the correct length is obtained. Refit the dowel to lock the plastic carrier in position.

Refitting

6 Refit the gear change lever by following the removal procedure in reverse.

Gear change lever housing

Note: *This procedure requires access to a non-standard type of socket, Saab tool no. 87 90 370, to allow removal of the housing retaining nuts. A similar tool can be fabricated by grinding down the sides of an old 12mm socket, to fit the triangular nut flats.*

Removal

7 Disconnect the battery negative cable and position it away from the terminal.
8 The operation may be simplified by removing the gear lever, as described in the previous sub-Section, but this is not mandatory.
9 With reference to Chapter 11, carry out the following:

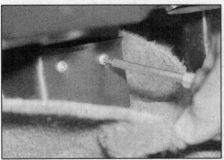

5.10a Remove screws and lift ventilation duct away from the underside of the facia

5.10b Disconnect rear seat heating ducting at joint below seat mounting member

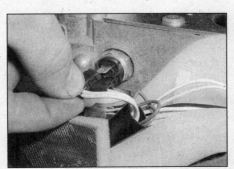

5.11a Unplug the wiring from the reversing light switch . . .

5.11b . . . and ignition switch at the connectors; label the cables to aid refitting

7A

7A•4 Manual transmission

5.12 Unscrew the lever housing retaining nuts and withdraw the bolts

5.13a Turn the housing over, then remove the screws . . .

5.13b . . . and lower the plastic cover panel from its underside

5.14 Disengage gear lever base from the selector rod boss, and remove the housing

Refitting

16 Refit the gear change lever housing by following the removal procedure in reverse. On completion check, and if necessary adjust, the alignment of the gear change linkage, as described in Section 6.

6 Gear change linkage - removal, refitting and adjustment

Removal

1 Working from Section 5, remove the gear change lever housing.
2 Ensure that the vehicle is parked on a level surface, then apply the handbrake and chock the rear wheels.
3 Raise the front of the vehicle and rest it securely on axle stands (see "Jacking and Vehicle Support").
4 Working under the engine bay, separate the two halves of the selector rod joint. On earlier models, this entails driving out the taper pin from the joint using a pin punch. Later models are fitted with a clamp bolt joint, located on the engine bay side of the bulkhead, which replaces the earlier taper pin type. On these models, slacken the clamp bolt and pull the two halves of the joint apart **(see illustration)**.
5 Draw selector rod into the cabin, taking care to avoid damaging the rubber gaiter at the bulkhead.

Refitting

6 Refit the selector rod by following the removal procedure in reverse. On completion, check and if necessary adjust the alignment of the gear change linkage, as described in the following sub-Section. **Note:** *The ignition switch must be turned to the 'unlocked' position before the selector rod is refitted, to ensure correct re-alignment of the locking mechanism.*

Adjustment - up to 1983 model year

7 Engage reverse gear and turn the ignition key to position 'L' (locked).
8 Check that the axial movement of the gear lever when moved back and forth is no more than 3.0 - 4.0 mm. If not, refer to Section 6 and loosen the lever housing retaining nuts, then move the housing forward or backward as necessary to bring the measurement within tolerance. Tighten the housing screws securely, after making the adjustment.

Adjustment - 1983 model year onwards

9 Disconnect the battery negative cable and position it away from the terminal.
10 Move both front seats towards the rear of the cabin, to the limit of their travel.
11 Refer to Chapter 11 and remove the bellows moulding from the front of the centre console.
12 Engage reverse gear and remove the ignition key from the cylinder.
13 Prise off the gear change lever gaiter, then remove the two console cover panel retaining screws beneath. Lift the front of the console cover panel and unplug the cabling for the interior lighting switches.
14 Insert the key into the ignition switch cylinder and release the gear change lever from the reverse position.
15 Remove the cover panel from the console, then remove the ashtray. Slacken and withdraw the four console retaining screws and push the console towards the rear of the cabin. Put the gear change lever into third gear.
16 At the alignment groove on the transmission casing, insert a 4 mm diameter rod (e.g. a twist drill bit) to lock the transmission in third gear. The alignment groove is located at the rear of the transmission, on the differential cover plate. When the rod is inserted into the groove, it should then be possible to slide it along into a drilling in the selector rod; peel back the rubber bellows where the selector rod enters the transmission casing, if neccesary **(see illustration)**.
17 From within the engine bay, slacken the clamp bolt on the selector rod joint. **Note:** *To improve access, it may be beneficial to raise the front of the vehicle and rest it on axle stands (see "Jacking and Vehicle Support").*
18 From inside the cabin, locate the gear change lever housing-to-selector rod alignment holes, at the front of the console

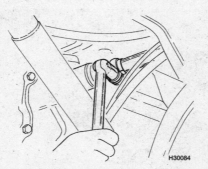

6.4 On later models, slacken clamp bolt and pull selector rod joint apart

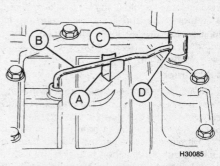

6.16 Method of locking transmission in third gear

A Alignment groove C Selector rod
B 4mm diameter rod D Hole
(with rod inserted)

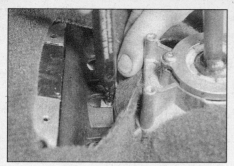

6.18 Insert a cross-head screwdriver to lock gear lever in third gear

7.1 Remove the screws retaining the metal baseplate

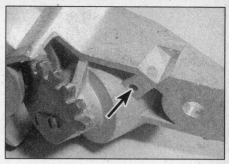

7.3 Drilling for lock cylinder pin, on left-hand side (base plate removed for clarity)

moulding. Insert a cross-head screwdriver with a 6 mm diameter shank into the holes, to lock the lever in the third gear position **(see illustration)**.
19 Tighten the selector rod clamp bolt securely.
20 Remove both the transmission casing and gear change lever alignment tools. Slide the console forward into position and refit the console retaining screws, tightening them securely.
21 Reconnect the cables to the interior lighting switches, then refit the console cover panel and retaining screws.
22 Check that all gears can be obtained smoothly and accurately.
23 Reconnect the battery negative cable.

7 Gear lever lock and ignition switch – removal and refitting

General

Located on the console between the front seats, adjacent to the gear change lever, the ignition switch and gear lever lock cylinder are both separate components, linked by a quadrant gear mechanism. When the ignition key is inserted into the lock cylinder and turned, the gear lever is released from its locked position (reverse). At the same time, the quadrant gear at the base of the lock cylinder gear turns the ignition switch spindle, enabling the vehicles auxiliary electrical, ignition and starting systems.

Lock cylinder
Removal

1 Refer to Section 6 and remove the gear change lever and housing. Unscrew the lower plastic cover, then remove the metal baseplate retaining screws **(see illustration)**.
2 Insert the key into the cylinder and turn it to a point between the 'L' and 'G' positions.
3 Turn the gear lever housing over, and identify the drilling for the lock cylinder pin, on the left hand side **(see illustration)**.
4 With the key set as described, the retaining pin can be driven in, by inserting a rod of suitable diameter.
5 Lift the lock cylinder out of the gear lever housing.

Refitting

6 Insert the key into the cylinder and turn it to a point midway between the 'L' and 'G' positions.
7 Press the lock cylinder pin into the drilling. Insert the cylinder into the gear lever housing, ensuring that the teeth of the quadrant gear engage with the slot in the lock cylinder.
8 Refer to Section 5 and refit the gear change lever housing.

Destructive removal (lost key or inoperative lock)

9 If the ignition key has been lost, or if for some reason the lock cannot be operated, the following procedure describes the removal of the lock cylinder without the use of the key.
Note: *The lock cylinder is effectively destroyed using this method of removal.*

10 Apply the handbrake. Move the gear change lever to the neutral position.
11 Unclip the lever gaiter from the surface of the console and fold it up over the lever knob.
12 Working from Chapter 11, remove the left hand front seat and and centre console.
13 Drill out the plug in the lock cylinder retaining pin drilling.
14 Using a 2 mm pin punch and hammer, drive in the cylinder retaining pin, then lift the lock cylinder out.

Ignition switch
Removal

15 Referring to Section 5, remove the gear change lever housing, then remove the lower plastic cover and metal baseplate.
16 Turn the housing over, to expose the heads of the upper cover plate retaining screws. Slacken and withdraw the screws, then lift off the cover plate **(see illustration)**.
17 Remove the two ignition switch retaining screws and withdraw the switch assembly **(see illustration)**.

Refitting

18 Before the switch can be refitted, it must be set in a reference condition to ensure correct alignment with the lock cylinder quadrant gear.
19 Turn the switch over, and insert a screwdriver into the slot on the end face of the spindle. Turn the spindle so that the punch mark on the edge of the switch end face lines up with the alignment arrow on the base of the switch body **(see illustration)**.
20 Turn the ignition key to the 'L' position.

7.16 Slacken and withdraw the screws, then lift off the cover plate

7.17 Remove the two ignition switch screws and withdraw the switch assembly

7.19 Punch mark on spindle end face aligns with the arrow on switch body

7.21 Fit the switch into the gear lever housing, ensuring that the lug (A) engages with the slot (B) in the lever housing

21 Fit the switch into the gear lever housing, ensuring that the alignment lug engages with the slot in the lever housing (see illustration).
22 Refit and tighten the switch retaining screws, then turn the ignition key back and forth to verify that the drive plate is correctly engaged with the teeth of the lock cylinder quadrant gear. Check the operation of the gear change lever lock mechanism.
23 Refit the gear lever housing cover plate and tighten the retaining screws.
24 Refer to Section 5 and refit the gear lever housing, ignition and reversing light switch cables.
25 Working from Chapter 11, refit the centre console, ventilation duct, carpet, door trim panels and seat.
26 Reconnect the battery negative cable and verify the operation of the ignition switch.

8 Manual transmission oil – draining and refilling

General information

1 The transmission is filled with the correct quantity and grade of oil at manufacture. The level must be checked regularly and if necessary topped up, in accordance with the maintenance schedule (see Chapter 1). In some markets outside the UK, there is a requirement to drain and renew the manual transmission oil at specified intervals; check your vehicles documentation and verify the service interval with your dealer if in doubt.

Draining

2 Drive the car sufficiently to warm the engine/transmission up to normal operating temperature; this will reduce the viscosity of the oil and speed up the draining process.
3 Park the car on level ground, switch off the ignition and chock the rear wheels (apply the handbrake on vehicles fitted with rear caliper actuating handbrake). For improved access, jack up the front of the car and support it securely on axle stands. **Note:** *The car must be lowered to the ground and parked on a level surface, to ensure accuracy when refilling and checking the oil level.*
4 Wipe clean the area around the filler plug and dipstick, which are situated on the cover plate, at the side of the transmission. Unscrew the plug from the casing and recover the sealing washer.
5 Position a container with a capacity of at least 3 litres (ideally with a large funnel) under the drain plug, which is located centrally, on the underside of the transmission – use a wrench to unscrew the plug from the casing.
6 Allow the all the oil to drain completely into the container. If the oil is still hot, take precautions against scalding. Clean the drain plug thoroughly, paying particular attention to the threads. Discard the original sealing washers; they should be always renewed whenever they are disturbed.

Refilling – routine maintenance

7 When the oil has drained out completely, clean the plug hole threads in the transmission casing. Fit a new sealing washer to the drain plug, coat the thread with sealing compound and screw it into the transmission casing, tightening it securely. If the car was raised for the draining operation, lower it to the ground.
8 When refilling the transmission, allow plenty of time for the oil level to settle completely before attempting to check it. Note that the car must be parked on a flat, level surface when checking the oil level. Use a funnel if necessary to maintain a gradual, constant flow and avoid spillage.
9 Refill the transmission with the specified grade and quantity of oil, then check the oil level using the dipstick. If the level is over the "MAX" graduation, refit the dipstick, then drive the car for a short distance so that the new oil is distributed fully around the transmission components. Re-check the level; if the level is still over the 'MAX' graduation, drain off a small quantity from the drain plug. Excessive oil in the transmission can cause stiff gear selection in cold conditions.
10 On completion, fit the filler plug with a new sealing washer, then coat its threads with thread sealing compound and refit it, tightening it securely.

Refilling – after overhaul

11 The procedure for refilling after overhaul is essentially the same as that for refilling during routine maintenance, except that about 0.3 litres (4 speed) / 0.4 litres (5 speed) of the total oil capacity should be poured into the primary gear casing before the transmission is refitted to the engine. This will protect the transmission components during the initial start-up period, before distribution of oil from the main casing has occurred.

9 Manual transmission – removal and refitting

The transmission can only realistically be removed from the vehicle whilst it is still bolted to the engine. Separation of the two within the confines of the engine bay will provide little gain and make the task unnecessarily complicated.; the procedure is therefore not recommended. Once the power unit has been removed, the transmission can then be detached from the engine on the bench and worked upon separately.

The removal and separation of the engine and gearbox assembly are described in Chapter 2 Part B.

10 Manual transmission overhaul – general information

The overhaul of a manual transmission is a complex (and often expensive) engineering task for the DIY home mechanic to undertake, which requires access to specialist equipment. It involves dismantling and reassembly of many small components, measuring clearances precisely and if necessary, adjusting them by the selection shims and spacers. Internal transmission components are also often difficult to obtain and in many instances, extremely expensive. Because of this, if the transmission develops a fault or becomes noisy, the best course of action is to have the unit overhauled by a specialist repairer or to obtain an exchange reconditioned unit.

Nevertheless, it is not impossible for the more experienced mechanic to overhaul the transmission if the special tools are available and the job is carried out in a deliberate step-by-step manner, to ensure that nothing is overlooked.

The tools necessary for an overhaul include internal and external circlip pliers, bearing pullers, a slide hammer, a set of pin punches, a dial test indicator and ideally, a hydraulic press. A large, sturdy workbench and bench vice will also be required.

During dismantling of the transmission, make careful notes of how each component is fitted to make reassembly easier and accurate.

Before dismantling the transmission, it will help if you have some idea of where the problem lies. Certain problems can be closely related to specific areas in the transmission which can make component examination and renewal easier. Refer to the Fault Diagnosis Section at the beginning of this manual for more information.

Chapter 7 Part B:
Automatic transmission

Contents

Automatic tranmission fluid - renewal . . . . .See Chapter 1	Selector cable - removal refitting adjustment .2
Automatic transmision overhaul - general information 9	Selector lever and housing - removal and refitting 5
Automatic transmission - removal and refitting 8	Selector lever lock and ignition switch - removal and refitting 7
Automatic transmission fluid - level check . . . . .See Chapter 1	Speedometer drive - removal and refitting 4
General information 1	Start inhibitor / reversing light switch - removal and refitting 6
Kickdown cable - adjustment 3	

Degrees of difficulty

Easy, suitable for novice with little experience		Fairly easy, suitable for beginner with some experience		Fairly difficult, suitable for competent DIY mechanic		Difficult, suitable for experienced DIY mechanic		Very difficult, suitable for expert DIY or professional	

Specifications

Designation	4 HP 18
Weight	65.5 kg approx. including torque converter
Fluid capacity, including fluid cooler	8.2 litres approx. including torque converter and oil cooler

Torque wrench settings	Nm	lbf ft
Selector lever to cable clamp nut	8	6
Transmission oil drain plug	6	4
Oil pressure gauge plug	6	4
Transmission lower cover plate retaining screws	11	8
Selector cable set screw	3	2

1 General information

The automatic transmission is of Borg-Warner Type 35 or Type 37 having three forward speeds and one reverse. The Type 37 transmission is a modified, strengthened version of the Type 35, fitted from 1979 model year. Only the Type 37 is covered in this Chapter.

The transmission is controlled by a cabin floor mounted, six position selector lever. The transmission operates in different modes, depending on the position of the selector lever.

In 'Park', the transmission is mechanically locked, thus preventing the roadwheels from turning. For this reason, 'Park' must only be selected when the vehicle has come to a complete standstill.

In 'Neutral', the transmission disconnects drive between the engine and the driven wheels, in the same manner as a manual transmission. This selection should not be made if the vehicle is moving. Apply the handbrake when parking with the transmission in this position.

In 'Drive', the transmission will automatically shift between the three forward gears according to road speed and accelerator position. First gear is always selected for a pull away from standstill and as the vehicle accelerates, higher ratio gears are sequentially selected at pre-defined road speed thresholds, to provide optimum comfort, fuel consumption and driveability.

To avoid transmission and drivetrain damage, the vehicle must be allowed to come to a standstill before selecting 'Reverse'.

The vehicle should be started in with either 'Park' or 'Neutral' selected; an electrical inhibitor switch prevents starting, if an inappropriate gear is currently selected.

The transmission shifting range can be limited to the first two, or even just first gear, if driving conditions require it. Selecting '2' allows automatic shifting between the first two gears, but upshift to third gear is prevented. If '2' is selected whilst the transmission is in third gear (with 'Drive' selected) the downshift to second takes place immediately. To avoid engine and transmission damage, do not select '2' at speeds above around 55 mph.

Selecting '1' locks the transmission in first gear, preventing upshift to higher ratio gears. This provides optimum performance and control when ascending or descending very steep hills, particularly when towing. It also prevents repeated automatic up and downshifts between first and second gear, which could otherwise overheat the transmission fluid. If '1' is selected whilst the transmission is in third or second gear (with 'Drive' or '2' selected) the downshift to first takes place sequentially when the road speed has dropped to the preset threshold for second gear, thus avoiding engine and transmission damage and sudden, unexpected engine braking.

To provide maximum acceleration, e.g. for overtaking, pressing the accelerator to the end of its travel, past the full throttle position, will cause the transmission to 'kickdown'. At this point, if the gear currently selected is not the optimum gear for maximum acceleration at the current road speed, the transmission will automatically downshift to a lower ratio gear. Selection of that gear will then be maintained either until the accelerator is released from the kickdown position, or until the maximum road speed for the gear is reached; upshift to a higher ratio gear will then occur.

7B•2 Automatic transmission

Due to the complexity of the transmission, major repairs and overhaul operations should be entrusted to a Saab dealer, who will be equipped with the necessary tools for fault diagnosis and repair. The information in this Chapter is therefore limited to descriptions of servicing operations that can be carried out without the need for specialised test or repair equipment.

2 Selector cable - removal, refitting and adjustment

Removal

1 Park the vehicle on a level surface, apply the handbrake (1988 model year onwards only), and chock the rear wheels.
2 Raise the front of the vehicle and rest it securely on axle stands (see "*Jacking and Vehicle Support*").
3 At the transmission casing, unscrew the selector cable retaining bolt from the transmission. Pull on the selector cable to draw the gear shift rod to its outermost position.
4 Slide back the spring tensioned sleeve and unhook the cable end fitting from the gear shift rod (see illustration).
5 Working inside the cabin, refer to Section 5 and remove the console panels and selector lever housing.
6 At the underside of the selector lever housing, extract the circlip from the end of the cable. Unscrew the clamp bolt, to free the cable from the selector lever (see illustration). Note the position of the cable in relation to the clamp, to provide guidance when the new cable is fitted.
7 Slacken the outer cable sheath lock nuts, then release the cable from the housing.
8 Working from Chapter 11, remove the trim panels from around the lower edge of the door apertures, then fold back the carpet to expose the centre heater duct.
9 Referring to Chapter 3, separate the ducting at the joint beneath the seat mounting member.
10 Release the existing selector cable from any securing clips, then pull the cable through the bulkhead into the cabin, taking care to avoid damaging the bulkhead grommet.

Refitting

11 Carefully guide the cable through bulkhead grommet and lay it in position.
12 At the transmission casing, slide back the spring loaded sleeve on the gear shift rod, then engage the cable end fitting with the corresponding fitting on the shift rod. Release the sleeve.
13 Fit the cable sheathing in place, then refit and tighten the retaining bolt.
14 Working inside the cabin, insert the end of the cable into the hole in the end of the selector lever, then refit the circlip.
15 Fit the cable sheathing into the selector lever housing, then tighten the lock nuts to secure it in place.
16 Reference the selector cable by pulling it into the lever housing, to the limit of its travel; this sets the transmission into 'PARK', Push the cable back into its sheathing by two stages; this sets the transmission to 'NEUTRAL'.
17 Refit the selector lever and housing to the floorpan, referring to Section 7 for guidance.
18 Move the selector lever to position 'N', then tighten the cable set screw to the specified torque. The set screw is located at the base of the selector lever housing, on the left hand side (see illustration).
19 Refit the heater duct, carpet and door aperture trim panels.
20 Perform a check of the selector mechanism adjustment, as described in the following sub-Section.

Adjustment

21 Move the selector lever to the 'N' position.
22 Depress the pawl button and apply light pressure to the lever; first in the forward direction (towards 'R'), then backwards (towards 'D'). Increasing resistance should be felt in both directions, as the gear selector valves and detent spring are actuated.
23 Hold the selector lever midway between the positions at which this resistance is felt, then release the pawl button. Note the position of the lever in relation to the gear indication plate.
24 If the lever is directly opposite the 'N' position, no further adjustment is necessary. Refit the lever housing console panels.

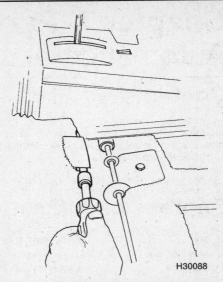

2.18 The set screw is located at the base of the selector lever housing, on the left-hand side

25 If the lever is not in line with the 'N' position on the indicator plate, adjustment is necessary. Depress the pawl button, then slacken the selector cable clamp bolt at the base of the selector lever. Release pawl button and move the lever to the 'N' position - positive location should be felt as the lever engages with the alignment notch in the gate. Tighten the cable clamp bolt to the specified torque.
26 Repeat the above check, this time at the 'D' position.
27 On completion, check the operation of the selection mechanism, before bringing the vehicle back into service.

3 Kickdown cable - removal, refitting and adjustment

Removal

1 Park the vehicle on a level surface, apply the handbrake (1988 model year onwards only), and chock the rear wheels.
2 Raise the front of the vehicle and rest it securely on axle stands (see "*Jacking and Vehicle Support*").
3 At the carburettor/throttle body (as applicable), disconnect the kickdown cable from the throttle spindle lever and mounting bracket.
4 Place a container with a capacity of at least 3 litres beneath the transmission, then unscrew the drain plug and drain the fluid.
5 Remove the retaining screws and lower the bottom cover from the front of the transmission, beneath the primary gearcase. Recover the gasket.
6 Remove the swarf collection magnet, oil strainer and level pipe.
7 Rotate the kickdown cam slightly, to allow

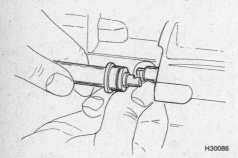

2.4 Slide back the sleeve and unhook the cable end fitting from the gear shift rod

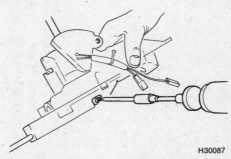

2.6 Unscrew the clamp bolt, to free the cable from the selector lever

Automatic transmission

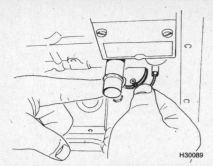

3.7 Rotate the kickdown cam slightly, to extract the inner cable nipple from it

3.12 Clamp the stop clip (arrowed) on the inner cable, so that it abuts the end of the threaded section of the outer cable

4 Speedometer drive - removal and refitting

Refer to the information in Chapter 7A.

5 Selector lever and housing - removal and refitting

Selector lever housing

Note: *This procedure requires access to a non-standard type of socket, Saab tool no. 87 90 370, to allow removal of the housing retaining nuts. However, a similar tool can be fabricated by grinding down the sides of an old 12mm socket, to fit the triangular nut flats.*

Removal

the inner cable nipple to be extracted from it **(see illustration)**.
8 Unscrew the outer cable nut from the transmission casing, then release it from the casing by pushing it out from inside the transmission. Recover the O-ring seal.
9 Withdraw the cable from the car.

Refitting

10 Insert the new cable into the transmission casing; fit a new O-ring seal to the threaded section of the cable outer. Tighten the retaining nut securely.
11 Working inside the transmission, connect the inner cable to the kickdown cam, making sure that the cable seats in the guide channel, and that the cable nipple slots into the hole in the cam.
12 With the outer cable pulled straight, draw the inner cable out until the kickdown cam just starts to rotate; at this point, clamp the stop clip on the inner cable, so that it abuts the end of the threaded section of the outer cable **(see illustration)**.
13 Connect the cable to the throttle disc at the carburettor/throttle (as applicable). Fit the cable outer into the mounting bracket.
14 With the accelerator pedal fully depressed, adjust the length of the outer cable, until the highest lobe of the kickdown cam is just in contact with the kickdown valve. It may be helpful to have an assistant depress the accelerator pedal whilst the adjustment is made.
15 Refit the cover plate to the underside of the transmission, together with a new gasket, then insert the retaining bolts and tighten them to the specified torque. Lower the vehicle to the ground.
16 Refer to the Specifications in Chapter 1 and refill the transmission with the specified grade and quantity of fluid.
17 Final adjustment of the downshift cable can only be made using a pressure gauge; if one is not available, the work should be entrusted to a Saab dealer.

Adjustment

18 Remove the line pressure plug from the front right-hand side of the transmission casing and connect a pressure gauge to it. Connect a hand held tachometer to the engine.
19 Raise the front of the vehicle and support it securely on axle stands (see "*Jacking and Vehicle Support*"). Chock the rear wheels, and on 1988 model year-onwards vehicles, apply the handbrake.
20 Select 'P'. Start the engine and check the engine idle speed using the tachometer. Verify that the speed is correct, according to the Specification in Chapter 4A. Adjust the idle speed, if necessary.
21 Disconnect the kickdown cable from the carburettor/throttle body (as applicable). Move the kickdown cable inner back and forth inside the outer cable, to verify that the throttle valve (which is operated in part by the kickdown valve) is not binding - the pressure reading at the gauge should vary accordingly as the cable is moved.
22 Slowly draw the kickdown cable inner from the outer to the the limit of its travel to obtain a maximum line pressure reading. Then allow the cable to retract to its original position and check that the line pressure drops to its original value. If the pressure does not drop to this value, or does not drop below 4.9 bar, the likelihood is that the throttle valve is faulty - the transmssion unit should be referred to a Saab dealer under these circumstances.
23 Reconnect the kickdown cable to the throttle spindle lever. Move the selector lever to 'D'.
24 Position the throttle spindle lever to relieve all load and tension from the kickdown cable, indicated by a minimum reading on the pressure gauge. Now adjust the kickdown cable outer, at the nut on the mounting bracket, to obtain an increase in the pressure reading of 0.1 bar.
25 Move the selector lever to 'P and check that the line pressure rises to between 4.2 and 4.9 bar.
26 Switch off the engine, remove the pressure gauge, and refit the plug at the transmission casing.
27 Road test the vehicle and check the operation of the kickdown mechanism.

1 Disconnect the battery negative cable and position it away from the terminal.
2 Refer to Chapter 11 and remove the left hand seat from the cabin area.
3 Move the selector lever to 'P', then remove the ignition key.
4 Prise the cap from the top of the selector lever knob then slacken and remove the locknut and knob.
5 Remove the lever indicator bezel.
6 Remove the bezel lamp and diffuser lens, refer to Chapter 12 for details.
7 Where fitted, remove the knob from the choke control cable; refer to Chapter 4 Part A.
8 Remove the two retaining screws from the upper edge of the console. Lift up the console, then reach underneath and disconnect the wiring from the interior lighting switches at the connectors; label the cables to aid refitting later.
9 Remove the ashtray from the rear of the console. Remove the remaining retaining screws and lift off the console.
10 Fold back the carpet from the left hand side of the selector lever housing, to expose the selector cable set screw.
11 Slacken the selector cable set screw, to free it from the selector lever.
12 Referring to Section 10 for guidance, unplug the wiring from the ignition switch at the connector.
13 Working from Sections 8 and 9 respectively, disconnect the wiring from the starter inhibitor switch and the reversing light switch. Label the cables and terminals carefully, to aid refitting later.
14 Where applicable, disconnect the cable from the choke control.
15 Using Saab special tool no. 87 90 370 (or the fabricated tool), slacken and remove the lever housing retaining screws.
16 Lift the lever housing away from the floorpan and turn it over; remove the screws and lift the cover panel from the underside.
17 Extract the circlip from the end of the selector cable, then withdraw the cable and remove the housing from the cabin.

7B•4 Automatic transmission

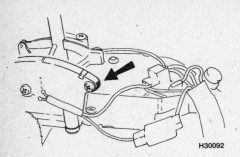

6.3 Location of the start inhibitor/reversing light switch on the selector lever housing

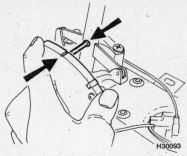

6.6 Rotate the switch body, until the alignment mark on the top surface lines up with the selector lever actuator pin

Refitting

18 Refitting is a reversal of removal, noting the following points:
(a) Referring to Section 2, refit and adjust the selector cable, then check the selector lever positioning.
(b) Reconnect all electrical cables, using the notes made during removal.
(c) Where applicable, refer to the relevant Section of Chapter 4A and refit the choke control knob.
(d) On completion, reconnect the battery negative cable and check the operation of all electrical components disturbed during the removal process, to verify that the cables have been reconnected correctly.

Selector lever

Removal

19 Refer to the previous sub-Section and remove the selector lever housing.
20 Using the key, turn the ignition switch to the 'D' position.
21 Remove the pin from the quadrant gear at the base of the ignition switch.
22 Slacken and withdraw the bolts from the base of the selector lever, then withdraw the lever from the housing.

Refitting

23 Refit the selector lever by following the reverse of the removal procedure.

6 Start inhibitor / reversing light switch - removal and refitting

Removal

1 Disconnect the battery negative cable and position it away from the terminal. Set the gear selector lever to the 'N' position.
2 Remove the gear selector console, as described in Chapter 11.
3 Unplug the wiring from the switch assembly at the connectors; label each connector carefully to ensure that it is reconnected correctly (see illustration).
4 Slacken the two mounting screws and lift the switch assembly away from the selector lever housing.

Refitting

5 Ensure that the selector lever is set to the 'N' position, then fit the switch assembly into position; engage the selector lever actuator pin with the slot in the side of the switch. Loosely fit the switch mounting screws.
6 Rotate the switch body, until the alignment mark on the top surface lines up with the selector lever actuator pin (see illustration).
7 Tighten the mounting screws securely, then restore the electrical connections, referring to the notes made during removal.
8 Refit the gear selector console, then reconnect the battery negative cable.
9 On completion, test the operation of the inhibitor switch and reversing light switch.

7 Selector lever lock and ignition switch - removal and refitting

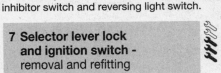

Refer to the information in Chapter 7A.

8 Automatic transmission - removal and refitting

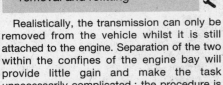

Realistically, the transmission can only be removed from the vehicle whilst it is still attached to the engine. Separation of the two within the confines of the engine bay will provide little gain and make the task unnecessarily complicated.; the procedure is therefore not recommended. Once the power unit has been removed, the transmission can then be detached from the engine and mounted on a stand or a work bench and worked upon separately.
Removal and separation of the engine and gearbox are described in Chapter 2B.

9 Automatic transmission overhaul - general information

In the event of a fault occurring, it will be necessary to establish whether the fault is electrical, mechanical or hydraulic in nature, before repair work can be contemplated. Diagnosis requires detailed knowledge of the transmissions operation and construction, as well as access to specialised test equipment, and so is deemed to be beyond the scope of this manual. It is therefore essential that problems with the automatic transmission are referred to a Saab dealer for assessment.
Note that a faulty transmission should not be removed before the vehicle has been assessed by a dealer, as fault diagnosis is carried out with the transmission in situ.

Chapter 8
Driveshafts

Contents

Driveshaft and outer CV joint - removal and refitting 2
Driveshaft gaiters - renewal 3
General description 1
Driveshaft inner joint - removal and refitting 4

Degrees of difficulty

Easy, suitable for novice with little experience	Fairly easy, suitable for beginner with some experience	Fairly difficult, suitable for competent DIY mechanic	Difficult, suitable for experienced DIY mechanic	Very difficult, suitable for expert DIY or professional

Specifications

Torque wrench settings

	Nm	lbf ft
Driveshaft nut:		
(Up to 1980 model year)	350	258
(1981 model year onwards)	300	221
Inner driveshaft joint flange bolts	25	18

1 General description

The driveshafts are of the solid type and are splined to needle roller type inner universal joints and constant velocity type outer joints. The inner shafts are splined to the transmission differential gears and supported in ball bearings located in the differential bearing housings. The outer shafts are splined to the front wheel hubs. On pre-1981 model year vehicles, the outer shafts are supported directly in double row ball bearings located in the steering knuckles. From 1981 model year onwards, the hubs are supported in the bearings and the shafts in the hubs; refer to Chapter 10 for greater detail of the hub assembly.

The inner joints incorporate tripod joints which are located in slotted cylindrical housings by needle roller bearing races. The outer joints incorporate a hub and six balls which move in slotted grooves, designed to provide constant velocity rotation at varying shaft angles (see illustration).

No routine maintenance is required, except to inspect the flexible driveshaft gaiters for damage or deterioration.

2 Driveshaft and outer joint - removal, overhaul and refitting

Removal

1 Turn the steering on full lock, then with the full weight of the vehicle on the suspension, place a suitable block of hardwood or metal between the front suspension upper wishbone and the body underframe. This is necessary to enable the lower control arm bolts to be removed without being under tension from the coil spring.

2 Prise the hub grease cap from the front wheel using a screwdriver and mallet.

3 Loosen the roadwheel nuts then unscrew and remove the driveshaft nut. **Note:** *If the driveshaft nut flange has been locked to the end of the driveshaft by peening, release it using a hammer and punch before attempting to unscrew it. The nut is tightened to a high torque, so if necessary fit an extension tube over the breaker bar handle to give additional leverage. Discard the driveshaft nut as a new one must be fitted on reassembly.*

4 Remove the washer/cone ring, as applicable, then discard it as a new item must be fitted on reassembly.

5 Chock the rear wheels, then raise the front of the vehicle, and support it securely on axle stands (see "Jacking and Vehicle Support"). Remove the roadwheel.

6 Working from Chapter 9, remove the brake caliper - note that there is no need to disconnect the hydraulic hose. Use a length of wire to tie the caliper to the suspension coil spring. **Note:** *On pre-88 model year vehicles where the handbrake acts on the front discs, it will be necessary to disconnect the handbrake cable from the caliper.*

7 Remove the brake disc with reference to Chapter 9.

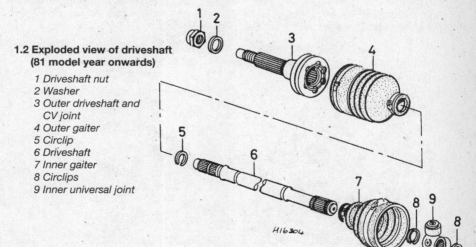

1.2 Exploded view of driveshaft (81 model year onwards)

1. Driveshaft nut
2. Washer
3. Outer driveshaft and CV joint
4. Outer gaiter
5. Circlip
6. Driveshaft
7. Inner gaiter
8. Circlips
9. Inner universal joint

2.8 Remove the large diameter clip from driveshaft inner joint, to release gaiter

8 Remove the large diameter clip from the driveshaft inner joint, to release the rubber gaiter (see illustration). The clip should be discarded once removed, as it is not re-usable.

9 Unscrew the nuts from the steering tie-rod end and upper control arm balljoint, then use a balljoint separator tool to release the tie-rod and upper balljoint; see Chapter 10 for details.

10 Still working from Chapter 10, unscrew and remove the bolts securing the lower balljoint to the lower wishbone.

11 Withdraw the driveshaft complete with the steering knuckle from the car. Ensure that the inner joint needle bearings and races do not slide off the tripod joint - use an elastic band or plastic bag to keep them in place.

12 On pre-1981 models, use a puller to remove the hub from the driveshaft, then support the steering knuckle and press or drive the driveshaft through the wheel bearing.

2.15 Use circlip pliers to expand the circlip inside the CV joint

2.16b The large circlip (arrowed) remains captive in the outer half of the CV joint

13 On vehicles from 1981 model year onwards, use a proprietary hub puller to press the driveshaft through the hub.

14 Release the rubber gaiter from the outer joint and slide it along the driveshaft.

15 Support the driveshaft with the outer joint downwards. Use circlip pliers to expand the circlip inside the CV joint (see illustration). Note: *On earlier models, it will be necessary to press down on the driveshaft to compress the spring washers, before the CV joint circlip can be expanded. These washers, together with their retaining circlip amd the spherical washer beneath should be discarded and **not** refitted; the components have been deleted from all subsequent vehicles.*

16 Withdraw the driveshaft from the CV joint. Note that the large circlip remains captive in the outer half of the CV joint (see illustrations).

Overhaul

17 Thoroughly clean the driveshaft splines, CV joint and tripod joint components with paraffin or a suitable solvent, taking care not to obliterate the alignment marks made during removal, the allow them to dry. Examine the CV joint components for wear and damage; in particular, check the balls and corresponding grooves for pitting and corrosion. If evidence of wear is visible, then the joint must be renewed. Note that the CV joint, balls and outboard driveshaft must be renewed as a matched set. Examine the tripod joint components for wear. Check that the three rollers are free to rotate without resistance and are not worn, damaged or corroded. The rollers are supported by arrays of needle

2.16a Withdraw the driveshaft from the CV joint

2.18 Packing the joint with the specified grease

bearings; wear or damage will be manifested as axial play in the rollers and/or roughness in rotation. If any such wear is discovered, the tripod joint must be renewed.

18 Commence reassembly by packing the joint with the specified grease, pushing it into the ball grooves and expelling any air that may be trapped underneath (see illustration). **Caution: Do not allow grease to come into contact with vehicle's paintwork, as discolouring may result.**

19 With the outer gaiter on the driveshaft, insert the shaft into the outer joint until the circlip engages the groove.

20 On pre-1981 model year vehicles, support the joint housing then locate the steering knuckle on the end of the driveshaft and use a metal tube on the wheel bearing inner race to drive the assembly fully onto the driveshaft. Locate the hub on the driveshaft splines and tap it up to the bearing. Fit the washer and nut loosely.

Refitting

21 On vehicles from 1981 model year onwards, lubricate the driveshaft splines with a molybdenum based grease then insert the shaft into the splined hub and tap in fully. Fit the new washer and nut loosely.

22 Pack the joint with more grease then fit the gaiter to the joint housing and secure with a new clip.

23 Check that the inner joint housing is packed with grease and that the needle bearings are in position on the tripod, then insert the driveshaft. (Remove the protective elastic band or plastic bag as applicable.)

24 Locate the lower balljoint in the lower control arm, insert the bolts (so that the bolt heads facing the rear of the vehicle) and tighten the nuts; refer to Chapter 10 for greater detail.

25 Still working fro Chapter 10, fit the tie-rod end and upper control arm balljoints to the steering knuckle and tighten the nuts.

26 Locate the rubber gaiter on the inner joint housing and secure with a new clip.

27 Refer to Chapter 9 and refit the brake disc, brake caliper and (where applicable) handbrake cable, with reference to Chapter 9.

28 Refit the roadwheel and lower the car to the ground.

29 Tighten the roadwheel nuts and the driveshaft nut to the specified torque. Where applicable, lock the driveshaft nut by peening the flange into the groove in the driveshaft .

30 Refit the hub cap then remove the hardwood or metal from beneath the suspension upper control arm.

3 Driveshaft gaiters - renewal

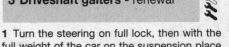

1 Turn the steering on full lock, then with the full weight of the car on the suspension place a suitable block of hardwood or metal between the front suspension upper control

Driveshafts 8•3

3.8 Fitting a new driveshaft gaiter

arm and the body underframe. This is necessary to enable the lower control arm bolts to be removed without being under tension from the coil spring.
2 Loosen the front roadwheel nuts, then chock the rear wheels,
3 Jack up the front of the car and support it on axle stands (see *"Jacking and Vehicle Support"*). Remove the relevant front roadwheel.
4 Unscrew the large diameter clip from the driveshaft inner joint, and release the rubber gaiter.
5 Working from Chapter 10, unscrew and remove the bolts securing the lower balljoint to the lower suspension wishbone, then pull out the steering swivel member and pivot it upwards. Support the member on an axle stand.
Caution: Do not lose any of the needle rollers from the inner joint tripod; use a plastic bag tightly secured with an elastic band to keep them in place.
6 Extract the circlip and remove the inner joint tripod, then remove the inner circlip (only fitted to early models).
7 Release the clips and slide the defective gaiter(s) from the driveshaft as necessary.
8 Wipe clean the driveshaft and slide on the new gaiters **(see illustration)**. Pack the outer gaiter with 80 grammes of molybdenum disulphide grease.
9 Fit the inner joint tripod and circlip(s).
10 Remove the protective bag and check that the needle rollers and races are in position on the inner joint tripod. Pack the inner gaiter with 60 grammes of grease.

11 Swivel the steering knuckle downwards and at the same time, locate the tripod and needle bearings in the inner joint housing.
12 Locate the lower balljoint in the lower control arm, insert the bolts (from the rear of the car) and tighten the nuts; see Chapter 10 for guidance.
13 Locate the gaiters on the joint housing(s) and secure with the new clips. If the clips are of the crimp variety, secure them in place by squeezing together the indentations in the tab with a pair of pinchers.
14 Refit the roadwheel and lower the car to the ground.
15 Tighten the roadwheel nuts and remove the hardwood or metal support from beneath the suspension upper control arm.

4 Driveshaft inner joint - removal and refitting

Removal

1 Remove the driveshaft complete with the steering knuckle as described in Section 2 paragraphs, but do not remove the driveshaft nut.
2 Place a suitable container beneath the rear of the transmission then unscrew the drain plug and drain the oil; refer to Chapter 7 Part A or B (as applicable) for reference. Clean the drain plug then refit and tighten it.
3 Mark the differential bearing housing in relation to the transmission casing then unscrew and remove the bolts. Disconnect the speedometer cable where applicable.

4.4a Withdrawing the housing assembly

4 Carefully tap the housing assembly free from the casing with a wooden mallet and withdraw it by hand. If the housing is tight, use a slide hammer in the slots provided on the side of the mounting flange to remove it. Recover the spring and plunger from the end of the inner driveshaft. Lift off the shims, noting their orientation and order of fitment **(see illustrations)**. Prise out the O-ring seal and discard it; a new one must be fitted on reassembly.
5 Extract the circlip from inside the housing, then support the housing and drive out the inner driveshaft using a soft metal drift **(see illustration)**.
6 Extract the circlip and pull the joint tripod from the end of the main driveshaft. If it is tight, use a puller to remove it **(see illustrations)**. **Note:** *On earlier models, a second circlip is fitted inboard of the tripod joint; this was deleted on later models.*
7 Clean all the components and examine them for wear and damage. In particular check the needle rollers, races, and tripod; referring to Section 2 for details - renew any components that appear worn. In addition, a new differential bearing housing oil seal must be fitted; refer to Chapter 7 Part A or B as appropriate.
8 Commence reassembly by locating the tripod joint on the main driveshaft and fitting the circlip.
9 Grease the tripod trunnions and fit the needle rollers followed by the races. Fill the races with grease.
10 Stand the inner driveshaft upright on the bench and use a metal tube on the bearing

4.4b Recover the spring . . .

4.4c . . . and plunger from the end of the inner driveshaft

4.4d Lift off the shims, noting their orientation and order of fitment

4.5 Extracting the circlip from inside the housing

inner race to drive the bearing housing fully onto the driveshaft **(see illustrations)**. Fit the circlip in the driveshaft groove.

11 Make sure that the spring and plunger are located in the end of the inner driveshaft, and the shims are in position correctly on the differential bearing housing. Fit a new O-ring seal to the housing.

12 Clean the mating faces then fit the housing and inner driveshaft to the transmission casing using a wooden mallet to tap it fully into position. Coat the threads with sealing compound, insert them and tighten them evenly in diagonal sequence to the specified torque.

13 Reconnect the speedometer cable where applicable.

14 Fill the transmission/final drive with oil with reference to Chapter 6 or 7.

15 Refit the driveshaft and steering knuckle as described in Section 2 paragraphs 22 to 29. As the driveshaft nut has not been disturbed there is no need to tighten it.

4.6a Extract the circlip and pull the joint tripod from the end of the main driveshaft

4.6b If it is tight, use a puller to remove it

4.10a Stand the inner driveshaft upright on the bench . . .

4.10b . . . and using a metal tube butted against the bearing inner race, drive the bearing housing fully onto the driveshaft

Chapter 9
Braking system

Contents

Anti-lock braking system (ABS) components - general information . .18
Anti-lock braking system (ABS) components - removal and refitting .19
Brake fluid level check .See "Weekly Checks"
Brake fluid - renewal . . . . .See Chapter 1
Brake light switch - adjustment, removal and refitting 17
Footbrake pedal - removal and refitting 16
Front brake caliper - removal, overhaul and refitting 5
Front brake disc - inspection, removal and refitting 7
Front brake pads - condition check . . . . .See Chapter 1
Front brake pads - renewal 3
General information 1
Handbrake 'ON' warning light switch - removal and refitting 15
Handbrake - adjustment check .See Chapter 1
Handbrake cables - removal, refitting and adjustment 13
Handbrake lever - removal and refitting 14
Hydraulic brake line and hoses - renewal .9
Hydraulic system - bleeding 2
Master cylinder - removal, overhaul and refitting 10
Rear brake caliper - removal, overhaul and refitting 6
Rear brake disc - inspection, removal and refitting 8
Rear brake pads - condition check . . . . .See Chapter 1
Rear brake pads - renewal 4
Vacuum servo unit - testing, removal and refitting 12
Vacuum servo unit non-return valve - removal, testing and refitting .11

Degrees of difficulty

| **Easy,** suitable for novice with little experience | | **Fairly easy,** suitable for beginner with some experience | | **Fairly difficult,** suitable for competent DIY mechanic | | **Difficult,** suitable for experienced DIY mechanic | | **Very difficult,** suitable for expert DIY or professional | |

Specifications

General
Brake system type and layout:
   Footbrake . . . . Diagonally split, dual hydraulic circuits, outboard discs/calipers fitted front and rear; ventilated front discs on certain models. Electronic ABS fitted as an option on certain models. Non-ABS models vacuum servo assisted, ABS models hydraulic servo assisted.
   Handbrake . . . . Dual cable, lever operated, acting on front discs (pre-1988 model year) or rear discs (1988 model year onwards)

Brake servo (non-ABS models)
Manufacturer . . . . Girling
Type . . . . Manifold vacuum assisted
Diameter 229mm

Master cylinder (non-ABS models)
Manufacturer . . . . Girling
Type . . . . Tandem cylinder
Cylinder internal diameter 22.2mm

Front brakes - pre-1988 model year
Manufacturer . . . . Girling

Brake discs:	Ventilated	Non-ventilated
Outside diameter	276mm	276mm
Thickness, new disc	20.0mm	12.7mm
Minimum after grinding	18.9mm	11.7mm
Maximum variation in disc thickness	0.015mm	0.015mm
Maximum runout	0.08 mm	0.08mm

Brake pad friction material minimum depth*: 1.0mm   1.0mm

Brake calipers:
   Type . . . . Sliding yoke, with direct and indirect pistons
   Cylinder diameter 54mm

* **Note:** *On pre-1983 model year vehicles, the inboard and outboard pads are not interchangeable.*

Rear brakes - pre-1988 model year
Manufacturer .... ATE
Brake discs:
   Outside diameter 267.5mm
   Thickness, new disc 10.5mm
   Minimum after grinding 9.5mm
Brake pad friction material minimum depth 1.0mm
Brake calipers:
   Type .... Fixed caliper, twin direct-acting pistons
   Cylinder diameter 30mm

Front brakes - 1988 model year onwards
Manufacturer .... Girling
Designation .... Colette 54
Brake discs:
   Outside diameter .... 278 mm
   Thickness, new disc .... 23.5 mm ± 0.2 mm
   Minimum thickness, after grinding .... 22.0 mm *
   Maximum variation in disc thickness .... 0.015 mm
   Maximum runout, disc fitted .... 0.08 mm
Note: Both surfaces of the disc must be turned/ground equally.
Brake pad friction material minimum depth .... 4.0 mm
Brake calipers:
   Type .... Sliding caliper, single piston
   Piston diameter .... 54 mm

Rear brakes - 1988 model year onwards
Manufacturer .... ATE
Brake discs:
   Outside diameter .... 258 mm
   Thickness, new disc .... 9.0 mm ± 0.2 mm
   Minimum thickness, after grinding .... 8.0 mm *
   Maximum variation in disc thickness .... 0.015 mm
   Maximum runout, disc fitted .... 0.08 mm
Note: Both surfaces of the disc must be turned/ground equally.
Brake pad friction material minimum depth .... 4.0 mm
Brake calipers:
   Type .... Sliding caliper, single piston
   Cylinder diameter .... 33 mm

Anti-lock Braking System (ABS)
Hydraulic unit:
   Make .... ATE
   Operating pressure, brake circuits .... 0 to 180 bar
Brake fluid reservoir:
   Fluid level indicator resistance .... 10 Ω (reservoir empty)
   ABS warning switch resistance .... 1 Ω (reservoir full)
Hydraulic pump:
   Operating pressure:
     Inlet side .... 0.1 to 1.0 Bar
     Delivery side .... 140 to 180 Bar
   Relief valve opening pressure .... 210 Bar
   Maximum continuous running time .... 2 minutes, followed by 10 minute pause
Wheel sensors:
   Resistance .... 800 to 1400Ω
   Sensor to toothed disc clearance .... 0.65 mm

Torque wrench settings

	Nm	lbf ft
Front caliper-to-steering swivel member, up to 87 model year	80	59
Front caliper-to-steering swivel member, from 88 model year	80	59
Rear caliper-to-axle upright, up to 87 model year	120	88
Rear caliper-to-axle upright, from 88 model year	45	33
Front hub centre nut (up to 81 model year)	350	258
Front hub centre nut (81 model year onwards)	300	221
Rear hub centre nut, early models (27mm across flats):		
Stage 1	49	36
Stage 2	Slacken nut completely	
Stage 3	3	2

Braking system 9•3

Torque wrench settings - continued

	Nm	lbf ft
Rear hub centre nut, later models (32mm across flats)	300	221
ABS accumulator retaining bolt	40	30
ABS hydraulic unit-to-bulkhead bolts	26	19
ABS pressure switch	23	17
ABS pump delivery hose unions	20	15

1 General information

Models with conventional braking system

Braking is achieved by a dual-circuit hydraulic system, assisted by a vacuum servo unit. All models have outboard discs fitted at the front and rear. On certain models, the front discs are ventilated to improve cooling and reduce brake fade.

The dual hydraulic circuits are diagonally split; one circuit operates the front right and rear left brakes, the other operates the front left and rear right brakes. This design ensures that at least 50% of the vehicles braking capacity will be available, should pressure be lost in one of the hydraulic circuits. Under these circumstances, the diagonal layout would prevent the vehicle from becoming unstable if the brakes are applied when only one circuit is operational.

Each front caliper houses two brake pads, one inboard and one outboard of the disc, actuated by a cylinder with one or two pistons, depending on the vehicles age. During braking, hydraulic pressure supplied to the caliper forces the (direct) piston along its cylinder and presses the inboard brake pad against the disc. The caliper body/yoke (or indirect piston, depending on the caliper type) reacts to this effort, bringing the outboard pad into contact with the disc. In this manner, equal pressure is applied to either side of the disc by the brake pads. When braking is ceased, the hydraulic pressure behind the piston drops and it is retracted back into the cylinder, releasing the inboard pad from the disc. The caliper body/yoke/indirect piston then slides back and releases the outboard pad.

On earlier models, the rear calipers incorporate one inboard and one outboard cylinder and piston, both of which act directly on the brake disc. On later models, single piston sliding calipers are fitted at the rear, see *Specifications* for details. The rear brake pad friction material surface area is smaller than that of the front brakes; the resulting difference in braking power between front and rear calipers prevents rear wheel lock-up during hard braking, eliminating the need for a hydraulic pressure regulating valve.

The cable operated handbrake is controlled via a cabin floor mounted lever. On earlier models, the cables actuate pushrods in the front brake calipers. These allow mechanical operation of the front brakes, bypassing the hydraulic system. The mechanism is automatically adjusted whenever the footbrake pedal is depressed, by means of a threaded sleeve, built into the direct-acting caliper piston. On later models, the handbrake cables actuate levers on the rear calipers.

The master cylinder converts footbrake pedal effort into hydraulic pressure. Its tandem construction incorporates two cylinders, one for each circuit, that operate in parallel. Each cylinder houses a piston and a corresponding return spring. The movement of the pistons along the cylinders causes brake fluid to flow through the brake lines in each circuit and transfer pressure from the brake pedal to the caliper pistons. The two (master cylinder) pistons are partially linked, an arrangement which allows equal pressure to be applied to all four calipers under normal operation and also allows full pedal effort to be transferred to the working circuit in the event of the other circuit failing, albeit with increased pedal travel.

A constant supply of brake fluid to the master cylinder is maintained by the brake fluid reservoir. The reservoir is semi-transparent to allow visual inspection of the fluid level and a screw-fit filler cap allows the level to to be topped up. A level detection switch is incorporated into the filler cap; this causes a warning lamp on the instrument panel to illuminate when the level of fluid in the reservoir becomes too low.

The vacuum servo unit uses engine manifold vacuum to amplify the effort applied to the master cylinder when the brake pedal is depressed.

Models with anti-lock braking system (ABS)

Available as an option on certain models from 1989 model year onwards, the anti-lock braking system prevents skidding which not only optimises stopping distances but allows full steering control to be maintained under maximum braking.

By electronically monitoring the speed of each roadwheel in relation to the other wheels, the system can detect when a wheel is about to lock-up, before control is actually lost. The brake fluid pressure applied to that wheel's brake caliper is then decreased and restored (or modulated) several times a second until control is regained. The system is split into three circuits giving control over each front wheel individually and both rear wheels together.

The system components comprise an Electronic Control Unit (ECU), four wheel speed sensors, a hydraulic unit, a hydraulic valve block, brake lines, a dedicated relay/fuse box and dashboard mounted warning lamps.

The hydraulic unit incorporates a tandem master cylinder which operates the two front brake calipers under normal braking, a valve block which modulates the pressure in the three brake circuits during ABS operation, an accumulator which provides a supply of highly pressurised brake fluid, a hydraulic pump to charge the accumulator, a servo cylinder which regulates the pressurised fluid supply from the accumulator to provide hydraulic power assistance (replacing the vacuum servo unit used in conventional braking systems), as well as pressure to operate the rear brakes, and a brake fluid reservoir.

The four wheel sensors are mounted on the wheel hubs. Each wheel has a rotating toothed disc mounted in the hub; the wheel speed sensors are mounted in close proximity to these discs. The teeth on the surface of the discs electrically excite the sensors, causing them to produce a voltage waveform whose frequency varies with the speed of the discs' rotation. These waveforms are transmitted to the ECU, which uses them to calculate the rotational speed of each wheel.

The fuse/relay box is mounted in the engine bay and houses fuses and relays for the ECU and a relay and fuse for the hydraulic pump.

The ECU has a self-diagnostic capability and will inhibit the operation of the ABS if a fault is detected, lighting the dashboard mounted warning lamp. The braking system will then revert to conventional, non-ABS operation. If the nature of the fault is not immediately obvious upon inspection, the vehicle *must* be taken to a Saab dealer, who will have the diagnostic equipment required to interrogate the ABS ECU electronically and pin-point the problem; refer to Section 19 for greater detail.

2 Hydraulic system - bleeding

 Warning: Hydraulic fluid is poisonous; thoroughly wash off spills from bare skin without delay. Seek immediate medical advice if any fluid is swallowed or gets into the eyes. Certain types of hydraulic fluid are inflammable and may ignite when brought into contact with hot components; when servicing any hydraulic system, it is safest to assume

that the fluid is inflammable and to take precautions against the risk of fire as though it were petrol that was being handled. Hydraulic fluid is an effective paint stripper and will also attack many plastics. If spillage occurs onto painted bodywork or fittings, it should be washed off immediately, using copious quantities of fresh water. It is also hygroscopic - i.e. it can absorb moisture from the air, which then renders it unusable. Old fluid may have suffered contamination and should never be re-used. When topping-up or renewing the fluid, always use the recommended grade, and ensure that it comes from a new sealed container.*

Models with conventional braking system

General

1 The correct operation of a hydraulic braking system relies on the fact that the fluid used in it is incompressible, otherwise the effort exerted at the brake pedal and master cylinder will not be fully transmitted to the brake calipers or wheel cylinders. The prescence of contaminants in the system will allow the fluid to compress - this results in a 'spongy' feel to the brakes and unpredictable performance, in the form of brake fade or at worst, brake failure. In addition, brake fluid deteriorates with age through oxidation and water absorption. This lowers its boiling point and may cause vapourisation due to the heat generated under hard braking, again affecting brake performance. For this reason, old or contaminated fluid must be renewed - this is achieved by bleeding the system. Similarly, if any of the hydraulic components in the braking system have been removed or disconnected, or if the fluid level in the master cylinder has been allowed to fall appreciably, it is inevitable that air will have been introduced into the system. The removal of this air from the hydraulic system is essential if the brakes are to function correctly; bleeding the system acheives this.
2 When refilling the system, use only clean, fresh fluid of the recommended type and grade; *never* re-use fluid that has already been bled from the system. Ensure sufficient fluid is available before starting work.

3 If there is any possibility of there being incorrect fluid in the system already, the brake components and circuits must be flushed completely with new fluid of the correct type and grade and new seals should be fitted throughout the system.
4 If hydraulic fluid has been lost from the system, or air has entered because of a leak, ensure that the fault is corrected before proceeding further.
5 Park the vehicle on level ground, switch off the engine and select first or reverse gear (manual transmission) or 'Park' (automatic transmission), then chock the wheels and release the handbrake.
6 Check that all pipes and hoses are secure, unions tight and bleed screws closed. Remove the dust caps and clean off all dirt from around the bleed screws.
7 Unscrew the master cylinder reservoir cap, and top the master cylinder reservoir up to the 'MAX' level line; refit the cap loosely, and remember to maintain the fluid level at least above the 'MIN' level line throughout the procedure, otherwise there is a risk of further air entering the system, as the level drops.
8 There are a number of one-man, do-it-yourself brake bleeding kits currently available from motor accessory shops. It is recommended that one of these kits is used whenever possible, as they greatly simplify the bleeding operation, and also reduce the risk of expelled air and fluid being drawn back into the system. If such a kit is not available, the basic (two-man) method must be used, which is described in detail below.
9 If a kit is to be used, prepare the vehicle as described previously, and follow the kit manufacturer's instructions, as the procedure may vary slightly according to the type being used. Generally, they are as outlined below in the relevant sub-section.
10 Whichever method is used, the same sequence must be followed (paragraphs 11 and 12) to ensure that the removal of all air from the system.

Bleeding sequence

11 It is possible to partially bleed the system, i.e. just one brake line and caliper. Providing fluid loss is kept to a minimum and air is not drawn into the system, it will not be necessary to bleed the other brake lines as well.

12 The system can be bled completely, if neccessary; the order in which the brakes lines are dealt with must always be as follows: **rear left-hand caliper - front right-hand caliper - rear right-hand caliper - front left-hand caliper**. Refer to Chapter 1 for the brake fluid renewal procedure.

Bleeding - basic (two-man) method

13 Obtain a clean glass jar, a suitable length of plastic or rubber tubing which is a tight fit over the bleed screw, and a ring spanner to fit the screw. Alternatively, a proprietary brake bleeding kit can be obtained. **Note:** *The help of an assistant will also be required.*
14 Remove the dust cap from the first caliper's bleed screw. Fit the spanner over the bleed screw and push the tube onto the bleed screw nipple. Place the other end of the tube in the jar and pour in sufficient fluid to cover the end of the tube **(see illustrations)**.
15 Throughout the procedure, keep an eye on the reservoir fluid level and ensure that it is maintained above the 'MIN' level line as the brakes are bled; top it up before starting if neccessary.
16 Have your assistant fully depress the brake pedal several times to build up pressure - then on the final downstroke, ask them to keep it depressed.
17 While pedal pressure is maintained, slacken the bleed screw (by turning it through approximately one turn) and allow the brake fluid to flow into the jar . Pedal pressure should be maintained throughout; follow the pedal down to the end of its travel if necessary, but do not release it. When the flow stops, tighten the bleed screw again, then have your assistant release the pedal slowly. Re-check the reservoir fluid level and top it up if neccessary.
18 If air is present in the brake lines, it will appear as bubbles in the expelled fluid. Repeat the steps given in the two previous paragraphs until the fluid emerging from the bleed screw is free from air bubbles. If the master cylinder has been drained and refilled and air is being bled from the first brake line in the sequence, allow approximately five seconds between cycles for the master cylinder passages to refill.
19 When no more air bubbles appear, tighten the bleed screw securely, remove the tube and spanner then refit the dust cap.
Caution: *Do not overtighten the bleed screw.*
20 Repeat the procedure on the remaining brake lines to be bled, until all air is removed from the system and the brake pedal feels firm again.

Bleeding - using a one-way valve kit

21 As their name implies, these kits consist of a length of tubing with a one-way valve fitted, to prevent expelled air and fluid being drawn back into the system; some kits include a translucent container, which can be positioned so that the air bubbles can be more easily seen flowing from the end of the tube.

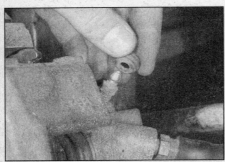

2.14a Remove the dust cap from the first caliper's bleed screw (later caliper shown)

2.14b Fit the spanner over the bleed screw and push the tube onto the nipple

22 The kit is connected to the bleed screw, which is then opened. The user returns to the driver's seat, depresses the brake pedal with a smooth, steady stroke and slowly releases it; this process is repeated until the expelled fluid is free of air bubbles.

23 Note that the use of these kits can simplify the bleed operation such, that it is very easy to neglect the level in the fluid reservoir. Ensure that it is maintained at least above the 'MIN' level line at all times, or air may be accidentally drawn into the system.

Caution: *When utilising one of the methods of bleeding described above, the master cylinder pistons are driven through unusually long strokes. The resulting stress on the piston seals can cause them to fail under these circumstances, although during normal operation, the seal would have been hydraulically sound. Avoid rapid, forceful pedal strokes, to preserve the piston seals as far as possible.*

Bleeding - using a pressure-bleeding kit

24 This type of kit is usually powered by the reservoir of pressurised air contained in a spare tyre. However, note that it will probably be necessary to reduce the tyre pressure to a level lower than normal, to avoid over-pressurising the bleeding kit; refer to the instructions supplied with the kit.

25 The method involves connecting a pressurised, fluid-filled container to the master cylinder reservoir. Bleeding can then be carried out simply by opening each bleed screw in turn, and allowing the fluid to flow out under moderate pressure until no more air bubbles can be seen in the expelled fluid.

26 This method has the advantage that the large reservoir of fluid provides an additional safeguard against air being drawn into the system during bleeding.

27 Pressure-bleeding is particularly effective when bleeding 'difficult' systems, or when bleeding the complete system at the time of routine fluid renewal.

All methods

28 When bleeding is complete, and firm pedal feel is restored, wash off any spilt fluid, tighten the bleed screws securely, and refit their dust caps (where applicable).

29 Check the hydraulic fluid level in the master cylinder reservoir; top it up if necessary.

30 Dispose of any hydraulic fluid that has been bled from the system; it cannot be re-used.

31 Check the feel of the brake pedal. If it feels at all spongy, it is probable that air is still present in the system; further bleeding will therefore be required. If the bleeding procedure has been repeated several times and brake feel has still not been restored, the problem may be caused by worn master cylinder seals - see Section 10 for a description of the master cylinder overhaul procedure.

Models with ABS

Caution: *The front wheel brake circuits must always be bled before the rear circuit.*

Front wheel brake circuits

32 Follow one of the methods described above, for non-ABS models. If both calipers are to be bled, bleed the right hand caliper before the left.

Rear wheel brake circuits

33 Top-up the level of fluid in the reservoir to the 'MAX' mark. Maintain the level at least above the 'MIN' mark throughout the bleeding process.

34 Fit one end of a length of tubing to the rear brake caliper bleed screw and immerse the other end of the tubing in brake fluid, contained in a clean jar.

35 Have an assistant turn the ignition switch to the second position, then depress and hold the brake pedal; this will power the hydraulic pump and pressurise the rear brake circuit.

36 Using a ring spanner, slacken the caliper bleed screw by about one turn and allow brake fluid to flow through the tube into the jar. Ensure that pedal pressure is maintained whilst the bleed screw is open.

Caution: *Do not allow the hydraulic pump to run for more than two minutes at a time. After this period, switch off the ignition and allow the pump to cool for ten minutes before restarting. Under no circumstances must the pump be allowed to run dry.*

37 Any air present in the system will be expelled as bubbles in the brake fluid. When no more bubbles can be seen escaping, tighten the bleed screw, release the brake pedal and switch off the ignition.

38 Clean off any excess brake fluid from around the bleed screw and refit the dust cap.

39 The above process can be repeated at the other rear wheel caliper, if necessary.

40 When the system has been bled, top-up the level of fluid in the reservoir to the 'MAX' mark and refit the cap.

3.5 Pull out the pad retaining pin

3 Front brake pads - renewal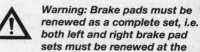

Warning: *Brake pads must be renewed as a complete set, i.e. both left and right brake pad sets must be renewed at the same time. Do not renew the pads on just one roadwheel, as unbalanced braking may occur, making the car unstable.*

Although more recent Saab brake pads do not contain asbestos, it is still wise to take safety precautions when cleaning the brake components. Do not use compressed air to blow out brake dust and debris - use a brush and avoid inhaling any of the airborne particles that may exude; wear an approved filtration mask. Use only proprietary brake cleaner fluid or methylated spirit to cleanse the brake components, DO NOT use petrol or any other petroleum-based product.

Pad removal - pre-1988 model year

1 Park the vehicle on a level surface, apply the handbrake and chock the rear wheels. Remove the wheel centre caps/trims and slacken the wheel bolts.

2 Raise the front of the vehicle, rest it securely on axle stands (see "*Jacking and Vehicle Support*") and remove the roadwheels.

3 Turn the brake disc so that one of the recesses on its edge is aligned with the disc pads.

4 Extract the spring clips from the ends of the pad retaining pin.

5 Pull out the pad retaining pin, then remove the damper spring plate **(see illustration)**. Use a hammer to tap out the pin if it is tight.

6 Withdraw the pads from both sides of the disc **(see illustration)**. If they are tight tap them lightly with a light mallet to release the accumulated dust and use grips on the backplates. Press the two pads away from the disc to retract the pistons back into the cylinders; the increased space will ease the removal of the pads.

Caution: *Keep an eye on the level in the brake fluid reservoir as you retract the*

3.6 Withdraw the pads from both sides of the disc

9•6 Braking system

3.8 Unscrew the upper and lower guide pin bolts (lower bolt shown)

3.9a Grasp the hydraulic body of the caliper and lift it away from the carrier

3.9b Hang the hydraulic body from the suspension coil spring

pads, to ensure that the displaced fluid does not cause it to overflow.

7 Remove the pads from the second caliper in the same manner.

Pad removal - 1988 model year onwards

8 At the first caliper, using a socket wrench and spanner, unscrew the upper and lower guide pin bolts **(see illustration)**.

9 Grasp the hydraulic body of the caliper and lift it away from the carrier, taking care not to strain brake hose or union. Hang the hydraulic body from the suspension coil spring using a length of wire, or a nylon cable-tie to avoid straining the brake hose **(see illustrations)**.

10 Draw out both brake pads; if they bind against the disc, apply pressure to the inboard pad with a pair of grips to retract the piston back into the caliper; the increased space will ease the removal of the pads.

Caution: Keep an eye on the level in the brake fluid reservoir as you retract the pad, to ensure that the displaced fluid does not cause it to overflow.

11 Remove the pads from the second caliper in the same manner.

Inspection - all vehicles

12 Douse the caliper, pads and disc with brake cleaning fluid and then brush away all traces of dust and dirt; observe the warning given at the beginning of this Section regarding the hazards of brake dust. Scrape any scale or rust from the disc outer edge.

13 Measure the depth of the friction material remaining on each pad. If any of the pads has worn down below its service limit (see *Specifications*), then the complete set of four pads (both wheels) must be renewed. Similarly, if any of the pads has been contaminated with grease or oil, it is not possible to clean and re-use it; the whole set must be renewed. If the pads have been contaminated, identify and rectify the cause before fitting new pads. If all the pads are still serviceable, clean them thoroughly using a brush (ideally a fine wire brush) and brake cleaning fluid. Pay particular attention to the metal backplate, where the pad contacts the caliper. Examine the surface of the friction material; carefully prise out any stones or grit that may have become embedded in it.

14 Apply a small quantity of anti-squeal/high-melting-point grease to the metal backing plates of the pads; *do not* allow any to come into contact with the friction material **(see illustration)**.

15 It is good practise to examine the condition of the brake discs when inspecting or renewing the pads; refer to Section 7.

16 Working from Section 5 examine the caliper piston seals for signs of leaking or deterioration, and the piston itself for signs of wear, damage or corrosion.

Pad refitting - pre-1988 model year

17 Apply a little high-melting-point grease to the sliding surfaces of the caliper yoke, moving the yoke in its groove to distribute the grease evenly.

18 Check the condition and security of the piston dust covers and retainers; renew them if they are in poor condition.

19 The self-adjusting handbrake mechanism must be reset, before the brake pads can be refitted. Engage a pair of circlip pliers, or a purpose-made tool, with the holes in the face of the inboard piston **(see illustration)**. Rotate the piston clockwise, whilst pressing the piston into the cylinder, until its face is flush with the surface of the cylinder.

Caution: Do not press in the piston any further than this, as the internal seal may be damaged.

20 Insert the disc pads between the caliper and disc. Note on pre-1983 model year vehicles the inner and outer pads are not interchangeable - the inner pad has one groove on its outer edge but the outer pad also has a groove on its inner edge.

21 Position the damper spring plate on the pads then insert the retaining pin through the pads and fit the spring clips.

22 Using a feeler blade, check that the clearance between the caliper handbrake lever and the yoke is no more than 0.5 mm. If necessary, adjust the handbrake, referring to Section 13 for details.

23 Depress the footbrake pedal several times, then pull up the handbrake lever through five 'clicks' of the ratchet mechanism and depress the pedal several more times. This will set the handbrake adjusting mechanism. Check the adjustment by applying the handbrake through two to four more 'clicks' and checking that the brake discs are held firmly.

24 Repeat the above procedure at the second caliper.

Pad refitting - 1988 model year onwards

25 At the first caliper, slide the pads into position, with the friction material facing the brake disc.

26 Fit the caliper hydraulic body into position on the carrier. If the pads bind against the disc, apply pressure to the inboard pad with a pair of grips to retract the piston back into the caliper.

Caution: Keep an eye on the level in the brake fluid reservoir as you retract the pad, to ensure that the displaced fluid does not cause it to overflow.

27 Coat the threads of the guide pin retaining bolts with locking compound, then refit and

3.14 Apply anti-squeal/high-melting-point grease to the pad backing plates

3.19 Engage a pair of circlip pliers with the holes in the face of the inboard piston

Braking system

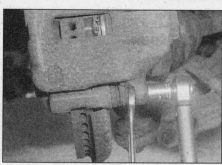

3.27 Refit the guide pin bolts and tighten them to the specified torque

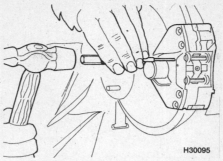

4.2 At the first caliper, use a pin punch to tap out the upper pad retaining pin

4.4 Withdraw the pads and backing plates from each side of the disc

tighten them to the specified torque (see illustration).

All models

28 Depress the brake pedal several times; this will pressurise the braking system and bring the pads into contact with the disc. If the pedal has a developed a spongy feel, air may have entered the system when the pads were being removed - refer to Section 2 for guidance in bleeding the system.
29 Refit the roadwheels, lower the vehicle to the ground and tighten the bolts to the correct torque. Refit the wheel trims/centre caps.
30 Top-up level of brake fluid in the reservoir to the 'MAX' mark and refit the filler cap, referring to "Weekly Checks" for guidance.

4 Rear brake pads - renewal

Note: *Refer to the warnings at the start of Section 3 before proceeding.*
1 Park the vehicle on a firm, level surface then chock the front wheels and select first gear (manual transmission) or 'Park' (automatic transmission) - do not apply the handbrake. Raise the rear of the vehicle, rest it securely on axle stands (see "Jacking and Vehicle Support") and remove both rear roadwheels

Pad removal - pre-1988 model year

2 At the first caliper, use a pin punch to tap out the upper pad retaining pin (see illustration).
3 Remove the damper spring plate then tap out the lower retaining pin.

4 Withdraw the pads and backing plates from each side of the disc (see illustration). If they are tight tap them lightly with a hammer to release the accumulated dust, then extract the pads from the caliper using grips on the pad backplates. Greater access can be gained by prising the two pads away from the disc, using water pump pliers to retract the pistons into the cylinders.

Caution: Keep an eye on the level in the brake fluid reservoir as you retract the pads, to ensure that the displaced fluid does not cause it to overflow.

Pad removal - 1988 model year onwards

5 At the first caliper, remove the dust plug from the adjusting screw drilling on the hydraulic body (greater access can be gained by unclipping the handbrake cable from the actuator lever - see Section 13). Using an

4.5a Removing dust plug from adjusting screw drilling on the hydraulic body

Allen key, turn the adjusting screw anticlockwise to the end of its travel - this will cause the piston to be retracted fully into the hydraulic body, drawing the brake pad away from the disc (see illustrations).
6 Prise out the two dust caps, then using a 7 mm hex bit and socket wrench, slacken and withdraw the caliper guide pins (see illustrations).
7 Using a screwdriver, lever the retaining spring clip off the caliper (see illustration).
8 Lift the hydraulic body away from the carrier, then (if not already removed) relieve the spring tension on the handbrake lever and release the handbrake cable from it.
9 Extract the brake pads from the hydraulic body. Note that the inboard pad must be prised away from the piston, as it is retained by an anti-rattle sping, attached to the pad backplate (see illustration).
10 Hang the hydraulic body casting from a

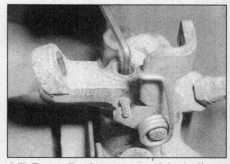

4.5b Turn adjusting screw anticlockwise to the end of its travel

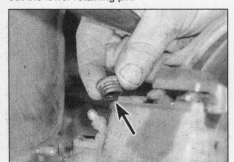

4.6a Prise out the two dust caps . . .

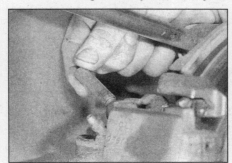

4.6b . . . then slacken and withdraw the caliper guide pins

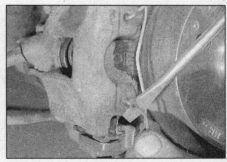

4.7 Using a screwdriver, lever the retaining spring clip off the caliper

9•8 Braking system

4.9 Inboard pad is retained by an anti-rattle spring (arrowed), attached to the backplate

4.10 Hang the hydraulic body casting from a rigid point on the vehicles' suspension

4.23a Fit the outboard pad to the caliper carrier . . .

rigid point on the vehicles' suspension, using cable-ties or wire. Do not allow it to dangle by the brake hose as this may cause damage to the hose **(see illustration)**.

11 Remove the pads from the caliper at the other roadwheel in the same manner.

Inspection - all models

12 Examine caliper piston seal and dust cover for signs of leaking or deterioration, and the piston itself for signs of wear, damage or corrosion. Douse the pads, disc and caliper with brake cleaning fluid, then clean off all traces of dust and dirt using a stiff brush. Prevent inhalation of any airborne dust by wearing an approved filtration mask.

13 Measure the depth of the friction material remaining on each pad. If any of the pads has worn down below its service limit (see Specifications), then the complete set of four pads (both roadwheels) must be renewed. Similarly, if any of the pads has been contaminated with grease or oil, it is not possible to clean and re-use it; the whole set must be renewed. If the pads have been contaminated, or have worn unevenly, identify and rectify the cause before fitting new pads.

14 If all the pads are still serviceable, clean them thoroughly using a brush (ideally a fine wire brush) and brake cleaning fluid. Pay particular attention to the metal backplate, where the pad contacts the carrier/piston. Examine the surface of the friction material; carefully prise out any fragments that have become embedded in it. Brush out the grooves if they are clogged

15 It is good practise to examine the condition of the brake discs when inspecting or renewing the pads; refer to Section 4 for guidance.

16 Apply a small quantity of anti-squeal grease to the metal backing plates of the pads; *do not* allow any to come into contact with the friction material.

Pad refitting - pre-1988 model year

17 Using a piece of wood, press the two pistons into their cylinders until they are flush with the inner faces of the caliper.
Caution: Keep an eye on the level in the brake fluid reservoir as you retract the pistons, to ensure that the displaced fluid does not cause it to overflow.

18 Insert the disc pads and backing plates each side of the disc. Note on early models the inner pad has one groove on its outer edge; the outer pad has an additional groove on its inner edge. On later models both pads are interchangeable.

19 Insert the upper retaining pin through the pads and tap it firmly into the caliper.

20 Hook the damper spring under the upper retaining pin then depress the lower part of the spring and insert the lower retaining pin. Tap the pin firmly into the caliper.

21 Repeat the above procedure at the second rear caliper, then refit the wheels and lower the car to the ground.

Pad refitting - 1988 model year onwards

22 With the hydraulic body removed, check that the guide pins slide freely in their bores, without excessive play.

23 Fit the outboard pad to the caliper carrier, press the inboard pad anti-rattle spring into the hollow of the piston in the hydraulic body, then fit the hydraulic body to the carrier **(see illustrations)**.

24 Refit the guide pins and using a 7 mm hex bit, tighten them to the correct torque. Refit the dust caps **(see illustration)**.
Caution: Do not lubricate the guide pins, as they will then tend to accumulate dust and debris, which will inhibit their operation.

25 Fit the spring clip into place, locating the ends into the holes provided in the hydraulic body casting.

4.23b . . . press the inboard pad anti-rattle spring into the piston, then fit the hydraulic body to the carrier

26 Using an Allen key, screw in the adjusting screw until the inboard pad is touching the brake disc, then back it off by half a turn.

27 Operate the handbrake lever against the return spring tension and reconnect the handbrake cable, then refer to Section 13 and adjust the operation of the handbrake.

All models

28 Depress the brake pedal several times; this will pressurise the braking system and bring the pads into contact with the disc. If the pedal has a developed a spongy feel, air may have entered the system when the pads were being removed - refer to Section 2 for guidance in bleeding the system.

29 Refit the roadwheels, lower the vehicle to the ground and tighten the bolts to the correct torque. Refit the wheel trims/centre caps.

30 Top-up level of brake fluid in the reservoir to the 'MAX' mark and refit the cap.

5 Front brake caliper - removal, overhaul and refitting

Removal

Pre-1988 model year

1 Remove the front brake pads from the caliper, as described in Section 3.

2 Disconnect the handbrake cable from the caliper; refer to Section 13 for more detail.

3 To allow the caliper to to be removed, the brake hose must be disconnected from it - this will entail some brake fluid loss. The

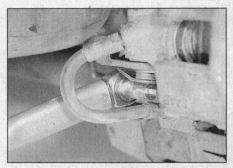

4.24 Refit the guide pins and using 7mm hex bit, tighten them to the correct torque

amount of fluid lost can be minimised by clamping the flexible brake hose using a proprietary hose clamp. These are designed to constrict hoses without pinching the walls and causing damage - they can be obtained from vehicle accessories shops at minimal cost. Disconnect and plug the rigid hydraulic pipe from the flexible hose at the support bracket.

Caution: The unprotected jaws of a G-clamp should not be used, as they may damage the hose, leading to premature failure.

4 Slacken and withdraw the two retaining bolts, then lift the caliper from the steering swivel member.

Overhaul

5 Clean the external surface of the caliper, then mount it in a bench vice; to protect the surface of the caliper, line the jaws of the vice with strips of wood or aluminium.
6 Release the return spring from the handbrake lever.
7 Remove the yoke from the caliper and withdraw the handbrake lever and return spring.
8 Remove the dust excluder and retaining ring from the indirect piston and then carefully apply air pressure from a bicycle pump at the fluid inlet port on the caliper, to eject the indirect piston.
9 Now press the pushrod in by hand and eject the direct piston from the cylinder.
10 Extract the seals and O-rings from the pistons and cylinder bore, taking great care not to scratch the surfaces of these components. Also, remove the two O-rings from the handbrake lever aperture.
11 Wash all components in clean hydraulic fluid or methylated spirit except the internal parts of the indirect piston otherwise the grease for the handbrake mechanism will be washed away.
12 Examine the surfaces of the pistons and cylinder bores for scoring or 'bright' wear areas. If these are evident, the entire caliper should be renewed. If the caliper appears to be in good condiiton, discard the old seals and obtain a caliper repair/service kit.
13 Commence reassembly by fitting the pushrod and handbrake lever O-rings to the indirect piston. Secure the handbrake lever O-rings with the special retainer, and lubricate them with brake grease.
14 Lubricate the cylinder bore with brake fluid and fit the piston seals in their grooves.
15 Fit the anchor plate to the pushrod and then insert the pushrod into the hole in the indirect piston. Ensure that the recess in the anchor plate comes immediately over the tension pin in the piston.
16 Wipe the indirect piston with a clean lint-free rag, dip it in clean hydraulic fluid and insert it into the caliper so that the recess for the yoke is in direct alignment with the groove in the caliper body.
17 Dip the direct piston into clean hydraulic

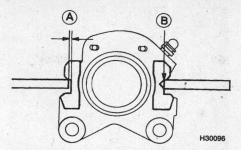

5.22 Yoke-to-caliper clearances (pre-1988)
A 0.15 - 0.30 mm   B Zero clearance

fluid and insert it into its cylinder, screwing together the piston and pushrod.
18 Screw and depress the two pistons until the edges of the dust excluder grooves are flush with the caliper body. Install the new dust excluders and their retaining rings.
19 Fit the spring and handbrake lever to the yoke, and lubricate the yoke sliding surfaces with grease.
20 Align the guide edges of the yoke with the grooves in the caliper body. Lift the handbrake lever and secure the end of its pivot pin in the hole in the indirect piston, making sure at the same time that the yoke engages in the recess in the indirect piston.
21 Install the handbrake lever return spring.
22 Check the yoke to caliper clearances are as shown in the accompanying illustration **(see illustration)**.

Refitting

23 Refit the caliper by following the removal procedure in reverse. Fit new locking plates to the caliper bolts, bending them over the head flats after tightening the bolts to the specified torque. On completion, carry out the following operations:
 a) Adjust the handbrake cables (Section 13).
 b) Bleed the hydraulic circuit (see Section 2).

1988 model year onwards

Removal

24 To allow the caliper to to be removed, the brake hose must be disconnected from it - this will entail some brake fluid loss. The amount of fluid lost can be minimised by clamping the flexible brake hose using a proprietary hose clamp. These are designed to constrict hoses without pinching the walls and causing damage - they can be obtained from vehicle accessories shops at minimal cost.

Caution: The unprotected jaws of a G-clamp should not be used, as they may damage the hose, leading to premature failure.

25 Clean the caliper in the area around the brake hose union with a clean rag. Slacken the hose union but do not attempt to unscrew it completely at this stage, as this may damage the pipe.
26 Remove the brake pads as described in Section 3.

27 Still working from Section 3, prise out the dust cap and unscrew the upper guide pin bolt using a spanner and socket wrench. The hydraulic body can then be lifted away from the carrier.
28 Unbolt the brake pipe support bracket from the caliper, then fully slacken and withdraw the pipe union from the caliper. Be prepared for some brake fluid loss - have a small container or a clean rag ready to catch spills. If the flow of fluid does not stop after the initial discharge, check the security and tightness of the hose clamp.
29 Wipe clean the end of the brake hose and the tapping in the hydraulic body and fit them with dust caps.

> **HAYNES HINT** *If dust caps are not available, cut the fingers out of an old rubber glove and stretch them over the open end of the brake pipe, securing them with elastic bands.*

30 Slacken and remove the retaining bolts, then lower the carrier away from the steering knuckle.

Overhaul

31 The cleanliness of the work area is of great importance when dismantling brake system components. Dirt entering the hydraulic system may adversely affect the systems performance and possibly cause failure. Select a clean, uncluttered surface to work on; laying a sheet of plain paper or card on the surface may help to keep everything clean and easily visible.
32 Clean off all traces of dirt and dust from the hydraulic body using a brush.

 Warning: Take great care to avoid inhaling the airborne dust.

33 Make a careful note of the orientation and order of assembly of all components, as the unit is dismantled. Laying the components out 'exploded' form is often an effective way of keeping them in order.
34 The piston can be removed from the cylinder by applying compressed air to the hydraulic hose tapping. Only low pressure is required, so that produced by a bicycle or foot pump will be sufficient. Remove the dust cap and couple the pump to the tapping with an old piece of vacuum or fuel hose to make a good seal - do not use anything that may damage the thread inside the tapping. Use a block of wood as padding to protect the end of the piston, as it is ejected under pressure.

 Warning: Brake fluid may be ejected under pressure - protect your eyes with safety goggles.

35 Carefully prise the dust cover out from the cylinder bore, leaving it attached to the piston - use a plastic instrument that will not damage the surface of the cylinder. Put the piston and dust cover to one side.

9•10 Braking system

36 Using the same plastic instrument, lever out the piston seal from its seat in the cylinder bore. Avoid scoring the inside of the bore or the seal seat.

37 Clean all components thoroughly, using only methylated spirit, isopropyl alcohol or clean hydraulic fluid as a cleaning agent. Do not use mineral-based solvents such as petrol or paraffin, as they will attack the hydraulic system's rubber components. Dry the components straight after cleaning, using compressed air or a clean, lint-free cloth. Use compressed air to blow the fluid passages clear. If reassembly is not going to be carried out immediately, remember to refit a dust cap to the brake hose tapping.

38 Examine all components closely and renew any that are worn or damaged. Check particularly the cylinder bore and piston; if they are scratched, worn or corroded in any way, they should be renewed (note that this means the renewal of the whole hydraulic body assembly). Check also the condition of the guide pins and their bores in the hydraulic body; both pins should be free from damage and corrosion and after cleaning, should form a reasonably tight sliding fit in their corresponding bores. Renew any component whose condition is dubious.

39 If the assembly is fit for further use, obtain an appropriate repair kit; the components are available from Saab dealers in various combinations.

40 Renew all rubber seals, dust covers and caps disturbed on dismantling, regardless of their apparent condition; the old items should *never* be re-used.

41 Before reassembly, ensure that all components are completely clean and dry.

42 Lubricate the new piston seal with the grease provided in the kit. Use the same grease to lubricate the new dust cap.

43 Fit the new piston seal into its seat in the cylinder; do not use any tools when doing this.

44 Fit the new dust cap to the piston; slide it over the end that contacts the brake pad, then pull it down to the other end of the piston.

45 With the hydraulic body laid on the work surface, offer up the piston to the cylinder bore. Starting at the lower edge, press the collar of the dust cap into the cylinder bore, working around its circumference until the collar is firmly seated all the way around.

46 The piston can now be pressed into the cylinder - use care as the piston engages with the new seal. Note that it may be necessary to remove the dust cap from the brake hose tapping as you do this, to allow the air trapped in the cylinder to escape.

Refitting

47 Reconnect the hydraulic body to the brake hose by threading the hose union into the tapping, holding the hose stationary and rotating the hydraulic body. Do not fully tighten the union at this stage.

48 Refit the carrier to the steering knuckle -

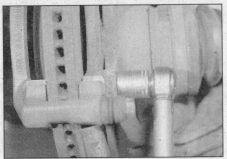

5.48 Tighten the carrier retaining bolts to the specified torque

apply a quantity of thread locking compound to the retaining bolts, then insert and tighten them to the specified torque **(see illustration)**.

49 Follow the brake pad refitting sequence at the end of Section 3, but do not refit the roadwheel at this stage.

50 With the brake pads fitted, tighten the brake pipe union to the correct torque.

51 Remove the brake pipe clamp tool, then refer to Section 2 and bleed the system to expel the air that will have entered the brake hose when the caliper was removed. If suitable precautions were taken to minimise fluid loss, it should only be necessary to bleed the system at that caliper.

52 Refit the roadwheel, lower the vehicle to the ground and tighten the bolts to the specified torque.

6 Rear brake caliper - removal, overhaul and refitting

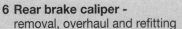

Pre-1988 model year

Removal

1 Remove the rear brake pads from the caliper, as described in Section 4.
2 Disconnect the hydraulic brake pipe from the caliper and plug the pipe to prevent loss of fluid.
3 Remove the two bolts and lift the caliper away from the rear axle.

Overhaul

4 Clean the external surfaces of the caliper, using brake cleaning fluid and a stiff brush.
5 Prise off the dust covers.
6 Using a bicycle pump in the hydraulic fluid aperture, force both pistons from their cylinders. Take care not to damage the pistons or cylinder bores; protect the piston faces with blocks of wood as they are ejected.
7 Prise the seals from the cylinder bores using a blunt, non-metallic instrument.
8 Wash all components in clean hydraulic fluid or methylated spirit.

Caution: Do not attempt to separate the two halves of the caliper.

9 Examine the surfaces of the pistons and cylinder bore for scoring or 'bright' wear areas. If these are evident, renewal of the

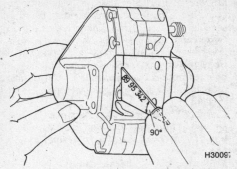

6.14 Using Saab tool 89 95 342 or a piece of card cut into a 20° triangle, position the cut-outs on the piston facing the bottom of the caliper

complete caliper will be neccessary. If the components are otherwise in good condition, discard the existing seals and dust caps. New seals and caps can be obtained in repair/overhaul kits, available from Saab dealers.

10 Commence reassembly by applying a coat of new brake fluid to the cylinder bores.
11 Fit the new seals to the cylinder bores.
12 Dip the inner surfaces of the pistons in brake fluid then insert them into their respective bores.
13 Fit the dust covers to the pistons, then press the pistons fully into the cylinders.
14 Using Saab tool 89 95 342 or a piece of card cut into a 20° triangle, position the cut-outs on the piston facing the bottom of the caliper **(see illustration)**.

Refitting

15 Refit the caliper by following the removal procedure in reverse. Fit new locking plates to the caliper bolts, bending them over the head flats after tightening the bolts to the specified torque. On completion, carry out the following operations:
a) Adjust the handbrake cables, as described in Section 13.
b) Bleed the hydraulic circuit (see Section 2).

1988 model year onwards

Removal

16 Clean the caliper in the area around the brake hose union with a clean rag. Slacken the hose union but do not attempt to unscrew it completely at this stage, or the hose will become twisted.
17 Remove the brake pads as described in Section 3.
18 Fit a brake hose clamp tool over the brake hose and tighten it. Hold the brake hose stationary with one hand and rotate the hydraulic body with the other, until the hose union unscrews from its tapping. Be prepared for some brake fluid leakage; have a small container or rag ready to catch spills. If the flow of fluid does not stop after the initial discharge, check the security and tightness of the hose clamp.

19 Wipe clean the end of the brake hose and the tapping in the hydraulic body and fit them with dust caps.

 HAYNES HiNT *If dust caps are not available, cut the fingers out of an old rubber glove and stretch them over the open brake pipe end, securing them with elastic bands.*

20 Slacken and remove the retaining bolts, then lower the carrier away from the axle upright.

Overhaul

Note: *Refer to the warnings at the start of Section 3 before proceeding.*

21 The cleanliness of the work area is of great importance when dismantling brake system components. Dirt entering the hydraulic system may adversely affect the systems performance and possibly cause failure. Select a clean, uncluttered surface to work on; laying a sheet of plain paper or card on the surface may help to keep things clean and easily visible.
22 Clean off all traces of dirt and dust from the hydraulic body using a brush.
23 Lay the hydraulic body on the work surface. Using a blunt screwdriver, lever out the guide pin spacer sleeves, together with the dust covers. Avoid scratching the internal surfaces of the spacer sleeves.
24 Lift the handbrake lever return spring off its pivot and put it to one side.
25 Prise off the dust cover retaining ring; avoid damaging the piston by using a plastic implement, or a small screwdriver with its blade wrapped in tape. Pull the dust cover away from the piston.
26 Using 4mm Allen key, rotate the adjusting screw clockwise to force the piston out of the cylinder. It may be necessary to remove the dust cap from the brake hose tapping at this point, to allow air into the cylinder as the piston is pushed out.
27 Lever the piston seal out of its seat using the plastic instrument; take care not to score the surface of the cylinder bore.
28 Clean all components thoroughly, using only methylated spirit, isopropyl alcohol or new hydraulic fluid as a cleaning agent. Do not use mineral-based solvents such as petrol or paraffin, as they will attack the hydraulic system's rubber components.
29 Dry the components immediately after cleaning; using compressed air or a clean, lint-free cloth. Use compressed air to blow the fluid passages clear . If reassembly is not going to be carried out immediately, remember to refit a dust cap to the brake hose tapping.
30 Examine all components closely and renew any that are worn or damaged. Check particularly the cylinder bore and piston; if they are scratched, worn or corroded in any way, they should be renewed (note that this means the renewal of the whole hydraulic body assembly). Check also the condition of the guide pins and their bores in the hydraulic body; both pins should be free from damage and corrosion and after cleaning, should form a reasonably tight sliding fit in their corresponding bores. If they cause the caliper to jam or stick, they may have distorted through overheating - check them for warpage against a straight edge. Renew any component whose condition is doubtful.
31 If the assembly is fit for further use, obtain an appropriate repair kit; the components are available from Saab dealers in various combinations.
32 Renew all rubber seals, dust covers and caps disturbed on dismantling as a matter of course; the old items should *never* be re-used.
33 Before reassembly, ensure that all components are completely clean and dry.
34 Lubricate the new piston seal and dust cap with the grease provided in the kit.
35 Fit the new piston seal into its seat in the cylinder - do not use any tools when doing this.
36 Fit the new dust cap to the piston: slide it over the end that contacts the brake pad, then pull it down to the other end of the piston.
37 Push the piston into the cylinder bore and retract it fully by turning the adjusting screw anticlockwise with a 4mm Allen key.
38 Push the skirt of the dust cap over the lip on the edge of the cylinder. Ensure that it is firmly seated, then fit the retaining ring.
39 Refit the handbrake lever return spring onto its pivot. Ensure that one end engages with the lever itself and the other braces against the casting.
40 Fit the two rubber bushes and guide pin spacer sleeves.

Refitting

41 Refit the carrier to the axle upright - apply a quantity of thread locking compound to the retaining bolts, then insert and tighten them to the specified torque.
42 Reconnect the hydraulic body to the brake hose by threading the hose union into the tapping (after removing the dust cap), holding the hose stationary and rotating the hydraulic body. Do not fully tighten the union at this stage.
43 Refer to Section 4 for details of fitting the brake pads and reconnecting the handbrake cable. Do not refit the roadwheel at this point.
44 Tighten the brake pipe union to the correct torque. Remove the brake pipe clamp tool, then refer to Section 2 and bleed the system to expel the air that will have entered the brake hose when the caliper was removed. If suitable precautions were taken to minimise fluid loss, it should only be necessary to bleed the system at that caliper.
45 Refer to Chapter 1 and adjust the operation of the handbrake.
46 Refit the roadwheel, lower the vehicle to the ground and tighten the bolts to the specified torque.

7 Front brake disc - inspection, removal and refitting

Inspection

Note: *Where brake disc backplates are fitted, it will be necessary to remove the brake caliper from the steering knuckle to allow an adequate inspection the discs rear surface. Disconnection of the hydraulic hose will not be neccessary, but the caliper must be hung from a convenient point on the suspension, to avoid straining the brake hose. Refer to Section 5 for details.*

1 Park the vehicle on a firm, level surface, then chock the rear wheels and apply the handbrake. Raise the front of the vehicle, rest it securely on axle stands (see "*Jacking and Vehicle Support*") and remove the front roadwheels.
2 Rotate the brake disc by hand and examine the whole of the surface area swept by the brake pads, on both sides of the disc. Typically, the surface will have a polished appearance, but should be free from heavy scoring. Smooth rippling is produced by normal operation and does not indicate excessive wear. Deep scoring and cracks, however, are indications of more serious damage in need of correction.
3 If deep scoring is discovered, it may be possible to have the disc reground to restore the surface, depending on the extent of the damage. Alternatively, it may be possible for a Saab dealer to use special abrasive blocks, temporarily fitted in place of the brake pads, to remove light scoring. To determine whether this is a feasible course of action, it will be necessary to measure the thickness of the disc, as described later.
4 Check the whole surface of the disc and for cracks, particularly around the roadwheel bolt holes. A cracked disc must be renewed.
5 Where applicbale, inspect the cooling vents between the two friction surfaces of the disc and clear out any traces of dirt, corrosion or brake dust; blocked air ways will impair the cooling efficiency and reduce brake performance. Use a piece of rag wrapped around a length of wire, soaked in brake cleaning fluid to clear the air ways. Do not use compressed air as this will propel the harmful brake dust into the air.
6 A ridge of rust and brake dust at the inner and outer edges of the disc, beyond the pad pontact area is normal - this can be scraped away quite easily using a stout screwdriver whilst rotating the disc.
7 Raised ridges caused the brake pads eroding the disc material, however, are an indication of excessive wear. If close examination reveals such ridges, the thickness of the disc must be measured, to assess whether it is still fit for use.
8 To measure the thickness of the disc, take readings at several points on the surface using a micrometer, in the area swept by the

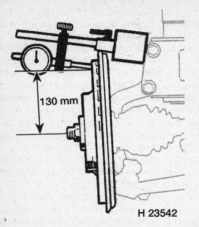

7.11 Brake disc runout - DTI gauge method

brake pads. Include any points where the disc has been scored; align the jaws of the micrometer with the deepst area of scoring, to get a true indication of the extent of the wear. Compare these measurements with the limits listed in Specifications. If the disc has worn below its minimum thickness, at any point, it must be renewed.

9 If the discs are suspected of causing brake judder, check the disc runout, using one of the following methods:-

Runout measurement - DTI gauge method

10 Refit the four roadwheel bolts, together with one M14 plain washer per bolt - this will ensure adequate disc to hub contact and whilst preventing the bolts from contacting the bearing housing. Tighten the bolts to 5 Nm (4 lb ft).
11 Clamp the DTI gauge to a stand and attach the stand, preferably via a magnetic base, to a fixed point on the suspension. Align the gauge so that its pointer rests upon the area of the disc swept by the brake pads, on an arc 130 mm from the centre of the hub (see illustration).
12 Zero the gauge and slowly rotate the disc through one revolution, observing the pointer movement. Note the maximum deflection recorded and compare the figure with that listed in Specifications.

Runout measurement - Feeler blade method

13 Use the gauges to measure the clearance between the disc and a convenient fixed point, such as the disc backplate. Rotate the disc and measure the variation in clearance at several points around the disc. Compare the maximum figure with that listed in Specifications.
14 If the disc runout is outside of tolerance, first check that the hub is not worn - refer to Chapter 10 for guidance. If the hub is in good condition, remove the disc (as described later in this Section), rotate it through 180° and refit it. This may improve the seating and eradicate the excessive runout.
15 If the runout is still unacceptable, then it may be possible to restore the disc by regrinding; consult your Saab dealer or a machine shop for a professional opinion - it may prove more economical to purchase a new disc. If the disc cannot be reground, then it must be renewed.
16 Measure the thickness of the disc at several points using a micrometer; compare the figures with the tolerance specified in Specifications.

Pre-1988 model year
Removal

17 To remove a front brake disc on pre-1982 99 models or pre-1981 900 models, temporarily refit two of the roadwheel bolts and slot a sturdy lever bar between them; use it to brace the wheel whilst the driveshaft nut is loosened and removed. Using a proprietary hub puller, withdraw the hub from the driveshaft then unbolt the brake disc.
18 Refitting is a reversal of removal; refer to Chapter 8 when fitting the driveshaft nut.
19 To remove a front brake disc on 99 models from 1982 onwards, or 900 models from 1981 onwards, unscrew the caliper mounting bolts and suspend the caliper from a convenient point on the suspension using a cable-tie or steel wire - take care not to strain the hydraulic hose. Remove the cross-head screws and withdraw the brake disc from the hub.

Refitting

20 Refit the brake disc by following the removal procedure in reverse.

1988 model year onwards
Removal

21 Mark the relationship between the disc and the hub with chalk or a marker pen. Slacken off the disc locating stud and retaining screw - these are on the same radius as the wheel bolt holes - but do not remove them at this stage (see illustration).
22 To allow the disc to be removed, the brake caliper must be unbolted from the hub assembly, but does not need to be dismantled - the brake pads and hydraulic hose can be left in place. Retract the inboard brake pad and piston back into the cylinder using a pair of water pump pliers or G-clamps - check that the displaced brake fluid does not overflow from the fluid reservoir.
23 Slide the hydraulic body along the guide pins so that both brake pads are clear of the disc. Referring to the relevant paragraphs of Section 5, remove the hydraulic body from the carrier; hang it from a rigid point on the suspension, using wire or a cable-tie. Do not allow it to dangle freely as this will strain the brake hose.
24 Still working from Section 5, remove the caliper carrier from the steering swivel member.
25 Remove the disc locating stud and retaining screw completely. Support the disc as you do this and lift it off as it becomes free (see illustration). If it sticks, tap the rear face lightly with a soft faced mallet to release it.
26 Remove the polished glaze from the surface of the disc with sand/emery paper. Use small, circular motions to avoid producing a directional finish on the surface.

Refitting

27 If a new disc is being fitted, remove the protective coating from the surface using an appropriate solvent.
28 Locate the disc on the hub so that the roadwheel bolt, retaining screw and locating stud holes are all correctly aligned; use the alignment marks made on removal. If the disc is being removed in an attempt to alter the seating and hence improve runout, turn the disc through 180° and then refit it.
29 Refit the locating stud and retaining screw, tightening them securely (see illustration).

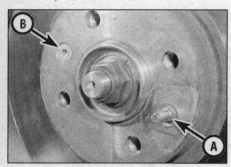

7.21 Disc locating stud (A) and retaining screw (B)

7.25 Removing the front brake disc

7.29 Tightening the locating stud using a 10 mm spanner

Braking system 9•13

8.7 Locate the disc on the hub, using the alignment marks made on removal

8.8 Refitting the rear disc locating stud

30 Refit the brake caliper carrier and hydraulic body, tightening the retaining bolts to the correct torque; refer to Section 5 for details.
31 Re-check the disc runout, using one of the methods described earlier.
32 Depress the brake pedal several times to advance the brake pads towards the disc.
33 Refit the roadwheel and lower the vehicle to the ground. Tighten the roadwheel bolts to the correct torque.

8 Rear brake disc - inspection, removal and refitting

Inspection

Note: *Where brake disc backplates are fitted, it will be necessary to remove the brake caliper from the axle upright to allow an adequate inspection the discs rear surface. Disconnection of the hydraulic hose will not be neccessary, but the caliper must be hung from a convenient point on the suspension, using a cable-tie or a length of wire, to avoid straining the brake hose. Refer to Section 6 for details.*

1 Park the vehicle on a level surface, select first gear (manual transmission) or 'PARK' (automatic transmission) and chock the front roadwheels; on earlier vehicles, fitted with a handbrake which operates on the front wheels, the handbrake should be applied. Remove the wheel trims/centre caps and slacken the roadwheel bolts.
2 Raise the rear of the vehicle, support it securely on axle stands (see "*Jacking and Vehicle Support*") and remove the roadwheels.
3 Refer to beginnining of Section 7, for guidance in the inspection of the brake disc - the information is applicable to both front and rear brake discs. Refer to Specifications for measurements and tolerances.

Removal

4 Working from Section 6, unbolt the brake caliper from the axle tube upright.
5 Mark the relationship between the brake disc and the hub assembly using chalk or a marker pen, then remove the retaining screws and lift off the brake disc. Note that on 1988 model year vehicles onwards, the disc is secured by means of one cross head screw and a locating stud.

Refitting

6 If a new disc is being fitted, remove the protective coating from the surface using an appropriate solvent.
7 Locate the disc on the hub so that the roadwheel bolt, retaining screw and locating stud holes are all correctly aligned, using the alignment marks made on removal **(see illustration)**. If the disc has been removed in an attempt to improve seating and runout, turn the disc through 180° and then refit it.
8 Refit the retaining screw (and where applicable, locating stud) and tighten them securely **(see illustration)**.
9 Refer to Section 6 and refit the brake caliper to the axle upright, tightening the retaining bolts to the correct torque.
10 Re-check the disc runout, using one of the methods described in Section 7.
11 Depress the brake pedal several times to advance the brake pads towards the disc.
12 On 1988 model year vehicles onwards, check and adjust the operation of the handbrake as described in Chapter 1.
13 Refit the roadwheel and lower the vehicle to the ground. Tighten the roadwheel bolts to the correct torque.

9 Hydraulic brake lines and hoses - removal and refitting

Note: *Before starting work, refer to the warning at the start of Section 2 concerning the hazards of working with hydraulic fluid.*
1 If any pipe or hose is to be renewed, minimise fluid loss as far as is possible. Isolate the section of the circuit to be worked on: flexible hoses can be sealed using a proprietary brake hose clamp, metal brake pipe unions can be plugged (if care is taken not to allow dirt into the system) or capped immediately they are disconnected. Place a small container or a wad of rag under any union that is to be disconnected, to catch any spilt fluid. Fluid loss can be further minimised by inserting a piece of polythene sheeting between the brake fluid reservoir and its filler cap. If a good seal is obtained, this will cause a vacuum in the reservoir which will prevent the fluid from escaping through the brake lines. Care should be taken to avoid damaging the filler cap thread, and to ensure that the polythene sheet is removed before the vehicle is brought back into service.
2 If a flexible hose is to be disconnected, unscrew the brake pipe union nut before removing hose from its mounting bracket.
3 To unscrew the union nuts, it is preferable to obtain a brake pipe spanner of the correct size; these are available from most large motor accessory shops. Failing this, a close-fitting open-ended spanner will be required, though if the nuts are tight or corroded, their flats may be rounded-off if the spanner slips. In this case, a self-locking wrench is often the only way to unscrew a stubborn union, but it follows that the pipe and the damaged nuts must be renewed on reassembly. Always clean a union and its surrounding area before disconnecting it, to reduce the risk of dirt entering the hydraulics. If disconnecting a component with more than one union, make a careful note of the connections before disturbing any of them.
4 If a brake pipe is to be renewed, it can be obtained, cut to length and with the union nuts and end flares in place, from Saab dealers. The pipe must then be bent into shape, following the line of the original pipe, before fitting it to the vehicle. Alternatively, most motor accessory shops can make up brake pipes from kits, but this requires very careful measurement of the original, to ensure that the replacement is of the correct length. Usually, the safest answer is to take the original to the shop as a pattern.
5 On refitting, do not overtighten the union nuts. It is not necessary to use excessive force to obtain a sound joint.
6 Ensure that the pipes and hoses are correctly routed, with no kinks and that they are properly secured in the clips or brackets provided. Ensure that they do not chafe against the vehicle's body or any other components; push the vehicle down on its suspension and turn the steering form lock to lock to check that the lines or pipes do not contact those components that are dynamic when the vehicle is being driven.
7 After fitting, bleed the hydraulic system (see Section 2). Thoroughly wash off any spilt fluid, and check carefully for fluid leaks before bringing the vehicle back into service.

10 Master cylinder - removal, overhaul and refitting

Removal - models with ABS

Note: *On models with ABS, the master cylinder is part of the hydraulic unit; refer to Sections 18 and 19.*

Removal - non-ABS models

1 Open the bonnet and place cloth beneath and around the master cylinder to protect the bodywork from brake fluid which may be spilt.

9•14 Braking system

Caution: *Spilled fluid must be washed off the bodywork immediately with cold water, otherwise the paintwork will be permanently damaged.*

2 Disconnect the wiring from the brake fluid reservoir filler cap, at the connector.
3 Disconnect the clutch master cylinder hose from the reservoir and plug the outlet
4 Make a note the brake lines' order of connection at the master cylinder, to aid refitting later, then unscrew the union nuts and disconnect the brake lines; plug the open ends to prevent fluid spillage. **Note:** *To improve access on later models, remove the retaining screws and lift off the cooling system expansion tank.*
5 Unscrew the mounting nuts and withdraw the master cylinder from the servo unit. Recover the O-ring seal and discard it; a new one must be used on refitting.
6 Remove the filler cap and tip the brake fluid into a container; dispose of it safely.

Overhaul

7 Select a clean flat worksurface upon which to carry out the overhaul. It may help to lay down a sheet of paper or card - this will aid visibility and absorb any brake fluid that drains from the dismantled components.
8 Clean all traces of dirt and brake fluid from the exterior of the master cylinder, then mount it in a vice, protecting the surface by lining the vice jaws with strips of wood or aluminium.
9 On earlier models, the fluid reservoir is retained by two pins that engage with lugs on the underside of the reservoir. Drive out the pins, using a hammer and pin punch.
10 Remove the reservoir from the master cylinder by gently rocking it from side to side; prise out the sealing rubbers.
11 Using a wooden dowel, push the primary piston into the cylinder, far enough to allow the the secondary plunger stop pin to be removed. Take care not to score the cylinder bore. Use wads of clean rag to absorb any brake fluid that may be ejected from the cylinder as you do this.
12 Allow the primary piston to protrude, under spring tension, then withdraw it from the cylinder. On earlier models, it will be neccessary to first extract the circlip from the mouth of the cylinder.

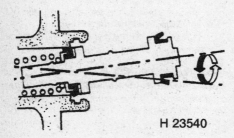

10.18 Insert the secondary piston assembly into the bore as shown

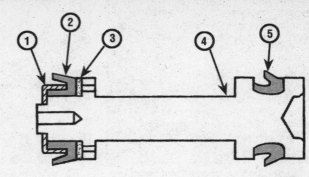

10.16 Secondary piston assembly

1 Spring seat
2 Piston seal
3 Washer
4 Piston
5 Piston seal

13 Remove the cylinder from the vice and carefully tap it on a piece of softwood to retrieve the secondary piston. As the components emerge, note their order of assembly and lay them in that order on a clean worksurface. If the piston proves difficult to extract, cautious application of compressed air to the brake pipe ports will force it out of the cylinder - pad the open end of the cylinder with a wad of rag, in case the pistons emeges faster than expected.
Caution: *Wear eye protection when using compressed air in this manner.*
14 Remove the sleeve and springs from the pistons, making a careful note of the components orientation and order of assembly. Remove the seals from the pistons using a flat non-metallic instrument.

 HAYNES HiNT *The end of nylon cable-tie makes a suitable tool.*

15 Clean all the components in methylated spirit and examine them for wear, damage and corrosion. In particular check the surfaces of the pistons and cylinder bore for scoring and corrosion - if such deterioration is discovered, it will be necessary to renew the complete master cylinder. Repair can only be considered if the wear is limited to the pistons and their seals; repair kits can be obtained from Saab dealers. Check that the inlet and outlet ports are clear of obstructions.
16 Dip the new seals in clean brake fluid and fit them to the pistons using the fingers only to manipulate them into position. Make sure that they are fitted the right way round, as illustrated **(see illustration)**.
17 Mount the cylinder in the vice and apply brake fluid to the bore.
18 Fit the spring to the secondary piston then insert the assembly into the bore, initially at an angle using a twisting motion, to avoid damage to the seals **(see illustration)**.
19 Depress the secondary piston with a wooden dowel and insert the stop pin in the reservoir inlet.
20 Fit the sleeve and spring to the primary piston and insert them into the bore again using a twisting motion to avoid damage to the seals.

21 Depress the primary piston against the spring tension and on later models, fit the circlip in the recess at mouth of the cylinder.
22 Locate the new sealing rubbers in the cylinder, lubricate them with clean brake fluid and then press the reservoir into position. Where applicable, drive in the two reservoir retaining pins.

Refitting

23 Fit a new O-ring seal to the mating surface bewteen the master cylinder and vacuum servo.
24 Fit the master cylinder over the mounting studs on the servo unit, then refit and tighten the nuts, observing specified torque wrench setting.
25 Fit the hydraulic pipes and tighten the union nuts, then fit the clutch master cylinder hose to the reservoir.
26 Connect the fluid level switch cable to the filler cap.
27 Fill the reservoir then top-up the fluid and refit the cap.
28 Bleed the entire hydraulic system at all four calipers, to purge all trapped air pockets - refer to Section 2.
29 Turn the ignition switch to the second position and observe the instrument panel, to verify the operation of the brake fluid level warning switch.

11 Vacuum servo unit non-return valve - removal, testing and refitting

Removal

1 Disconnect the vacuum hose from the servo unit non-return valve.
2 Withdraw the valve from its rubber sealing grommet, using a twisting motion. Extract the grommet from the servo housing.

Testing

3 Examine the check valve for signs of damage, and renew if necessary. The valve may be tested by blowing through it in both directions. Air should flow through the valve in one direction only - when blown through from the *servo unit* side of the valve. Renew the valve if this is not the case.

Braking system 9•15

4 Examine the rubber sealing grommet and flexible vacuum hose for signs of damage or deterioration, and renew as necessary.

Refitting

5 Fit the sealing grommet into position in the servo unit.
6 Carefully ease the non-return valve into position, taking great care not to displace or damage the grommet. Reconnect the vacuum hose to the valve and, where applicable, fit its retaining clip.
7 On completion, start the engine and check the non return valve-to-servo unit connection for signs of air/vacuum leaks, which would be indicated by a hissing sound, whilst the engine is running.

12 Vacuum servo unit - removal and refitting

Removal

1 To remove the servo unit, first remove the master cylinder as described in Section 8.
2 Remove the lower facia panel and where fitted, the centre console - refer to Section 11 for details.
3 Unscrew the lower facia panel retaining screws and lift out the panel.
4 Disconnect the vacuum hose from the servo unit.
5 Move the wiring harness from the servo unit.
6 On left-hand drive 99 models, disconnect the pullrod from the intermediate lever and disconnect the wiring from the stop-light switch. Unscrew the mounting nuts and withdraw the servo together with the bracket. Unbolt the bracket from the servo unit.
7 On all other models, extract the circlip and remove the pivot pin from the servo pushrod. Unscrew the mounting nuts and withdraw the servo unit.
Caution: Get an assistant to support the servo unit, to prevent it from dropping into the engine bay as the last nut is removed.
8 If the servo unit is to be refitted, the felt air filters should be renewed. Cut a slit in the centres of the new filters, using a sharp blade, to enable them to be located them over the pushrod.

Refitting

9 Refit the vacuum servo unit by following the removal procedure in reverse; make reference to Section 10 when refitting the master cylinder.
10 To check the operation of the servo unit, depress the footpedal and keep it depressed whilst starting the engine. The footpedal should move towards the floor, proving that the sevo unit is providing assistance. To repeat the test, the pedal must first be depressed several times with the engine switched off, to dissipate the residual vacuum in the unit.

13.8 Handbrake cable connection to front brake caliper (pre-88 model year vehicles)

13 Handbrake cables - removal, refitting and adjustment

Pre-1988 model year

Removal

1 Working from Chapter 11, unbolt the driver's seat from the floorpan.
2 Remove the sill scuff plates from the bottom of the door aperture, then pull the carpet away from the heater ducts.
3 Remove the screws and withdraw the gear lever cover - on earlier models, take care not to damage the ignition switch lamp.
4 Remove the air ducts, then unscrew the adjustment nut from the handbrake lever. Note that the two cables cross each other in front of the lever, hence the right hand nut adjusts the left hand cable and vice versa.
5 Remove the clip securing the two cables to the floor.
6 Chock the vehicles rear wheels, then jack up the front of the car, support it securely on axle stands (see "*Jacking and Vehicle Support*") and remove the relevant front wheel.
7 Remove the screws and withdraw the cable bush from the inner wheel arch.
8 Relieve the spring tension on the brake caliper handbrake lever, then unhook the inner cable from it **(see illustration)**. Withdraw the outer cable, and remove the rubber gaiter.
9 Withdraw the entire handbrake cable from beneath the engine compartment.

Refitting

10 Refit the cable by following the removal procedure in reverse.

Adjustment

11 Fully apply the handbrake lever several times, then fully release it. Using a feeler blade check that the clearance between the bottom of the handbrake lever on the brake caliper and the yoke is 0.019 in (0.5 mm). If this is not the case, turn the appropriate adjusting nut on the rear of the handbrake lever, until the correct clearance is obtained **(see illustration)**.
12 Verify that the clearance is identical at the opposite front caliper, then refit the

13.11 Cable adjustment nuts at the rear of the handbrake lever (pre-1988 model)

roadwheels and lower the car to the ground. Finally tighten the roadwheel bolts to the specified torque.

1988-1990 model year

Removal

13 Ensure that the vehicle is parked on a flat surface; select first gear (manual gearbox) or 'PARK' (automatic transmission) and chock the front roadwheels; do not apply the handbrake.
14 Working from Chapter 11, separate the centre console from the gear lever console by unclipping the plastic bellows.
15 Prise the gearchange lever gaiter off its mounting frame and fold it up over the lever knob. Models with automatic transmission: prise off the gear selection indicator plate and rotate it around the selector lever shaft, to expose the console retaining screws beneath.
16 Again referring to Chapter 11, remove the centre console, to expose the handbrake cable terminations beneath.
17 Unscrew the adjusting nuts from the end of the relevant handbrake cable, to release it from the handbrake lever.
18 Prise off the locking plate and withdraw the cable from the console area.
19 Tilt the rear seat cushion forwards, then remove the locking pins from the two dowels that secure the leading edge of the cushion. Withdraw the dowels and remove the seat cushion.
20 Where applicable, remove the retaining screws and lift off the rear door aperture sill scuff plates.
21 Roll back the carpet to expose the handbrake cable clamp bracket on the floor pan. Slacken the retaining screws and free the relevant cable from the clamp bracket.
22 Remove the centre caps and slacken the rear roadwheel bolts.
23 At the rear roadwheel, slacken the wheel bolts, then raise the rear of the vehicle and rest it securely on axle stands (see "*Jacking and Vehicle Support*"). Remove the roadwheel.
24 At the suspension lower arm, remove the retaining screws and detach the handbrake cable bracket.
25 At the caliper, relieve the spring tension on the handbrake lever, then unhook the cable inner from it.

9•16 Braking system

26 From inside the cabin, slowly withdraw the cable through the floorplan, prising out the rubber sealing grommet.

Refitting

27 Refit the cable by reversing the removal procedure. Before refitting the gear lever centre console, refer to the cable adjustment procedure, described in the following sub-Section.

Adjustment

28 If a new cable has been fitted, apply and release the handbrake lever several times, to eliminate the initial stretching the new cable inner.
29 With the rear of the vehicle raised and supported on axle stands, use a spanner to unscrew the protective cap from adjusting screw drilling, at the rear of the brake caliper.
30 Using an Allen key, turn the adjusting screw clockwise to the end of its travel (so that the inboard brake pad contcats the disc) then turn it back anticlockwise through one half turn.
31 Check at this point that the rear wheels are free to rotate, without any sign of binding between the brake disc and pad. If this is not the case, turn the adjusting screw anticlockwise through a further quarter of a turn. If the binding cannot eliminated, refer to Section 6 of this Chapter and check the condition of the brake caliper.
32 Refit the protective cap to the caliper adjusting screw drilling.
33 Relieve the spring tension on the caliper handbrake lever by hand, then insert a 1.0 mm feeler blade between it and the end-stop pin on the caliper body.
34 With the gear lever centre console removed (Chapter 11), tighten the relevant handbrake cable adjusting nut until the correct clearance is at the caliper handbrake lever is obtained, indicated by the feeler blade.
35 Verify that the clearance is correct at the opposite rear caliper.
36 Refit the gear lever/centre console, referring to Chaper 11 for guidance. Refit the gear lever gaiter/selector lever indicator plate, as appropriate.
37 Refit the roadwheels, lower the vehicle to the ground, then tighten the roadwheel bolts to the specified torque.
38 Check that the handbrake can be applied securely, within approximately 4 or 5 'clicks' of the lever ratchet mechanism.

1990 model year onwards

Removal

39 Ensure that the vehicle is parked on a flat surface; select first gear (manual gearbox) or 'PARK' (automatic transmission) and chock the front roadwheels; do not apply the handbrake.
40 Working from Chapter 11, separate the centre console from the gear lever console by detaching the bellows moulding. Prise the gearchange lever gaiter off its mounting frame

13.46 Unscrew the retaining nut(s) from the end of the handbrake cable (arrowed)

and fold it up over the lever knob. Models with automatic transmission: prise off the gear selection indicator plate and rotate it around the selector lever shaft, to expose the console retaining screws beneath .
41 Still working from Chapter 11, remove the left hand front seat., then remove the gear lever/centre console.
42 Tilt the rear seat cushion forwards, then remove the locking pins from the two dowels that secure the leading edge of the cushion. Withdraw the dowels and remove the seat cushion.
43 Where applicable, remove the retaining screws and lift off the rear door aperture sill scuff plates.
44 Roll back the carpet and sound insulating material to expose the metal shield over handbrake cable clamp bracket on the floor pan. Remove the retaining screws and remove the shield , then slacken the fixings and free the relevant cable from the clamp bracket.
45 On vehicles equipped with ABS, cut through the nylon tie that secures the wheel sensor to the handbrake cable.
46 Ensure that the handbrake lever is in the released position, then unscrew the retaining nut from the end of the handbrake cable **(see illustration)**.
47 Prise off the locking plate and withdraw the cable from the console area.
48 At the rear roadwheel, slacken the wheel bolts, then raise the rear of the vehicle and rest it securely on axle stands (see "*Jacking and Vehicle Support*"). Remove the roadwheel.
49 At the brake caliper, relieve the spring tension on the handbrake lever, then unhook the cable inner from it **(see illustration)**. Withdraw the cable outer and inner through the guide loop.
50 At the suspension lower arm, remove the retaining screws and detach the handbrake cable bracket.
51 From inside the cabin, slowly withdraw the cable through the floorplan, prising out the rubber sealing grommet.

Refitting

Note: *The right hand handbrake cable is 20mm longer than the left.*
52 On the new cable, slacken the locknut away from the adjustment sleeve, so that the

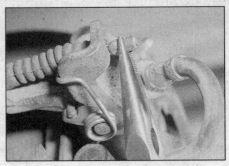

13.49 At the brake caliper, unhook the cable inner from the handbrake lever

cable sheathing can be inserted into the end of the adjustment sleeve. Re-tighten the locknut by three turns, to secure the sheathing in place.
53 Refit the remainder of the cable by reversing the removal procedure, but before refitting the rear seat cushion, refer to the cable adjustment procedure, as described in the following sub-Section.

Adjustment

54 If a new cable has been fitted, apply and release the handbrake lever several times, to eliminate the initial stretching the new cable inner.
55 With the rear seat cushion raised (or removed), use a screwdriver to prise the adjustment sleeve and locknut apart .
56 Partially relieve the spring tension on the caliper handbrake lever, then insert a 2.0 mm feeler blade between it and the end-stop pin on the caliper body, using the spring tension to trap the gauge in position **(see illustration)**.
57 Screw the adjusting sleeve locknut against the sleeve, until the correct clearance at the caliper handbrake lever is obtained, indicated by the feeler blade dropping out **(see illustration)**.
58 Apply and release the handbrake lever several times, to settle the adjustment mechanism. Using the feeler blade, check that the clearance at the brake caliper handbrake lever remains between 0.5 mm and 2.0 mm.
59 Verify that the clearance is identical at the opposite rear caliper.
60 Refit and lower the rear seat cushion, referring to Chaper 11 for guidance. Refit the

13.56 Setting handbrake lever clearance at the end-stop pin on the caliper body

Braking system 9•17

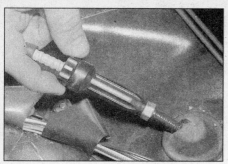

13.57 Screw the adjusting sleeve locknut against the sleeve, until correct clearance at the caliper handbrake lever is obtained

gear lever gaiter/selector lever indicator plate, as appropriate.
61 Refit the roadwheels, lower the vehicle to the ground, then tighten the roadwheel bolts to the specified torque.
62 Check that the handbrake can be applied securely, within approximately 4 or 5 'clicks' of the lever ratchet mechanism.

14 Handbrake lever - removal, refitting and adjustment

Pre 1988 model year

Removal

1 Park the vehicle on a level surface, select first gear (manual gearbox) or 'PARK' (automatic transmission) and chock the rear wheels.
2 Remove the cover from the rear of the handbrake lever and unscrew the cable adjusting nuts; refer to Section 13 for greater detail.
3 Extract the circlip and withdraw the pivot pin.
4 Withdraw the handbrake lever from the vehicle.

Refitting

5 Refit the handbrake lever by reversing the removal procedure. On completion, adjust the handbrake cables as described in Section 13.
6 On pre-1983 models the distance from the end of the lever to the front of the pushbutton with the handbrake fully applied should be

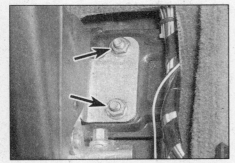

14.10 Handbrake lever retaining nuts (post-1988 model)

between 6.0 and 10.0 mm - if necessary screw the pushbutton in or out as required. On post 1983 models, with the handbrake fully applied, screw the pushbutton in to the end of its travel, then back it off by up to one turn.
7 Check that the instrument panel warning lamp illuminates with the handbrake lever applied within two or three clicks of the rachet mechanism; if necessary adjust the position of the switch, referring to Section 15 for details.

Post 1988 model year

Removal

8 Park the vehicle on a level surface, select first gear (manual transmission) or 'PARK' (automatic transmission) and chock the rear wheels.
9 Remove the cover from the rear of the handbrake lever and unscrew the cable retaining nuts; refer to Section 13 for greater detail. Extract the cables from the drawbar.
10 Remove the retaining nuts and lift the lever assembly away from the floorpan **(see illustration)**.

Refitting

11 Refit the handbrake lever by following the removal procedure in reverse. On completion, refer to Section 13 and adjust the handbrake cables.
12 With the handbrake fully applied, screw the pushbutton in to the end of its travel, then back it off by up to one turn.
13 Check that the instrument panel warning lamp illuminates with the handbrake lever applied within two or three clicks of the rachet mechanism; if necessary adjust the position of the switch, referring to Section 15 for details.

15 Handbrake 'ON' warning light switch - removal, testing and refitting

1 The handbrake 'ON' warning light switch is mounted on a bracket, which is screwed to the floorpan. It is operated by the handbrake cable drawbar. Access to it can be gained by removing the centre console (where applicable); refer to Chapter 11 for details.
2 To remove the switch, ensure that the ignition switch is turned to the 'OFF' position, then unplug the wiring connector from the rear of the switch body.
3 Squeeze together the metal tangs at the side of the switch body and extract the switch from the bracket **(see illustration)**.
4 The operation of the switch can be tested by applying the probes of an electrical continuity tester across the metal switch body and the connector terminal and pressing the plunger in and out. A healthy switch will register an open circuit when the plunger is pressed in.
5 The switch can be refitted by reversing the removal procedure.

15.3 Squeeze together the metal tangs at the side of the switch body, and extract the switch from the bracket

16 Footbrake pedal - removal and refitting

General

1 The brake pedal pivots on a mounting bracket, which is bolted to the vehicles bulkhead. On left hand drive models, the pedal is connected directly to the input push rod at the brake servo unit. On right hand drive models, the pedal is connected to a drive rod, which communicates the pedals movements across the bulkhead to a link-rod and slave lever arrangement, mounted directly behind the brake servo unit (or hydraulic unit, on vehicles with ABS). The input push rod passes through the bulkhead and is pinned to final slave lever in the link-rod arrangement.

Removal

2 Park the car on level ground, and apply the handbrake.
3 Working from Chapter 11, remove the steering column lower cover panel, the lower facia panel assembly, the ashtray and where fitted, the centre console.
4 Remove the lower air duct from under the facia, referring to Chapter 3 for details.
5 Unhook the return spring from the pedal.
6 Remove the retaining screws and lower the brake light switch bracket from the pedal bracket.
7 Extract the split pin and remove the washer and pivot pin securing the servo pushrod/linkage to the pedal. Note on right-hand drive models, the pedal shaft is extended to the left-hand side of the car by means of a coupling rod.
8 On right-hand drive models remove the circlip and washer from the left-hand end of the shaft, unscrew the right-hand bracket nuts, and withdraw the pedal and shaft.
9 On left-hand drive models unscrew the locknut, remove the pivot bolt, and withdraw the pedal.

Refitting

10 Refitting is a reversal of removal; where applicable on right hand drive vehicles, adjust the length of the servo pushrod so that the brake pedal is the same height as the clutch pedal, when both are at rest.

9•18 Braking system

17.1 Location of brake light switch (arrowed)

17 Brake light switch - removal, refitting and testing

900 models

1 The brake light switch is clipped to a mounting plate which, on left hand drive vehicles is screwed to the footpedal assembly. On right hand drive vehicles, it is screwed to the auxiliary pedal bracket, which is located on left hand side of the bulkhead below the facia. **(see illustration)** Note: *On vehicles fitted with cruise control, two vacuum switches are also clipped to the same mounting plate as the brake light switch - see Chapter 4 Part B for greater detail.*

2 To gain access to the switch, first refer to Chapter 11 and remove the facia lower panel.

3 Ensure that the ignition switch is turned to the 'OFF' position then unplug the electrical connector from the rear of the switch body.

4 Squeeze together the plastic tangs at the side of the switch to release it from the mounting plate.

90/99 models

5 The stop-light switch is located in the engine compartment, and is activated by the servo pushrod.

6 Ensure that the ignition switch is turned to the 'OFF' position, then unplug the electrical connector from the switch body.

7 Unscrew the switch from its mounting.

All models

8 The operation of the switch can be tested by applying the probes of an electrical continuity tester across the metal switch body and the connector terminal and pressing the plunger in and out. A healthy switch will register an open circuit when the plunger is pressed in.

9 The switch can be refitted by reversing the removal procedure.

10 On completion, verify the operation of the switch by turning the ignition switch to the second position and depressing the brake pedal; the brake lights should illuminate when the pedal has been depressed by 10 mm or more.

18 Anti-lock braking system (ABS) components - general information

1 The Anti-lock Braking System (ABS) is managed by an Electronic Control Unit (ECU), which has the capacity to monitor the status and condition of all the components in the system, including itself. In the ECU detects a fault, it responds by shutting down the ABS and illuminating the dashboard mounted ABS warning lamp. Under these circumstances, conventional non-ABS braking is maintained. If the nature of the fault detected is such that it may interfere with the operation of the conventional braking system aswell, the ECU also illuminates the normal brake warning lamp.

2 As a consequence, if the ABS warning lamps indicate that all is not well, it is very difficult to diagnose problems with the system, without the equipment and expertise to electronically 'interrogate' the ECU. Hence, this Section is limited firstly to a list of the basic checks that should be carried out, to establish the integrity of the system, (e.g. is there enough brake fluid?, is anything leaking? etc). Section 19 is limited to a description of the removal and refitting of the major components only.

3 If the cause of the fault cannot be immediately identified using the check list described, the *only* course of action open is to take the vehicle to a Saab dealer for examination. Dedicated test equipment is needed to interrogate ABS ECU, to determine the nature and incidence of the fault. For safety reasons, owners are strongly advised against attempting to diagnose complex problems with the ABS using standard workshop equipment.

Basic fault finding checks

Brake fluid level

4 Check the brake fluid level with the vehicle parked on a flat surface and the accumulator fully charged. To ensure that the accumulator is fully charged, switch on the ignition and listen to the hydraulic pump, which is part of the hydraulic unit assembly. The sound made by the pump will be quite audible; when it stops the accumulator will be fully charged. Switch off the ignition and without touching the brake pedal, check the brake fluid level in the reservoir. If neccessary, top it up to the 'MAX' mark with brake fluid of the correct grade.

5 If the level is exceptionally low, there may be leak somewhere in the hydraulic sysem. Refer to Chapter 1 and carry out a check of the brake hoses and pipes throughout the vehicle. If no leaks are apparent, remove each roadwheel in turn and check for leaks at the brake calipers pistons - refer to Sections 5 and 6 for details of the front and rear caliper layout.

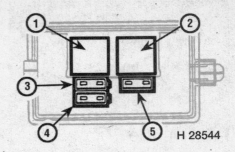

18.6 ABS fuses and relays
1 Hydraulic pump motor relay
2 ECU relay
3 ECU fuse, 10A
4 ECU fuse, 30A
5 Hydraulic pump motor fuse, 30A

Fuses and relays

6 The fuses (and relays) for the ABS are situated in a separate protective box, in the engine bay adjacent to the brake fluid reservoir. **(see illustration)**. Remove the cover and pull out each fuse, one by one. Visually check the fuse filament; if it is difficult to see whether or not it has blown, use a multimeter or continuity tester to check the electrical continuity of the fuse. If any of the fuses are blown, *do not* fit a new one, until the fault that caused the fuse to blow has been found and rectified; have the vehicle inspected by a Saab dealer and inform him of your findings.

Caution: Fuses are fitted to protect equipment and people from damage; a blown fuse is an indication of a problem. Simply fitting a new fuse, or even bypassing the old one before the problem is corrected will only serve to increase the risk of damage.

7 The system relays can be found in the same protective box as the fuses. In general, relays are difficult to test conclusively without an electrical specification. However, the metal contacts inside a relay can usually be felt (and often heard) to open or close as it operates - if the relay in question does behave in this way when the ignition switch is turned on, but fails to operate the circuit, it may be faulty. It should be noted that this is not a conclusive test; substitution with a known good relay *of the same type* is a preferable method of verifying the components operation. If the relay is suspected of being faulty, it can be renewed by pulling it out of its socket - noting its orientation - and pushing in a new unit.

Note: *Some of the heavier duty relays are secured with one or more retaining screws.*

Electrical connections and earthing points

8 The engine bay is a hostile environment for electrical connections and even the best seals can sometimes be penetrated. Water, chemicals and air will induce corrosion on the

Braking system 9•19

connector's contacts and prevent good continuity, sometimes intermittently. Disconnect the battery negative cable, then check the security and condition of all connectors at the ABS hydraulic unit, situated in the engine bay on the rear bulkhead. Unplug each connector and examine the contacts inside.

9 Clean contacts that are found to be dirty or corroded using a proprietary contact cleaning oil, obtainable in most garages and car accessories shops. Avoid scraping the contacts clean with a blade, as this will accelerate corrosion later. Use a piece of lint-free cloth in conjunction with the cleaning oil to produce a clean, shiny contact surface that will result in good electrical continuity.

10 In addition, check the security and condition of the electrical earthing point, on the front of the hydraulic unit.

19 Anti-lock braking system (ABS) components - removal and refitting

1 Due to the complexity of the anti-lock braking system and the fact that specialised test equipment is needed to diagnose faults with it, this Section is limited to a description of the removal and refitting of major components only. Section 18 contains a list of the basic checks that should be carried out, to establish the integrity of the ABS system and to help identify any prominent faults.

⚠ **Warning: After working on any part of the braking system, test the vehicle exhaustively before bringing it back into service. Make sure that all warning lamps extinguish, check all disturbed joints and unions for leaks and top-up the fluid level in the reservoir to the 'MAX' mark. Finally, repeatedly check that the braking system is capable of stopping the vehicle normally, before taking it out onto the public highway.**

Electronic Control Unit (ECU)

General

Caution: The ECU contains components that are sensitive to the levels of static electricity generated by a person during normal activity. Once the multiway harness connector has been unplugged, the exposed ECU pins can freely conduct stray static electricity to these components, damaging or even destroying them - the damage will be invisible and may not manifest itself immediately. Expensive repairs can be avoided by observing the following basic handling rules:

a) *Handle a disconnected ECU by its outer case only; do not allow fingers or tools to come into contact with the pins.*
b) *When carrying an ECU around, "ground" yourself from time to time, by touching an earthed metal object, such as an unpainted water pipe; this will discharge any static that may have built up.*
c) *Do not leave the ECU unplugged from its connector for any longer than is absolutely necessary.*

Removal

2 The ECU is located in the engine bay, on the of the inside the left hand wing.
3 Disconnect the battery negative cable and position it away from the terminal.
4 Remove the retaining screws and lift the ECU away from the mounting bracket.
5 Unplug the ECU harness connector.

Refitting

6 The ECU can be refitted by reversing the removal procedure.

Wheel sensors - general

7 Both the front and rear ABS sensors have two fixings built into the sensor housing; a bolt to secure the sensor to the wheel hub assembly and a set screw to adjust the protrusion depth of the sensor tip. The clearance between the tip of the sensor and the toothed sensor wheel is critical to the operation of the ABS system. One method of checking this clearance is to use a feeler blade, however access to the rear of the wheel hub is restricted, making accurate setting difficult. The preferred method is to attach a self-adhesive spacer disc, available at minimal cost from a Saab dealer, to the end of the sensor tip to set the clearance. Once the sensor is secured in place, the spacer is worn away by the toothed sensor wheel, leaving the correct air-gap. The refitting procedure details this part of the operation.

Front wheel sensors

Removal

8 Disconnect the battery negative cable and position it away from the terminal.
9 Locate the harness connectors in the engine bay, on the bulkhead directly behind the cylinder head. They are bayonet type connectors that are released by squeezing together the finger grips and twisting the two halves of the connector apart.
10 Ensure that the vehicle is parked on a level surface, apply the handbrake and chock the rear wheels. Raise the front of the vehicle, resting it securely on axle stands (see "*Jacking and Vehicle Support*") and remove the front roadwheel(s).
11 Where applicable, raise the air conditioning coolant pipe clips by slackening the pipe clips, to allow the sensor cable connector to be passed underneath. Carefully guide the sensor cable and connector through the grommet in the wheel arch.
12 Slacken and withdraw the sensor bolt **(see illustration)**, extract the cable from the bracket at the side of the upper suspension arm and remove the sensor from the hub assembly, together with its protective sleeve.

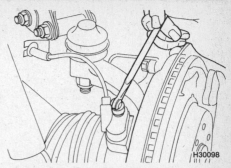

19.12 Slacken and withdraw the wheel sensor retaining bolt (front wheel).

Refitting

13 Slacken the set screw and slide the adjusting sleeve off the old wheel sensor. Clean the sleeve thoroughly; lubricate it lightly with a smear of anti-seize grease and fit it to the new sensor.
14 Ensure that all traces of the old fibre spacer are removed from the sensor tip, using a fine wire brush and a rag moistened with a suitable solvent. Once clean, peel the backing film from the rear of the new 0.65mm thick fibre spacer and press it onto the sensor tip.
15 Slowly turn the brake disc by hand and using a fine wire brush, clean off any traces of dirt and old fibre spacer from the toothed disc.
16 Insert the sensor into the drilling in the hub assembly and tighten the retaining bolt.
17 Gently push the spacer against the surface of the toothed disc and tighten the set screw.

Caution: Do not rotate the toothed disc until the sensor has been secured in position, or the disc teeth may gouge the fibre spacer, giving an incorrect sensor clearance.

18 Alternatively, if a fibre spacer is not available, insert a feeler blade between the sensor tip and the toothed disc; and set the clearance to the value given in Specifications.
19 Pass the sensor connector and cable through the grommet in the wheel arch, then secure the cable to the mounting bracket on the suspension upper arm.
20 Refit the remainder of the components by reversing the removal sequence. On completion, turn the steering from lock to lock and push the vehicle up and down on its suspension, to check that the sensor cable cannot chafe against any of the adjacent components.

Rear wheel sensors

Removal

21 Disconnect the battery negative cable and position it away from the terminal.
22 Tilt the rear seat forward to expose the rear wheel sensor cables, running adjacent to the handbrake cables, along floor pan.
23 Unplug the sensor cable at the connector. Note that the connectors are of the bayonet

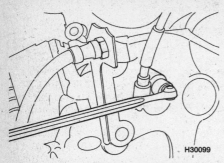

19.27 Slacken and withdraw the wheel sensor retaining bolt (rear wheel)

type, that are released by squeezing together the finger grips and twisting the two halves of the connector apart. Snip through any cable-ties that are securing the cable in position.

24 Ensure that the vehicle is parked on a level surface then chock the front wheels and select first gear (manual transmission) or 'PARK' (automatic transmission). Raise the rear of the vehicle, rest it securely on axle stands (see "*Jacking and Vehicle Support*") and remove the right or left hand rear roadwheel, as appropriate.

25 Extract the sensor cable from the clip and carefully guide the connector through the grommet in the floorplan.

26 At the rear suspension lower arm, remove the retaining screws and lift off the handbrake cable guide bracket, then release the wheel sensor cable from the guide on the suspension arm.

27 Slacken and withdraw the sensor retaining bolt **(see illustration)**, then extract the sensor from the hub assembly.

Refitting

28 Slacken the set screw and slide the adjusting sleeve off the old wheel sensor. Clean the sleeve thoroughly; lubricate it lightly with a smear of anti-seize grease and fit it to the new sensor.

29 Ensure that all traces of the old fibre spacer are removed from the sensor tip, using a fine wire brush and a rag moistened with a suitable solvent. Once clean, peel the backing film from the rear of the new 0.65mm thick fibre spacer and press it onto the sensor tip.

30 Slowly turn the brake disc by hand and using a fine wire brush, clean off any traces of dirt and old fibre spacer from the toothed disc.

31 Insert the sensor into the drilling in the hub assembly and tighten the retaining bolt.

32 Gently push the spacer against the surface of the toothed disc and tighten the set screw.

Caution: *Do not rotate the toothed disc until the sensor has been secured in position, or the disc teeth may gouge the fibre spacer, giving an incorrect sensor clearance.*

33 Alternatively, if a fibre spacer is not available, insert a feeler blade between the sensor tip and the toothed disc, and set the clearance to the value given in Specifications.

34 Pass the sensor connector and cable through the grommet in the wheel arch, then secure the cable in the guide on the suspension lower arm. Refit the handbrake cable guide bracket.

35 Refit the remainder of the components by reversing the removal sequence. On completion, push the vehicle up and down on its suspension, to check that the sensor cable cannot chafe against any of the adjacent components.

Fluid reservoir

Removal

Caution: *The braking system must be depressurised before the fluid reservoir can be removed. This can be achieved by depressing and releasing the brake pedal in excess of 20 times, with the ignition switched off, until firm resistance at the pedal is felt.*

36 Park the vehicle on a level surface and apply the handbrake firmly. Disconnect the battery negative cable and position it away from the terminal.

37 Wipe the surfaces around the hose connections clean of dirt, to prevent contamination as the hoses are disconnected.

38 Working from Chapter 4, remove the air inlet scoop from the side of the engine bay.

39 With reference to Chapter 3, unbolt the cooling system expansion tank from the side of the engine bay and move it to one side; it is not necessary to disconnect the coolant hoses from it.

40 Siphon out as much brake fluid as possible from the reservoir, using an old poultry baster or pipette. Note that it will not be possible to remove all the fluid from the reservoir.

41 Unplug the wiring for the fluid level switch at the connector. The connector has a metal locking ring which must be splayed open at the sides, to release it.

42 On vehicles with manual transmission, attach a hose clamp to the clutch master cylinder supply hose, then disconnect the end from the fluid reservoir.

43 At the base of the reservoir, slacken and withdraw the retaining screw, using an allen key.

44 Carefully prise the reservoir away from the master cylinder, allowing the plastic ports on the underside of the reservoir to pull out of the rubber grommets. Recover the O-ring and spacer from the rearward connection, noting their order of fitment.

45 Pull the hydraulic pump supply pipe off the port on the base of the reservoir.

46 Extract the rubber grommets from the ports on the master cylinder and discard them; new items must be fitted on reassembly.

Refitting

47 Refit the reservoir by following the removal procedure in reverse, noting the folllowing points:

a) *Ensure that the spacer and O-ring are fitted to the rearward reservoir port in the correct order.*
b) *Fit new rubber grommets into the master cylinder to reservoir connection ports.*
c) *Refill the system with correct grade and quantity of brake fluid, then bleed the entire braking system; refer to Section 2 of this Chapter for details.*
d) *On vehicles with manual transmission, bleed the clutch hydraulic system; refer to Chapter 6 for details.*
e) *On completion, check all disturbed joints and unions for leaks. Verify that the brake and ABS warning lamps extinguish after the ignition has been turned on. Finally, check that the braking system functions correctly, before bringing the vehicle back into service.*

Hydraulic unit

General

48 The hydraulic unit incorporates a tandem master cylinder which operates the two front brake calipers under normal braking, a valve block which modulates the pressure in the three brake circuits during ABS operation, an accumulator which provides a supply of highly pressurised brake fluid, a hydraulic pump to charge the accumulator, a servo cylinder which regulates the pressurised fluid supply from the accumulator to provide hydraulic power assistance (replacing the vacuum servo unit used in conventional braking systems), aswell as pressure to operate the rear brakes, and a brake fluid reservoir **(see illustration)**.

Removal

Caution: *The braking system must be depressurised before the fluid reservoir can be removed. This can be achieved by depressing and releasing the brake pedal in excess of 20 times, with the ignition switched off, until firm resistance at the pedal is felt.*

49 Park the vehicle on a level surface and apply the handbrake firmly. Disconnect the battery negative cable and position it away from the terminal.

50 The hydraulic unit is mounted on the right hand side of the engine bay, at the bulkhead. Wipe the surfaces around the hose connections clean of dirt, to prevent contamination as the hoses are disconnected.

51 Working from Chapter 11, remove the steering column lower cover panel, the lower facia panel assembly, the ashtray and where fitted, the centre console.

52 Remove the lower air duct from under the facia, referring to Chapter 3 for details.

53 Unclip the sound insulation padding from behind the pedal assembly.

54 Detach the left hand defroster duct from the heater unit and remove it.

55 Referring to Section 16, separate the hydraulic unit pushrod from the brake pedal (left hand drive vehicles) / slave lever (right hand drive vehicles).

Braking system 9•21

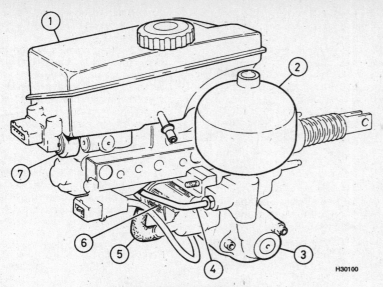

19.48 ABS hydraulic unit

1 Brake (and clutch) fluid reservoir
2 Accumulator
3 Hydraulic pump
4 Pump delivery hose
5 Pump inlet hose
6 Hydraulic pump motor
7 Main valve

56 Working from Chapter 4, remove the air inlet scoop from the side of the engine bay.
57 With reference to Chapter 3, unbolt the cooling system expansion tank from the side of the engine bay and move it to one side; it is not necessary to disconnect the coolant hoses from it.
58 Siphon out as much brake fluid as possible from the reservoir, using an old poultry baster or pipette. Note that it will not be possible to remove all the fluid from the reservoir.
59 Unplug the electrical connections for the pressure switch, main valve, fluid level indicator and pump motor from the hydraulic unit; label each connector to aid refitting later.
Note: *The connectors have metal locking rings which must be splayed open at the sides, to release them.*
60 Unbolt the steady bracket from the front of the hydraulic unit. Remove the bolt and disconnect the earth cable from the bracket. Unplug the sensor cable connector and move it to one side; label it to aid refitting later.
61 On vehicles with manual transmission, attach a hose clamp to the clutch master cylinder supply hose, then disconnect the end of the hose from the fluid reservoir.
62 Slacken the unions and disconnect all four rigid brake pipes from the side of the hydraulic unit, including the larger bore return pipe. Plug the open ports in the hydraulic unit and the ends of the brake pipes, then label them to aid refitting later. Position a wad of rags underneath the unions, to absorb any brake fluid spillage.
63 In the cabin area, remove the four retaining nuts from the hydraulic unit mounting bolts, that protrude through the bulkhead, behind the pedal assembly. Have an assistant support the hydraulic unit inside the engine bay, to prevent it dropping as the retaining nuts are removed.
64 Lift the hydraulic unit out of the engine bay; recover the rubber bellows from the input pushrod and peel the gasket from the bulkhead, if it has stuck in position.

Refitting

65 The hydraulic unit can be refitted by reversing the removal procedure, noting the following points:
a) Observe the specified torque wrench setting, when refitting the retaining nuts at the bulkhead.
b) Refit the brake pipes according to the notes made during removal and tighten the unions securely - refer to Section 9 for guidance.
c) Refit the electrical connectors according to the notes made during removal.
d) Refill the system with correct grade and quantity of brake fluid, then bleed the entire braking system; refer to Section 2 of this Chapter for details.
e) On vehicles with manual transmission, bleed the clutch hydraulic system; refer to Chapter 6 for details.
f) On completion, check all disturbed joints and unions for leaks. Verify that the brake and ABS warning lamps extinguish after the ignition has been turned on. Finally, check that the braking system functions correctly, before bringing the vehicle back into service. It may be prudent to have the system checked by a Saab dealer at the earliest opportunity.

Valve block

Removal

Caution: *The braking system must be depressurised before the fluid reservoir can be removed. This can be achieved by depressing and releasing the brake pedal in excess of 20 times, with the ignition switched off, until firm resistance at the pedal is felt.*

66 Park the vehicle on a level surface and apply the handbrake firmly. Disconnect the battery negative cable and position it away from the terminal.
67 Wipe the surfaces around the hose connections clean of dirt, to prevent contamination as the hoses are disconnected.
68 Working from Chapter 4, remove the air inlet scoop from the side of the engine bay.
69 With reference to Chapter 3, unbolt the cooling system expansion tank from the side of the engine bay and move it to one side; it is not necessary to disconnect the coolant hoses from it.
70 Siphon out as much brake fluid as possible from the reservoir, using an old poultry baster or pipette. Note that it will not be possible to remove all the fluid from the reservoir.
71 Unplug the electrical cabling from the valve block at the connector.
72 Slacken the unions and disconnect all brake pipes from the valve block (three on the underside, four on the side - including the larger bore return pipe). Plug the open ends of the pipes and the ports on the valve block, then label them to aid correct refitting later.
Caution: *To avoid straining the brake pipes, slacken the connections at the hydraulic unit, to allow some freeplay.*
73 Cut through the plastic tie that secures the wiring loom to the valve block.
74 Remove the three retaining nuts and lift the valve block out of the engine bay.

Refitting

75 The hydraulic unit can be refitted by reversing the removal procedure, noting the following points:
a) Refit the brake pipes according to the notes made during removal and tighten the unions securely - refer to the general information given in Section 9 for guidance.
b) Refill the system with correct grade and quantity of brake fluid, then bleed the entire braking system; refer to Section 2 of this Chapter for details.
c) On completion, check all disturbed joints and unions for leaks. Verify that the brake and ABS warning lamps extinguish after the ignition has been turned on. Finally, check that the braking system functions correctly, before bringing the vehicle back into service.

Notes

Chapter 10
Suspension and steering systems

Contents

Front hub assembly and wheel bearings - removal and refitting 4
Front suspension anti-roll bar - removal and refitting 5
Front suspension coil spring - removal and refitting 6
Front suspension wishbones - removal and refitting 3
Front suspension ball joints - testing and renewal 7
Front suspension damper - removal, testing and refitting 2
General description 1
Power assisted steering hydraulic system - draining,
  refilling and bleeding 18
Power assisted steering servo pump - removal and refitting 17
Power steering fluid level check . . . . .See Chapter 1
Rear axle tube - removal and refitting 16
Rear hub assembly - removal and refitting 15
Rear suspension anti-roll bar - removal and refitting 11
Rear suspension coil spring - removal and refitting 10
Rear suspension damper - removal and refitting 9
Rear suspension Panhard rod - removal and refitting 12
Rear suspension torque arm - removal and refitting 13
Rear suspension lower arm - removal and refitting 14
Steering and suspension check . . . . .See Chapter 1
Steering gear (manual) - removal and refitting 19
Steering rack (power assisted) - removal and refitting 20
Steering rack rubber gaiters - renewal 21
Steering swivel member - removal and refitting 8
Steering wheel - removal and refitting 22
Steering column - removal and refitting 23
Steering column intermediate shaft - removal and refitting 24
Tyre checks . . . . .See "Weekly Checks"
Wheel alignment and steering angles -checking and adjustment . . .25

Degrees of difficulty

Easy, suitable for novice with little experience		Fairly easy, suitable for beginner with some experience		Fairly difficult, suitable for competent DIY mechanic		Difficult, suitable for experienced DIY mechanic		Very difficult, suitable for expert DIY or professional	

Specifications

General
Front suspension layout . . . . Independent with unequal length, asymmetric double wishbones, coil springs and telescopic dampers. Anti-roll bar fitted to certain models
Rear suspension layout . . . . Dead beam axle supported in Watts linkage with trailing lower arm/spring link, leading upper arm/torque bar, Panhard rod, coil springs and gas/oil dampers. Anti-roll bar fitted to certain models
Steering . . . . Rack and pinion, hydraulic power assistance on certain models
Wheelbase:
  Up to 82 model year . . . . 2525 mm
  82 model year onwards . . . . 2517 mm

Coil springs *
Front suspension:
  Total number of coils per spring (except sports chassis) 8.25
    Sports chassis (silver/bronze colour coded) 6.5
  Free length (class 1/2):
    Dark green/light green colour coded . . . . 373 mm
    Yellow/red colour coded . . . . 380 mm
    Pink/brown colour coded . . . . 388 mm
    Black/white colour coded . . . . 373 mm
    Silver/bronze colour coded . . . . 301 mm
Rear suspension:
  Total number of coils per spring 9
  Free length (class 1/2):
    Black colour coded . . . . 311 mm
    Green colour coded . . . . 308 mm
    Silver/bronze colour coded . . . . 293 mm

* **Note:** *Coil springs on the same axle must be of the same class; this maintains a uniform ride height across the vehicle.*

Road wheels

	Aluminium alloy	Steel
Maximum radial runout	0.5 mm	1.0 mm
Maximum lateral runout	0.5 mm	1.0 mm

Wheel alignment (vehicle unladen)

Rear:
- Toe-in .... 4 ± 2 mm, 1 - 3mm / side (measured between rims)
- Camber .... -0.50 ± 0.25°

Front:
- Toe-in:
  - Standard chassis .... 2 ± 1mm (measured between rims)
  - Sports chassis .... 1.5 ± 0.5 mm
- Camber:
  - Up to 91 model year (standard chassis) .... +0.50 ± 0.50 °
  - Up to 91 model year (sports chassis) .... +0.25 ± 0.25 °
  - 91 model year onwards (standard chassis) ...... -0.25 ± 0.50 °
  - 91 model year onwards (sports chassis) .... +0.25 ± 0.25 °
- Castor:
  - Manual steering .... +1 ± 0.50°
  - PAS .... +2 ± 0.50°
  - Sports chassis .... +2 ± 0.25°
- King pin inclination .... 11.5 ± 1°

Steering angle:
- Outer wheel .... 20°
- Inner wheel .... 20.75 ± 0.50°

Tyre pressures
Refer to the end of "Weekly Checks".

Balljoints

Track rods:
- Wear limits in track rod ends:
  - Axial play .... 2 mm
  - Radial play .... 1 mm
- Steering swivel member:
  - Axial play .... 2 mm
  - Radial play .... 1 mm
- Steering inner ball joints:
  - Axial play .... 2 mm
  - Radial play .... 1 mm

Manual steering

- Turns, lock to lock .... 4.2 (EMS 3.5)
- Damping yoke cover plate clearance .... 0.05 - 0.15 mm
- Shim thicknesses .... 0.13mm, 0.19mm and 0.25mm

Power assisted steering (PAS)

- Steering wheel turns, lock to lock 3.6
- Servo fluid capacity .... 75 cl
- Hydraulic servo pump: .... Belt tension:
  - New belt .... 400-490 N
  - After adjustment .... 290-330 N
  - Minimum .... 220 N

Torque wrench settings

	Nm	lbf ft
Steering wheel centre nut	26	19
Steering column universal joint pinch bolt	30	22
Steering rack mounting bolts	70	52
Steering rack inner ball joint (adjustable)	47	35
Steering rack inner ball joint (non-adjustable)	120	89
Steering gear yoke cover plate bolts	18	13
Track rod end locknut	70	52
Track rod end to track arm	55	41
PAS rack mounting bolts	70	52
PAS hydraulic hose fittings	28	21
PAS pinion locknut	40	30
PAS damper yoke locknut	80	59
PAS inner balljoint	90	66

Steering and suspension

Torque wrench settings - continued

	Nm	lbf ft
PAS track rod end locknut	70	52
PAS track rod end to track arm	55	41
Hub centre nut, front (pre-81 model year)	350	258
Hub centre nut, front (81 model year on)	300	221
Hub centre nut, rear (27 mm across flats):		
Stage 1	49	36
Stage 2	Slacken nut completely	
Stage 3	3	2
Hub centre nut, rear (32 mm across flats)	300	221
Wheel nuts (pre 88 model year)	100	74
Wheel bolts (88 model year onwards)	120	89
Wheel locating stud (88 model year onwards)	28	21
Steering swivel member to upper and lower ball joints	45	33
Lower wishbone to ball joint bolts	50	37
Upper wishbone to ball joints	80	59
Upper wishbone bearing bracket mounting bolts	50	37
Lower wishbone bearing bracket mounting bolts	85	63
Upper wishbone pivot bolt nuts	65	48
Lower wishbone pivot bolt nuts	80	59
Rear suspension torque arm mounting bolts (dry bushes)	60	44
Rear suspension torque arm mounting bolts (lubricated bushes)	30	22
Panhard rod mounting bolts	60	44

1 General description

A fully independent front suspension system is fitted, utilising unequal length, asymmetric double wishbones, coil springs, telescopic dampers; an anti-roll bar is fitted to certain models. The coil springs are mounted between the upper wishbones and the inner wings, with pivoted lower seats that prevent the springs from deforming under compression. The dampers are mounted between the lower wishbones and the inner wings and can be removed separately from the coil springs. The anti-roll bar is bolted to the lower wishbones and is supported by rubber bushes, mounted in brackets that are bolted to the vehicles' subframe. Bracketed rubber bushes are also employed as flexible mountings for the front suspension wishbones, at the inner wings and subframe. Bolt-on balljoints couple the wishbones to the top and bottom of the steering swivel members.

The front wheel hubs and outer drive shafts are journalled in the steering swivel member in double row ball bearing races. The outboard driveshafts are secured by a single nut and thrustwasher. A modified (but mechanically similar) hub assembly was fitted from 1981 model year onwards. The wheel bearings can be renewed on all variants.

The rear suspension is arranged in a Watts linkage configuration, employing a lightweight, rigid, tubular axle located by upper and lower suspension arms - the trailing lower arms also serving as carriers for the coil springs. The leading upper arms provide positive axle location, opposing the torsional reaction caused when the rear wheel brakes are applied, without limiting suspension travel. On certain models, a rear anti-roll bar is fitted. Lateral axle location is achieved by means of a Panhard rod, bolted between the axle tube and the vehicles underside. The anti-roll bar and dampers are anchored on the lower trailing arms.

The rear hub carriers are mounted at either end of the axle tube, housing the stub axles onto which the wheel hubs are mounted. Three different types of rear wheel hub may be fitted; according to the vehicles build date; earlier versions are secured with a 27 mm A/F nut and incorporate double row tapered roller bearing races; later versions up to 1988 model year are secured with a 32m A/F nut and incorporate double row ball bearing races. The latter variant is a sealed unit; the wheel bearings are non-serviceable and can only be renewed as an integral part of the hub assembly. An updated but mechanically simillar version of this design, with an outboard seal, was fitted from 1988 model year.

A rack and pinion steering system is fitted to all variants. Effort applied at the steering wheel is transmitted to the steering rack pinion by the steering column. The rotation of the pinion causes the rack to move linearly, extending and retracting the track rods, which connect the ends of the steering rack to the steering swivel members, by means of bolt-on ball joints. This movement then alter the steering angle of the roadwheels.

Power assisted steering (PAS) is fitted on certain models. The power assisted steering rack contains a hydraulic ram, actuated by pressurised hydraulic fluid supplied by the PAS pump. When the steering wheel is turned, a control valve mounted in line with the steering pinion directs fluid to the appropriate side of the hydraulic ram, amplifying the effort produced at the pinion and assisting the movement of the steering rack. The PAS hydraulic pump is mounted externally on the engine block and is driven by the auxiliary drivebelt.

The two-stage steering column is coupled by universal joints and incorporates a telescopic inner shaft and collapsible outer column, which houses the steering column bearings. The design and mounting position of the steering column are such that, in the event of a head-on collision, it will be deflected away from the driver and also absorb impact by crumpling longitudinally.

2 Front suspension damper - removal, testing and refitting

 Warning: To preserve the handling characteristics of the vehicle, BOTH front dampers (left and right hand) should renewed as a pair.

Removal

1 Jack up the front of the car and support it on axle stands (see "Jacking and Vehicle Support") Chock the rear wheels and remove the relevant front wheel.

2 Position a trolley jack beneath the outer end of the lower wishbone and raise the suspension to relieve the pressure on the damper mountings. **Note:** *To improve access to the left hand damper top mounting, refer to Chapter 4 and unbolt the coolant expansion tank from the body.*

3 Unscrew the damper mounting nuts and withdraw the unit from the car. On 99 models,

10•4 Suspension and steering

2.3a 900 model front damper upper . . .

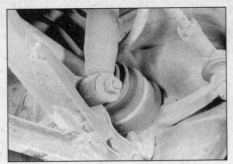

2.3b . . . and lower mountings

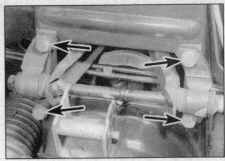

3.5 Upper wishbone bearing mounting bolts (arrowed)

the upper mounting bolt is located in the wheel housing; on 900 models it is located in the engine compartment **(see illustrations)**. Recover the washers and mounting rubbers, noting their order of fitment. **Note:** *It will be necessary to hold the lower of the two top mounting nut still as the upper lock nut is slackened - a thin, open ended spanner will be needed for this.*

 HAYNES HiNT *Use an open-ended bicycle spanner, which will be much thinner than an automotive spanner.*

Testing

4 Mount the damper vertically in a vice and move the upper section through its full stroke. The resistance should be even throughout the stroke - if not, renew the damper.

Refitting

5 Refitting is a reversal of the removal procedure. Where hydraulic dampers are being fitted, first bleed the air from the damper by operating it through its full stroke several times while mounted vertically in a vice. Refer to Section 9, when disposing of the pneumatic (gas filled) damper units.

3 Front suspension wishbones - removal and refitting

Upper wishbone

Removal

1 If removing the left-hand side upper wishbone, first refer to Chapter 2B and remove the engine and transmission.
2 Remove the front coil spring (Section 6).
3 On 99 models remove the damper with reference to Section 2.
4 Support the lower wishbone with a trolley jack, then unscrew and remove the bolt securing the upper balljoint and spring seat to the upper wishbone. Note which way around the bolt heads are fitted.
5 Unscrew the bolts from the wishbone inner bearing brackets, then pivot the wishbone up and withdraw it from the vehicle **(see illustration)**. Where shims are fitted beneath the bearing brackets, note their locations and order of fitment and ensure that they are refitted in the same manner.
6 Mark the fitted position of the bearing brackets, then unscrew the nuts and remove the washers and brackets.
7 Using a long bolt, metal tube, and washers press the rubber bushes from the brackets.
8 Clean all the components and check them for damage and distortion. Renew the parts as necessary. Obtain new rubber bushes.
9 Dip the new rubber bushes in soapy water then press them into the brackets using the method described in paragraph 7.

Refitting

10 Fit the brackets to the wishbone together with the washers and nuts. Position the brackets at an angle of 52° (99 models) or 62° (900 models) to the wishbone, then tighten the nuts to the specified torque. Note that the wishbone is asymmetrical **(see illustrations)**; ensure that is fitted the correct way around.

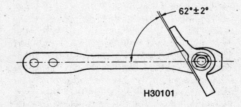

3.10a Position the bearing brackets at 62° (900 models) to the control arm

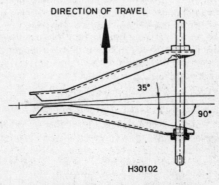

3.10b The wishbone is asymmetrical; ensure it is fitted the correct way round

11 Fit the together with the shims, and insert and tighten the bolts.
12 Locate the upper balljoint and spring seat on the wishbone then insert and tighten the bolts.
13 On 99 models, refit the damper with reference to Section 2.
14 Refit the front coil spring as described in Section 6.
15 Refit the engine/transmission if applicable as described in Chapter 1.
16 Check and if necessary adjust the front wheel alignment.

Lower wishbone

Removal

17 Jack up the front of the car and support it on axle stands (see *"Jacking and Vehicle Support"*). Chock the rear wheels and remove the front wheel.
18 Position a trolley jack under the outer end of the lower wishbone and raise the suspension until a hardwood or metal block can be inserted beneath the upper wishbone and the bodyframe, to relieve the coil spring pressure. Lower the jack.
19 Unscrew the damper lower mounting nut and disconnect the damper, referring to Section 2 for guidance.
20 Unscrew and remove the bolts securing the lower balljoint to the lower wishbone. Note which way round the bolt heads are fitted.
21 Unscrew the nuts from the wishbone inner bearing brackets and lower the wishbone from the underframe **(see illustration)**. If necessary, remove the upper brackets.

3.21 Lower wishbone rear bearing bracket

Steering and suspension 10•5

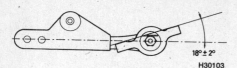

3.26 Position the brackets at an angle of 18° to the wishbone

22 Mark the bearing brackets for position then unscrew the nuts and remove the washer and brackets.
23 Using a long bolt, metal tube, and washers press the rubber bushes from the brackets.
24 Clean all the components and check them for damage and distortion. Renew the parts as necessary and obtain new rubber bushes.
25 Dip the new rubber bushes in soapy water then press them into the brackets using the method described above.

Refitting

26 Fit the brackets to the wishbone together with the washers and nuts. Position the brackets at an angle of 18° to the wishbone then tighten the nuts to the specified torque (see illustration).
27 Fit the wishbone to the underframe then insert and tighten the bolts.
28 Locate the balljoint in the lower wishbone then insert and tighten the bolts.
29 Re-connect the damper then fit the washer and nut, and tighten the nut.
30 Raise the front suspension with a trolley jack and remove the wood or metal block.
31 Refit the front wheel and lower the car to the ground.
32 Check and if necessary adjust the front wheel alignment.

4 Front hub assembly and wheel bearings - removal and refitting

All vehicles

1 Refer to Section 8 and remove the steering swivel member from the vehicle.
Note: *The wheel bearings are an interference fit in the steering swivel member casting. Ideally, they should be removed using a hydraulic press. If access to such a press is impossible, the bearings can be driven out using a mallet and a section of metal tubing as a drift. The tubing should be of such a diameter that it bears squarely on the surface of the bearing and does not damage the bore of bearing housing.*

Vehicles up to 1980 model year
Removal

2 With the driveshaft, hub and brake disc removed, the wheel bearing retaining circlip must be removed using a pair of circlip pliers. If the circlip is at all corroded or damaged, renew it.

3 Press the bearing races from the steering swivel member. **Note:** *The bearings cannot be reused, once removed.*
4 Inspect the bearing housing for damage or corrosion. Do not attempt to restore a badly corroded surface with emery cloth; renew the steering swivel member if necessary.
5 Pack the new bearing with a quantity of molybdenum disulphide based grease. Smear a small amount on the surface of the bearing housing.
6 Press the bearing into the steering swivel member and refit the cirlip.
7 Press the driveshaft into the bearing in the steering swivel member, then press on the inner bearing race, over the driveshaft.
8 Press the hub and brake disc assembly onto the driveshaft splines, then fit the thrustwasher and nut - do not attempt to tighten the nut at this point.

Vehicles from 1981 to 1987 model year

9 Press the hub out of the steering swivel member. **Note:** *Once the hub has been removed, the wheel bearings cannot be reused.*
10 If the inner bearing race has recesses machined into it, it may removed using a universal puller. Otherwise, it should be driven out using a mallet and cold chisel.
11 Remove the retaining circlips from either side of the bearing, then press the bearing out of the steering swivel member. Renew the circlips if they appear at all corroded or damaged.
12 Inspect the bearing housing for damage or corrosion. Do not attempt to restore a badly corroded surface with emery cloth; renew the steering swivel member if necessary.
13 Pack the new bearing with a quantity of molybdenum disulphide-based grease. Smear a small amount on the surface of the bearing housing.
14 Fit the circlip to the inboard side of the bearing housing in the steering swivel member.
15 Press the new bearing into its housing, until it bears against the cirlip.
16 Fit the circlip to the outboard side of the bearing housing.
17 Press the hub squarely into the bearing.
18 Press the inner bearing race into position.
19 Refer to Section 8 and refit the steering swivel member to the vehicle.

Vehicles from 1988 model year onwards

20 Press the hub out of the steering swivel member. **Note:** *Once the hub has been removed, the wheel bearings cannot be reused.*
21 Use a universal puller to draw the bearing inner race off the hub (see illustration).
22 Remove the circlips from either side of the bearing in the steering swivel member, then press out the bearing (see illustration).

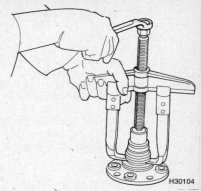

4.21 Use a universal puller to draw the bearing inner race off the hub

23 Clean all components thoroughly, then inspect them for evidence of wear or corrsion; pay particular attention to the bearing housing recesses.
24 Fit the circlip to the inboard side of the bearing housing in the steering swivel member.
25 Pack the new bearing with a quantity of molybdenum disulphide based grease. Smear a small amount on the surface of the bearing housing, then press the bearing into its housing, until it bears against the cirlip.
26 Fit the circlip to the outboard side of the bearing housing.
27 Press the hub squarely into the bearing, then press the inner bearing race into position.

All vehicles

28 Refer to Section 8 and refit the steering swivel member to the vehicle.

5 Front suspension anti-roll bar - removal and refitting

Removal

1 Park the vehicle on a level surface, apply the hanbrake and chock the rear wheels.
2 Push the vehicle down on its suspension and insert a block of wood, approximately 25cm wide by 25cm high between the suspension upper wishbone and the inner

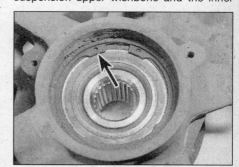

4.22 Remove the circlips (driveshaft side arrowed) from either side of the bearing in the steering swivel member

10•6 Suspension and steering

wing - this will relieve the load placed on the suspension components by the coil spring, when the vehicle is raised.
3 Raise the front of the vehicle, rest it securely on axle stands (see "*Jacking and Vehicle Support*") and remove the roadwheels.
4 Unbolt the anti-roll bar link from the lower suspension wishbone, then unbolt the link from the end of the anti-roll bar.
5 Working underneath the floorpan, slacken and withdraw the U-clamp bracket bolts. Recover the rubber bushes, and where fitted, the shims.
6 Working from Section 20 or 21, as applicable, unbolt the steering rack from the subframe and move it towards the front of the vehicle.
7 Rotate the anti-roll bar so that the angled ends are pointing upwards, then manipulate it out through the wheelarch.

Refitting

8 Refit the anti-roll bar by following the removal procedure in reverse, noting the following points:
  a) Inspect all rubber bushes carefully; renew them if they appear worn or damaged.
  b) Lubricate the bushes with a suitable rubber lubricant, Saab recommend Molycote 1-4382.

6 Front coil spring - removal and refitting

1 Jack up the front of the car and support it on axle stands (see "*Jacking and Vehicle Support*"). Chock the rear wheels and remove the relevant front wheel.
2 Using a proprietary spring compressor tool, compress the coil spring to give approximately 1.5 in (38 mm) clearance at the upper seat **(see illustration)**. If necessary, extra clearance can be obtained by supporting the lower wishbone with a trolley jack, disconnecting the damper lower mounting, then lowering the jack.

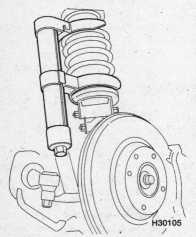

6.2 Using a proprietary spring compressor tool to compress the coil spring

3 Prise the upper steel cone from the wheel housing if necessary, then withdraw it together with the coil spring from the car.
4 Remove the steel cone and rubber ring from the top of the coil spring then release the spring compressor tool.
5 Clean the components including the spring pivot plate, and examine them for wear and damage. Renew them as necessary and obtain a new rubber ring. The spring pivot plate bush can be renewed if necessary.
6 Compress the coil spring with the compressor tool, and locate the rubber ring and steel cone on the top.
7 Locate the spring on the pivot plate then gradually release the compressor, at the same time guiding the steel cone into the wheel housing. If the pivot plate has been removed, the bolt should be left loose until the compressor is removed; it can then be tightened.
8 Re- connect the damper lower mounting, if applicable.
9 Refit the front wheel and lower the car to the ground.

7 Front suspension ball joints - testing and renewal

Track rod ends

99 models

1 Wear in the track rod end balljoint can be checked by attempting to turn the roadwheel alternately on each lock while noting any play in the balljoint. Moderate wear is compensated for by a split bearing within the balljoint, so if any play is detectable, the track rod end must be renewed. In addition, check the condition of the rubber seal covering the joint; if it is split or perished, it must be renewed.
2 Jack up the front of the car and support it on axle stands (see "*Jacking and Vehicle Support*"). Chock the rear wheels and remove the relevant front wheel.
3 Turn the steering wheel to give better access then unscrew the track rod end nut, and use a ball joint splitter to release the tie-rod from the steering swivel member.
4 Loosen the locknut and unscrew the track rod end noting how many turns are required to remove it.
5 Screw the new track rod end onto the track rod to the previously noted position, by screwing it in by the same number of turns. Engage the track rod end with the steering swivel member, then fit the nut and tighten it to the specified torque.
6 Fit the front wheel and lower the car to the ground.
7 Check and adjust the front wheel alignment as described in Section 26.

900 models

8 When measuring the tie-rod end fitting length after adjusting the toe-in setting, measure the distance between the inner face

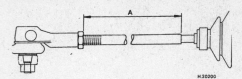

7.9 Note distance A when checking or adjusting front wheel toe-in
A = 100 mm max (manual steering)
A = 125 mm max (power assisted steering)

of the balljoint locknut and the outer edge of the bellows location groove. Unclip the outer end of the bellows and slide them inwards towards the steering gear to enable the measurement to be taken.
9 Note when checking or adjusting the front wheel toe-in, the distance A **(see illustration)** must not exceed the specified maximum. This applies to both the left- and right-hand side track rods. In addition, the maximum difference between the set distance on each side must not exceed 2.0 mm.

Upper and lower wishbone balljoints

10 Jack up the front of the car and support it on axle stands (see "*Jacking and Vehicle Support*"). Chock the rear wheels and remove the front wheels.
11 Position a trolley jack under the outer end of the lower wishbone and raise the suspension until a hardwood or metal block can be inserted beneath the upper wishbone and the bodyframe. Lower the jack.
12 Using water pump pliers or a pair of grips, compress the upper and lower balljoints in turn and check that the endfloat does not exceed that given in *Specifications*. Using a lever against the wishbones check that the side-to-side movement of the balljoint does not exceed that given in *Specifications*. If necessary renew the balljoints as follows.
13 Unscrew the damper lower mounting nut and disconnect the damper.
14 Unscrew the balljoint nut and use a balljoint separator tool to release the balljoint from the steering knuckle. Support the steering knuckle on an axle stand when removing the upper balljoint.
15 Unbolt the balljoint from the wishbone noting which way the bolt heads are fitted.
16 Fit the new balljoint(s) by following the removal procedure in reverse, but use new self-locking nuts on the wishbone bolts. Tighten the nuts to the specified torque.

8 Steering swivel member - removal and refitting

Vehicles up to 1980 model year

Removal

1 Press the front of the vehicle down on its suspension and insert a block of metal or

Steering and suspension 10•7

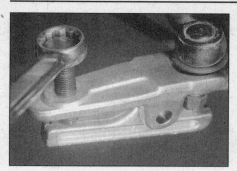

8.17 Separating the track rod end from steering swivel member

8.18a Remove the retaining bolts . . .

8.18b . . . then draw the upper . . .

8.18c . . . and lower ball joints away from the wishbones

8.19 Withdrawing the steering swivel member and hub assembly

hardwood between the upper wishbone and the bodyframe; this will relieve the suspension components of the load placed on them by the coil spring.
2 Remove the hub grease cap, using a chisel and mallet, then slacken the hub centre nut.
3 Slacken the roadwheel nuts, then chock the rear wheels and raise the front of the vehicle, resting it securely on axle stands (see "*Jacking and Vehicle Support*"). Remove the roadwheel.
4 Referring to Chapter 9, remove the brake caliper assembly from the steering swivel member, then unbolt the brake disc from the hub.
5 Using a proprietary hub puller, remove the hub from the end of the outboard driveshaft.
6 At the inboard driveshaft universal joint, remove the metal clip and slide the rubber gaiter off the driver cup; refer to Chapter 8.
7 Remove the nut securing the track-rod end to the steering swivel member, then using a ball-joint splitter, separate the joint.
8 Slacken and remove the bolts that secure the upper and lower ball joints to the suspension wishbones.
9 Withdraw the steering swivel member from the wheel arch, allowing the inner dirveshaft universal joint to separate.
10 Mount the assembly in the jaws of a vice and remove the hub centre nut. Then using a soft-faced mallet, tap the driveshaft out of the wheel bearing.
11 For details of wheel bearing removal, refer to Section 4.

Refitting
12 Refit the steering swivel member by following the removal procedure in reverse. Ensure that the hub nut is tightened to the specified torque.

Post 1980 model year vehicles
13 Press the front of the vehicle down on its suspension and insert a block of metal or hardwood between the upper wishbone and the bodyframe; this will relieve the suspension components of the load placed on them by the coil spring.
14 Remove the hub grease cap, using a cold chisel and mallet, then slacken and remove the hub centre nut, slacken the roadwheel nuts, then chock the rear wheels and raise the

front of the vehicle, resting it securely on axle stands (see "*Jacking and Vehicle Support*"). Remove the roadwheel.
15 Referring to Chapter 9, remove the brake caliper assembly from the steering swivel member, then unbolt the brake disc from the hub. Vehicles with ABS: refer to Chapter 9 and remove the wheel sensor.
16 Separate the track rod end from the steering swivel member using a ball joint splitter **(see illustration)**.
17 Remove the retaining bolts, then draw the upper and lower ball joints away from the wishbones. Make a note the orientation of the bolts, to aid refitting later **(see illustrations)**.
18 Carefully draw the steering swivel member and hub assembly away from the suspension wishbones and driveshaft **(see illustration)**. If the driveshaft splines stick in the hub, tap it through with a mallet.
19 Refer to Section 4 for details of the hub and wheel bearing removal.

Refitting
20 Refit the steering swivel member by following the removal procedure in reverse. On completion, ensure that the hub centre nut is tightened to the specified torque.

9 Rear suspension damper – removal, testing and refitting

Warning: To preserve safe handling characteristics, BOTH rear dampers (left- and right-hand) should renewed as a pair.

Hydraulic
1 Jack up the rear of the car and support the bodyframe with axle stands (see "*Jacking and Vehicle Support*"). Apply the handbrake and remove the rear wheel.
2 Position a trolley jack beneath the rear end of the trailing link, and raise it slightly to relieve the tension on the damper bolts.
3 Slacken and remove the nuts from the top of the damper and recover the upper mounting rubber and washer.
4 Unscrew and remove the lower mounting bolt, and withdraw the damper from the car.
5 Mount the damper in a vice. Move the upper section through its full stroke a few times. The resistance should be even throughout the stroke - if not, fit a new damper.
6 Refitting is a reversal of removal, however first bleed the air from the damper by operating it through its full stroke several times while mounted vertically in a vice.

Pneumatic
7 Jack up the rear of the car and support the bodyframe with axle stands (see "*Jacking and Vehicle Support*"). Apply the handbrake and remove the rear wheel.
8 Place an additional axle stand under the rear axle near the damper to be removed.
9 Place a trolley jack below the rear end of the trailing link. Raise it slightly to relieve the tension on the damper bolts **(see illustration)**.
10 Slacken and remove the nuts from the top of the damper and recover the upper mounting rubber and washer **(see illustration)**.

10•8 Suspension and steering

9.9 Support the rear end of the trailing link, to relieve the tension on the damper bolts

9.10 Slacken and remove the nuts from the top of the damper

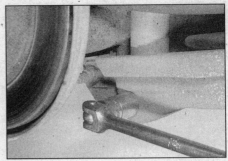

9.11 Unscrew and remove the damper lower mounting bolt

11 Unscrew and remove the damper lower bolt, then remove the bolt securing the trailing link to the rear axle (see illustration).
12 Lower the trailing link and withdraw the damper.
13 The gas charge in the damper is under very high pressure and therefore the unit should not be heated or damaged in any way, but disposed of safely.
14 Refit the damper by folowing the removal procedure in reverse.

10 Rear coil spring -
removal and refitting

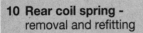

1 Jack up the rear of the car and support the bodyframe with axle stands. Apply the handbrake and remove the rear wheel.
2 Position a trolley jack beneath the trailing link directly under the damper mounting and raise it slightly, to relieve the load on the suspension components.
3 Unscrew and remove the damper lower mounting bolt, then remove the bolt that secures the suspension lower arm to the rear axle (see illustration).
4 Gradually lower the trailing link and withdraw the coil spring, together with the upper and lower seats.
5 On models with auxiliary pneumatic springs, unscrew the valve and release the air, then remove the unit from the trailing link.
6 Check the coil spring and seats for wear and damage, and check the free length with the dimension given in the Specifications. Renew it if necessary.

7 Refitting is a reversal of removal. Where applicable, the auxiliary pneumatic springs should be inflated to a pressure of 2.0 bar.

11 Rear suspension anti-roll bar -
removal and refitting

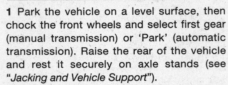

1 Park the vehicle on a level surface, then chock the front wheels and select first gear (manual transmission) or 'Park' (automatic transmission). Raise the rear of the vehicle and rest it securely on axle stands (see "Jacking and Vehicle Support").
2 Unscrew and remove the damper bottom retaining bolt and allow the anti-roll bar and link arm to hang down as shown (see illustration). Repeat this procedure on the opposite side.
3 Undo the forward mounting nuts each side and lower the anti-roll bar at the rear. Compress the leading edge mounting and withdraw the anti-roll bar.
4 Refit in the reverse order of removal. Renew the rubber bushes if they appear worn or perished. When fitting the rear mounting bolts, use a large flat bladed screwdriver as a lever to align the anti roll bar with the lower arm bolt hole.
5 When fitted, do not overtighten the bolts, as the rubber bushes may be damaged. Rock the vehicle on its suspension and check that the anti-roll bar does not contact the rear suspension lower arm.

12 Rear suspension Panhard rod -
removal and refitting

1 Jack up the rear of the car and support the bodyframe with axle stands (see "Jacking and Vehicle Support).
2 Slacken and withdraw the mounting bolts, then remove the Panhard rod.
3 If the rubber bushes appear worn or damaged, they should be renewed: Construct a bush extractor using a long bolt, metal tubing and washers. Brace the tube against the Panhard rod boss, then thread the bolt through the centre of the bush and fit washers that will bear on the bush but pass through the boss. Screw a nut onto the end of the bolt, and tighten it to draw the bush out of the Panhard rod boss.
4 Lubricate the new bushes by dipping them in soapy water to aid fitting.
5 Refitting is a reversal of removal, but do not tighten the Panhard rod bolts until the weight of the car is back on the suspension.

13 Rear suspension torque arm -
removal and refitting

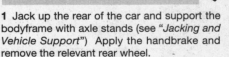

1 Jack up the rear of the car and support the bodyframe with axle stands (see "Jacking and Vehicle Support") Apply the handbrake and remove the relevant rear wheel.
2 Unscrew and remove the mounting bolts and remove the torque arm. Note which way round the bolts are fitted (see illustration).

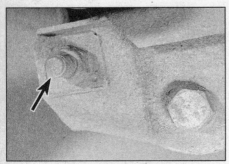

10.3 Remove the lower arm-to-rear axle bolt (arrowed)

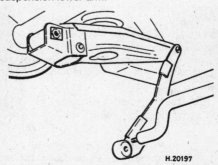

11.2 Remove the damper bottom bolt and allow the anti-roll bar to hang down

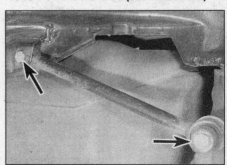

13.2 Remove the mounting bolts (arrowed) and remove the torque arm.

Steering and suspension

3 If the rubber bushes appear worn or damaged, they should be renewed. Construct a bush extractor using a long bolt, metal tubing and washers. Brace the tube against the torque arm boss, then thread the bolt through the centre of the bush and fit washers that will bear on the bush but pass through the boss. Screw a nut onto the end of the bolt, and tighten it to draw the bush out of the torque arm boss.
4 Lubricate the new bushes by dipping them in soapy water to aid fitting.
5 Refitting is a reversal of removal, but delay tightening the mounting bolts until the full weight of the car is on the suspension.

14 Rear suspension lower arm - removal and refitting

1 Remove the rear coil spring as described in Section 10, and where applicable lower the trailing edge of the anti-roll bar away from lower arm; refer to Section 11 for guidance.
2 Unscrew and remove the nuts from the bolts securing the leading edge of lower arm to the vehicles underside (see illustration), then lower the arm away from the vehicle.
3 Check the condition of the rubber bushes on the rear axle and renew them if necessary; refer to the information given at the end of Section 16.
4 Refitting is a reversal of removal, but delay tightening the rubber bush mounting bolts until the full weight of the car rests on the suspension.

15 Rear wheel hub assembly - removal and refitting

Removal

All vehicles

1 Refer to Chapter 9 and remove the rear brake caliper and disc.
2 Prise off the hub grease cap, using a stout screwdriver (see illustration).
3 Where applicable, lever up the locking collar, then unscrew the hub nut and remove the washer (see illustration). Note: *On post 1981 model year vehicles, bear in mind that the hub nut is tightened to a very high torque.*
4 Pull the hub from the stub axle - use a universal puller if necessary (see illustration).

Pre-1981 model year vehicles only

5 Note that on vehicles from 1981 model year onwards, the wheel bearings and oil seal form an integral part of the hub, therefore if the wheel bearings are worn, the entire hub must be renewed.
6 On earlier models with renewable bearings, prise the oil seal from the hub and remove both bearing inner races.
7 Support the hub on a block of wood to protect it, then using a soft metal drift inserted in the machined recesses provided, drive out

14.2 Remove the nuts (arrowed) from the leading edge of lower arm

the bearing outer races. Take care to avoid damaging the bearing housing.
8 Clean the components in a suitable solvent, then dry them thoroughly and check them for signs of wear and damage. Note: *The oil seal must be renewed as a matter of course.* In particular, check the surfaces of the tapered rollers and bearing race; if wear is evident, the bearings must be renewed.
9 Drive the inner races into the hub using a section of metal tubing of suitable diameter.
10 Half-fill the space between the races with molybdenum disulphide-based grease.
11 Fit the inner bearing inner race, then smear a little of the grease on the lip of the oil seal and using a block of wood, carefully drive it into the hub until it is flush with the surface of the hub.
12 Clean the stub axle and smear a little grease on the sealing surface, then fit the hub over it. **Note:** *Fit a new hub nut and thrustwasher as a matter of course.*

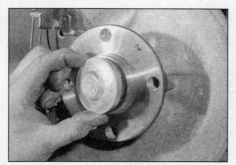

15.2 Prise off the hub grease cap

15.4 Removing the hub from the stub axle

13 Fit the inner race to the outer bearing, then fit the hub washer and nut.

All vehicles

14 Tighten the hub nut, observing the specified torque settings (see illustration). On pre-1988 model year vehicles, lock the hub nut by peening the collar into the machined groove in the stub axle.
15 Obtain a new grease cap and fill it with grease. Using a light mallet, tap it into position on the hub.
16 Refit the brake disc, caliper and roadwheel (Chapter 9). Observe the correct torque when tightening the roadwheel bolts.

16 Rear axle tube - removal and refitting

1 Jack up the rear of the car and support the bodyframe with axle stands (see "Jacking and Vehicle Support") Apply the handbrake and remove the rear wheels.
2 Unscrew the brake fluid reservoir filler cap and tighten it down onto a piece of polythene sheeting in order to reduce the loss of fluid in the subsequent procedure. Alternatively fit hose clamps to the two rear brake hoses.
3 Identify the brake hoses for position then unscrew the unions and disconnect them from the rear axle. Plug the hoses.
4 Working on each side of the vehicle in turn, jack up the suspension lower arms, then unscrew and remove the damper lower mounting bolts. Lower the jack and remove the coil springs; refer to Section 10.
5 Unscrew and remove the bolts securing the

15.3 Removing the rear hub nut

15.14 Tighten the hub nut, observing the correct torque setting

16.10 View of the rear axle upright and suspension components

torque arms to the rear axle uprights; refer to Section 13 for guidance.
6 Unbolt the Panhard rod from the rear axle and bodyframe; refer to Section 12.
7 Unscrew and remove the bolts securing the suspension lower arms to the rear axle, and withdraw the rear axle from under the vehicle.
8 If necessary, the suspension lower arm rubber bushes may be removed from the rear axle, using tubular drift to drive them out.
9 Lubricate the new bushes by dipping them in soapy water to aid fitting.
10 Refit the axle by following the removal procedure in reverse, but delay tightening the mounting bolts until the full weight of the car rests on the suspension (see illustration). Note that the bolt securing the Panhard rod to the bodyframe should be inserted from the rear of the vehicle. On completion, bleed the brake hydraulic system (see Chapter 9).

17 Power assisted steering servo pump - removal and refitting

Removal
1 Clean the servo pump particularly around the hose connections.

99 models
2 Siphon the fluid from the reservoir.
3 Drain the cooling system with reference to Chapter 2 and disconnect the hose between the expansion tank and water pump.
4 Disconnect the hydraulic supply and delivery lines from the pump.
5 Unbolt the pump from the bracket and engine mounting, release the drivebelt, and withdraw the servo pump. If necessary remove the pulley and pump mounting.

900 models
6 Disconnect the return hose from the pump and drain the fluid into a suitable container.
7 Remove the alternator (see Chapter 5).
8 Remove the adjusting links for the alternator and servo pump.
9 On early models loosen the screws and remove the drivebelt then move the pump to one side and remove the adjusting link. Disconnect the pressure hose then remove the mounting bolt and withdraw the pump.

10 On later models, withdraw the pump together with the adjustment link.

Refitting
99 models
11 Fit the mounting to the pump so that the clamp bolt is aligned with the delivery outlet and the lug is 49.0 mm from the centre of the pulley.
12 Refit the pump using a reversal of the removal procedure and adjust the drivebelt tension so that it can be moved between 10.0 and 15.0 mm midway between the pulleys under firm thumb pressure. Fill the cooling system with reference to Chapter 3. Fill and bleed the hydraulic system.

900 models
13 Refitting is a reversal of removal, but adjust the drivebelt tension so that it can be moved between 0.4 and 0.6 in (10.0 and 15.0 mm) midway between the pulleys under firm thumb pressure. Fill and bleed the hydraulic system. Refit the alternator (see Chapter 5).

18 Power assisted steering (PAS) hydraulic system - draining, refilling and bleeding

Draining
16 valve engine
1 Remove the retaining screws and lift the PAS fluid reservoir from the bodywork.
2 Slacken the hose clip, then disconnect the fluid return hose from the reservoir; be prepared for some fluid loss. Plug the port in the reservoir to minimise leakage and prevent the ingress of dirt.
3 Insert the hose end into a receptacle having a capacity of at least 1.0 litre.

8 valve engine
4 Slacken the clip that secures the fluid return hose to the port on the PAS pump, then disconnect the hose from the pump. Plug the open hose to minimise fluid loss and prevent the ingress of dirt.
5 Connect a length of hose to PAS pump return port, then insert the other end into a suitable (at least 1.0 litre capacity) container.

All vehicles
6 Start the engine and allow the fluid to be drained into the container. Turn the steering from lock to lock twice to ensure that all the fluid is purged.
7 Stop the engine when fluid ceases to flow into the receptacle; do not allow the PAS pump to run dry for any longer than is necessary.
8 Reconnect the fluid hoses to thier original ports and tighten the hose clips.

Refilling
9 Fill the hydraulic system via the fluid reservoir filler cap with the specified grade and quantity of fluid.

Bleeding
10 Ensure that the vehicle is parked on a level surface, then chock the rear wheels, raise the front of the vehicle and rest it securely on axle stands (see "Jacking and Vehicle Support).
11 With the front wheels clear of the ground, turn the steering from lock to lock three or four times.
12 Lower the vehicle to the ground, then start the engine and turn the steering (now with power assistance) from lock to lock three or four times.
13 Check the fluid level in the reservoir and top it up if necessary.

19 Steering gear (manual) - removal and refitting

Removal
1 Jack up the front of the car and support it on axle stands (see "Jacking and Vehicle Support"). Chock the rear wheels and remove the front wheels.
2 Working in the engine compartment, mark the intermediate shaft in relation to the steering rack pinion then unscrew and remove the pinch-bolt.
3 With reference to Section 7, unscrew the track rod end nuts and use a ball joint splitter to release the track rod from the steering arms.

99 models
4 Remove the lower facia panel and prise the intermediate shaft bellows from the body.
5 Unbolt the steering column from the bulkhead then pull the intermediate shaft from the pinion. Tie the column to one side, taking care not to strain the wiring.

All models
6 Unscrew the steering rack mounting bolts; recover the nuts and washers.
7 Move the whole assembly to the side (disconnect the intermediate shaft on 900 models) then lower the steering rack between the bodyframe members.

Refitting
8 Refitting is a reversal of removal, but tighten the nuts and bolts to the specified torque and finally check and adjust the front wheel alignment, as described in Section 25.

20 Steering rack (power-assisted) - removal and refitting

1 Drain the fluid from the PAS system, as described in Section 18.
2 Unbolt the servo pump from its mounting.
3 Jack up the front of the car and support it on axle stands (see "Jacking and Vehicle Support"). Chock the rear wheels and remove the front wheels.

Steering and suspension 10•11

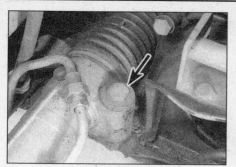

20.9a Power assisted steering rack left-hand . . .

20.9b . . . and right-hand mounting bolt (arrowed)

22.4 Remove the centre pad from the steering wheel

4 Support the engine/gearbox with a hoist or trolley jack then remove the left-hand side engine mounting.
5 In the engine bay, at the bulkhead, mark the steering column intermediate shaft in relation to the steering rack pinion then unscrew and then remove the pinch-bolt.
6 Disconnect the pipe clips from the front suspension crossmember.
7 Unscrew the track rod end nuts and use a separator tool to release them from the steering swivel member.
8 On pre-1988 model year vehicles, remove the left-hand side handbrake cable from the yoke and wheel housing, and the right-hand handbrake cable clip from the steering gear.
9 Unbolt the steering gear from the bodyframe noting the location of the spacers and bracket (see illustrations).
10 Disconnect the hydraulic feed and return pipes from the servo valve on the steering gear. Plug the pipes and apertures to prevent the ingress of debris.
11 Move the whole assembly to the side and disconnect the steering column intermediate shaft, then lower the PAS rack assembly between the bodyframe members.

Refitting

12 Refitting is a reversal of removal, noting the following points:
 a) Ensure that no foreign matter enters the hydraulic lines and apertures as the hoses are being refitted. Fit new seals to the hydraulic lines if necessary
 b) Tighten all fixings to the specified torque wrench settings.
 c) Check and adjust the front wheel alignment as described in Section 25.
 d) Adjust the servo pump drivebelt as described in Section 17.
 e) Refill and bleed the hydraulic system, as described in Section 18.

21 Steering rack rubber gaiters - renewal

1 The gaiters can easily be removed with the steering rack in place. Thoroughly clean the area around the gaiter first of all, to prevent dirt from contaminating the inner joints.

2 Refer to Section 7 and disconnect the track rod end from the steering swivel member. Slacken the locknut then unscrew the track rod end from the track rod.
3 Remove the clips and slide the gaiter off the end of the track rod end.
4 Pack the new gaiter with the specified quantity and grade of grease, then fit it over the track rod, securing it in place with new clips.
5 Refit the track rod end to the track rod and steering swivel member, then with reference to Sections 7 and 25, check and if necessary adjust the wheel alignment.

22 Steering wheel - removal and refitting

1 Set the front wheels in the straight-ahead position.
2 Disconnect the battery negative lead.
3 Refer to Section 11 and remove the steering column lower shroud.
4 Remove the centre pad from the steering wheel (see illustration). To do this on the standard steering wheel, remove the screws located behind the steering wheel and where applicable prise out the centre badge. Disconnect the wiring for the horn buttons.
5 Hold the steering wheel stationary and unscrew the retaining nut. Remove the washer.
6 Mark the steering wheel in relation to the inner column then pull off the steering wheel. Do not knock or jar the steering wheel in an attempt to free it, otherwise the collapsible

23.2 Remove the steering gear pinion-to-intermediate shaft pinch-bolt (arrowed)

steering column outer cage may be distorted - if possible use an extractor to pull the wheel off the steering column splines.

Refitting

7 Refit the steering wheel by following the removal procedure in reverse; tighten the centre nut to the specified torque.

23 Steering column - removal and refitting

Removal

1 Remove the steering wheel as described in Section 22.
2 Working in the engine compartment mark the intermediate shaft in relation to the steering gear pinion then unscrew and remove the pinch-bolt (see illustration).
3 Unscrew the bolts and withdraw the column lower cover and lower facia panel; refer to Chapter 11 for guidance.
4 Remove the direction indicator and wiper combination switch (refer to Chapter 12).
5 Prise the column rubber boot from the bulkhead.
6 Unscrew the mounting bracket bolts (see illustration) and withdraw the steering column from the car, at the same time disconnecting the intermediate shaft from the steering gear pinion.

Refitting

7 Refit the steering column , but do not fully tighten the mounting bracket bolts until all are inserted. Make sure that the intermediate

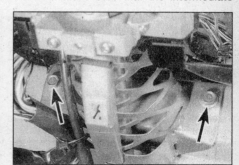

23.6 Steering column outer cage mounting bolts (arrowed)

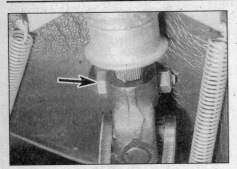

24.2 Remove steering column inner shaft-to-intermediate shaft pinch bolt (arrowed)

shaft pinch-bolt engages the groove in the steering gear pinion. Suitable adhesive should be applied to the groove in the rubber boot before fitting it to the bulkhead. Seal the mounting holes in the bulkhead with waterproof sealing compound.

24 Steering column intermediate shaft- removal and refitting

1 Remove the steering column (Section 23).
2 Mark the intermediate shaft in relation to the inner column then unscrew and remove the pinch-bolt and slide off the intermediate shaft (see illustration).
3 To renew the rubber bellows prise it over the sealing ring and pull it off the universal joint. A tapered fork is now required in order to ease the new bellows over the universal joint - Saab tool No 89 95 813 can be used or a home made tool from a suitable plastic bottle. Lubricate the tool with petroleum jelly to prevent damage to the bellows.

Refitting

4 Refit the intermediate shaft by following the removal procedure in reverse.

25 Wheel alignment- checking and adjustment

Definitions

1 A vehicle's steering and suspension geometry is defined in four basic settings - all angles are expressed in degrees (toe settings are also expressed as a measurement); the steering axis is defined as an imaginary line drawn through the axis of the suspension strut, extended where necessary to contact the ground.
2 Camber is the angle between each wheel and a vertical line drawn through its centre and tyre contact patch, when viewed from the front or rear of the car. Positive camber is when the wheels are tilted outwards from the vertical at the top; negative camber is when they are tilted inwards.
3 Camber is not adjustable, and is given for reference only; while it can be checked using a camber checking gauge, if the figure obtained is significantly different from that specified, the vehicle must be taken for careful checking by a professional, as the fault can only be caused by wear or damage to the body or suspension components.
4 Castor is the angle between the steering axis and a vertical line drawn through each roadwheel centre and tyre contact patch, when viewed from the side of the car. Positive castor is when the steering axis is tilted so that it contacts the ground ahead of the vertical; negative castor is when it contacts the ground behind the vertical.
5 Castor is not adjustable, and is given for reference only; while it can be checked using a castor checking gauge, if the figure obtained is significantly different from that specified, the vehicle must be taken for careful checking by a professional, as the fault can only be caused by wear or damage to the body or suspension components.
6 Steering axis inclination/SAI - also known as kingpin inclination/KPI - is the angle between the steering axis and a vertical line drawn through each roadwheel centre and tyre contact patch, when viewed from the front or rear of the car.
7 SAI/KPI is not adjustable, and is given for reference only.
8 Toe is the difference, viewed from above, between lines drawn through the wheel centres and the car's centre-line. 'Toe-in' is when the wheels point inwards, towards each other at the front, while 'toe-out' is when they splay outwards from each other at the front.
9 The front wheel toe setting is adjusted by screwing the balljoints in or out of their track-rods, to alter the effective length of the track-rod assemblies.
10 Rear wheel toe setting is not adjustable, and is given for reference only. While it can be checked, if the figure obtained is significantly different from that specified, the vehicle must be taken for careful checking by a professional, as the fault can only be caused by wear or damage to the body or suspension components.

Checking - general

11 Due to the special measuring equipment necessary to check the wheel alignment, and the skill required to use it properly, the checking and adjustment of these settings is best left to a Saab dealer or similar expert. Note that most tyre-fitting shops now possess sophisticated checking equipment.
12 For accurate checking, the vehicle must be at the kerb weight, i.e. unladen and with a full tank of fuel, and the ride height must be correct.
13 Before starting work, check first that the tyre sizes and types are as specified, then check the tyre pressures and tread wear, the roadwheel run-out, the condition of the hub bearings, the steering wheel free play, and the condition of the front suspension components (Chapter 1). Correct any faults found.
14 Park the vehicle on level ground, check that the front roadwheels are in the straight-ahead position, then rock the rear and front ends to settle the suspension. Release the handbrake, and roll the vehicle backwards approximately 1 metre, then forwards again, to relieve any stresses in the steering and suspension components.

Toe setting - checking and adjusting

Front wheel toe setting

15 The front wheel toe setting is checked by measuring the distance between the front and rear inside edges of the roadwheel rims. Proprietary toe measurement gauges are available from motor accessory shops.
16 Prepare the vehicle as described in paragraphs 12 to 14 above.
17 A tracking gauge must now be obtained. Two types of gauge are available, and can be obtained from motor accessory shops. The first type measures the distance between the front and rear inside edges of the roadwheels, as described previously, with the car stationary. The second type, known as a scuff plate, measures the actual position of the contact surface of the tyre in relation to the road surface, with the vehicle in motion. This is achieved by pushing or driving the front tyre over a plate, which then moves slightly according to the scuff of the tyre, and shows this movement on a scale. Both types have their advantages and disadvantages, but either can give satisfactory results if used correctly and carefully. Alternatively, a tracking gauge can be fabricated from a length of steel tubing, suitably cranked to clear the engine and gearbox assembly, with a set-screw and a locknut at one end.
18 Many tyre specialists will also check toe settings free, or for a small charge.
19 Make sure that the steering is in the straight-ahead position when taking measurements.
20 If adjustment is found to be necessary, clean the ends of the track-rods in the areas of the adjustment pin locknuts.
21 Slacken the locknuts (one at the inner and outer end of each adjustment pin), and use a pair of grips to turn each track-rod by equal amounts in the same direction. Only turn each track rod by a quarter of a turn at a time, before re-checking.
22 Check that the track-rod end balljoints are centralised, and not forced to the limit of movement in any direction.
23 When adjustment is correct, tighten the locknuts.
24 Check that the track-rod lengths are equal, and that the steering wheel spokes are in the straight-ahead position.

Rear wheel toe setting

25 The procedure for checking the rear toe setting is same as described for the front in paragraph 17. The setting is not adjustable - see paragraph 10.

Chapter 11
Bodywork and fittings

Contents

Body exterior fittings - removal and refitting 22	General information 1
Bonnet and front grille - removal and refitting 8	Hinge and lock lubrication ....See Chapter 1
Bonnet lock - removal and refitting 10	Interior trim panels- removal and refitting 25
Bonnet release cable - removal and refitting 9	Maintenance - bodywork and underframe 2
Central locking servo motors - removal and refitting 17	Maintenance - upholstery and carpets 3
Centre console - removal and refitting 26	Major body damage - repair 5
Door - removal, refitting and adjustment 11	Minor body damage - repair 4
Door handle and lock components - removal and refitting 13	Rear bumper - removal and refitting 7
Door inner trim panel - removal and refitting 12	Seat belt components - removal and refitting 24
Door window glass and regulator - removal and refitting 14	Seats - removal and refitting 23
Electric window components - removal and refitting 18	Sunroof assembly - removal and refitting 21
Exterior door mirrors - removal and refitting 19	Tailgate and support struts/boot lid - removal and refitting ...15
Facia assembly - removal and refitting 27	Tailgate/bootlid lock components - removal and refitting 16
Front bumper - removal and refitting 6	Windscreen and fixed windows - general information 20

Degrees of difficulty

Easy, suitable for novice with little experience	**Fairly easy,** suitable for beginner with some experience	**Fairly difficult,** suitable for competent DIY mechanic	**Difficult,** suitable for experienced DIY mechanic	**Very difficult,** suitable for expert DIY or professional

1 General information

The vehicle body is constructed from overlapping pressed steel sections that are either spot welded or seam welded together, depending on the position of the joint and the stresses it is expected to withstand, both in normal use and in the event of a collision. The overall rigidity of the body is increased by the use of stiffening beams built into the body panels and steel flanges in the window and door openings.

The body underside is coated with polyester underseal and an anti-corrosion compound. This treatment provides protection against the elements and also serves as an effective sound insulation layer. The cabin, luggage area and engine compartment are also lined with felting and other sound insulating materials, to provide further noise damping.

Certain models are fitted with electric windows front and rear. The window glass is raised and lowered by an electric motor, linked by a cable to the window regulator mechanism. A master switch panel is mounted in the centre console, from which all windows can be operated and locked. In addition, individual switch panels are mounted on the rear door trim panels.

Central locking is fitted to certain models and is actuated from the drivers door lock. It operates the locks on all four doors and the tailgate/boot. The lock mechanisms are actuated by servo motor units that can be separated from the lock assemblies and renewed individually; refer to Section 17 for details.

2 Maintenance - bodywork and underframe
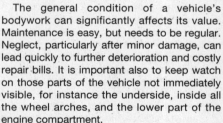

The general condition of a vehicle's bodywork can significantly affects its value. Maintenance is easy, but needs to be regular. Neglect, particularly after minor damage, can lead quickly to further deterioration and costly repair bills. It is important also to keep watch on those parts of the vehicle not immediately visible, for instance the underside, inside all the wheel arches, and the lower part of the engine compartment.

The basic maintenance routine for the bodywork is washing - preferably with a lot of water, from a hose. This will remove all the loose solids which may have stuck to the vehicle. It is important to flush these off in such a way as to prevent grit from scratching the finish. The wheel arches and underframe need washing in the same way, to remove any accumulated mud which will retain moisture and tend to encourage rust. Strange as it may seem, the best time to clean the underframe and wheel arches is in wet weather, when the mud is thoroughly wet and soft. In very wet weather, the underframe is usually cleaned of large accumulations automatically, and this is a good time for inspection.

Periodically, except on vehicles with a wax-based underbody protective coating, it is a good idea to have the whole of the underframe of the vehicle steam-cleaned, engine compartment included, so that a thorough inspection can be carried out to see what minor repairs and renovations are necessary. Steam-cleaning is available at many garages, and is necessary for the removal of the accumulation of oily grime, which sometimes is allowed to become thick in certain areas. If steam-cleaning facilities are not available, there are some excellent grease solvents available, which can be brush-applied; the dirt can then be simply hosed off. Note that these methods should not be used on vehicles with wax-based underbody protective coating, or the coating will be removed. Such vehicles should be inspected annually, preferably just prior to Winter, when the underbody should be washed down, and any damage to the wax coating repaired using an Undershield. Ideally, a completely fresh coat should be applied. It would also be worth considering the use of such wax-based protection for injection into door panels, sills, box sections, etc, as an additional safeguard against rust damage, where such protection is not provided by the vehicle manufacturer.

After washing paintwork, wipe off with a chamois leather to give an unspotted clear finish. A coat of clear protective wax polish like the many excellent wax polishes, will give added protection against chemical pollutants in the air. If the paintwork sheen has dulled or oxidised, use a cleaner/polisher combination to restore the brilliance of the shine. This requires a little effort, but such dulling is

11•2 Bodywork and fittings

usually caused because regular washing has been neglected. Care needs to be taken with metallic paintwork, as special non-abrasive cleaner/polisher is required to avoid damage to the finish. Always check that the door and ventilator opening drain holes and pipes are completely clear, so that water can be drained out. Brightwork should be treated in the same way as paintwork. Windscreens and windows can be kept clear of the smeary film which often appears, by the use of proprietary glass cleaner. Never use any form of wax or other body or chromium polish on glass.

3 Maintenance - upholstery and carpets

Mats and carpets should be brushed or vacuum-cleaned regularly, to keep them free of grit. If they are badly stained, remove them from the vehicle for scrubbing or sponging, and make quite sure they are dry before refitting. Seats and interior trim panels can be kept clean by wiping with a damp cloth. If they do become stained (which can be more apparent on light-coloured upholstery), use a little liquid detergent and a soft nail brush to scour the grime out of the grain of the material. Do not forget to keep the headlining clean in the same way as the upholstery. When using liquid cleaners inside the vehicle, do not over-wet the surfaces being cleaned. Excessive damp could get into the seams and padded interior, causing stains, offensive odours or even rot. If the inside of the vehicle gets wet accidentally, it is worthwhile taking some trouble to dry it out properly, particularly where carpets are involved.

Warning: Do not leave oil or electric heaters inside the vehicle for this purpose.

4 Minor body damage - repair

Repairs of minor scratches in bodywork

If the scratch is very superficial, and does not penetrate to the metal of the bodywork, repair is very simple. Lightly rub the area of the scratch with a paintwork renovator, or a very fine cutting paste, to remove loose paint from the scratch, and to clear the surrounding bodywork of wax polish. Rinse the area with clean water.

Apply touch-up paint to the scratch using a fine paint brush; continue to apply fine layers of paint until the surface of the paint in the scratch is level with the surrounding paintwork. Allow the new paint at least two weeks to harden, then blend it into the surrounding paintwork by rubbing the scratch area with a paintwork renovator or a very fine cutting paste. Finally, apply a wax polish.

Where the scratch has penetrated right through to the metal of the bodywork, causing the metal to rust, a different repair technique is required. Remove any loose rust from the bottom of the scratch with a penknife, then apply rust-inhibiting paint to prevent the formation of rust in the future. Using a rubber or nylon applicator, fill the scratch with bodystopper paste. If required, this paste can be mixed with cellulose thinners, to provide a very thin paste which is ideal for filling narrow scratches. Before the stopper-paste in the scratch hardens, wrap a piece of smooth cotton rag around the top of a finger. Dip the finger in cellulose thinners, and quickly sweep it across the surface of the stopper-paste in the scratch; this will ensure that the surface of the stopper-paste is slightly hollowed. The scratch can now be painted over as described earlier in this Section.

Repairs of dents in bodywork

When deep denting of the vehicle's bodywork has taken place, the first task is to pull the dent out, until the affected bodywork almost attains its original shape. There is little point in trying to restore the original shape completely, as the metal in the damaged area will have stretched on impact, and cannot be reshaped fully to its original contour. It is better to bring the level of the dent up to a point which is about 3 mm below the level of the surrounding bodywork. In cases where the dent is very shallow anyway, it is not worth trying to pull it out at all. If the underside of the dent is accessible, it can be hammered out gently from behind, using a mallet with a wooden or plastic head. Whilst doing this, hold a suitable block of wood firmly against the outside of the panel, to absorb the impact from the hammer blows and thus prevent a large area of the bodywork from being 'belled-out'.

Should the dent be in a section of the bodywork which has a double skin, or some other factor making it inaccessible from behind, a different technique is called for. Drill several small holes through the metal inside the area - particularly in the deeper section. Then screw long self-tapping screws into the holes, just sufficiently for them to gain a good purchase in the metal. Now the dent can be pulled out by pulling on the protruding heads of the screws with a pair of pliers.

The next stage of the repair is the removal of the paint from the damaged area, and from an inch or so of the surrounding 'sound' bodywork. This is accomplished most easily by using a wire brush or abrasive pad on a power drill, although it can be done just as effectively by hand, using sheets of abrasive paper. To complete the preparation for filling, score the surface of the bare metal with a screwdriver or the tang of a file, or alternatively, drill small holes in the affected area. This will provide a really good 'key' for the filler paste.

To complete the repair, see the Section on filling and respraying.

Repairs of rust holes or gashes in bodywork

Remove all paint from the affected area, and from an inch or so of the surrounding 'sound' bodywork, using an abrasive pad or a wire brush on a power drill. If these are not available, a few sheets of abrasive paper will do the job most effectively. With the paint removed, you will be able to judge the severity of the corrosion, and therefore decide whether to renew the whole panel (if this is possible) or to repair the affected area. New body panels are not as expensive as most people think, and it is often quicker and more satisfactory to fit a new panel than to attempt to repair large areas of corrosion.

Remove all fittings from the affected area, except those which will act as a guide to the original shape of the damaged bodywork (eg headlamp shells etc). Then, using tin snips or a hacksaw blade, remove all loose metal and any other metal badly affected by corrosion. Hammer the edges of the hole inwards, in order to create a slight depression for the filler paste.

Wire-brush the affected area to remove the powdery rust from the surface of the remaining metal. Paint the affected area with rust-inhibiting paint; if the back of the rusted area is accessible, treat this also.

Before filling can take place, it will be necessary to block the hole in some way. This can be achieved by the use of aluminium or plastic mesh, or aluminium tape.

Aluminium or plastic mesh, or glass-fibre matting, is probably the best material to use for a large hole. Cut a piece to the approximate size and shape of the hole to be filled, then position it in the hole so that its edges are below the level of the surrounding bodywork. It can be retained in position by several blobs of filler paste around its periphery.

Aluminium tape should be used for small or very narrow holes. Pull a piece off the roll, trim it to the approximate size and shape required, then pull off the backing paper (if used) and stick the tape over the hole; it can be overlapped if the thickness of one piece is insufficient. Burnish down the edges of the tape with the handle of a screwdriver or similar, to ensure that the tape is securely attached to the metal underneath.

Bodywork repairs - filling and respraying

Before using this Section, see the Sections on dent, deep scratch, rust holes and gash repairs.

Many types of bodyfiller are available, but generally speaking, those proprietary kits which contain a tin of filler paste and a tube of resin hardener are best for this type of repair, which can be used directly from the tube. A wide, flexible plastic or nylon applicator will be found invaluable for imparting a smooth and well-contoured finish to the surface of the filler.

Mix up a little filler on a clean piece of card or board - measure the hardener carefully (follow the maker's instructions on the pack), otherwise the filler will set too rapidly or too slowly. Alternatively, a no-mix filler can be used straight from the tube without mixing, but daylight is required to cure it. Using the applicator, apply the filler paste to the prepared area; draw the applicator across the surface of the filler to achieve the correct contour and to level the surface. As soon as a contour that approximates to the correct one is achieved, stop working the paste - if you carry on too long, the paste will become sticky and begin to 'pick-up' on the applicator. Continue to add thin layers of filler paste at 20-minute intervals, until the level of the filler is just proud of the surrounding bodywork.

Once the filler has hardened, the excess can be removed using a metal plane or file. From then on, progressively-finer grades of abrasive paper should be used, starting with a 40-grade production paper, and finishing with a 400-grade wet-and-dry paper. Always wrap the abrasive paper around a flat rubber, cork, or wooden block - otherwise the surface of the filler will not be completely flat. During the smoothing of the filler surface, the wet-and-dry paper should be periodically rinsed in water. This will ensure that a very smooth finish is imparted to the filler at the final stage.

At this stage, the 'dent' should be surrounded by a ring of bare metal, which in turn should be encircled by the finely 'feathered' edge of the good paintwork. Rinse the repair area with clean water, until all of the dust produced by the rubbing-down operation has gone.

Spray the whole area with a light coat of primer, - this will show up any imperfections in the surface of the filler. Repair these imperfections with fresh filler paste or bodystopper, and once more smooth the surface with abrasive paper. If bodystopper is used, it can be mixed with cellulose thinners, to form a really thin paste which is ideal for filling small holes. Repeat this spray-and-repair procedure until you are satisfied that the surface of the filler, and the feathered edge of the paintwork, are perfect. Clean the repair area with clean water, and allow to dry fully.

The repair area is now ready for final spraying. Paint spraying must be carried out in a warm, dry, windless and dust-free atmosphere. This condition can be created artificially if you have access to a large indoor working area, but if you are forced to work in the open, you will have to pick your day very carefully. If you are working indoors, dousing the floor in the work area with water will help to settle the dust which would otherwise be in the atmosphere. If the repair area is confined to one body panel, mask off the surrounding panels; this will help to minimise the effects of a slight mis-match in paint colours. Bodywork fittings (eg chrome strips, door handles etc) will also need to be masked off. Use genuine masking tape, and several thicknesses of newspaper, for the masking operations.

Before commencing to spray, agitate the aerosol can thoroughly, then spray a test area (an old tin, or similar) until the technique is mastered. Cover the repair area with a thick coat of primer; the thickness should be built up using several thin layers of paint, rather than one thick one. Using 400 grade wet-and-dry paper, rub down the surface of the primer until it is really smooth. While doing this, the work area should be thoroughly doused with water, and the wet-and-dry paper periodically rinsed in water. Allow to dry before spraying on more paint.

Spray on the top coat, again building up the thickness by using several thin layers of paint. Start spraying in the centre of the repair area, and then, using a circular motion, work outwards until the whole repair area and about 2 inches of the surrounding original paintwork is covered. Remove all masking material 10 to 15 minutes after spraying on the final coat of paint.

Allow the new paint at least two weeks to harden, then, using a paintwork renovator or a very fine cutting paste, blend the edges of the paint into the existing paintwork. Finally, apply wax polish.

Plastic components

With the use of more and more plastic body components by the vehicle manufacturers (eg bumpers, spoilers, and in some cases major body panels), rectification of more serious damage to such items has become a matter of either entrusting repair work to a specialist in this field, or renewing complete components. Repair of such damage by the DIY owner is not really feasible, owing to the cost of the equipment and materials required for effecting such repairs. The basic technique involves making a groove along the line of the crack in the plastic, using a rotary burr in a power drill. The damaged part is then welded back together, using a hot air gun to heat up and fuse a plastic filler rod into the groove. Any excess plastic is then removed, and the area rubbed down to a smooth finish. It is important that a filler rod of the correct plastic is used, as body components can be made of a variety of different types (eg polycarbonate, ABS, polypropylene).

Damage of a less serious nature (abrasions, minor cracks etc) can be repaired by the DIY owner using a two-part epoxy filler repair material. Once mixed in equal proportions, this is used in similar fashion to the bodywork filler used on metal panels. The filler is usually cured in twenty to thirty minutes, ready for sanding and painting.

If the owner is renewing a complete component himself, or if he has repaired it with epoxy filler, he will be left with the problem of finding a suitable paint for finishing which is compatible with the type of plastic used. At one time, the use of a universal paint was not possible, owing to the complex range of plastics encountered in body component applications. Standard paints, generally speaking, will not bond to plastic or rubber satisfactorily, but a type spraymatch paint, to match any plastic or rubber finish, can be obtained from dealers. However, it is now possible to obtain a plastic body parts finishing kit which consists of a pre-primer treatment, a primer and coloured top coat. Full instructions are normally supplied with a kit, but basically, the method of use is to first apply the pre-primer to the component concerned, and allow it to dry for up to 30 minutes. Then the primer is applied, and left to dry for about an hour before finally applying the special-coloured top coat. The result is a correctly-coloured component, where the paint will flex with the plastic or rubber, a property that standard paint does not normally posses.

5 Major body damage - repair

Where serious damage has occurred, or large areas need renewal due to neglect, it means that complete new panels will need welding-in, and this is best left to professionals. If the damage is due to impact, it will also be necessary to check completely the alignment of the bodyshell, and this can only be carried out accurately by a Saab dealer using special jigs. If the body is left misaligned, it is primarily dangerous, as the car will not handle properly, and secondly, uneven stresses will be imposed on the steering, suspension and possibly transmission, causing abnormal wear, or complete failure, particularly to such items as the tyres.

6 Front bumper - removal and refitting

Removal

1 On 99 models unscrew the vertical Allen bolt from beneath each mounting bracket.
2 On 900 models, first remove the spoiler (where applicable), then unscrew the vertical through-bolt from beneath each mounting (see illustration).

6.2 On 900 models, unscrew the two through-bolts from beneath the mountings

11•4 Bodywork and fittings

7.1 Unscrew the nuts from the bumper mounting brackets

8.3 Mark the position of the hinge halves, then loosen the bolts on the hinge bar

8.6 Grille retaining screw (900 models)

3 Ease the bumper off its mountings and remove it from the vehicle.
4 Refitting is a reversal of removal.

7 Rear bumper - removal and refitting

Removal

1 Unscrew the nuts from the left and right hand bumper mounting brackets **(see illustration)**.
2 Ease the bumper away from its mounting brackets and remove it from the vehicle.

Refitting

3 Refitting is a reversal of removal.

8 Bonnet and front grille - removal and refitting

Bonnet

Removal

1 Pull the bonnet release handle (located beneath the left-hand side of the facia panel) and fully open the bonnet.
2 Disconnect the windscreen washer hose from the jet ports, then unclip it from the retainers.
3 On 99 models, loosen the pivot bolts at the top of the hinges. On 900 models, use a pencil to mark the relative position of the two halves of the hinge, then loosen the retaining bolts on the hinge bar. **(see illustration)**.
4 With the help of an assistant, take the weight of the bonnet, then remove the pivot bolts and withdraw the bonnet from the car.

Refitting

5 Refitting is a reversal of removal, but check that the bonnet aligns with the surrounding bodywork and that it locks correctly. If adjustment is required on 99 models, loosen the lower hinge bolts and reposition the bonnet within the elongated slots. On 900 models, remove the front radiator grille and loosen the hinge bolts on the cross-panel. The catch can be adjusted by inserting shims (maximum thickness 5.0 mm), and the rubber

stops on the front of the bonnet should be adjusted so that they support the bonnet, without rattling in the closed position.

Grille

Removal

6 To remove the radiator grille on 900 models remove the three screws from the cross-panel then lift the grille from the holes in the front valance **(see illustration)**.
7 To remove the radiator grille on 99 models remove the seven screws (1 on top and 6 from the front) and withdraw the grille, at the same time disconnecting the headlight washer tubes.

Refitting

8 Refitting is a reversal of removal.

9 Bonnet release cable - removal and refitting

Removal

1 Refer to Section 8 and remove the front grille.
2 Working underneath the bonnet lock, remove the stop nipple from the end of the release cable.
3 Work around the engine bay and remove the release cable outer from the retaining clips.
4 Referring to Section 27, extract the fixings and lower the sound insulation panel away from the under side of the steering column/facia.

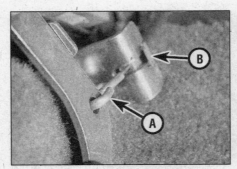

9.5 Bonnet release handle assembly
A Cable inner   B Cable outer spring clip

5 At the release handle, prise off the spring clip and release the cable outer from the mounting bracket. Unhook the cable inner from the handle **(see illustration)**.
6 At the bulkhead, prise out the rubber grommet, then pull the entire cable through into the cabin area.

Refitting

7 Refit the cable by reversing the removal process. Adjust both cable stop at the bonnet lock, to ensure that the lock release correctly when the handle is operated. Check also that when the handle is at rest, the bonnet is held securely in closed position by the lock; pull up firmly at each corner of the bonnet to assess this. If necessary, adjust the striker pin length by slackening the locknut and then screwing the pin in or out as necessary.

10 Bonnet lock - removal and refitting

Removal

1 Refer to Section 8 and remove the front grille.
2 Disconnect the release cable from the lock mechanism by removing the stop nipple.
3 Slacken the and withdraw the lock retaining bolts from underneath the front crossmember **(see illustration)**.

Refitting

4 Refit the bonnet lock by following the removal procedure in reverse.

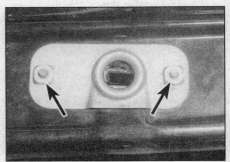

10.3 Withdraw the lock retaining bolts from underneath the front crossmember

Bodywork and fittings 11•5

11.3 Slacken and withdraw the door hinge retaining bolts

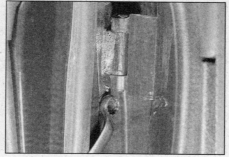

11.9 Mark the position of the hinge through-bolts, then unscrew them

11 Door - removal, refitting and adjustment

Front doors

Removal

1 Mark the fitted position of both hinges on the front pillar with a pencil.
2 Where applicable, peel back the rubber seal and unplug the electrical connector at the leading edge of the door.
3 Starting at the lower hinge, slacken and withdraw the hinge retaining bolts (see illustration).
4 Get an assistant to support the door as the last bolts are removed, then carefully draw the hinges out of the door frame.

Refitting

5 Refitting is a reversal of removal. Close the door and check that it is positioned correctly within its aperture and that it is flush with the surrounding bodywork.

Adjustment

6 To adjust the flush fit of the door, loosen the hinge bolts on the front pillar and move the door in the slotted holes. Re-tighten the bolts when the correct fit is achieved. **Note: Later models have a grub Allen screw that allows hinge location during factory assembly; this must be removed and discarded to allow adjustment; the remaining hole should then be filled.**
7 To adjust the position of the door within its aperture, first remove the door trim panel (see Section 12), then loosen the adjustment screws through the access holes at the front edge of the door. Move the door as required and retighten the screws when the correct fit is achieved. If necessary, check and adjust the alignment of the door lock striker plate; refer to Section 13 for details.

Rear doors

Removal

8 Remove the trim from the lower section of the centre pillar.
9 Mark the position of the hinge through-bolts then unscrew them, starting at the lower hinge (see illustration).
10 As the upper hinge is released, get an assistant to support the door and remove it from the vehicle.

Refitting

11 Refitting is a reversal of removal. Close the door and check that it is positioned correctly within its aperture and that it is flush with the surrounding bodywork. To adjust the door loosen the hinge bolts on the door and reposition it as necessary. Check that the door striker supports and retains the door in the correct position - if necessary loosen the screws and reposition it.

12 Door inner trim panel - removal and refitting

Front door

Removal

1 Disconnect the battery negative cable and position it away from the terminal. Fully close the window and on models with manual windows, note the position of the window winder handle. Remove the handle by pressing out the cover disc and removing the screw (see illustrations).
2 Unscrew and remove the door locking knob (see illustration).
3 Where applicable, remove the screws from the door sill strip.
4 Remove the armrest front and rear retaining screws and lift it away from the door (see illustrations).
5 Prise the bezel away from the door latch release handle (see illustration).
6 Work around the outside of the trim panel and unfasten the plastic retaining studs; using a flat bladed screwdriver, turn each stud through 90° until it is felt to disengage.

12.1a Remove the handle by pressing out the cover disc . . .

12.1b . . . and removing the screw

12.2 Unscrew and remove the door locking knob

12.4a Remove the armrest front . . .

12.4b . . .and rear retaining screws

11•6 Bodywork and fittings

12.5 Prise the bezel away from the door latch release handle

12.8 With all the fasteners released, ease the trim panel away from the door

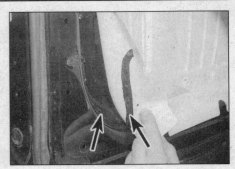
12.9 Cut through the adhesive sealant (arrowed) with a sharp blade

7 On later models, the lower edge of the trim panel is secured with flange bolts; remove these and recover the washers.
8 When all the fasteners have been released, ease the trim panel away from the door, guiding the lock release rod through the hole in the top if the panel **(see illustration)**.
9 Remove the protective plastic panel by pulling one corner away from the door, then working around the panel and cutting through the adhesive sealant with a sharp blade **(see illustration)**.

Refitting
10 Refitting is a reversal of removal, but apply a liquid locking compound to the threads of the locking knob.

Rear door
11 The procedure for removing the rear door trim panel is essentially the same as that for the front door. On models with electric rear windows, prise the electric window switch from the trim panel and unplug the cabling at the connector **(see illustration)**.

13 Door handle and lock components - removal and refitting

Interior door handles
Removal
Note: *This procedure is applicable to both front and rear doors.*
1 Ensure that the window is fully closed. On vehicles fitted with electric windows, disconnect the battery negative cable and position it away from the terminal.
2 Remove the door trim panel (Section 12).
3 Remove the four door handle screws (three in the rear door) **(see illustration)**.

4 At the door lock mechanism, unhook the linkrod from the back of the actuator lever.
5 Withdraw the handle assembly, together with the link-rod, from the door aperture **(see illustrations)**.

Refitting
6 Refit the handle by following the removal procedure in reverse.

Exterior door handles
Removal
7 Ensure that the window is fully closed. On vehicles fitted with electric windows, disconnect the battery negative cable and position it away from the terminal.
8 Remove the door trim panel (Section 12).
9 Working inside the door aperture, remove the screw that secures the leading edge of the exterior door handle to the door; recover the plastic spacer **(see illustrations)**.

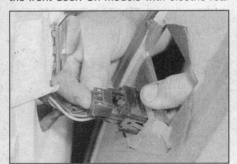
12.11 Prise electric window switch from the trim panel and unplug the connector

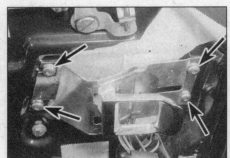
13.3 Remove the four door handle screws (arrowed) - front door shown

13.5a Withdraw the handle assembly and link-rod from the door aperture

13.5b Link rod connection at the rear of the handle

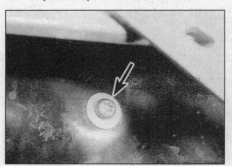
13.9a Remove the exterior door handle leading edge screw . . .

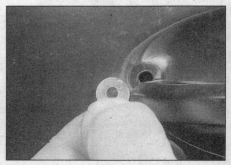
13.9b . . . and recover the plastic spacer

Bodywork and fittings 11•7

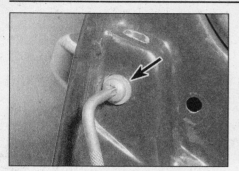

13.10a Remove the screw that secures the trailing edge of the handle to the door

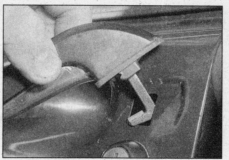

13.10b Unhook the actuator lever from the internal lock mechanism, and withdraw the handle from the door

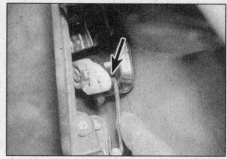

13.14 Disconnect the link rod from the lock cylinder plastic arm (arrowed)

10 At the rear edge of the door, remove the screw that secures the trailing edge of the handle to the door. Unhook the actuator lever from the internal lock mechanism and withdraw the handle from the door **(see illustrations)**.

Refitting
11 Refit the handle by following the removal procedure in reverse. Ensure that the lug at the end of the handle lever engages with the corresponding lug on the internal lock mechanism.

Lock cylinders
Removal
12 Ensure that the window is fully closed. On models with electric windows, disconnect the battery negative cable and position it away from the terminal.
13 Remove the door trim panel (Section 12).
14 Working inside the door aperture, disconnect the link rod from the lock cylinder plastic arm **(see illustration)**.
15 Using a pair of grips, extract the metal spring clip from the rear of the cylinder body, then withdraw the lock cylinder from the door **(see illustrations)**.

Refitting
16 Refitting is a reversal of removal. Should a new lock be required, an identical lock can be obtained from a Saab dealer if the key serial number is quoted.

Lock mechanism - front door (two-door models)
Removal
17 Fully shut the window, then with reference to Section 12, remove the door trim.
18 Turn the lock to the closed position then remove the cross-head retaining screws.
19 On the link rod type, disconnect the link from the lock cylinder and lock. On the crank rod type, remove the clip and disconnect the link from the lock and crank rod.
20 Unscrew the outer door handle screws as necessary and withdraw the lock from the door. If necessary completely remove the outer door handle. Pull the protective plastic bag from the lock mechanism.

Refitting
21 Refitting is a reversal of removal, but lubricate the moving parts with multi-purpose grease.

Lock mechanism - front door (four-door models)
Removal
22 Unsure that the window is fully closed. On vehicles fitted with central locking and/or electric windows, disconnect the battery negative cable and position it away from the terminal.
23 Refer to Section 12 and remove the door trim panel.
24 On vehicles fitted with central locking, refer to Section 17 and disconnet the servo motor link rod from the lock mechanism.
25 Refer to the *Interior Door Handles* removal description earlier in this Section and disconnect the interior handle link rod from the lock mechanism.
26 Refer to the *Lock Cylinder* removal description earlier in this Section and disconect the lock cylinder link rod from the lock mechanism.
27 With the lock release knob removed, unhook the link rod from the clip on the lock mechanism and withdraw the rod from the door **(see illustration)**.
28 At the rear edge of the door, remove the lower mounting screw from the window guide channel. Carefully bend the lower edge towards the front of the door, to allow lock mechanism to be removed.
29 Using the key, turn the lock to the closed position. Then at the rear edge of the door, slacken and withdraw the lock mechanism mounting screws **(see illustration)**.

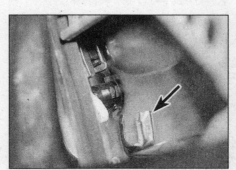

13.15a Extract metal spring clip (arrowed) from the rear of the cylinder body . . .

13.15b . . . then withdraw the lock cylinder from the door

13.27 Unhook the lock release link rod, and withdraw the rod from the door

13.29 At the rear edge of the door, slacken the lock mechanism mounting screws

11•8 Bodywork and fittings

13.30a Withdraw the lock mechnism past the window regulator and out of the door

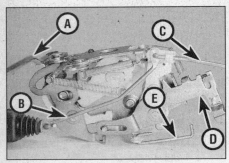

13.30b Lock mechanism connections (front door lock, 4-door model shown)

A Link rod to interior door handle
B Link rod to central locking servo motor
C Link rod to lock release knob
D Exterior door handle release
E Link rod to lock cylinder

30 Withdraw the lock mechnism into the door, then manoeuvre it past the window regulator and out of the door **(see illustrations)**. Recover the padded gasket from the lock to door mating surface and pull off the protective plastic bag.

Refitting

31 Refit the lock mechanism by following the removal procedure in reverse, noting the following points:
a) Grease all moving components with multi purpose grease.
b) Ensure that the latch release lug engages with the corresponding lug on the exterior door handle; refer to the 'Exterior Door Handle' removal description earlier in this Section.
c) Check that the lock and its asscciated components function correctly when reassenbled, before the door trim panel is refitted.

Lock mechanism - rear door

Removal

32 Fully shut the window then remove the door trim as described in Section 12.
33 Turn the lock to the closed position then remove the cross-head retaining screws from the lock.
34 Disconnect the locking knob link.

35 Unscrew the outer door handle front screw and loosen the rear screw on the edge of the door.
36 Remove the screws from the interior door handle and disconnect the front end of the link.
37 Push the lock into the door, disconnect the link, then withdraw the lock. Recover the padded gasket from the lock to door mating surface and pull off the protective plastic bag.
38 If necessary, remove the outer door handle and where applicable, the central locking electric motor.

Refitting

39 Refitting is a reversal of removal, but lubricate the moving parts with multi-purpose grease and refit the protective plastic bag.

14 Door window glass and regulator - removal and refitting

Front door window glass

Removal

1 Refer to Section 12 and remove the door trim panel.
2 Remove the inner weatherstrip from the bottom of the window opening.
3 Unscrew the regulator-to-window channel screws and rotate the window so that its front edge is downwards.
4 On early models, prise off the plastic runner.
5 Withdraw the window from the door.
6 If necessary, remove the channel and rubber pad from the window, noting their fitted positions.

Refitting

7 Refitting is a reversal of removal, but if the channel and rubber pad have been removed, ensure that they are refitted in their original positions.

Rear door window glass

Removal

8 Remove the door trim (see Section 12).
9 Remove the weatherstrip from the bottom of the window opening.
10 Push out the quarter-light window from the inside and remove the moulding.
11 Remove the regulator-to-window channel screws and lower the window.
12 Extract the rubber moulding from the rear channel then remove the channel screws. The bottom screw is reached by removing the rubber plug.
13 Move the channel to the rear then lift the window from the inside of the door.
14 If necessary remove the channel and rubber pad from the window, noting their fitted positions.

Refitting

15 Refitting is a reversal of removal, but if the channel and rubber pad have been removed, ensure that they are refitted in their original positions.

Window regulator assembly - electric

Removal

16 Disconnect the battery negative cable and position it away from the terminal.
17 Remove the door trim (see Section 12).
18 Disconnect the wiring from the electric window motor at the cable.
19 Remove the screw from the window channel, then remove the window.
20 Unbolt the regulator and withdraw it through the aperture in the door.
21 Remove the protective foil from the electric motor then unbolt the motor from the regulator.

Refitting

22 Refitting is a reversal of removal.

Door window regulators - manual, front

Removal

23 Remove the door window glass as described earlier in this Section.
24 Remove the regulator mounting bolts and withdraw it through the aperture in the door.

Refitting

25 Refitting is a reversal of removal.

Door window regulators - manual, rear

Removal

26 Fully open the window then remove the trim panel as described in Section 12.
27 Unscrew the regulator screws at the crank pivot and the bottom of the door frame.
28 Unbolt the window retainer, then withdraw the regulator through the aperture in the door taking care not to scratch the window.

Refitting

29 Refitting is a reversal of removal.

15 Tailgate and support struts / bootlid - removal and refitting

Tailgate

Removal

Caution: It is essential that the help of an assistant is enlisted during this operation.

1 Disconnect the battery negative cable and position it away from the terminal.
2 Remove the tailgate interior trim panel, as described in Section 25.
3 Unplug the all electrical cabling to the exterior lamps and where applicable the central locking servo motor. Label each connector to avoid confusion on refitting.

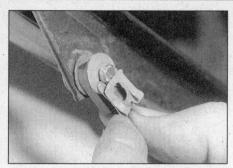

15.11 Prise the clips from the joints at the upper ends of the support struts

4 If the original tailgate is to be refitted, tie string to the ends of the cables, to pull them back into position later.
5 Lay a dust sheet on the rear edge of the roof. Prise out the rubber grommets and withdraw the cables from the tailgate, laying them on the dustsheet to avoid scratching the paintwork.
6 Support the weight of the tailgate using a stout prop of appropriate length. Prise off the retaining clips, then detach the upper ends of the support struts from the spigots on the tailgate.
7 At each hinge, slacken and remove the two grub screws, then remove the main retaining bolts and separate the two halves of the hinge. Maintain firm support of the tailgate to prevent it from tilting.
8 With the help of your assistant, lift off the tailgate and lower it to ground, resting it on a dust sheet to protect the edges. If the original tailgate is to be refitted, leave the drawstrings tails in place, with enough slack to allow the cabling to be pulled back through.

Refitting
9 Refit the tailgate by reversing the removal procedure. If the tailgate is being renewed, transfer all serviceable components from the old tailgate to the new one before fitting it. Close the tailgate and check the operation of the catch/lock mechanism; if excessive force is required to engage it correctly or if the catch appears loose when the tailgate is closed, then refer to Section 16 and adjust the alignment of the latch mechanism.

Support struts
Removal
10 Support the weight of the tailgate using a stout prop of suitable length.
11 Prise the retaining clips from the joints at the upper ends of the support struts, then carefully lever the joint off the spigot on the tailgate (see illustration).
12 Repeat this operation at the lower end of the struts, then remove them from the vehicle.

 Warning: *Ensure the struts are fully extended by keeping the tailgate propped open, or they may expand suddenly when released from their mountings.*

Refitting
13 Refitting is a reversal of removal.

Bootlid
Removal
14 Open the boot lid and mark the position of the hinges with a pencil.
15 With the help of an assistant unscrew the bolts/nuts and withdraw the boot lid from the car. Disconnect the wiring where necessary.

Removal
16 Refitting is a reversal of removal, but make sure that the boot lid is central within its aperture. If necessary loosen the bolts and reposition the boot lid - the striker plate can also be adjusted in the same way.

16 Tailgate/bootlid lock components - removal and refitting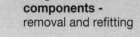

Lock and cylinder
99 saloon models
1 With the boot lid open, remove the two screws and withdraw the lock mechanism.
2 Remove the screws and withdraw the handle from the boot lid.
3 Prise off the plastic disc and remove the spring and push button.
4 Extract the circlip and remove the lever ring and torsion spring.
5 With the key inserted withdraw the cylinder from the sleeve.
6 Refitting is a reversal of removal.

16.19 Prise the lid from the plastic cover

16.21 Remove the screws and withdraw the cylinder from inside the tailgate

900 4-door saloon models
Conventional locking
7 Open the boot lid then pull the clip from the lock cylinder.
8 rise the link from the cylinder lever and withdraw the cylinder.
9 Unscrew the bolts and withdraw the lock Remove the handle from between the inner and outer panels.
10 Refitting is a reversal of removal.
Central locking
11 Disconnect the battery negative lead.
12 Open the boot lid and disconnect the wiring to the electric motor.
13 Unbolt and remove the electric motor.
14 Remove the lock and cylinder as described above.
15 Disconnect the link rod from the link.
16 Refitting is a reversal of removal.

Removal - 900 5-door models
17 Disconnect the battery negative cable and position it away from the terminal.
18 Open the tailgate, then refer to Section 25 and remove the tailgate trim panel.
19 Remove the retaining bar, then prise the lid from the plastic cover (see illustration).
20 Unhook the exterior handle link rod from the lock cylinder plastic arm, then unhook the central locking servo link rod from the same point (see illustration).
21 Remove the retaining screws and withdraw the cylinder from inside the tailgate (see illustration).
22 Remove the retaining nuts and prise the exterior handle away from the tailgate (see illustration)

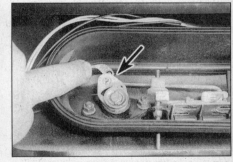

16.20 Unhook the central locking servo link rod from the lock cylinder plastic arm

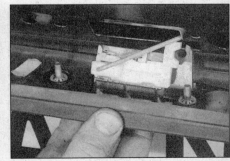

16.22 Remove the retaining nuts and prise the exterior handle away from the tailgate

11•10 Bodywork and fittings

16.25 Slacken and withdraw the latch mechanism retaining screws

Refitting

23 Refit the lock components by following the removal procedure in reverse. Ensure that the push rod from the latch mechanism engages with the lug on the exterior handle actuator lever.

Latch mechanism

24 Refer to the above sub-Section and disconnect the latch mechanism link rod from the lock mechanism.
25 Mark the fitted position of the latch mechanism in relation to its retaining screws. Slacken and withdraw the retaining screws, then lower the latch mechanism away from the tailgate (see illustration).
26 Note that the link rod has an adjustment nut to allow the relationship between the rod and the exterior handle actuator lever to be altered.
27 Refit is the reverse of removal. If much force is required to engage the tailgate latch correctly, or if the latch appears loose when the tailgate is closed, the latch alignment can be adjusted as required by slackening the retaining screws and moving the mechanism up and down on its slooted mounting holes.

17 Central locking servo motors - removal and refitting

Door lock servo motors

Note: *This procedure is applicable to both front and rear doors.*

Removal

1 Disconnect the battery negative cable and position it away from the terminal.
2 Remove the door trim panel (Section 12).
3 With reference to Section 13, disconnect the servo motor link rod from the lock mechanism.
4 Unplug the motor cabling at the connector (see illustration).
5 Slacken and withdraw the retaining screws and remove the motor from the door (see illustrations). Ensure that the plastic protective bag remains intact.

Refitting

6 Refit the servo motor by reversing the removal procedure.

Tailgate/boot lock servo motor

Removal

7 Disconnect the battery negative cable and position it away from the terminal.
8 Refer to Section 25 and remove the tailgate inner trim panel.
9 With reference to Section 16, disconnect the servo motor link rod from the lock mechanism.
10 Unplug the motor cabling at the connector.
11 Slacken and withdraw the retaining screws and remove the motor from the tailgate (see illustration).

Refitting

12 Refit the servo motor by reversing the removal procedure.

18 Electric window components - removal and refitting

Regulators and motors

1 Refer to Section 14.

Switches

2 Refer to Chapter 12.

19 Exterior door mirrors - removal and refitting

Mirror glass - manual mirrors

1 To remove the mirror, remove the handle (manual type) plastic clip, and bezel then unscrew the bolts and withdraw the mirror from the door. Refitting is a reversal of removal.
2 Tilt the mirror down and use a wide blade screwdriver to prise out the mirror midway along the top edge. Slide the mirror up against the stop then press down the stop with a screwdriver and remove the glass. Refitting is a reversal of removal. Lubricate the joint with grease and press on the middle of the glass.

Mirror glass - electric mirrors

3 Using the mirror control switch, tilt the mirror glass up to the full extent of its travel.
4 Insert a flat bladed screwdriver into the hole in the underside of the mirror housing and engage the blade with slotted plastic wheel (see illustration).

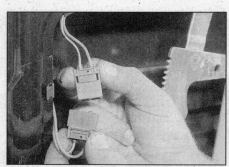
17.4 Unplug the door lock motor cabling at the connector

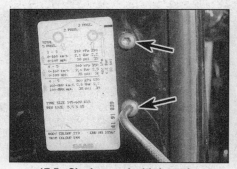
17.5a Slacken and withdraw the retaining screws (arrowed) . . .

17.5b . . . and remove the motor from the door

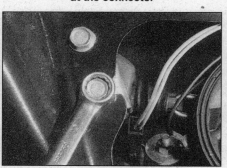
17.11 Slacken the retaining screws and remove the motor from the tailgate

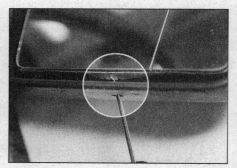
19.4 Insert a screwdriver in to the hole in the underside of the mirror housing

Bodywork and fittings 11•11

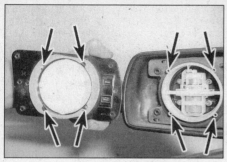

19.8 Ensure the plastic lugs on the back of the mirror are lined up with the cut-outs in the inside edge of the wheel

19.11 Releasing a mirror trim panel press stud fixing centre rod with a pin-punch

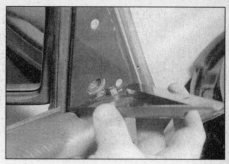

19.12 Unclip the plastic trim panel from inside the door, to expose the mirror mounting screws

5 Turn the wheel two 'clicks' to the right to bring the third slot in line with the centre of the hole in the housing.

6 Lift the mirror glass out of the housing.

7 If the mirror glass is being removed due to breakage, remember to brush out any fragments of broken glass from inside the mirror housing and check that none are wedged in the position adjustment mechanism.

8 When refitting the mirror glass, ensure that the plastic slotted wheel is positioned such that plastic lugs on the back of the mirror are lined up with the cut-outs in the inside edge of the wheel (see illustration). Fit the mirror glass to the housing and, using a screwdriver inserted into the hole in the underside of the mirror, turn the slotted wheel two clicks to the left to secure the glass in place. On completion, check the operation of the electric position adjustment.

Mirror assembly - manual mirrors

9 Remove the handle plastic clip and bezel then unscrew the retaining bolts and withdraw the mirror from the door. Refitting is a reversal of removal.

Mirror assembly - electric mirrors

10 Disconnect the battery negative cable and position it away from the terminal.

11 Release the press stud fixings by pushing in the centre rod with a pin-punch and then prising them out of the plastic cover (see illustration).

12 Unclip the plastic trim panel from inside the door, to expose the mirror mounting screws (see illustration).

13 Slacken and remove the two mounting screws (see illustration).

14 Lift the mirror assembly away from the door slightly, then unplug the cabling at the connector (see illustration).

15 Recover the rubber gasket from the mirror and inspect it for wear or damage; renew it if necessary (see illustration).

Refitting

16 Refit the mirror assembly by reversing the removal procedure. Refer to the previous sub-section for details of refitting the mirror glass.

17 Note that to refit the press stud fixings, the centre rod must be pulled out before the stud is fitted into its mounting hole (see illustration).

20 Windscreen and fixed windows - general information

The renewal of areas of fixed glass is a complex, messy and time-consuming task, which is deemed to be beyond the scope of the home mechanic; without the benefit of extensive practice, it is difficult to attain a secure, waterproof fit. Furthermore, the task carries a high risk of accidental breakage - this applies especially to the laminated glass windscreen. In view of this, owners are strongly advised to entrust work of this nature to a Saab dealer or one of the many specialist windscreen fitters.

21 Sunroof assembly - removal and refitting

General information

Due to the complexity of the sunroof mechanism, considerable expertise is required to repair, renew or adjust the sunroof components successfully. Removal of the sunroof first requires that the headlining be removed, which is a tedious operation, not to be undertaken lightly. Therefore it is recommended that all problems related to the sunroof are referred to a Saab dealer.

19.13 Slacken and remove the two mounting screws

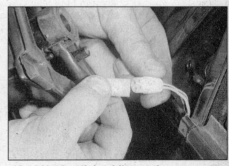

19.14 Unplug the cabling at the connector

19.15 Recover the rubber gasket from the mirror and inspect it for wear or damage

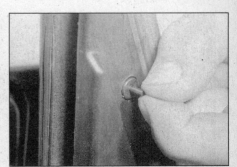

19.17 The press stud centre rod must be pulled out before the stud is refitted

11•12 Bodywork and fittings

22 Body exterior fittings - removal and refitting

Badges and trim mouldings

Removal

1 Side trim panels, rubbing strips, bonnet, bootlid and tailgate emblems are all secured in place by a combination of press studs and adhesive tape.
2 To remove the fittings from the bodywork, select an implement to use as a lever that will not damage the paintwork, such as a plastic spatula, or a filling knife wrapped in PVC tape.
3 Insert the lever between the top edge of the fitting and the bodywork and carefully prise it away, to release the press stud fixings; if more than one stud is used, start at one end of the fitting and work along its length, releasing the studs one by one.
4 Progressively pull the lower edge of the fitting away from the bodywork, allowing the adhesive tape to peel off.
5 Clean the bodywork surface to remove all traces of dirt and the remains of any adhesive tape.

Refitting

6 Peel the backing strip from the new fitting. Offer it up to its mounting position, top edge first, and press the stud fixings into their holes. Smooth the lower edge of the fitting into place, then press down on it firmly to ensure that the tape adheres along its whole length.

23 Seats - removal and refitting

Front seats

Removal

1 On pre-1981 models set the seat to its middle height position then disconnect the wiring for the heating element and seat belt warning system. Push back the catches and raise the front of the seat until the rear brackets are disengaged. The seat can then be removed complete with the rails.
2 On 1981 on models slide the seat fully to the rear and remove the bolts that secure the front of the seat rails to the mounting member. Slide the seat to the front and remove the two rear mounting nuts in the same manner. Note: The mounting bolts screw into nuts that are held in captive cages - ensure that these do not fall inside the seat mounting member as the bolts are removed (see illustrations).
3 Tilt the seat back, then disconnect the wiring for the seat heating elements and seat belt warning lamp switches. Remove the seat from the cabin area.

Refitting

4 Refit the front seats by following the removal procedure in reverse.

Rear seat back rests

Rmoval

5 Fold the rear seat cushions forward, then tilt the back rest fully forwards.
6 Remove the spring clips from the hinge brackets on both sides; a length of welding rod bent into a hook will aid removal. Then using a pair of grips, draw out the plastic discs (see illustrations).
7 Push the whole back rest to one side, then lift the hinge pin out of the bracket and plastic retaining clip (see illustration). Repeat this operation at the opposite side.
8 Lift the seat back rest out of the cabin area.

Refitting

9 Refit the seat back rest by reversing the removal procedure.

Rear seat cushion

Removal

10 Tilt the seat cushion forwards, then remove the circlips from the ends of the hinge pins (see illustration).
11 Slide cushion to one side, disengage the hinge pins then lift the cushion out of the cabin.

Refitting

12 Refit the seat back rest by reversing the removal procedure.

24 Seat belt components - general

General

1 Access to the front seat belt components is gained by removing the front seats/centre console, or the B-pillar trim panel, as applicable.

23.2a Slide the seat forwards and remove the rear seat-to-mounting bolts

23.2b Seat mounting captive nuts

23.6a Remove the spring clips from the hinge brackets on both sides . . .

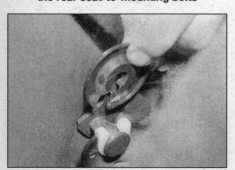

23.6b . . . then draw out the plastic discs

23.7 Lift the hinge pin out of the bracket and plastic retaining clip

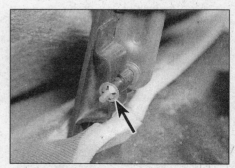

23.10 Remove the circlips (arrowed) from the ends of the hinge pins

Bodywork and fittings 11•13

2 Access to the rear seat belt components is gained by removing the load space trim panel and/the rear seats, as applicable.
3 All mountings must be re-tightened to the specified torque, once disturbed.
4 Seat belts that have been used during a collision will stretch by up to 60 mm, and must be renewed.

25 Interior trim panels - removal and refitting

A-pillar trim panels
1 Open the relevant front door and prise the rubber sealing strip from the door aperture at the A-pillar.

90/99 models
2 Remove the clips and unscrew the panel from the pillar.
3 Refitting is a reversal of removal.

900 models
4 Working from the headlining down, grasp the panel firmly and progressively ease it away from the pillar, allowing the press studs beneath to disengage one at a time.
5 To refit, offer the panel up to its mounted position and apply firm pressure over each press stud until it engages. Press the door aperture sealing strip back into place.

B-pillar trim panels
6 The B-pillar trims form the cover for the seat belt inertia reel assemblies. Prise out the door aperture sealing strips, then remove the press-stud fixings (see illustration).

Headlining
7 The headlining is clipped to the roof panel and can only be lowered once all fittings such as the grab handles, sun visors, sunroof (where fitted), pillar trim panels, and interior lamps have been removed. The weatherstrip for door, and where applicable tailgate and sunroof, apertures will also have to be prised away from the bodywork.
8 Note that the removal of the headlining requires considerable skill and experience if it is to be carried out without damage, and is therefore best entrusted to an expert.

Tailgate interior trim panel
9 Open the tailgate and remove the internal trim panel fixings studs by twisting each one anticlockwise through a quarter turn.
10 Where applicable, remove the retaining screws and lift off the internal grab handle.
11 Carefully lower the interior trim panel away from the tailgate.
12 To refit, reverse the removal procedure.

Load space side trim panels
13 Remove retaining screws and lift out the load space floor panel.
14 Remove the plastic caps from the studs that secure the trim panel to the rear of the load space then prise the rubber sealing strip away from the edge of the tailgate aperture.
15 Working from the tailgate aperture, carefully draw the panel away from the side of the load space. At the underside of the parcel shelf support/speaker bracket, use a screwdriver to release the plastic tangs from the cut-outs in the trim panel.
16 Carefully ease the trim panel away from the bodywork and lift it out of the vehicle.

26 Centre console - removal and refitting

Removal
1 Disconnect the battery negative cable and position it away from the terminal.
2 On models with manual transmission, peel the gear lever gaiter off the mounting frame. On models with automatic transmission, prise out the selector indicator plate. Extract the flexble bellows from the front of the console (see illustrations).
3 Remove the ashtray module from the rear of the centre console, then slide both front seats fully forward.
4 Remove the mounting screws from the rear of the console (see illustration).
5 At the gear/selector lever aperture, remove the mounting screws at the front of the console (see illustration).
6 Where applicable, refer to Chapter 12 and remove the switches for electric windows and interior lighting.
7 Select reverse gear and remove the ignition key from the lock cylinder.
8 Lift the console away from the floorpan slighlty and unplug the ignition switch illumination cabling. Re-insert the igntion key from underneath the console, then move the gear/selector lever towards the front of the vehicle; this gives enough clearance to allow the console to be lifted clear of the gear lever housing and out of the cabin.
9 Recover the diffuser lense from the ignition lock cylinder.
10 Remove the retaining screws and lift out the stowage bin (see illustration).
11 Pull out the trim panel (see illustration).

Refitting
12 Refit the centre console by following the removal procedure in reverse.

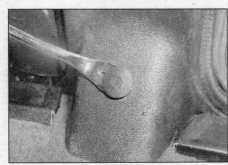

25.6 Removing the B-pillar press-stud fixings

26.2a On manual gearbox models, peel the gear lever gaiter off the mounting frame

26.2b Extract the flexble bellows from the front of the console

26.4 Remove the mounting screws from the rear of the console

26.5 Remove the mounting screws at the front of the console

11•14 Bodywork and fittings

26.10 Lift out the stowage bin

26.11 Pull out the trim panel

12 Where applicable, disconnect the control shaft from the heater unit so that it remains connected to the contol knob on the facia panel; refer to Chapter 4 for greater detail.
13 Disconnect the ducting from the vents at the side of the switch panel.
14 Ensure that nothing remains connected to the switch panel, then withdraw it from the facia completely (see illustration).

Refitting

15 Refitting is a reversal of removal, noting the following points:
a) Ensure all vacuum and electrical connections are remade securely, according to the notes made on removal.
b) Refit the four through-bolts in their correct holes; if they are mixed up, they could be pushed through the facia top.
c) On completion, check the operation of all controls and switches.

27 Facia assembly - removal and refitting

Lower facia panel

Removal

1 Remove the centre console (Section 26).
2 Pull out the ashtray drawer, then remove the screws and lower the mounting frame away from the lower facia panel.
3 Remove the screw located behind the ashtray assembly (see illustration).
4 Open the bonnet, then working inside the wing cavity at the bulkhead slacken and withdraw the retaining screws on either side of the vehicle (see illustration).
5 Extract the stud fixing and remove the lower cover panel from the underside of the steering column (see illustrations).
6 Lower the panel away from the facia, ensuring that all the cabling and ducting is released from the mounting clips before removing it completely (see illustration).

Refitting

7 Refitting is a reversal of removal.

Facia switch panel

Removal

8 Disconnect the battery negative cable and position it away from the terminal.
9 Remove the steering wheel (Chapter 10).
10 Working at the lower edge of the facia panel, remove the four through-bolts; note the fitted position of each bolt as they are all of different length (see illustration).
11 Tilt the panel back slighltly, then working at the rear of the panel, unplug the cabling for each electrical control, in turn. Label each connector carefully, to aid refitting later. Unlug and label the vacuum control hoses from the panel in the same manner.

Upper facia panel

Removal

16 Remove the facia switch panel as described in paragraphs 8 to 14.
17 Remove the two speaker/demister grilles from the top of the facia. Undo the retaiing screw now exposed by each removed grille.
18 Undo and remove the screws located below the glove compartment.
19 Lift out the upper facia panel.

Refitting

20 Refitting is a reversal of removal.

27.3 Remove the mounting screw located behind the ashtray assembly

27.4 Slacken the facia screws on either side of the car (right-hand side shown)

27.5 Removing the lower cover panel from the underside of the steering column

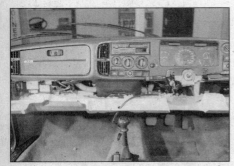

27.6 Lower the panel away from the facia

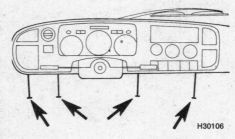

27.10 At the lower edge of the facia panel, remove the four through-bolts (arrowed)

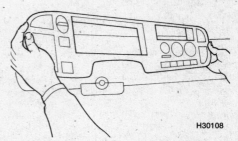

27.14 Withdraw the switch panel from the facia

Chapter 12
Body electrical systems

Contents

Anti - theft alarm - general information 17
Cigarette lighter - removal and refitting 10
Electrical fault finding - general information 2
Exterior light bulbs - removal and refitting 7
Exterior light units - removal and refitting 6
Fuses and relays - general information 3
General information and precautions 1
Headlight beam adjustment - general information 8
Heated front seat components - general information 18
Horn components - removal and refitting 12
In car entertainment components - general information 15
Interior light bulbs - removal and refitting 5
Instrument panel - removal and refitting 9
Radio aerial - removal and refitting 16
Speedometer drive cable - renewal 11
Switches and controls - removal and refitting 4
Windscreen, and headlight wiper arm - removal and refitting 13
Windscreen, and headlight wiper motor and linkage -
    removal and refitting 14

Degrees of difficulty

| **Easy,** suitable for novice with little experience | **Fairly easy,** suitable for beginner with some experience | **Fairly difficult,** suitable for competent DIY mechanic | **Difficult,** suitable for experienced DIY mechanic | **Very difficult,** suitable for expert DIY or professional |

Specifications

Bulbs (typical)   **Watts**
Headlamps 60/55
Rear direction indicators 21
Brake lights 21
Reversing lights 21
Rear fog lights 21
Front direction indicators 21
Daytime driving lights / parking lights 21/5
Tail /brake lights 21/5
Courtesy lights 10
Load space illumination 10
Number plate illumination 5
Rear view mirror mounted light 5
Glove box illumination 5
Switch illumination 1.2
Warning lights 1.2
Ashtray illumination 1.2
Charge warning lamp 2
Heater/ventilation control illumination 2
Cigarette lighter illumination 2
Instrument illumination 3
Direction indicator repeaters 5

Fuses - 99 models

No	Rating (A)	Circuit protected
1	5	Left-hand rear tail and parking lights
2	5	Right-hand rear tail and town lights
3	8	Horn, reversing lights
4	8	Wiper washer, instruments (electric rear view mirror)
5	8 (to 1981) 16 (1982 on)	Electrically heated seat, town lights
6	16	Passenger compartment fan
7	16	Radiator fan (extra lights)
8	16	Heated rear window
9	8	Cigarette lighter, clock, interior lighting
10	16 (1979 and 1982) 8 (1980 and 1981)	Fuel pump
11	5 (to 1981) 8 (1982 on)	Direction indicators, hazard flashers
12	5 (to 1981) 8 (1982 on)	Brake lights
13	3	Headlight wipers (in-line)

Fuses - 900 models (up to 1991 model year)

No	Rating (A)	Circuit protected
1	8	Right-hand main beam
2	8	Left-hand main beam
3	8	Right-hand low beam
4	8	Left-hand low beam
5	16	Radiator fan
6	16	Heated rear window
7	5	Interior lighting
8	16	Fuel pump
9	8	Hazard warning lights
10	5	Brake lights
11	16 (to 1979) 8 (1980 on)	Air conditioning Rear fog lights
12	5	Right-hand parking lights
13	5	Left-hand parking lights
14	8	Horn
15	8	Headlight wipers
16	16	Electrically heated seat
17	25	Heater fan
18	8	Spare/air conditioning
19	8	Instrument panel lights
20	8	Direction indicators
21	8	Windscreen wipers
22	8	Day driving lights

Fuses - 900 models (1991 model year onwards)

Main fusebox

No	Rating (A)	Circuit protected
1	10	Lambda sensor pre-heating
2	Spare	
3	15	Ignition system, EGR modulator valve
4	20	Daytime driving lights
5	15	Windscreen wipers, headlight wipers, seat belt reminder lamp
6	30	Air conditioning
7	15	Direction indicators, tachometer, warning lamps: shift-up, battery charging, check engine, oil pressure, SRS
8	10	Headlight wipers, electric mirrors, cruise control, EZK test connector
9	30	Ventilation fan
10	10	APC system, headlight alignment.
11	Spare	
12	Spare	
13	20	Reversing lights, cigarette lighter
14	15	Right hand headlamp full beam
15	15	Left hand headlamp full beam, and full beam warning lamp

Body electrical systems

Fuses - 900 models (1991 model year onwards) - continued

Main fusebox

No	Rating (A)	Circuit protected
16	15	Right hand headlamp, dipped beam
17	15	Left hand headlmap, dipped beam
18	10	Right hand parking light, right hand tail light, number plate illumination
19	10	Left hand parking light, left hand tail light
20	Spare	
21	15	Rear fog lamps and warning lamp
22	10	Fuel system, temperature gauge, handbrake warning lamp, footbrake and ABS warning lamp, fuel gauge, speed sensor
23	10	Switch and glovebox illumination
24	Spare	
25	30	Radiator cooling fan
26	25	Horn
27	15	Hazard warning lights
28	15	Clock
29	Spare	
30	20	Fuel pump
31	15	Brake lights

Auxiliary fusebox (under rear seat)

No	Rating (A)	Circuit protected
1	20	Convertible top operation
2	20	Interior lighting, EXH warning lamp, front seat heaters, seat belt and ignition key warning buzzers
3	20	Rear window and rear view mirrors heaters
4	10	Interior lighting
5	20	Burglar alarm
6	20	Radio, electric aerial
7	30	Electric front windows
8	30	Electric rear windows and sunroof
9	30	Electric front seat adjustment, left hand
10	30	Electric front seat adjustment, right hand

Relays - 99 models

1. Headlight
2. Starter inhibitor (automatic models)
3. Safety relay (CI system)
4. Fuel pump (CI system)
5. Ignition switch
6. Radiator fan
7. Parking/town lights
8. Windscreen wiper delay
9. Lights
10. Heated rear window
11. Ignition system service outlet
12. Spare

Relays - 900 models (to 1990)

A. Lights
B. Lights
C. Heated rear window
D. Lambda
E. Ignition switch
F. Reversing light
G. Fuel pump
H. Starter interlock
I. Air conditioning
J. Radiator fan

Relays - 900 models (1991 on)

A. Headlights
B. Headlights
C. Headights
D. Additional fog lamps (US and CA only)
E. Ignition switch
F. Daytime driving lights (CA only)
G. Radiator fan
H. Air conditioning compressor
I. -
J. Air conditioning radiator fan
K. Horn

1 General information and precautions

1 This Chapter covers all electrical systems not directly related to the vehicles powertrain. The starting, charging and ignition systems are covered in Chapter 5, Parts A and B.
2 Refer to the precautions listed in "Safety First!", before commencing any of the operations detailed in this Chapter.
Caution: Electronic Control Units (ECUs) contain components that are sensitive to the levels of static electricity generated by a person during normal activity. Once the harness connector has been unplugged, the exposed ECU connector pins can freely conduct stray static electricity to these components, damaging or even destroying them - the damage will be invisible and may not manifest itself immediately. Expensive repairs can be avoided by observing these handling rules:

a) Handle a disconnected ECU by its case only; do not allow fingers or tools to come into contact with the pins.
b) When carrying an ECU around, "ground" yourself from time to time, by touching a metal object such as an unpainted water pipe, this will discharge any potentially damaging static that may have built up.
c) Do not leave the ECU unplugged from its connector for any longer than is absolutely necessary.

2 Electrical fault-finding - general information

1 **Note:** *Refer to the precautions given in 'Safety first!' and in Section 1 of this Chapter before starting work. The following tests relate to testing of the main electrical circuits, and should not be used to test delicate electronic circuits (such as anti-lock braking systems), particularly where an electronic control unit (ECU) is used.*

General

1 A typical electrical circuit consists of an electrical component, any switches, relays, motors, fuses, fusible links or circuit breakers related to that component, and the wiring and connectors which link the component to both the battery and the chassis. To help to pinpoint a problem in an electrical circuit, wiring diagrams are included at the end of this manual.

2 Before attempting to diagnose an electrical fault, first study the appropriate wiring diagram, to obtain a more complete understanding of the components included in the particular circuit concerned. The possible sources of a fault can be narrowed down by noting whether other components related to the circuit are operating properly. If several components or circuits fail at one time, the problem is likely to be related to a shared fuse or earth connection.

3 Electrical problems usually stem from simple causes, such as loose or corroded connections, a faulty earth connection, a blown fuse, a melted fusible link, or a faulty relay (refer to Section 3 for details of testing relays). Visually inspect the condition of all fuses, wires and connections in a problem circuit before testing the components. Use the wiring diagrams to determine which terminal connections will need to be checked, in order to pinpoint the trouble-spot.

4 The basic tools required for electrical fault-finding include a circuit tester or voltmeter (a 12-volt bulb with a set of test leads can also be used for certain tests); a self-powered test light (sometimes known as a continuity tester); an ohmmeter (to measure resistance); a battery and set of test leads; and a jumper wire, preferably with a circuit breaker or fuse incorporated, which can be used to bypass suspect wires or electrical components. Before attempting to locate a problem with test instruments, use the wiring diagram to determine where to make the connections.

5 To find the source of an intermittent wiring fault (usually due to a poor or dirty connection, or damaged wiring insulation), an integrity test can be performed on the wiring, which involves moving the wiring by hand, to see if the fault occurs as the wiring is moved. It should be possible to narrow down the source of the fault to a particular section of wiring. This method of testing can be used in conjunction with any of the tests described in the following sub-Sections.

6 Apart from problems due to poor connections, two basic types of fault can occur in an electrical circuit - open-circuit, or short-circuit.

7 Open-circuit faults are caused by a break somewhere in the circuit, which prevents current from flowing. An open-circuit fault will prevent a component from working, but will not cause the relevant circuit fuse to blow.

8 Short-circuit faults are caused by a 'short' somewhere in the circuit, which allows the current flowing in the circuit to 'escape' along an alternative route, usually to earth. Short-circuit faults are normally caused by a breakdown in wiring insulation, which allows a feed wire to touch either another wire, or an earthed component such as the bodyshell. A short-circuit fault will normally cause the relevant circuit fuse to blow.

Caution: A short-circuit that occurs in the wiring between a circuit's battery supply and its fuse will not cause the fuse in that particular circuit to blow. This part of the circuit is unprotected - bear this in mind when fault-finding on the vehicle's electrical system.

Finding an open-circuit

9 To check for an open-circuit, connect one lead of a circuit tester or voltmeter to either the negative battery terminal or a known good earth.

10 Connect the other lead to a connector in the circuit being tested, preferably nearest to the battery or fuse.

11 Switch on the circuit, bearing in mind that some circuits are live only when the ignition switch is moved to a particular position.

12 If voltage is present (indicated either by the tester bulb lighting or a voltmeter reading, as applicable), this means that the section of the circuit between the relevant connector and the battery is problem-free.

13 Continue to check the remainder of the circuit in the same fashion.

14 When a point is reached at which no voltage is present, the problem must lie between that point and the previous test point with voltage. Most problems can be traced to a broken, corroded or loose connection.

Finding a short-circuit

15 To check for a short-circuit, first disconnect the load(s) from the circuit (loads are the components which draw current from a circuit, such as bulbs, motors, heating elements, etc).

16 Remove the relevant fuse from the circuit, and connect a circuit tester or voltmeter to the fuse connections.

17 Switch on the circuit, bearing in mind that some circuits are live only when the ignition switch is moved to a particular position.

18 If voltage is present (indicated either by the tester bulb lighting or a voltmeter reading, as applicable), this means that there is a short-circuit.

19 If no voltage is present, but the fuse still blows with the load(s) connected, this indicates an internal fault in the load(s).

Finding an earth fault

20 The battery negative terminal is connected to 'earth' - the metal of the engine/transmission and the car body - and most systems are wired so that they only receive a positive feed, the current returning via the metal of the car body. This means that the component mounting and the body form part of that circuit. Loose or corroded mountings can therefore cause a range of electrical faults, ranging from total failure of a circuit, to a puzzling partial fault. In particular, lights may shine dimly (especially when another circuit sharing the same earth point is in operation), motors (eg wiper motors or the radiator cooling fan motor) may run slowly, and the operation of one circuit may have an apparently-unrelated effect on another. Note that on many vehicles, earth straps are used between certain components, such as the engine/transmission and the body, usually where there is no metal-to-metal contact between components, due to flexible rubber mountings, etc.

21 To check whether a component is properly earthed, disconnect the battery, and connect one lead of an ohmmeter to a known good earth point. Connect the other lead to the wire or earth connection being tested. The resistance reading should be zero; if not, check the connection as follows.

22 If an earth connection is thought to be faulty, dismantle the connection, and clean back to bare metal both the bodyshell and the wire terminal or the component earth connection mating surface. Be careful to remove all traces of dirt and corrosion, then use a knife to trim away any paint, so that a clean metal-to-metal joint is made. On reassembly, tighten the joint fasteners securely; if a wire terminal is being refitted, use serrated washers between the terminal and the bodyshell, to ensure a clean and secure connection. When the connection is remade, prevent the onset of corrosion in the future by applying a coat of petroleum jelly or silicone-based grease, or by spraying on (at regular intervals) a proprietary ignition sealer.

3 Fuses and relays - general

Fuses

1 Fuses are designed to break an electrical circuit when a predetermined current limit is reached, in order to protect the components and wiring which could be damaged by excessive current flow. Any excessive current flow will be due to a fault in the circuit, usually a short-circuit (see Section 2).

2 The main fuses are located in the fusebox, which is mounted in the engine bay, inside the left hand wing cavity. The fuse box also acts as a distribution panel for the vehicles wiring harnesses.

3 To gain access to the fusebox, open the bonnet and unclip the transparent plastic lid.

4 On later models, additional fuses are housed in the auxiliary fuse box, which is mounted underneath the rear seat **(see illustration)**. On certain models, fuses specific to the ABS system are housed in a dedicated fusebox, mounted in the engine bay.

Body electrical systems 12•5

3.4 Location of auxiliary fusebox, under rear seat (later models)

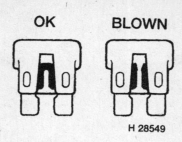

3.7 A blown fuse can be recognised by its melted or broken wire

5 To remove a fuse, first ensure that the relevant circuit is switched off. If in doubt, disconnect the battery negative cable to isolate the supply.
6 Using the plastic tool provided in the fusebox, pull the fuse from its socket.
7 Inspect the fuse from the side, through the transparent plastic body - a blown fuse can be recognised by its melted or broken wire **(see illustration)**.
8 Spare fuses are provided in the blank terminal positions in the fusebox.
9 Before renewing a blown fuse, trace and rectify the cause, and always use a fuse of the correct rating. Never substitute a fuse of a higher rating, or make temporary repairs using wire or metal foil - more serious damage, or even fire, could result.
10 Note that the fuses are colour-coded, as described below; refer to the wiring diagrams for details of the fuse ratings and the circuits protected.

Colour	Rating
Orange	5A
Red	10A
Blue	15A
Yellow	20A
Clear or White	25A
Green	30A

Relays

11 A relay is an electrically-operated mechanical switch, which has the following properties:
a) A relay can switch a heavy current remotely from the circuit in which the current is flowing, therefore allowing the use of lighter-gauge wiring and switch contacts.
b) A relay can receive more than one control input, unlike a mechanically operated switch.
c) Relays are available with internal timing components, to provide a time delay function - for example, the intermittent wiper relay and direction indicator flasher module.

12 The main relays are mounted in the fusebox; see *Fuses* above.
13 Additional relays are located in the auxiliary fusebox, underneath the rear seat; see *Fuses* above.

14 Relays specific to the ABS system are housed in a dedicated fusebox, located in the engine bay. In addition, relays specific to particular systems are mounted adjacent to the components they are supplying.
15 If a circuit or system that is controlled by a relay develops a fault and the performance of the relay is in doubt, power up the system in question and *in general*, if the relay is functioning it should be possible to hear it 'click' as it is energises. If this is found to be the case, then it is probable that the fault lies with the systems components or wiring. If the relay cannot be heard to energise, then either the relay is not receiving a main supply or switching voltage, or the relay itself is faulty. Verification can be carried out by the substitution of a known good unit, but be careful - while some relays are identical in appearance and operation, others look similar but perform different functions - ensure that the substitute relay is of exactly the same type.
16 To remove a relay, first ensure that the relevant circuit is switched off. The relay can then simply be pulled out from the socket, and pushed back into position. Note that some of the more heavy duty relays may be secured in position by a retaining screw.

4 Switches and controls - removal and refitting

Caution: *To eliminate the risk of causing accidental short circuits, disconnect the battery negative cable and position it away from the terminal, before attempting any of the following operations.*

Ignition switch and gear/selector lever lock

1 Refer to the information given in Chapter 7A or B, as applicable.

Steering column combination switches

Removal

2 Refer to Chapter 10 and remove the steering wheel, then refer to Chapter 11 and remove the steering cloumn upper and lower cowling panels.
3 Remove the retaining screws and slide out the switch body **(see illustrations)**.
4 Unplug cabling at the connector.

Refitting

5 Refit the switches by following the removal procedure in reverse.

Mirror adjustment controls

Removal

6 Using a small flat bladed screwdriver padded with insulating tape, prise the switch and bezel out of the facia and unplug the cabling at the connector.

Refitting

7 Plug in the connector securely and push the switch housing back into the aperture in the facia, until the locking lugs are felt to engage.

Fog/driving light switches, heated rear window switch, hazard light switch

8 These switches are encased in modules of a standard size and can be removed in a similar manner.
9 Refer to Chapter 11 and remove the facia switch panel.
10 Reach behind the panel and squeeze the tangs at the side of the switch body together, then pull the switch out of the panel.
11 Unplug the cabling at the connector, labelling it to aid refitting later.
12 Refit the switch by reversing the removal procedure.

Seat belt warning switches

13 Switches in the seat belt buckles activate the "Fasten Seatbelts" warning panel when

4.3a Remove the retaining screws . . .

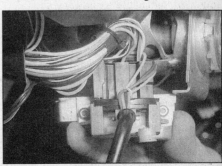

4.3b . . . and slide out the switch body

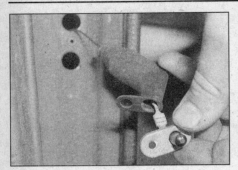

4.16 Removing the courtesy light switch from the door pillar

4.18 Using a screwdriver to prise the switch panel from the centre console

Reversing light switch

22 Refer to the information given in Chapter 7A or B as applicable

5 Interior light bulbs – removal and refitting

Caution: *To eliminate the risk of causing accidental short circuits, disconnect the battery negative cable and position it away from the terminal, before attempting any of the following operations.*

Forward courtesy lights

Facia light

1 Using a small, flat bladed screwdriver, prise the plastic lens from the facia light unit to expose the bulb **(see illustration)**.
2 Unclip the bulb from the spring loaded contacts **(see illustration)**.
3 Refitting is a reversal of removal.

Centre light

4 Using a small, flat bladed screwdriver, prise the plastic lens from the centre light unit to expose the bulb **(see illustration)**.
5 Unclip the bulb from the spring loaded contacts.
6 Refitting is a reversal of removal.

Load space courtesy light

7 Remove the retaining screws and lift off the lens **(see illustration)**.
8 Prise bulb from the spring contacts **(see illustration)**.

4.19 Release the clip and unplug the switch wiring connector(s)

4.21 Removing the interior lighting switch

the vehicle is started, if the belts have not been fastened. However, they are not serviceable components and can only be renewed as part of the stalk/buckle assembly.

Courtsey light door switches

14 The switches that operate the courtesy lights as the doors open are located on the leading edges of the door frame, adjacent to the hinges.
15 Open the relevant door and prise the rubber gaiter from the switch.
16 Remove the retaining screw, then using a small flat bladed screwdriver padded with insulating tape, prise the switch from the door pillar **(see illustration)**. Disconnect the wiring connector as it becomes accessible.
Tip: Tie a length of string to the wiring and tape it to the frame, to prevent it falling back into the door pillar.
17 Refitting is a reversal of removal; ensure that the rubber gaiter is correctly seated on the switch, to prevent moisture ingress.

Centre console-mounted switches

Electric window and sunroof switches

18 Using a suitable flat-bladed screwdriver to carefully prise the switch panel from the centre console; at the same time reach through the handbrake lever aperture and press on the switch panel from below to ease its release **(see illustration)**.
19 Release the clip and unplug the switch wiring connector(s) **(see illustration)**.
20 Refit the switch panel by reversing the removal procedure.

Interior lighting switch

21 Use a suitable flat-bladed screwdriver to carefully prise the switch panel from the centre console **(see illustration)**.

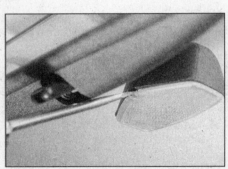

5.1 Prise the plastic lens from the facia light unit to expose the bulb

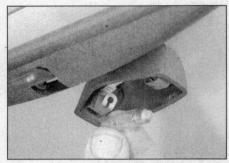

5.2 Unclip the bulb from the spring-loaded contacts

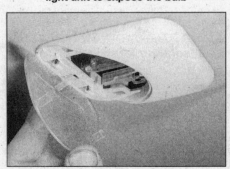

5.4 prise the plastic lens from the centre light unit to expose the bulb

5.7 Remove the retaining screws and lift off the lens

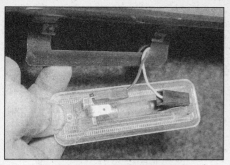
5.8 Prise load space light bulb from the spring contacts

6.3a Remove the headlight unit lower . . .

6.3b . . . and upper retaining screws

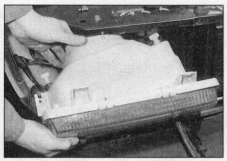
6.3c Removing the headlight unit

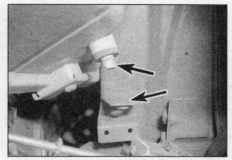
6.5 Locating peg (arrowed) engages with the lug (arrowed) on the car body

6.9 The direction indicator/side light unit is secured by two screws at the trailing edge

Instrument panel illumination

9 Remove the instrument panel, as described in Section 9 of this Chapter.
10 The panel illumination bulbs are fitted to the rear of the unit by bayonet fixings. To remove them, turn the body through one quarter turn until it is felt to disengage.

6 Exterior light units - removal and refitting

Headlight units

900 models

Removal

1 Disconnect the battery negative lead
2 Remove the headlight bulb (see Section 7).
3 Remove the direction indicator light unit as described later in this Section. Where applicable, lower the wiper blade then remove the retaining screws and withdraw the headlamp, guiding the cabling through the aperture in the body panel (see illustrations).
4 The lens may be removed by prising off the metal clips. The reflector may then also be removed.

Refitting

5 Refitting is a reversal of removal, ensure that the locating peg at the rear of the light unit engages with the lug on the body (see illustration).

99 models

Removal

6 To remove the headlamp on 99 models, first remove the bulb then close the bonnet without locking it and remove the radiator grille.
7 Remove the screws and withdraw the headlamp from the frame. Release the adjusting screws and remove the frame from the headlamp.

Refitting

8 Refitting is a reversal of removal.

Indicator/side light units

9 The direction indicator/side light unit is secured by two screws at the trailing edge (see illustration) and a hinge fixing at the leading edge.
10 To remove it, slacken the two screws and pivot the unit away from the wing.

Tail light units

11 Working in the load space, remove the tail light unit retaining nuts (see illustration).

12 Lift the unit away from the wing and unplug the cabling at the connectors, labelling them to aid refitting later (see illustration).
13 Refitting is a reversal of removal.

High level brake lights

14 The high level brake light unit is held in position by a spring clip on 5 door models and a metal bracket on 4/2 door models.
15 Refer to Section 7 and remove the light bulb. With the rear cover removed, unplug the cabling from the bulb terminals; label the cables to aid refitting.
16 Release the light unit from its mountings, then guide the cabling through the entry hole and remove the unit from the vehicle.
17 Refitting is a reversal of removal.

Direction indicator side repeaters

18 Prise the unit from the wing using a small flat bladed screwdriver - pad the blade with

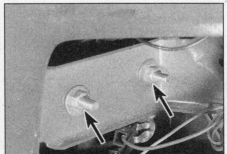
6.11 Remove the tail light unit retaining nuts

6.12 Lift the unit away from the wing

12•8 Body electrical systems

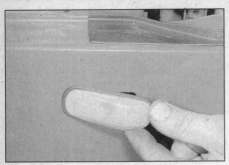

6.18 Removing a direction indicator side repeater unit

PVC tape to protect the paintwork **(see illustration)**.
19 Refit the unit by pressing it firmly back into the wing - ensure that the rubber gasket lies flat and forms a good seal.

7 Exterior light bulbs - removal and refitting

Headlights

1 Working in the engine bay, remove the cap from the rear of the headlight unit. Pull the connector from the bulb, then depress the bulb retainer and turn it anti-clockwise, and withdraw the bulb. On later models, squeeze together the ends of the spring clip to release the bulb from the headlight unit **(see illustration)**.

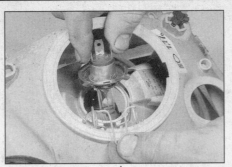

7.1 Removing the headlight bulb on later models

Caution: *Do not touch the glass bulb with bare fingers, as the deposits left on the surface will evaporate and tarnish the headlight unit reflector.*
2 Refitting is a reversal of removal, but make sure that the lugs on the bulb flange are correctly located in the headlamp.

Direction indicator/side lights

3 Refer to Section 6 and remove the light unit from the wing.
4 The bulbs are a bayonet fit in the rear of the light unit reflector **(see illustration)**.

Tail lights

5 Remove the retaining screws and lift the lens from the rear of the tail light unit **(see illustration)**.
6 The bulbs are a bayonet fit in the centre of the reflector **(see illustration)**.

High level brake lights

7 Unclip the cover from the rear of the light unit.
8 The bulb is a bayonet fit in its holder.

Direction indicator side repeaters

9 Refer to Section 6 and remove the light unit from the wing.
10 The bulb is a push fit in its holder **(see illustration)**.

Number plate illumination

11 Remove the retaining screws, then lower the lens and prise out the bulb from the spring contacts **(see illustrations)**.
12 Refitting is a reversal of removal.

8 Headlight beam adjustment - general information

1 It is recommended that the headlamp alignment is carried out by a Saab dealer using beam setting equipment. However, in an emergency the following procedure will provide an acceptable light pattern.
2 Position the car on a level surface with tyres correctly inflated approximately 5.0 metres in front of, and at right-angles to, a wall or garage door.
3 Draw a vertical line on the wall corresponding to the centre line of the car. The position of the line can be ascertained by marking the centre of the front and rear screens with crayon then viewing the wall from the rear of the car.

7.4 Removing a direction indicator/side light bulb

7.5 Remove the retaining screws and lift the lens from the rear of the tail light unit

7.6 The tail light bulbs are a bayonet fit in the centre of the reflector

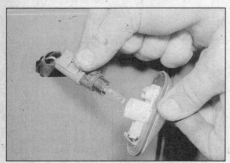

7.10 Direction indicator side repeater bulb removal

7.11a Remove the number plate light screws, then lower the lens . . .

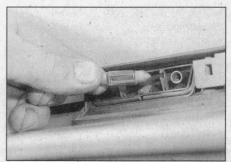

7.11b . . . and prise out the bulb from the spring contacts

Body electrical systems 12•9

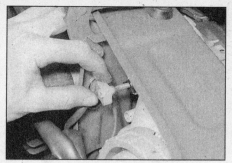

8.6 Manual headlight adjustment on 900 models

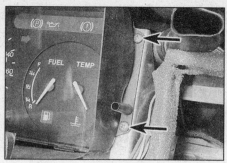

9.10a Remove the retaining screws . . .

9.10b . . . and withdraw the instrument panel from the facia

4 Measure the distance between the headlamp centres and their height above the ground, then mark the positions on the wall.
5 Switch the headlamps on main beam and check that the areas of maximum illumination coincide with the marks on the wall. On dipped beam the area of maximum illumination should be 50 mm below the centre marks.
6 If adjustment is necessary turn the adjustment screws on the headlamp rim (99 models) or on the rear of the headlamp (900 models) until the setting is correct (see illustration). On 900 models insert a screwdriver in one of the bonnet hinge holes to hold the bonnet half open so that the screws can be reached when adjusting the headlamps.

9 Instrument panel - removal and refitting

1 Disconnect the battery negative lead.

99 models

Removal

2 Remove the three screws under the instrument panel, then slide the safety padding from the clips on the facia.
3 Remove the retaining screws then withdraw the instrument panel so that the speedometer cable and wiring can be disconnected.
4 Withdraw the instrument panel from the facia.

Refitting

5 Refitting is a reversal of removal.

900 models

Removal

6 Remove the steering wheel as described in Chapter 11.
7 Refer to Chapter 11 and remove the facia switch panel.
8 Tilt the facia panel back to give sufficient room to remove the instrument panel.
9 Prise out the defroster/speaker grille and disconnect the wiring multi-plugs and speedometer cable, label all disconnected cables, controls and vacuum hoses carefully to aid refitting later.

10 Remove the retaining screws and withdraw the instrument panel from the facia (see illustrations).

Refitting

11 Refitting is a reversal of removal.

10 Cigarette lighter - removal and refitting

Removal

1 Disconnect the battery negative cable and position it away from the terminal.
2 Refer to Chapter 11 and remove the facia switch panel.
3 Unplug the cabling from the rear of the cigarette lighter, then remove the spring clip and push the unit out through the front of the of the facia panel.

Refitting

4 Refit the unit by reversing the removal procedure.

11 Speedometer drive cable - renewal

Removal

1 Jack up the front of the car and support on axle stands (see "Jacking and Vehicle Support"). Chock the rear wheels.
2 Unscrew the collar from the left-hand rear of the gearbox and disconnect the speedometer cable.
3 On 99 models remove the instrument panel as described in Section 13, then remove the grommet from the bulkhead and withdraw the speedometer cable.
4 On 900 models prise out the defroster/speaker grille, and bend back the cable entry spring at the bulkhead. Disconnect the cable from the instrument panel, release it from the grommet and withdraw it into the engine compartment.

Refitting

5 Refitting is a reversal of removal.

12 Horn components - removal and refitting

Horn unit(s)

Removal

1 On 99 models the left-hand horn is located in the engine compartment next to the bumper bracket, and the right-hand horn unit is located outside beneath the bumper. On 900 models, both unit are mounted in the engine compartment.
2 To remove a horn unit, first disconnect the battery negative lead then disconnect the wiring from the horn terminal.
3 Unscrew the bracket mounting bolt and withdraw the horn from the car.

Refitting

4 Refitting is a reversal of removal.

Switches

Removal

5 Refer to Chapter 10 and remove the padding from the steering wheel.
6 Unclip the horn button from the steering wheel spoke, to expose the switch contacts (see illustration).
7 Unplug the cabling then lift the contacts from the plastic moulding.

Refitting

8 Refitting is a reversal of removal.

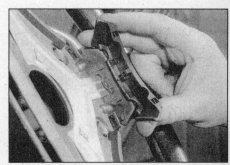

12.6 Unclip the horn button from the steering wheel spoke, to expose the switch contacts

13.2 Flip up the plastic cover at the base of the wiper arm, to expose the retaining nut

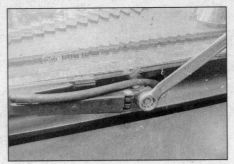

13.7 Slacken and remove the retaining nut and lift off the headlight wiper arm

14.9 Windscreen wiper motor (900 models)

13 Windscreen and headlight wiper arms - removal and refitting

Winscreen wiper arm

1 Note the position on the windscreen at which wiper blades settle, when in the parked position. Mark this position on the screen with a strip of masking tape.
2 Flip up the hinged plastic cover at the base of the wiper arm, to expose the retaining nut (see illustration).
3 Slacken and remove the nut, then lift off the wiper arm.
4 When refitting, ensure that the splined drive spindle engages with the wiper arm mounting hole such that the blade falls in the same position on the windscreen as before, as marked by the strip of masking tape. Tighten the retaining nut and flip down the plastic cover.

Headlamp wiper arm

5 Note the position on the headlight lens at which wiper blade settles, when in the parked position. Mark this position on the lens with a strip of masking tape.
6 Flip up the hinged plastic cover at the base of the wiper arm, to expose the retaining nut.
7 Hold the drive spindle stationary with a pair of grips, then slacken and remove the retaining nut and lift off the wiper arm (see illustration).
8 When refitting, ensure that the drive spindle engages with the wiper arm mounting hole such that the blade falls in the same position on the headlight lens as before, as marked by the strip of masking tape. Tighten the retaining nut securely and flip down the plastic cover.

14 Windscreen and headlight wiper motor and linkage - removal and refitting

Windscreen wiper

1 Remove the wiper arms and blades.
2 Disconnect the battery negative lead.

99 models

3 Working in the engine compartment, unscrew the nut securing the steel tube to the wiper motor.
4 Unbolt the wiper motor strap and detach the wiring. Remove the motor, pulling the flexible cable from the steel tube.
5 Unscrew the bolts from the bottom of the spindle housings and detach the cable tubes.
6 Unscrew the spindle nuts and remove the housings noting the location of the washers and rubber spacers.
7 Refitting is a reversal of removal, but leave the housing bolts loose while inserting the cable.

900 models

8 Prise the rubber grommets from the spindles.
9 Unscrew the linkage mounting bolts and disconnect the wiring from the motor (see illustration). There are two bolts by the spindles and two bolts on the bracket.
10 Extract the circlip from the motor crank and disconnect the linkage rod, then unbolt the motor from the bracket.
11 Refitting is a reversal of removal.

Headlight wiper

12 Disconnect the battery negative lead.

99 models (except Turbo)

13 Remove the radiator grille (Chapter 12) and prise the linkage from the motor crank. Remove the crank.
14 Remove the battery (Section 2) and disconnect the wires from the wiper motor.
15 Unbolt the wiper motor from the fan cover and withdraw it from the car.
16 Unhook the spring securing the bushes to the front sheet, then unbolt the cover and linkage.
17 Refitting is a reversal of removal, but lubricate the recesses in the front sheet with grease, and apply locking fluid to the wiper motor crank securing screw. If necessary adjust the parked position of the wipers by altering the length of the linkage rod. Also check the tension of the cord and adjust if necessary.

99 models (Turbo)

18 Prise up the spindle cover, unscrew the nut, and remove the wiper arm from the spindle.
19 Remove the damper housing on the left-hand side or the heat shield on the right-hand side.
20 Unscrew the wiper mounting bolts from the headlamp surround.
21 Disconnect the wiring and withdraw the wiper motor from the car.
22 Refitting is a reversal of removal.

900 models

23 Remove the headlamp as described in Section 6.
24 Prise off the cap, unscrew the nut, and remove the wiper arm from the spindle.
25 Remove the mounting screws, disconnect the wiring, and withdraw the wiper motor (see illustration). Disconnect the washer tube and remove the wiper blade (photos).
26 Refitting is a reversal of removal.

15 In-car entertainment components - general information

The make and type of radio/casette or compact disc player fitted depends on the age and specification of the vehicle, as does the method of removal. For specific instructions regarding the removal of the unit, refer to the manufacturers documentation supplied with the vehicle, or see advice from you Saab dealer.

14.25 Headlight wiper motor (900 models)

16 Radio aerial - removal and refitting

Removal

1 Switch off the radio and ensure that aerial is returned to its fully retracted position. Where an electric aerial is fitted, disconnect the battery negative cable and position it away from the terminal.
2 Refer to Chapter 11 and remove the inner trim trim panel from the left hand side of the load space.
3 Remove the locknut from the top of the aerial, then lift off the grommet.
4 At the mounting bracket, remove the upper and lower retaining screws and lift out the aerial assembly.
5 Disconnect the coaxial cable (and on electric aerials the power cable) from the aerial assembly.

Refitting

6 Refit the aerial assembly by reversing the removal procedure.

17 Anti-theft alarm - general information

At the time of writing, very little information was available on the alarm. It is recommended that any problems or queries with the system should be referred to a Saab dealer.

18 Heated front seat components - general information

Certain models are fitted with thermostatically regulated heated front seats. Individual control switches are provided for each seat, which allow the heating element temperature to be set to one of three levels, or switched off completely.

Two heating elements are fitted to each seat; one in the backrest and one in the seat cushion. Access to the heating elements can only be gained by carefully removing the upholstery from the seat - this is a specialised operation which should be entrusted to a Saab dealer.

Wiring diagrams overleaf

12•12 Wiring diagrams

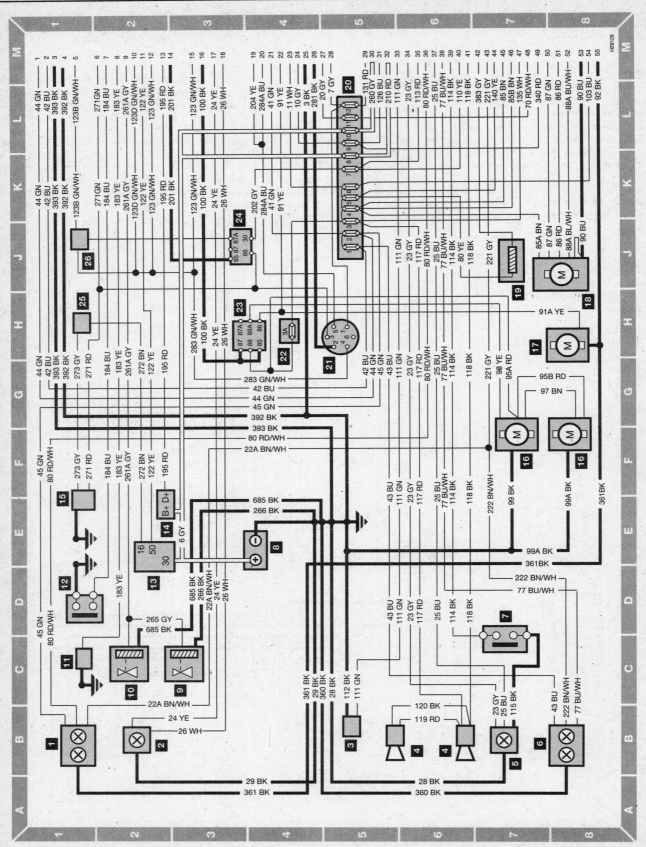

Typical 99 models - part 1

Wiring diagrams 12•13

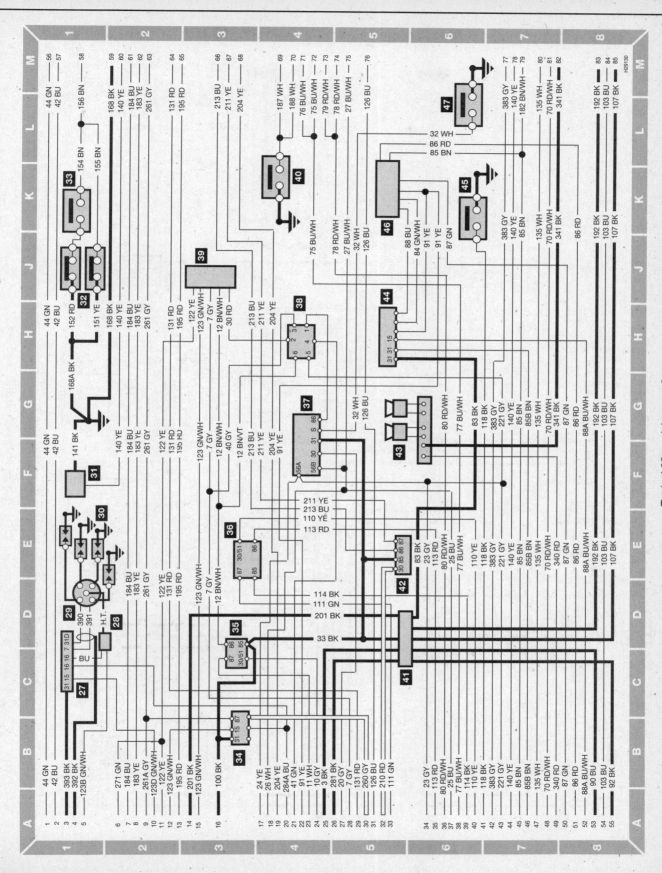

Typical 99 models - part 2

12•14 Wiring diagrams

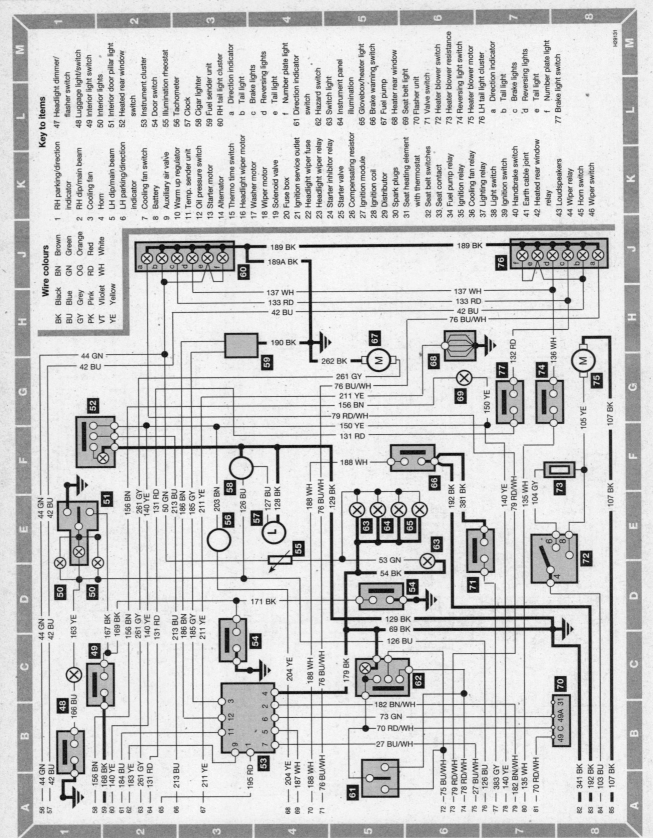

Typical 99 models - part 3

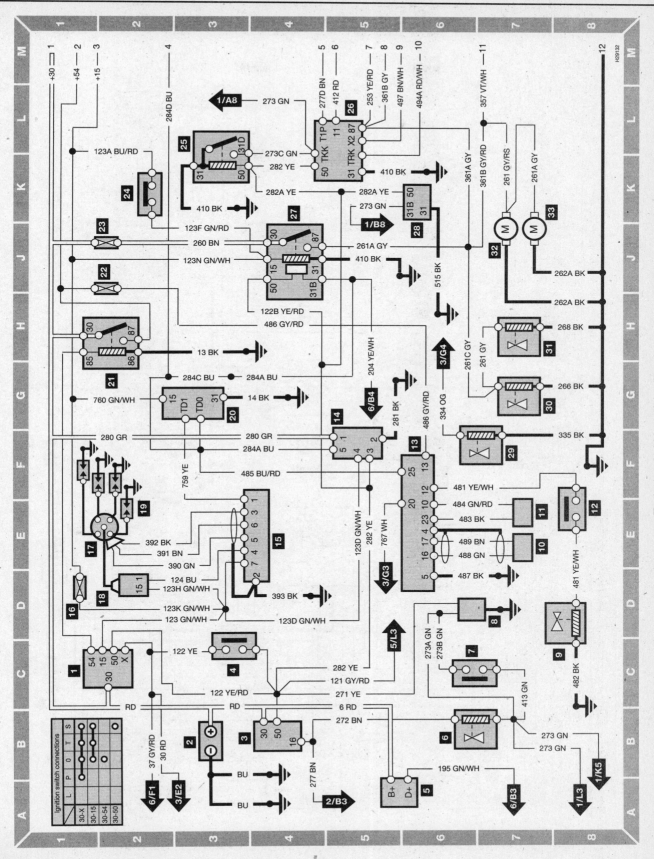

Typical early 900 models - part 1

12•16 Wiring diagrams

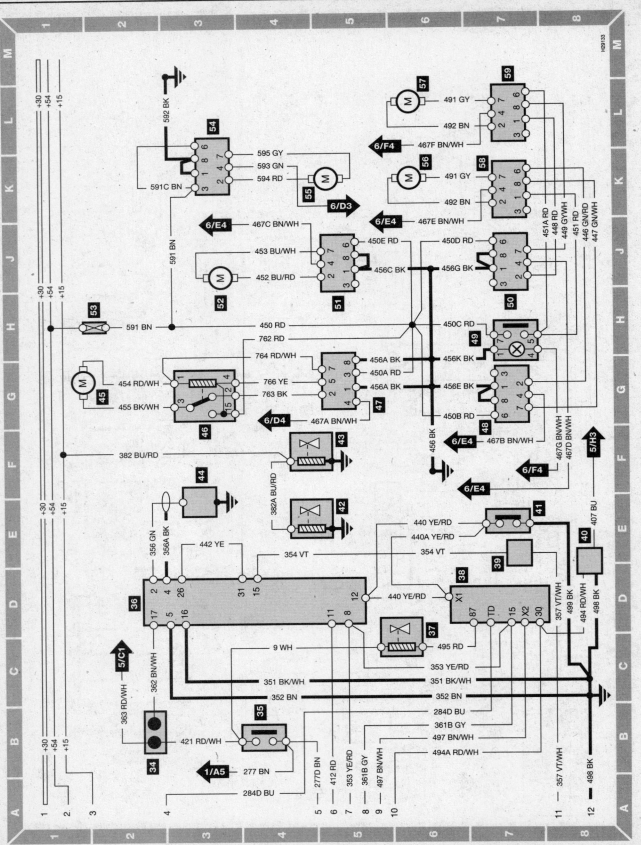

Typical early 900 models - part 2

Wiring diagrams 12•17

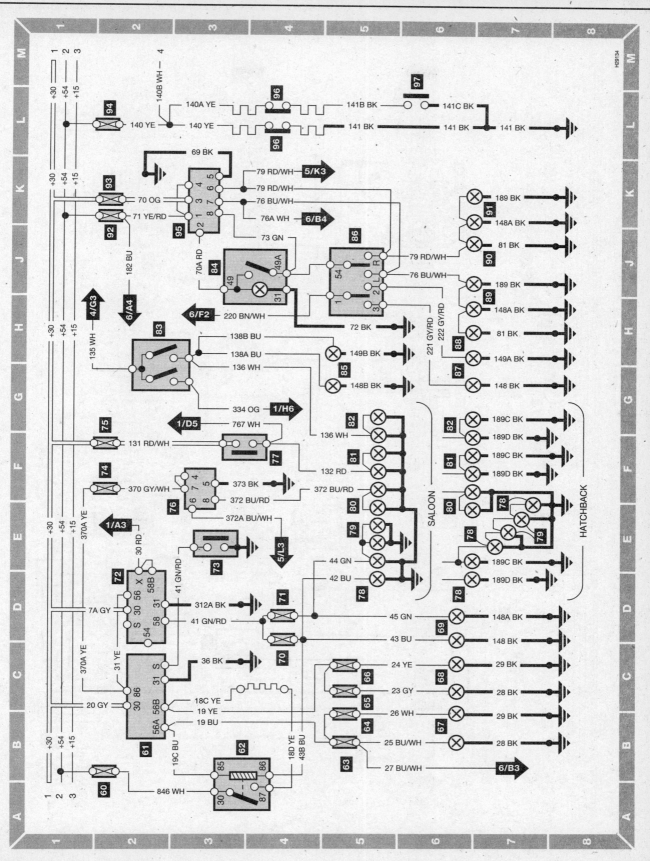

Typical early 900 models - part 3

12•18 Wiring diagrams

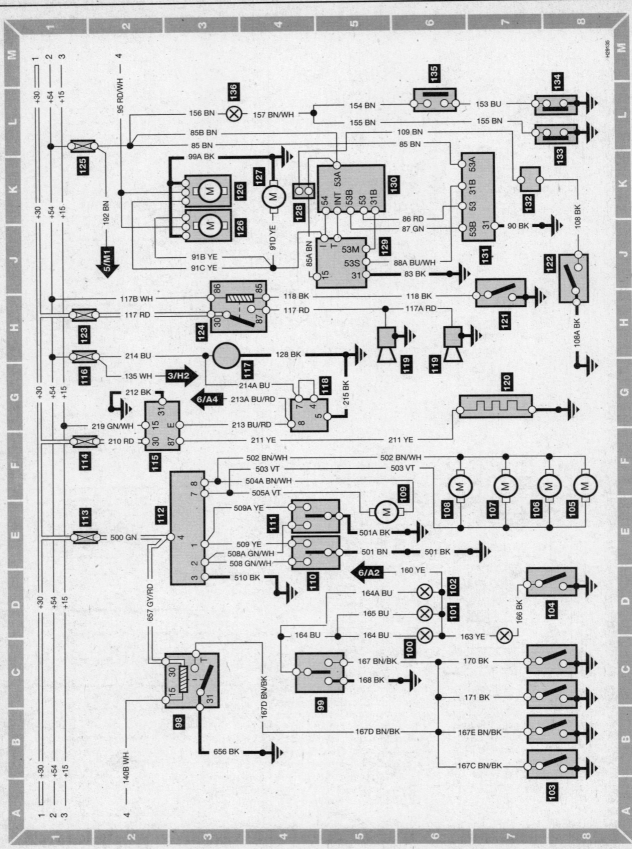

Typical early 900 models - part 4

Wiring diagrams 12•19

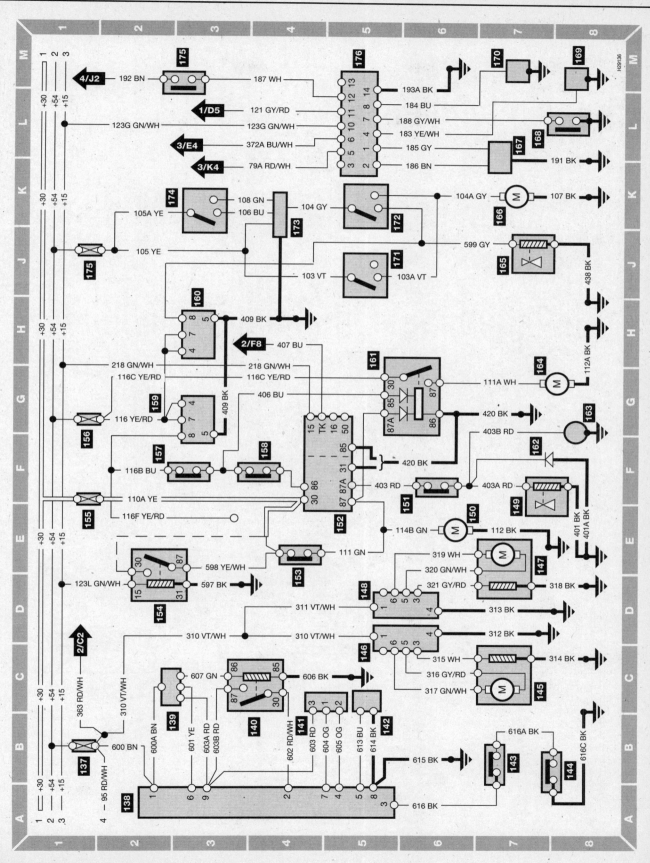

Typical early 900 models - part 5

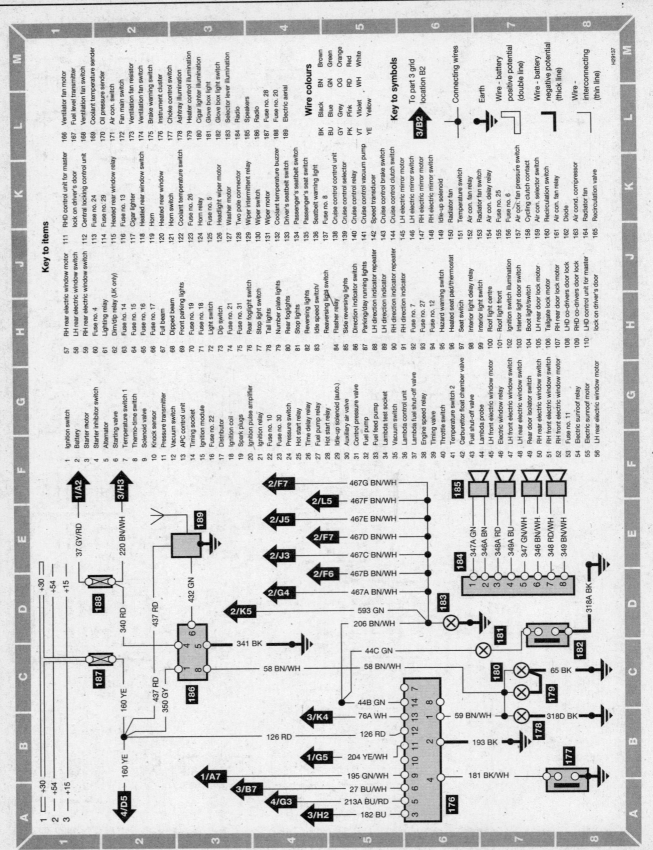

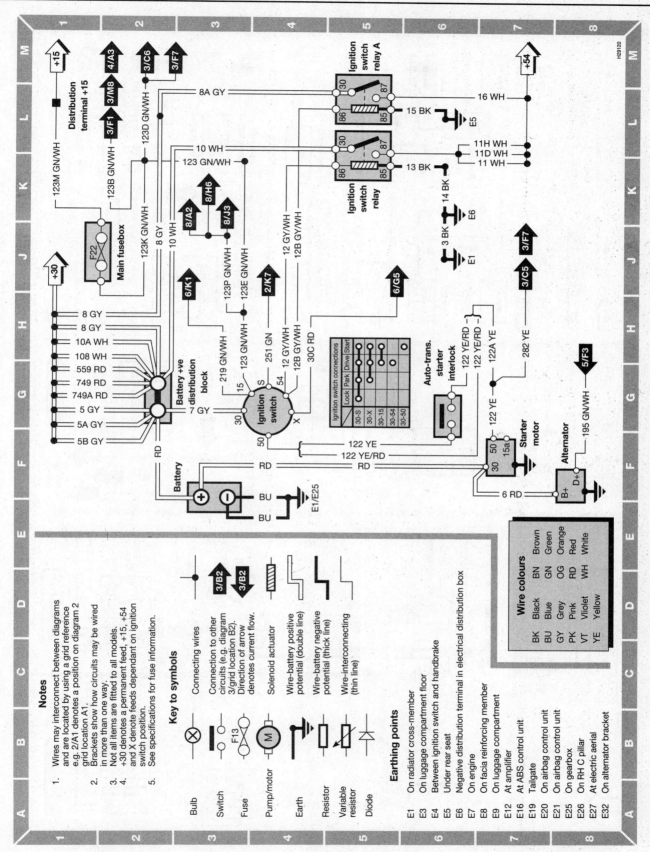

Diagram 1 : Typical later 900 models - notes, key to symbols, earthing points, starting, charging and power supply

12•22 Wiring diagrams

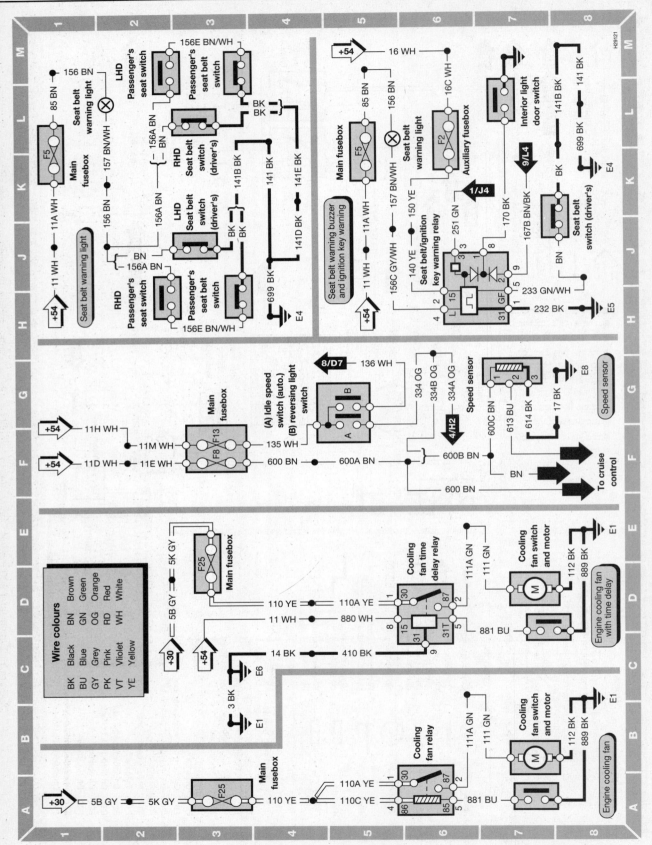

Diagram 2 : Typical later 900 models – cooling fan, speed sensor, seat belt buzzer and ignition key warning

Wiring diagrams 12•23

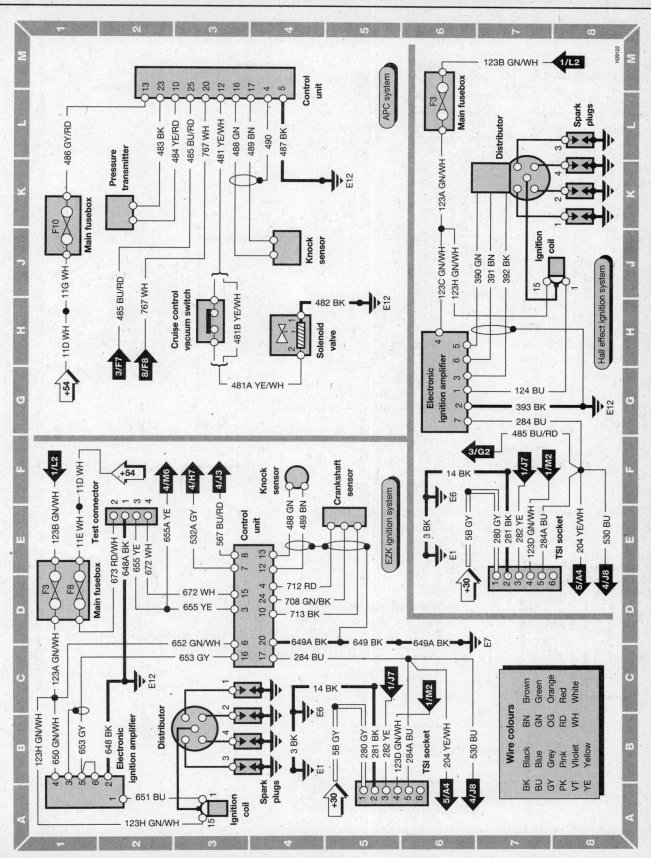

Diagram 3 : Typical later 900 models – APC and ignition systems

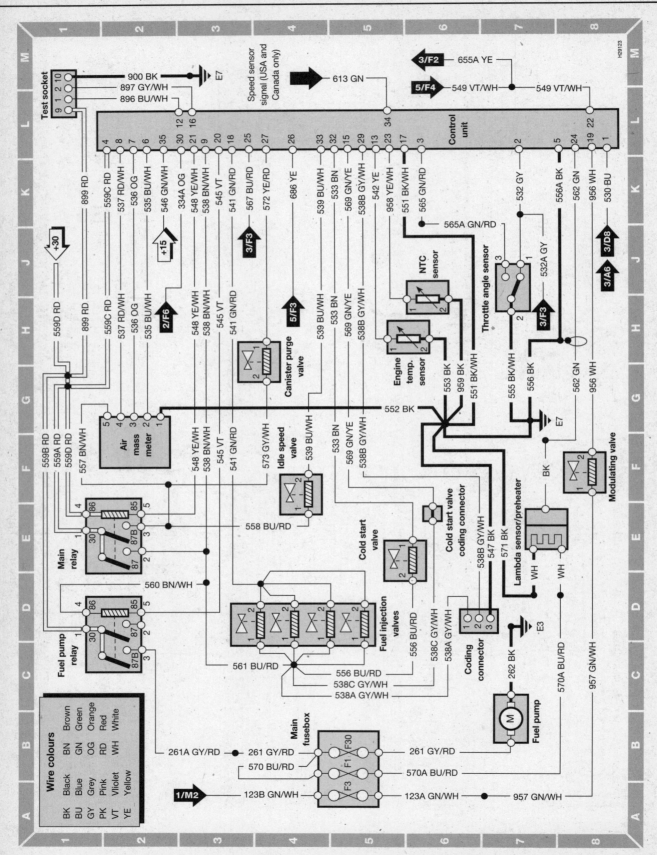

Diagram 4 : Typical later 900 models - typical fuel injection system

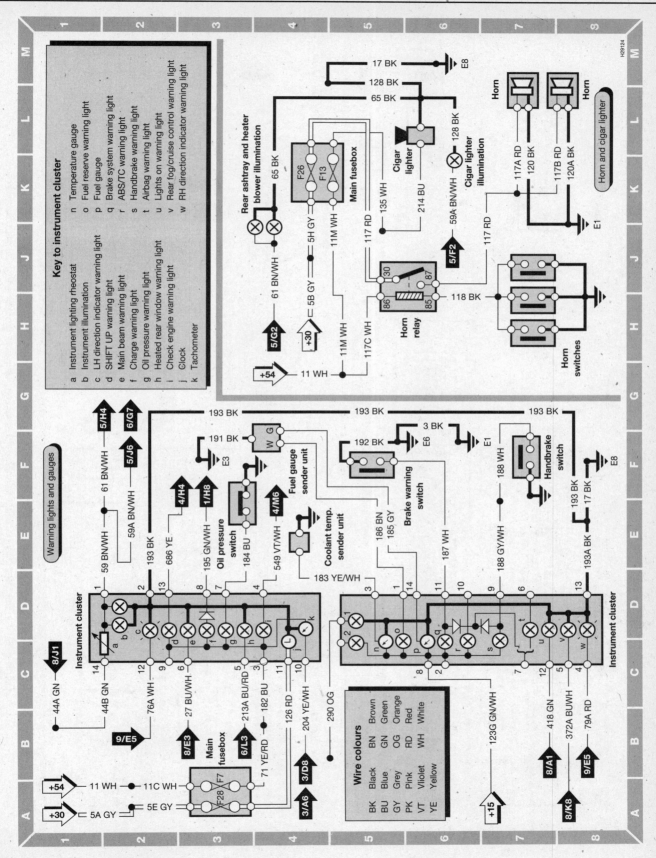

Diagram 5 : Typical later 900 models - warning lights/gauges, horn and cigar lighter

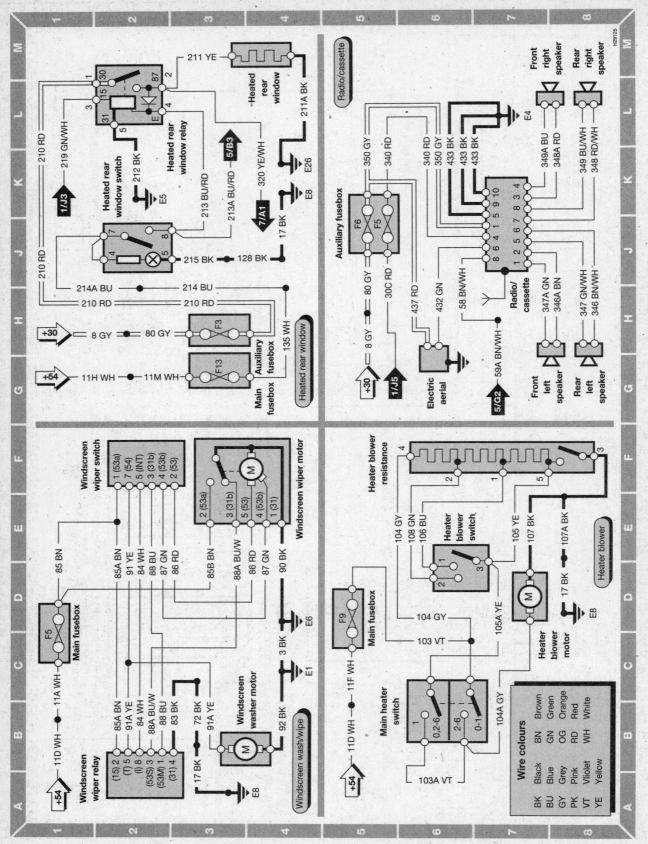

Diagram 6 : Typical later 900 models – wash/wipe, heater blower, heated rear window and radio/cassette

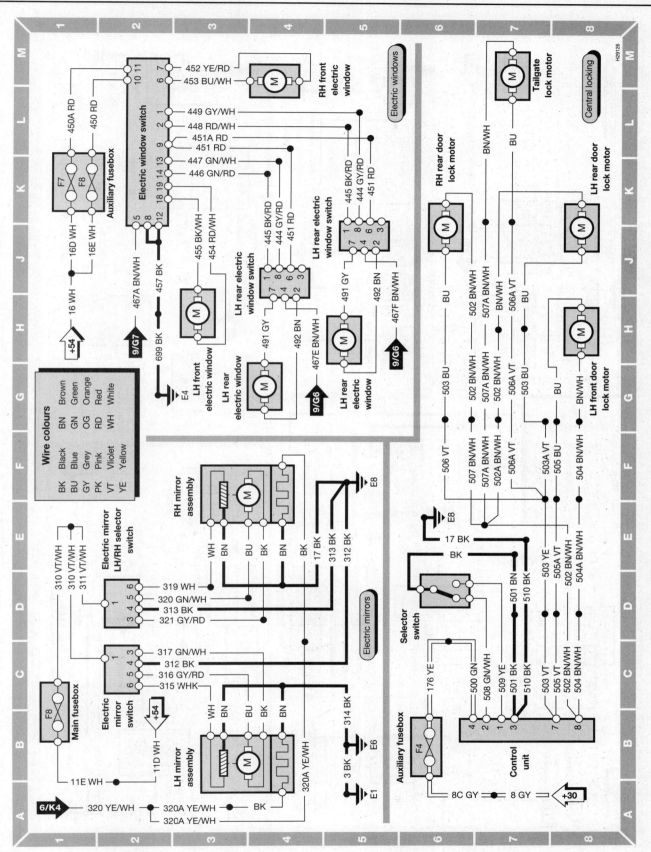

Diagram 7 : Typical later 900 models - central locking, electric mirrors and windows

12•28 Wiring diagrams

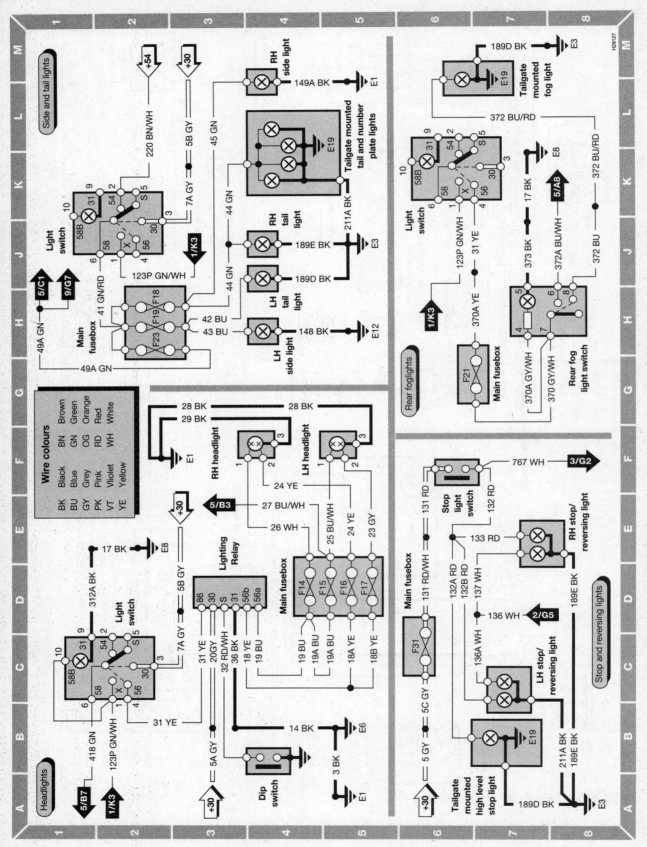

Diagram 8 : Typical later 900 models - exterior lighting

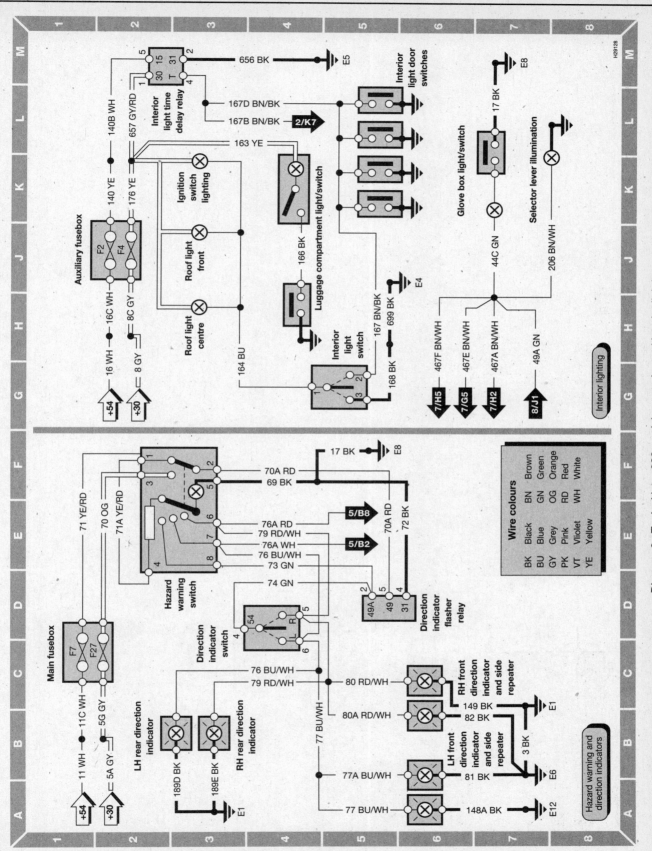

Diagram 9 : Typical later 900 models - exterior and interior lighting

Notes

Reference   REF•1

Dimensions and Weights . . . . **REF•1**	Radio/cassette Anti-theft System - precaution . . **REF•5**
Conversion Factors . . . . **REF•2**	Tools and Working Facilities . . . . **REF•6**
Buying Spare Parts . . . . **REF•3**	MOT Test Checks . . . . **REF•8**
Vehicle Identification . . . . **REF•3**	Fault Finding . . . . **REF•12**
General Repair Procedures . . . . **REF•4**	Glossary of Technical Terms . . . . **REF•20**
Jacking and Vehicle Support . . . . **REF•5**	Index . . . . **REF•25**

Dimensions and Weights
Note: *All figures are approximate, and may vary according to model. Refer to manufacturer's data for exact figures.*

Dimensions
Overall length - 99 models:
    Saloon - early 1979 . . . . 4420 mm
    Saloon - late 1979 on . . . . 4477 mm
    Combi Coupe . . . . 4530 mm
Overall length - 900 models:
    Saloon and Combi Coupe . . . . 4687 to 4739 mm
Overall width:
    All models . . . . 1690 mm
Overall height (at curb weight):
    99 models . . . . 1440 mm
    900 models . . . . 1420 mm

Weights
99 models:
    1979 Saloon . . . . 1200 to 1250 kg
    1979 Combi Coupe . . . . 1230 to 1320 kg
    1980 Saloon . . . . 1200 to 1240 kg
    1981 Saloon . . . . 1190 to 1230 kg
    1982 on models . . . . 1170 to 1205 kg
900 models:
    1979 Combi Coupe . . . . 1200 to 1280 kg
    1980 Combi Coupe . . . . 1185 to 1295 kg
    1981 on models . . . . 1210 to 1440 kg
Maximum roof rack load . . . . 100 kg
Maximum trailer weight:
    With brakes . . . . 1500 kg
    Without brakes . . . . 500 kg

Conversion Factors

Length (distance)
Inches (in)	x 25.4	= Millimetres (mm)	x 0.0394	=	Inches (in)
Feet (ft)	x 0.305	= Metres (m)	x 3.281	=	Feet (ft)
Miles	x 1.609	= Kilometres (km)	x 0.621	=	Miles

Volume (capacity)
Cubic inches (cu in; in^3)	x 16.387	= Cubic centimetres (cc; cm^3)	x 0.061	=	Cubic inches (cu in; in^3)
Imperial pints (Imp pt)	x 0.568	= Litres (l)	x 1.76	=	Imperial pints (Imp pt)
Imperial quarts (Imp qt)	x 1.137	= Litres (l)	x 0.88	=	Imperial quarts (Imp qt)
Imperial quarts (Imp qt)	x 1.201	= US quarts (US qt)	x 0.833	=	Imperial quarts (Imp qt)
US quarts (US qt)	x 0.946	= Litres (l)	x 1.057	=	US quarts (US qt)
Imperial gallons (Imp gal)	x 4.546	= Litres (l)	x 0.22	=	Imperial gallons (Imp gal)
Imperial gallons (Imp gal)	x 1.201	= US gallons (US gal)	x 0.833	=	Imperial gallons (Imp gal)
US gallons (US gal)	x 3.785	= Litres (l)	x 0.264	=	US gallons (US gal)

Mass (weight)
Ounces (oz)	x 28.35	= Grams (g)	x 0.035	=	Ounces (oz)
Pounds (lb)	x 0.454	= Kilograms (kg)	x 2.205	=	Pounds (lb)

Force
Ounces-force (ozf; oz)	x 0.278	= Newtons (N)	x 3.6	=	Ounces-force (ozf; oz)
Pounds-force (lbf; lb)	x 4.448	= Newtons (N)	x 0.225	=	Pounds-force (lbf; lb)
Newtons (N)	x 0.1	= Kilograms-force (kgf; kg)	x 9.81	=	Newtons (N)

Pressure
Pounds-force per square inch (psi; lbf/in^2; lb/in^2)	x 0.070	= Kilograms-force per square centimetre (kgf/cm^2; kg/cm^2)	x 14.223	=	Pounds-force per square inch (psi; lbf/in^2; lb/in^2)
Pounds-force per square inch (psi; lbf/in^2; lb/in^2)	x 0.068	= Atmospheres (atm)	x 14.696	=	Pounds-force per square inch (psi; lbf/in^2; lb/in^2)
Pounds-force per square inch (psi; lbf/in^2; lb/in^2)	x 0.069	= Bars	x 14.5	=	Pounds-force per square inch (psi; lbf/in^2; lb/in^2)
Pounds-force per square inch (psi; lbf/in^2; lb/in^2)	x 6.895	= Kilopascals (kPa)	x 0.145	=	Pounds-force per square inch (psi; lbf/in^2; lb/in^2)
Kilopascals (kPa)	x 0.01	= Kilograms-force per square centimetre (kgf/cm^2; kg/cm^2)	x 98.1	=	Kilopascals (kPa)
Millibar (mbar)	x 100	= Pascals (Pa)	x 0.01	=	Millibar (mbar)
Millibar (mbar)	x 0.0145	= Pounds-force per square inch (psi; lbf/in^2; lb/in^2)	x 68.947	=	Millibar (mbar)
Millibar (mbar)	x 0.75	= Millimetres of mercury (mmHg)	x 1.333	=	Millibar (mbar)
Millibar (mbar)	x 0.401	= Inches of water (inH$_2$O)	x 2.491	=	Millibar (mbar)
Millimetres of mercury (mmHg)	x 0.535	= Inches of water (inH$_2$O)	x 1.868	=	Millimetres of mercury (mmHg)
Inches of water (inH$_2$O)	x 0.036	= Pounds-force per square inch (psi; lbf/in^2; lb/in^2)	x 27.68	=	Inches of water (inH$_2$O)

Torque (moment of force)
Pounds-force inches (lbf in; lb in)	x 1.152	= Kilograms-force centimetre (kgf cm; kg cm)	x 0.868	=	Pounds-force inches (lbf in; lb in)
Pounds-force inches (lbf in; lb in)	x 0.113	= Newton metres (Nm)	x 8.85	=	Pounds-force inches (lbf in; lb in)
Pounds-force inches (lbf in; lb in)	x 0.083	= Pounds-force feet (lbf ft; lb ft)	x 12	=	Pounds-force inches (lbf in; lb in)
Pounds-force feet (lbf ft; lb ft)	x 0.138	= Kilograms-force metres (kgf m; kg m)	x 7.233	=	Pounds-force feet (lbf ft; lb ft)
Pounds-force feet (lbf ft; lb ft)	x 1.356	= Newton metres (Nm)	x 0.738	=	Pounds-force feet (lbf ft; lb ft)
Newton metres (Nm)	x 0.102	= Kilograms-force metres (kgf m; kg m)	x 9.804	=	Newton metres (Nm)

Power
Horsepower (hp)	x 745.7	= Watts (W)	x 0.0013	=	Horsepower (hp)

Velocity (speed)
Miles per hour (miles/hr; mph)	x 1.609	= Kilometres per hour (km/hr; kph)	x 0.621	=	Miles per hour (miles/hr; mph)

Fuel consumption*
Miles per gallon (mpg)	x 0.354	= Kilometres per litre (km/l)	x 2.825	=	Miles per gallon (mpg)

Temperature

Degrees Fahrenheit = (°C x 1.8) + 32   Degrees Celsius (Degrees Centigrade; °C) = (°F - 32) x 0.56

*It is common practice to convert from miles per gallon (mpg) to litres/100 kilometres (l/100km), where mpg x l/100 km = 282

Buying Spare Parts

Spare parts are available from many sources, including maker's appointed garages, accessory shops, and motor factors. To be sure of obtaining the correct parts, it will sometimes be necessary to quote the vehicle identification number. If possible, it can also be useful to take the old parts along for positive identification. Items such as starter motors and alternators may be available under a service exchange scheme - any parts returned should always be clean.

Our advice regarding spare part sources is as follows:

Officially-appointed garages

This is the best source of parts which are peculiar to your car, and which are not otherwise generally available (eg badges, interior trim, certain body panels, etc). It is also the only place at which you should buy parts if the vehicle is still under warranty.

Accessory shops

These are very good places to buy materials and components needed for the maintenance of your car (oil, air and fuel filters, spark plugs, light bulbs, drivebelts, oils and greases, brake pads, touch-up paint, etc). Components of this nature sold by a reputable shop are of the same standard as those used by the car manufacturer.

Besides components, these shops also sell tools and general accessories, usually have convenient opening hours, charge lower prices, and can often be found not far from home. Some accessory shops have parts counters where the components needed for almost any repair job can be purchased or ordered.

Motor factors

Good factors will stock all the more important components which wear out comparatively quickly, and can sometimes supply individual components needed for the overhaul of a larger assembly (eg brake seals and hydraulic parts, bearing shells, pistons, valves, alternator brushes). They may also handle work such as cylinder block reboring, crankshaft regrinding and balancing, etc.

Tyre and exhaust specialists

These outlets may be independent, or members of a local or national chain. They frequently offer competitive prices when compared with a main dealer or local garage, but it will pay to obtain several quotes before making a decision. When researching prices, also ask what "extras" may be added - for instance, fitting a new valve and balancing the wheel are both commonly charged on top of the price of a new tyre.

Other sources

Beware of parts or materials obtained from market stalls, car boot sales or similar outlets. Such items are not invariably sub-standard, but there is little chance of compensation if they do prove unsatisfactory. In the case of safety-critical components such as brake pads, there is the risk not only of financial loss but also of an accident causing injury or death.

Second-hand components or assemblies obtained from a car breaker can be a good buy in some circumstances, but this sort of purchase is best made by the experienced DIY mechanic.

Vehicle Identification

Modifications are a continuing and unpublicised process in vehicle manufacture, quite apart from major model changes. Spare parts lists are compiled upon a numerical basis, the individual vehicle identification numbers being essential to correct identification of the component concerned.

When ordering spare parts, always give as much information as possible. Quote the car model, year of manufacture, body and engine numbers, as appropriate.

The *Vehicle Identification Number (VIN)* plate is located on top of the right-hand side front wheel arch except on early 99 models where it is located on the left-hand side **(see illustration)**. It is also punched into the body in the luggage compartment or beneath the rear seat cushion.

The *engine number* is on the front left-hand side of the cylinder block **(see illustration)**.

The *transmission number* is situated on the top of the primary gear casing **(see illustration)**.

The *body trim and colour codes* are either on the right or left-hand side of the engine compartment, or on later models in the luggage compartment.

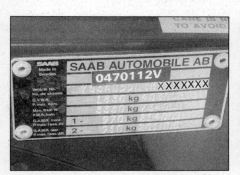

Vehicle identification number (VIN) plate

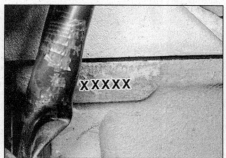

Engine number on an H-type engine

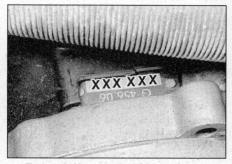

Transmission number (as seen from the front of the car)

General Repair Procedures

Whenever servicing, repair or overhaul work is carried out on the car or its components, it is necessary to observe the following procedures and instructions. This will assist in carrying out the operation efficiently and to a professional standard of workmanship.

Joint mating faces and gaskets

When separating components at their mating faces, never insert screwdrivers or similar implements into the joint between the faces in order to prise them apart. This can cause severe damage which results in oil leaks, coolant leaks, etc upon reassembly. Separation is usually achieved by tapping along the joint with a soft-faced hammer in order to break the seal. However, note that this method may not be suitable where dowels are used for component location.

Where a gasket is used between the mating faces of two components, ensure that it is renewed on reassembly, and fit it dry unless otherwise stated in the repair procedure. Make sure that the mating faces are clean and dry, with all traces of old gasket removed. When cleaning a joint face, use a tool which is not likely to score or damage the face, and remove any burrs or nicks with an oilstone or fine file.

Make sure that tapped holes are cleaned with a pipe cleaner, and keep them free of jointing compound, if this is being used, unless specifically instructed otherwise.

Ensure that all orifices, channels or pipes are clear, and blow through them, preferably using compressed air.

Oil seals

Oil seals can be removed by levering them out with a wide flat-bladed screwdriver or similar tool. Alternatively, a number of self-tapping screws may be screwed into the seal, and these used as a purchase for pliers or similar in order to pull the seal free.

Whenever an oil seal is removed from its working location, either individually or as part of an assembly, it should be renewed.

The very fine sealing lip of the seal is easily damaged, and will not seal if the surface it contacts is not completely clean and free from scratches, nicks or grooves. If the original sealing surface of the component cannot be restored, and the manufacturer has not made provision for slight relocation of the seal relative to the sealing surface, the component should be renewed.

Protect the lips of the seal from any surface which may damage them in the course of fitting. Use tape or a conical sleeve where possible. Lubricate the seal lips with oil before fitting and, on dual-lipped seals, fill the space between the lips with grease.

Unless otherwise stated, oil seals must be fitted with their sealing lips toward the lubricant to be sealed.

Use a tubular drift or block of wood of the appropriate size to install the seal and, if the seal housing is shouldered, drive the seal down to the shoulder. If the seal housing is unshouldered, the seal should be fitted with its face flush with the housing top face (unless otherwise instructed).

Screw threads and fastenings

Seized nuts, bolts and screws are quite a common occurrence where corrosion has set in, and the use of penetrating oil or releasing fluid will often overcome this problem if the offending item is soaked for a while before attempting to release it. The use of an impact driver may also provide a means of releasing such stubborn fastening devices, when used in conjunction with the appropriate screwdriver bit or socket. If none of these methods works, it may be necessary to resort to the careful application of heat, or the use of a hacksaw or nut splitter device.

Studs are usually removed by locking two nuts together on the threaded part, and then using a spanner on the lower nut to unscrew the stud. Studs or bolts which have broken off below the surface of the component in which they are mounted can sometimes be removed using a stud extractor. Always ensure that a blind tapped hole is completely free from oil, grease, water or other fluid before installing the bolt or stud. Failure to do this could cause the housing to crack due to the hydraulic action of the bolt or stud as it is screwed in.

When tightening a castellated nut to accept a split pin, tighten the nut to the specified torque, where applicable, and then tighten further to the next split pin hole. Never slacken the nut to align the split pin hole, unless stated in the repair procedure.

When checking or retightening a nut or bolt to a specified torque setting, slacken the nut or bolt by a quarter of a turn, and then retighten to the specified setting. However, this should not be attempted where angular tightening has been used.

For some screw fastenings, notably cylinder head bolts or nuts, torque wrench settings are no longer specified for the latter stages of tightening, "angle-tightening" being called up instead. Typically, a fairly low torque wrench setting will be applied to the bolts/nuts in the correct sequence, followed by one or more stages of tightening through specified angles.

Locknuts, locktabs and washers

Any fastening which will rotate against a component or housing during tightening should always have a washer between it and the relevant component or housing.

Spring or split washers should always be renewed when they are used to lock a critical component such as a big-end bearing retaining bolt or nut. Locktabs which are folded over to retain a nut or bolt should always be renewed.

Self-locking nuts can be re-used in non-critical areas, providing resistance can be felt when the locking portion passes over the bolt or stud thread. However, it should be noted that self-locking stiffnuts tend to lose their effectiveness after long periods of use, and should be renewed as a matter of course.

Split pins must always be replaced with new ones of the correct size for the hole.

When thread-locking compound is found on the threads of a fastener which is to be re-used, it should be cleaned off with a wire brush and solvent, and fresh compound applied on reassembly.

Special tools

Some repair procedures in this manual entail the use of special tools such as a press, two or three-legged pullers, spring compressors, etc. Wherever possible, suitable readily-available alternatives to the manufacturer's special tools are described, and are shown in use. In some instances, where no alternative is possible, it has been necessary to resort to the use of a manufacturer's tool, and this has been done for reasons of safety as well as the efficient completion of the repair operation. Unless you are highly-skilled and have a thorough understanding of the procedures described, never attempt to bypass the use of any special tool when the procedure described specifies its use. Not only is there a very great risk of personal injury, but expensive damage could be caused to the components involved.

Environmental considerations

When disposing of used engine oil, brake fluid, antifreeze, etc, give due consideration to any detrimental environmental effects. Do not, for instance, pour any of the above liquids down drains into the general sewage system, or onto the ground to soak away. Many local council refuse tips provide a facility for waste oil disposal, as do some garages. If none of these facilities are available, consult your local Environmental Health Department, or the National Rivers Authority, for further advice.

With the universal tightening-up of legislation regarding the emission of environmentally-harmful substances from motor vehicles, most current vehicles have tamperproof devices fitted to the main adjustment points of the fuel system. These devices are primarily designed to prevent unqualified persons from adjusting the fuel/air mixture, with the chance of a consequent increase in toxic emissions. If such devices are encountered during servicing or overhaul, they should, wherever possible, be renewed or refitted in accordance with the vehicle manufacturer's requirements or current legislation.

Note: It is antisocial and illegal to dump oil down the drain. To find the location of your local oil recycling bank, call this number free.

Jacking and Vehicle Support

The jack supplied with the vehicle tool kit should only be used for changing the roadwheels - see *"Wheel changing"* at the front of this manual. When carrying out any other kind of work, raise the vehicle using a hydraulic (or "trolley") jack, and always supplement the jack with axle stands positioned under the vehicle jacking points at the front and rear of the sills on each side of the car. When using a trolley jack, the front of the car can be raised beneath the engine compartment crossmember, and the rear raised beneath the reinforced bracket immediately behind the fuel tank.

To raise the front of the vehicle, position the trolley jack head beneath the reinforced subframe for the engine compartment. **Do not** jack the vehicle under the sump, or under any of the steering or suspension components, or they may be damaged.

To raise the rear of the vehicle, position the jack head beneath the reinforced member immediately behind the fuel tank. **Do not** jack the vehicle under the rear axle.

The jack supplied with the vehicle locates in the jacking points positioned at the front and rear of the body sills on each side of the car. Ensure that the jack head is correctly engaged before attempting to raise the vehicle **(see illustrations)**.

Never work under, around, or near a raised vehicle, unless it is adequately supported in at least two places.

Axle stands can be placed under, or adjacent to, the jacking point

Vehicle jack located in sill jacking point

Radio/cassette Anti-theft System - precaution

On later models, the radio/cassette unit fitted as standard equipment by Saab has a built-in security code, to deter thieves. If the power source to the unit is cut, the anti-theft system will activate. Even if the power source is immediately reconnected, the radio/cassette unit will not function until the correct security code has been entered. Therefore, if you do not know the correct security code for the radio/cassette unit, **do not** disconnect the battery negative terminal of the battery, or remove the radio/cassette unit from the vehicle.

To enter the correct security code, follow the instructions provided with the radio/cassette player handbook.

If an incorrect code is entered, the unit will become locked, and cannot be operated.

If this happens, or if the security code is lost or forgotten, seek the advice of your Saab dealer.

Tools and Working Facilities

Introduction

A selection of good tools is a fundamental requirement for anyone contemplating the maintenance and repair of a motor vehicle. For the owner who does not possess any, their purchase will prove a considerable expense, offsetting some of the savings made by doing-it-yourself. However, provided that the tools purchased meet the relevant national safety standards and are of good quality, they will last for many years and prove an extremely worthwhile investment.

To help the average owner to decide which tools are needed to carry out the various tasks detailed in this manual, we have compiled three lists of tools under the following headings: *Maintenance and minor repair*, *Repair and overhaul*, and *Special*. Newcomers to practical mechanics should start off with the *Maintenance and minor repair* tool kit, and confine themselves to the simpler jobs around the vehicle. Then, as confidence and experience grow, more difficult tasks can be undertaken, with extra tools being purchased as, and when, they are needed. In this way, a *Maintenance and minor repair* tool kit can be built up into a *Repair and overhaul* tool kit over a considerable period of time, without any major cash outlays. The experienced do-it-yourselfer will have a tool kit good enough for most repair and overhaul procedures, and will add tools from the *Special* category when it is felt that the expense is justified by the amount of use to which these tools will be put.

Maintenance and minor repair tool kit

The tools given in this list should be considered as a minimum requirement if routine maintenance, servicing and minor repair operations are to be undertaken. We recommend the purchase of combination spanners (ring one end, open-ended the other); although more expensive than open-ended ones, they do give the advantages of both types of spanner.

- Combination spanners:
  Metric - 8 to 19 mm inclusive
- Adjustable spanner - 35 mm jaw (approx.)
- Spark plug spanner (with rubber insert)
- Spark plug gap adjustment tool
- Set of feeler gauges
- Brake bleed nipple spanner
- Screwdrivers:
  Flat blade - 100 mm long x 6 mm dia
  Cross blade - 100 mm long x 6 mm dia
- Combination pliers
- Hacksaw (junior)
- Tyre pump
- Tyre pressure gauge
- Oil can
- Oil filter removal tool
- Fine emery cloth
- Wire brush (small)
- Funnel (medium size)

Repair and overhaul tool kit

These tools are virtually essential for anyone undertaking any major repairs to a motor vehicle, and are additional to those given in the *Maintenance and minor repair* list. Included in this list is a comprehensive set of sockets. Although these are expensive, they will be found invaluable as they are so versatile - particularly if various drives are included in the set. We recommend the half-inch square-drive type, as this can be used with most proprietary torque wrenches.

The tools in this list will sometimes need to be supplemented by tools from the *Special* list:

- Sockets (or box spanners) to cover range in previous list (including Torx sockets)
- Reversible ratchet drive (for use with sockets)
- Extension piece, 250 mm (for use with sockets)
- Universal joint (for use with sockets)
- Torque wrench (for use with sockets)
- Self-locking grips
- Ball pein hammer
- Soft-faced mallet (plastic/aluminium or rubber)
- Screwdrivers:
  Flat blade - long & sturdy, short (chubby), and narrow (electrician's) types
  Cross blade – Long & sturdy, and short (chubby) types
- Pliers:
  Long-nosed
  Side cutters (electrician's)
  Circlip (internal and external)
- Cold chisel - 25 mm
- Scriber
- Scraper
- Centre-punch
- Pin punch
- Hacksaw
- Brake hose clamp
- Brake/clutch bleeding kit
- Selection of twist drills
- Steel rule/straight-edge
- Allen keys (inc. splined/Torx type)
- Selection of files
- Wire brush
- Axle stands
- Jack (strong trolley or hydraulic type)
- Light with extension lead

Sockets and reversible ratchet drive

Valve spring compressor

Clutch plate alignment set

Spline bit set

Piston ring compressor

Tools and Working Facilities

Special tools

The tools in this list are those which are not used regularly, are expensive to buy, or which need to be used in accordance with their manufacturers' instructions. Unless relatively difficult mechanical jobs are undertaken frequently, it will not be economic to buy many of these tools. Where this is the case, you could consider clubbing together with friends (or joining a motorists' club) to make a joint purchase, or borrowing the tools against a deposit from a local garage or tool hire specialist. It is worth noting that many of the larger DIY superstores now carry a large range of special tools for hire at modest rates.

The following list contains only those tools and instruments freely available to the public, and not those special tools produced by the vehicle manufacturer specifically for its dealer network. You will find occasional references to these manufacturers' special tools in the text of this manual. Generally, an alternative method of doing the job without the vehicle manufacturers' special tool is given. However, sometimes there is no alternative to using them. Where this is the case and the relevant tool cannot be bought or borrowed, you will have to entrust the work to a dealer.

☐ Valve spring compressor
☐ Valve grinding tool
☐ Piston ring compressor
☐ Piston ring removal/installation tool
☐ Cylinder bore hone
☐ Balljoint separator
☐ Coil spring compressors (where applicable)
☐ Two/three-legged hub and bearing puller
☐ Impact screwdriver
☐ Micrometer and/or vernier calipers
☐ Dial gauge
☐ Stroboscopic timing light
☐ Dwell angle meter/tachometer
☐ Universal electrical multi-meter
☐ Cylinder compression gauge
☐ Hand-operated vacuum pump and gauge
☐ Clutch plate alignment set
☐ Brake shoe steady spring cup removal tool
☐ Bush and bearing removal/installation set
☐ Stud extractors
☐ Tap and die set
☐ Lifting tackle
☐ Trolley jack

Buying tools

Reputable motor accessory shops and superstores often offer excellent quality tools at discount prices, so it pays to shop around.

Remember, you don't have to buy the most expensive items on the shelf, but it is always advisable to steer clear of the very cheap tools. Beware of 'bargains' offered on market stalls or at car boot sales. There are plenty of good tools around at reasonable prices, but always aim to purchase items which meet the relevant national safety standards. If in doubt, ask the proprietor or manager of the shop for advice before making a purchase.

Care and maintenance of tools

Having purchased a reasonable tool kit, it is necessary to keep the tools in a clean and serviceable condition. After use, always wipe off any dirt, grease and metal particles using a clean, dry cloth, before putting the tools away. Never leave them lying around after they have been used. A simple tool rack on the garage or workshop wall for items such as screwdrivers and pliers is a good idea. Store all normal spanners and sockets in a metal box. Any measuring instruments, gauges, meters, etc, must be carefully stored where they cannot be damaged or become rusty.

Take a little care when tools are used. Hammer heads inevitably become marked, and screwdrivers lose the keen edge on their blades from time to time. A little timely attention with emery cloth or a file will soon restore items like this to a good finish.

Working facilities

Not to be forgotten when discussing tools is the workshop itself. If anything more than routine maintenance is to be carried out, a suitable working area becomes essential.

It is appreciated that many an owner-mechanic is forced by circumstances to remove an engine or similar item without the benefit of a garage or workshop. Having done this, any repairs should always be done under the cover of a roof.

Wherever possible, any dismantling should be done on a clean, flat workbench or table at a suitable working height.

Any workbench needs a vice; one with a jaw opening of 100 mm is suitable for most jobs. As mentioned previously, some clean dry storage space is also required for tools, as well as for any lubricants, cleaning fluids, touch-up paints etc, which become necessary.

Another item which may be required, and which has a much more general usage, is an electric drill with a chuck capacity of at least 8 mm. This, together with a good range of twist drills, is virtually essential for fitting accessories.

Last, but not least, always keep a supply of old newspapers and clean, lint-free rags available, and try to keep any working area as clean as possible.

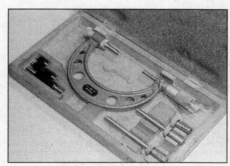
Micrometer set

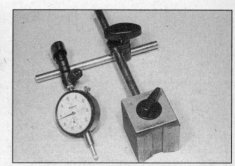
Dial test indicator ("dial gauge")

Stroboscopic timing light

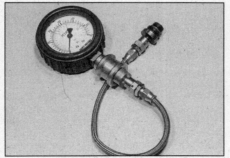
Compression tester

Stud extractor set

MOT Test Checks

This is a guide to getting your vehicle through the MOT test. Obviously it will not be possible to examine the vehicle to the same standard as the professional MOT tester. However, working through the following checks will enable you to identify any problem areas before submitting the vehicle for the test.

Where a testable component is in borderline condition, the tester has discretion in deciding whether to pass or fail it. The basis of such discretion is whether the tester would be happy for a close relative or friend to use the vehicle with the component in that condition. If the vehicle presented is clean and evidently well cared for, the tester may be more inclined to pass a borderline component than if the vehicle is scruffy and apparently neglected.

It has only been possible to summarise the test requirements here, based on the regulations in force at the time of printing. Test standards are becoming increasingly stringent, although there are some exemptions for older vehicles. For full details obtain a copy of the Haynes publication Pass the MOT! (available from stockists of Haynes manuals).

An assistant will be needed to help carry out some of these checks.

The checks have been sub-divided into four categories, as follows:

1 Checks carried out **FROM THE DRIVER'S SEAT**

2 Checks carried out **WITH THE VEHICLE ON THE GROUND**

3 Checks carried out **WITH THE VEHICLE RAISED AND THE WHEELS FREE TO TURN**

4 Checks carried out on **YOUR VEHICLE'S EXHAUST EMISSION SYSTEM**

1 Checks carried out FROM THE DRIVER'S SEAT

Handbrake

☐ Test the operation of the handbrake. Excessive travel (too many clicks) indicates incorrect brake or cable adjustment.

☐ Check that the handbrake cannot be released by tapping the lever sideways. Check the security of the lever mountings.

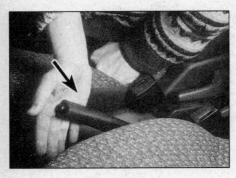

Footbrake

☐ Depress the brake pedal and check that it does not creep down to the floor, indicating a master cylinder fault. Release the pedal, wait a few seconds, then depress it again. If the pedal travels nearly to the floor before firm resistance is felt, brake adjustment or repair is necessary. If the pedal feels spongy, there is air in the hydraulic system which must be removed by bleeding.

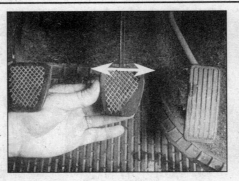

☐ Check that the brake pedal is secure and in good condition. Check also for signs of fluid leaks on the pedal, floor or carpets, which would indicate failed seals in the brake master cylinder.

☐ Check the servo unit (when applicable) by operating the brake pedal several times, then keeping the pedal depressed and starting the engine. As the engine starts, the pedal will move down slightly. If not, the vacuum hose or the servo itself may be faulty.

Steering wheel and column

☐ Examine the steering wheel for fractures or looseness of the hub, spokes or rim.

☐ Move the steering wheel from side to side and then up and down. Check that the steering wheel is not loose on the column, indicating wear or a loose retaining nut. Continue moving the steering wheel as before, but also turn it slightly from left to right.

☐ Check that the steering wheel is not loose on the column, and that there is no abnormal movement of the steering wheel, indicating wear in the column support bearings or couplings.

Windscreen and mirrors

☐ The windscreen must be free of cracks or other significant damage within the driver's field of view. (Small stone chips are acceptable.) Rear view mirrors must be secure, intact, and capable of being adjusted.

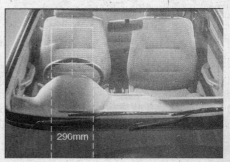

MOT Test Checks   REF•9

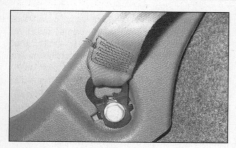

Seat belts and seats

Note: *The following checks are applicable to all seat belts, front and rear.*

☐ Examine the webbing of all the belts (including rear belts if fitted) for cuts, serious fraying or deterioration. Fasten and unfasten each belt to check the buckles. If applicable, check the retracting mechanism. Check the security of all seat belt mountings accessible from inside the vehicle.

☐ The front seats themselves must be securely attached and the backrests must lock in the upright position.

Doors

☐ Both front doors must be able to be opened and closed from outside and inside, and must latch securely when closed.

2 Checks carried out WITH THE VEHICLE ON THE GROUND

Vehicle identification

☐ Number plates must be in good condition, secure and legible, with letters and numbers correctly spaced – spacing at (A) should be twice that at (B).

☐ The VIN plate (A) and homologation plate (B) must be legible.

Electrical equipment

☐ Switch on the ignition and check the operation of the horn.

☐ Check the windscreen washers and wipers, examining the wiper blades; renew damaged or perished blades. Also check the operation of the stop-lights.

☐ Check the operation of the sidelights and number plate lights. The lenses and reflectors must be secure, clean and undamaged.

☐ Check the operation and alignment of the headlights. The headlight reflectors must not be tarnished and the lenses must be undamaged.

☐ Switch on the ignition and check the operation of the direction indicators (including the instrument panel tell-tale) and the hazard warning lights. Operation of the sidelights and stop-lights must not affect the indicators - if it does, the cause is usually a bad earth at the rear light cluster.

☐ Check the operation of the rear foglight(s), including the warning light on the instrument panel or in the switch.

Footbrake

☐ Examine the master cylinder, brake pipes and servo unit for leaks, loose mountings, corrosion or other damage.

☐ The fluid reservoir must be secure and the fluid level must be between the upper (A) and lower (B) markings.

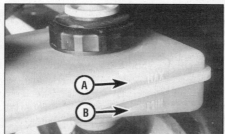

☐ Inspect both front brake flexible hoses for cracks or deterioration of the rubber. Turn the steering from lock to lock, and ensure that the hoses do not contact the wheel, tyre, or any part of the steering or suspension mechanism. With the brake pedal firmly depressed, check the hoses for bulges or leaks under pressure.

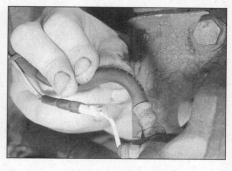

Steering and suspension

☐ Have your assistant turn the steering wheel from side to side slightly, up to the point where the steering gear just begins to transmit this movement to the roadwheels. Check for excessive free play between the steering wheel and the steering gear, indicating wear or insecurity of the steering column joints, the column-to-steering gear coupling, or the steering gear itself.

☐ Have your assistant turn the steering wheel more vigorously in each direction, so that the roadwheels just begin to turn. As this is done, examine all the steering joints, linkages, fittings and attachments. Renew any component that shows signs of wear or damage. On vehicles with power steering, check the security and condition of the steering pump, drivebelt and hoses.

☐ Check that the vehicle is standing level, and at approximately the correct ride height.

Shock absorbers

☐ Depress each corner of the vehicle in turn, then release it. The vehicle should rise and then settle in its normal position. If the vehicle continues to rise and fall, the shock absorber is defective. A shock absorber which has seized will also cause the vehicle to fail.

REF•10 MOT Test Checks

Exhaust system

☐ Start the engine. With your assistant holding a rag over the tailpipe, check the entire system for leaks. Repair or renew leaking sections.

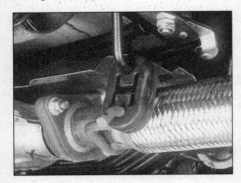

3 Checks carried out WITH THE VEHICLE RAISED AND THE WHEELS FREE TO TURN

Jack up the front and rear of the vehicle, and securely support it on axle stands. Position the stands clear of the suspension assemblies. Ensure that the wheels are clear of the ground and that the steering can be turned from lock to lock.

Steering mechanism

☐ Have your assistant turn the steering from lock to lock. Check that the steering turns smoothly, and that no part of the steering mechanism, including a wheel or tyre, fouls any brake hose or pipe or any part of the body structure.

☐ Examine the steering rack rubber gaiters for damage or insecurity of the retaining clips. If power steering is fitted, check for signs of damage or leakage of the fluid hoses, pipes or connections. Also check for excessive stiffness or binding of the steering, a missing split pin or locking device, or severe corrosion of the body structure within 30 cm of any steering component attachment point.

Front and rear suspension and wheel bearings

☐ Starting at the front right-hand side, grasp the roadwheel at the 3 o'clock and 9 o'clock positions and shake it vigorously. Check for free play or insecurity at the wheel bearings, suspension balljoints, or suspension mountings, pivots and attachments.

☐ Now grasp the wheel at the 12 o'clock and 6 o'clock positions and repeat the previous inspection. Spin the wheel, and check for roughness or tightness of the front wheel bearing.

☐ If excess free play is suspected at a component pivot point, this can be confirmed by using a large screwdriver or similar tool and levering between the mounting and the component attachment. This will confirm whether the wear is in the pivot bush, its retaining bolt, or in the mounting itself (the bolt holes can often become elongated).

☐ Carry out all the above checks at the other front wheel, and then at both rear wheels.

Springs and shock absorbers

☐ Examine the suspension struts (when applicable) for serious fluid leakage, corrosion, or damage to the casing. Also check the security of the mounting points.

☐ If coil springs are fitted, check that the spring ends locate in their seats, and that the spring is not corroded, cracked or broken.

☐ If leaf springs are fitted, check that all leaves are intact, that the axle is securely attached to each spring, and that there is no deterioration of the spring eye mountings, bushes, and shackles.

☐ The same general checks apply to vehicles fitted with other suspension types, such as torsion bars, hydraulic displacer units, etc. Ensure that all mountings and attachments are secure, that there are no signs of excessive wear, corrosion or damage, and (on hydraulic types) that there are no fluid leaks or damaged pipes.

☐ Inspect the shock absorbers for signs of serious fluid leakage. Check for wear of the mounting bushes or attachments, or damage to the body of the unit.

Driveshafts (fwd vehicles only)

☐ Rotate each front wheel in turn and inspect the constant velocity joint gaiters for splits or damage. Also check that each driveshaft is straight and undamaged.

Braking system

☐ If possible without dismantling, check brake pad wear and disc condition. Ensure that the friction lining material has not worn excessively, (A) and that the discs are not fractured, pitted, scored or badly worn (B).

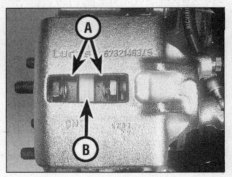

☐ Examine all the rigid brake pipes underneath the vehicle, and the flexible hose(s) at the rear. Look for corrosion, chafing or insecurity of the pipes, and for signs of bulging under pressure, chafing, splits or deterioration of the flexible hoses.

☐ Look for signs of fluid leaks at the brake calipers or on the brake backplates. Repair or renew leaking components.

☐ Slowly spin each wheel, while your assistant depresses and releases the footbrake. Ensure that each brake is operating and does not bind when the pedal is released.

MOT Test Checks

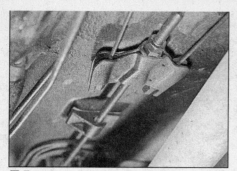

☐ Examine the handbrake mechanism, checking for frayed or broken cables, excessive corrosion, or wear or insecurity of the linkage. Check that the mechanism works on each relevant wheel, and releases fully, without binding.

☐ It is not possible to test brake efficiency without special equipment, but a road test can be carried out later to check that the vehicle pulls up in a straight line.

Fuel and exhaust systems

☐ Inspect the fuel tank (including the filler cap), fuel pipes, hoses and unions. All components must be secure and free from leaks.

☐ Examine the exhaust system over its entire length, checking for any damaged, broken or missing mountings, security of the retaining clamps and rust or corrosion.

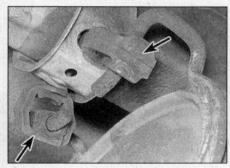

Wheels and tyres

☐ Examine the sidewalls and tread area of each tyre in turn. Check for cuts, tears, lumps, bulges, separation of the tread, and exposure of the ply or cord due to wear or damage. Check that the tyre bead is correctly seated on the wheel rim, that the valve is sound and

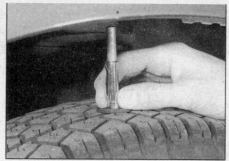

properly seated, and that the wheel is not distorted or damaged.

☐ Check that the tyres are of the correct size for the vehicle, that they are of the same size and type on each axle, and that the pressures are correct.

☐ Check the tyre tread depth. The legal minimum at the time of writing is 1.6 mm over at least three-quarters of the tread width. Abnormal tread wear may indicate incorrect front wheel alignment.

Body corrosion

☐ Check the condition of the entire vehicle structure for signs of corrosion in load-bearing areas. (These include chassis box sections, side sills, cross-members, pillars, and all suspension, steering, braking system and seat belt mountings and anchorages.) Any corrosion which has seriously reduced the thickness of a load-bearing area is likely to cause the vehicle to fail. In this case professional repairs are likely to be needed.

☐ Damage or corrosion which causes sharp or otherwise dangerous edges to be exposed will also cause the vehicle to fail.

4 Checks carried out on YOUR VEHICLE'S EXHAUST EMISSION SYSTEM

Petrol models

☐ Have the engine at normal operating temperature, and make sure that it is in good tune (ignition system in good order, air filter element clean, etc).

☐ Before any measurements are carried out, raise the engine speed to around 2500 rpm, and hold it at this speed for 20 seconds. Allow the engine speed to return to idle, and watch for smoke emissions from the exhaust tailpipe. If the idle speed is obviously much too high, or if dense blue or clearly-visible black smoke comes from the tailpipe for more than 5 seconds, the vehicle will fail. As a rule of thumb, blue smoke signifies oil being burnt (engine wear) while black smoke signifies unburnt fuel (dirty air cleaner element, or other carburettor or fuel system fault).

☐ An exhaust gas analyser capable of measuring carbon monoxide (CO) and hydrocarbons (HC) is now needed. If such an instrument cannot be hired or borrowed, a local garage may agree to perform the check for a small fee.

CO emissions (mixture)

☐ At the time or writing, the maximum CO level at idle is 3.5% for vehicles first used after August 1986 and 4.5% for older vehicles. From January 1996 a much tighter limit (around 0.5%) applies to catalyst-equipped vehicles first used from August 1992. If the CO level cannot be reduced far enough to pass the test (and the fuel and ignition systems are otherwise in good condition) then the carburettor is badly worn, or there is some problem in the fuel injection system or catalytic converter (as applicable).

HC emissions

☐ With the CO emissions within limits, HC emissions must be no more than 1200 ppm (parts per million). If the vehicle fails this test at idle, it can be re-tested at around 2000 rpm; if the HC level is then 1200 ppm or less, this counts as a pass.

☐ Excessive HC emissions can be caused by oil being burnt, but they are more likely to be due to unburnt fuel.

Diesel models

☐ The only emission test applicable to Diesel engines is the measuring of exhaust smoke density. The test involves accelerating the engine several times to its maximum unloaded speed.

Note: *It is of the utmost importance that the engine timing belt is in good condition before the test is carried out.*

☐ Excessive smoke can be caused by a dirty air cleaner element. Otherwise, professional advice may be needed to find the cause.

REF•12 Fault Finding

Engine
- [ ] Engine fails to rotate when attempting to start
- [ ] Engine rotates, but will not start
- [ ] Engine difficult to start when cold
- [ ] Engine difficult to start when hot
- [ ] Starter motor noisy or rough in engagement
- [ ] Engine starts, but stops immediately
- [ ] Engine idles erratically
- [ ] Engine misfires at idle speed
- [ ] Engine misfires throughout the driving speed range
- [ ] Engine hesitates on acceleration
- [ ] Engine stalls
- [ ] Engine lacks power
- [ ] Engine backfires
- [ ] Oil pressure warning light illuminated with engine running
- [ ] Engine runs-on after switching off
- [ ] Engine noises

Cooling system
- [ ] Overheating
- [ ] Overcooling
- [ ] External coolant leakage
- [ ] Internal coolant leakage
- [ ] Corrosion

Fuel and exhaust systems
- [ ] Excessive fuel consumption
- [ ] Fuel leakage and/or fuel odour
- [ ] Excessive noise or fumes from exhaust system

Clutch
- [ ] Pedal travels to floor - no pressure or very little resistance
- [ ] Clutch fails to disengage (unable to select gears)
- [ ] Clutch slips (engine speed increases, with no increase in vehicle speed)
- [ ] Judder as clutch is engaged
- [ ] Noise when depressing or releasing clutch pedal

Manual transmission
- [ ] Noisy in neutral with engine running
- [ ] Noisy in one particular gear
- [ ] Difficulty engaging gears
- [ ] Jumps out of gear
- [ ] Vibration
- [ ] Lubricant leaks

Automatic transmission
- [ ] Fluid leakage
- [ ] Transmission fluid brown, or has burned smell
- [ ] General gear selection problems
- [ ] Transmission will not downshift (kickdown) with accelerator fully depressed
- [ ] Engine will not start in any gear, or starts in gears other than Park or Neutral
- [ ] Transmission slips, shifts roughly, is noisy, or has no drive in forward or reverse gears

Driveshafts
- [ ] Clicking or knocking noise on turns (at slow speed on full-lock)
- [ ] Vibration when accelerating or decelerating

Braking system
- [ ] Vehicle pulls to one side under braking
- [ ] Noise (grinding or high-pitched squeal) when brakes applied
- [ ] Excessive brake pedal travel
- [ ] Brake pedal feels spongy when depressed
- [ ] Excessive brake pedal effort required to stop vehicle
- [ ] Judder felt through brake pedal or steering wheel when braking
- [ ] Brakes binding
- [ ] Rear wheels locking under normal braking

Suspension and steering systems
- [ ] Vehicle pulls to one side
- [ ] Wheel wobble and vibration
- [ ] Excessive pitching and/or rolling around corners, or during braking
- [ ] Wandering or general instability
- [ ] Excessively-stiff steering
- [ ] Excessive play in steering
- [ ] Lack of power assistance
- [ ] Tyre wear excessive

Electrical system
- [ ] Battery will only hold a charge for a few days
- [ ] Ignition/no-charge warning light remains illuminated with engine running
- [ ] Ignition/no-charge warning light fails to come on
- [ ] Lights inoperative
- [ ] Instrument readings inaccurate or erratic
- [ ] Horn inoperative, or unsatisfactory in operation
- [ ] Windscreen/tailgate wipers inoperative, or unsatisfactory in operation
- [ ] Windscreen/tailgate washers inoperative, or unsatisfactory in operation
- [ ] Electric windows inoperative, or unsatisfactory in operation
- [ ] Central locking system inoperative, or unsatisfactory in operation

Introduction

The vehicle owner who does his or her own maintenance according to the recommended service schedules should not have to use this section of the manual very often. Modern component reliability is such that, provided those items subject to wear or deterioration are inspected or renewed at the specified intervals, sudden failure is comparatively rare. Faults do not usually just happen as a result of sudden failure, but develop over a period of time. Major mechanical failures in particular are usually preceded by characteristic symptoms over hundreds or even thousands of miles. Those components which do occasionally fail without warning are often small and easily carried in the vehicle.

With any fault-finding, the first step is to decide where to begin investigations. Sometimes this is obvious, but on other occasions, a little detective work will be necessary. The owner who makes half a dozen haphazard adjustments or replacements may be successful in curing a fault (or its symptoms), but will be none the wiser if the fault recurs, and ultimately may have spent more time and money than was necessary. A calm and logical approach will be found to be more satisfactory in the long run. Always take into account any warning signs or abnormalities that may have been noticed in the period preceding the fault - power loss, high or low gauge readings, unusual smells, etc - and remember that failure of components such as fuses or spark plugs may only be pointers to some underlying fault.

The pages which follow provide an easy-reference guide to the more common problems which may occur during the operation of the vehicle. These problems and their possible causes are grouped under

Fault Finding REF•13

headings denoting various components or systems, such as Engine, Cooling system, etc. The Chapter and/or Section which deals with the problem is also shown in brackets. Whatever the fault, certain basic principles apply. These are as follows:

Verify the fault. This is simply a matter of being sure that you know what the symptoms are before starting work. This is particularly important if you are investigating a fault for someone else, who may not have described it very accurately.

Don't overlook the obvious. For example, if the vehicle won't start, is there fuel in the tank? (Don't take anyone else's word on this particular point, and don't trust the fuel gauge either!) If an electrical fault is indicated, look for loose or broken wires before digging out the test gear.

Cure the disease, not the symptom. Substituting a flat battery with a fully-charged one will get you off the hard shoulder, but if the underlying cause is not attended to, the new battery will go the same way. Similarly, changing oil-fouled spark plugs for a new set will get you moving again, but remember that the reason for the fouling (if it wasn't simply an incorrect grade of plug) will have to be established and corrected.

Don't take anything for granted. Particularly, don't forget that a "new" component may itself be defective (especially if it's been rattling around in the boot for months), and don't leave components out of a fault diagnosis sequence just because they are new or recently-fitted. When you do finally diagnose a difficult fault, you'll probably realise that all the evidence was there from the start.

Engine

Engine fails to rotate when attempting to start
- ☐ Battery terminal connections loose or corroded (Chapter 1).
- ☐ Battery discharged or faulty (Chapter 5).
- ☐ Broken, loose or disconnected wiring in the starting circuit (Chapter 5).
- ☐ Defective starter solenoid or switch (Chapter 5).
- ☐ Defective starter motor (Chapter 5).
- ☐ Starter pinion or flywheel/driveplate ring gear teeth loose or broken (Chapter 2A or 5).
- ☐ Engine earth strap broken or disconnected (Chapter 2A).

Starter motor turns engine slowly
- ☐ Partially-discharged battery (recharge, use jump leads, or push start) (Chapter 5).
- ☐ Battery terminals loose or corroded (Chapter 1).
- ☐ Battery earth to body defective (Chapter 5).
- ☐ Engine earth strap loose (Chapter 2A).
- ☐ Starter motor (or solenoid) wiring loose (Chapter 5).
- ☐ Starter motor internal fault (Chapter 5).

Engine rotates, but will not start
- ☐ Fuel tank empty.
- ☐ Battery discharged (engine rotates slowly) (Chapter 5).
- ☐ Battery terminal connections loose or corroded (Chapter 1).
- ☐ Ignition components damp or damaged (Chapter 1 and 5).
- ☐ Broken, loose or disconnected wiring in the ignition circuit (Chapters 1 and 5).
- ☐ Worn, faulty or incorrectly-gapped spark plugs (Chapter 1).
- ☐ Faulty choke or carburettor (Chapters 1 and 4A).
- ☐ Fuel injection system fault (Chapter 4B).
- ☐ Major mechanical failure (eg broken timing chain) (Chapter 2B).

Engine difficult to start when cold
- ☐ Battery discharged (Chapter 5).
- ☐ Battery terminal connections loose or corroded (Chapter 1).
- ☐ Worn, faulty or incorrectly-gapped spark plugs (Chapter 1).
- ☐ Faulty choke or carburettor (Chapters 1 and 4A).
- ☐ Fuel injection system fault (Chapter 4B).
- ☐ Other ignition system fault (Chapters 1 and 5).
- ☐ Low cylinder compressions (Chapter 2A).

Engine difficult to start when hot
- ☐ Air filter element dirty or clogged (Chapter 1).
- ☐ Faulty choke or carburettor (Chapters 1 and 4A).
- ☐ Fuel injection system fault (Chapter 4B).
- ☐ Low cylinder compressions (Chapter 2A).

Starter motor noisy or rough in engagement
- ☐ Starter pinion or flywheel/driveplate ring gear teeth loose or broken (Chapter 2A or 5).
- ☐ Starter motor mounting bolts loose or missing (Chapter 5).
- ☐ Starter motor internal components worn or damaged (Chapter 5).

Engine starts, but stops immediately
- ☐ Loose or faulty electrical connections in the ignition circuit (Chapters 1 and 5).
- ☐ Vacuum leak at the carburettor, throttle body or inlet manifold (Chapters 4A or 4B).
- ☐ Faulty carburettor (Chapter 4A).
- ☐ Fuel injection system fault (Chapter 4B).

Engine idles erratically
- ☐ Incorrectly-adjusted idle speed (Chapter 4A or 4B).
- ☐ Air filter element clogged (Chapter 1).
- ☐ Vacuum leak at the carburettor, throttle body, inlet manifold or associated hoses (Chapter 4A or 4B).
- ☐ Worn, faulty or incorrectly-gapped spark plugs (Chapter 1).
- ☐ Uneven or low cylinder compressions (Chapter 2A).
- ☐ Camshaft lobes worn (Chapter 2A).
- ☐ Faulty carburettor (Chapter 4A).
- ☐ Fuel injection system fault (Chapter 4B).

Engine misfires at idle speed
- ☐ Worn, faulty or incorrectly-gapped spark plugs (Chapter 1).
- ☐ Faulty spark plug HT leads (Chapter 1).
- ☐ Vacuum leak at the carburettor, throttle body, inlet manifold or associated hoses (Chapter 4A or 4B).
- ☐ Faulty carburettor (Chapter 4A).
- ☐ Fuel injection system fault (Chapter 4B).
- ☐ Distributor cap cracked or tracking internally (Chapter 1).
- ☐ Uneven or low cylinder compressions (Chapter 2A).
- ☐ Disconnected, leaking, or perished crankcase ventilation hoses (Chapter 4C).

Engine misfires throughout the driving speed range
- ☐ Fuel filter choked (Chapter 1).
- ☐ Fuel pump faulty, or delivery pressure low (Chapter 4A or 4B).
- ☐ Fuel tank vent blocked, or fuel pipes restricted (Chapter 4A or 4B).
- ☐ Vacuum leak at the carburettor, throttle body, inlet manifold or associated hoses (Chapter 4A or 4B).
- ☐ Worn, faulty or incorrectly-gapped spark plugs (Chapter 1).
- ☐ Faulty spark plug HT leads (Chapter 1).
- ☐ Distributor cap cracked or tracking internally (Chapter 1).
- ☐ Faulty ignition coil (Chapter 5).
- ☐ Uneven or low cylinder compressions (Chapter 2A).
- ☐ Faulty carburettor (Chapter 4A).
- ☐ Fuel injection system fault (Chapter 4B).

Fault Finding

Engine - continued

Engine hesitates on acceleration
☐ Worn, faulty or incorrectly-gapped spark plugs (Chapter 1).
☐ Vacuum leak at the carburettor, throttle body, inlet manifold or associated hoses (Chapter 4A or 4B).
☐ Faulty carburettor (Chapter 4A).
☐ Fuel injection system fault (Chapter 4B).

Engine stalls
☐ Vacuum leak at the carburettor, throttle body, inlet manifold or associated hoses (Chapter 4A or 4B).
☐ Fuel filter choked (Chapter 1).
☐ Fuel pump faulty, or delivery pressure low (Chapter 4A or 4B).
☐ Fuel tank vent blocked, or fuel pipes restricted (Chapter 4A or 4B).
☐ Faulty carburettor (Chapter 4A).
☐ Fuel injection system fault (Chapter 4B).

Engine lacks power
☐ Fuel filter choked (Chapter 1).
☐ Fuel pump faulty, or delivery pressure low (Chapter 4A or 4B).
☐ Uneven or low cylinder compressions (Chapter 2A).
☐ Worn, faulty or incorrectly-gapped spark plugs (Chapter 1).
☐ Vacuum leak at the carburettor, throttle body, inlet manifold or associated hoses (Chapter 4A or 4B).
☐ Faulty carburettor (Chapter 4A).
☐ Fuel injection system fault (Chapter 4B).
☐ Faulty turbocharger, where applicable (Chapter 4B).
☐ Brakes binding (Chapters 1 and 9).
☐ Clutch slipping (Chapter 6).

Engine backfires
☐ Vacuum leak at the carburettor, throttle body, inlet manifold or associated hoses (Chapter 4A or 4B).
☐ Faulty carburettor (Chapter 4A).
☐ Fuel injection system fault (Chapter 4B).

Oil pressure warning light illuminated with engine running
☐ Low oil level, or incorrect oil grade ("*Weekly Checks*").
☐ Faulty oil pressure sensor (Chapter 2A).
☐ Worn engine bearings and/or oil pump (Chapter 2A or 2B).
☐ Excessively high engine operating temperature (Chapter 3).
☐ Oil pressure relief valve defective (Chapter 2A).
☐ Oil pick-up strainer clogged (Chapter 2B).

Note: *Low oil pressure in a high-mileage engine at tickover is not necessarily a cause for concern. Sudden pressure loss at speed is far more significant. In any event, check the gauge or warning light sender before condemning the engine.*

Engine runs-on after switching off
☐ Excessive carbon build-up in engine (Chapter 2A or 2B).
☐ Excessively high engine operating temperature (Chapter 3).

Engine noises

Pre-ignition (pinking) or knocking during acceleration or under load
☐ Ignition timing incorrect/ignition system fault (Chapters 1 and 5).
☐ Incorrect grade of spark plug (Chapter 1).
☐ Incorrect grade of fuel (Chapter 1).
☐ Vacuum leak at carburettor, throttle body, inlet manifold or associated hoses (Chapter 4A or 4B).
☐ Excessive carbon build-up in engine (Chapter 2A or 2B).
☐ Faulty carburettor (Chapter 4A).
☐ Fuel injection system fault (Chapter 4B).

Whistling or wheezing noises
☐ Leaking inlet manifold or throttle body gasket (Chapter 4A or 4B).
☐ Leaking exhaust manifold gasket (Chapter 4A or 4B).
☐ Leaking vacuum hose (Chapters 4A, 4B and 9).
☐ Blowing cylinder head gasket (Chapter 2A).

Tapping or rattling noises
☐ Worn valve gear, timing chain, camshaft or hydraulic tappets (Chapter 2A).
☐ Incorrect valve clearances - B201 engine (Chapter 1)
☐ Ancillary component fault (water pump, alternator, etc) (Chapters 3, 5, etc).

Knocking or thumping noises
☐ Worn big-end bearings (regular heavy knocking, perhaps less under load) (Chapter 2B).
☐ Worn main bearings (rumbling and knocking, perhaps worsening under load) (Chapter 2B).
☐ Piston slap (most noticeable when cold) (Chapter 2B).
☐ Ancillary component fault (water pump, alternator, etc) (Chapters 3, 5, etc).

Fault Finding   REF•15

Cooling system

Overheating
- [ ] Auxiliary drivebelt broken - or incorrectly adjusted (Chapter 1).
- [ ] Insufficient coolant in system ("*Weekly Checks*").
- [ ] Thermostat faulty (Chapter 3).
- [ ] Radiator core blocked, or grille restricted (Chapter 3).
- [ ] Electric cooling fan or thermostatic switch faulty (Chapter 3).
- [ ] Pressure cap faulty (Chapter 3).
- [ ] Ignition timing incorrect, or ignition system fault (Chapters 1 and 5).
- [ ] Inaccurate temperature gauge sender unit (Chapter 3).
- [ ] Airlock in cooling system (Chapter 1).

Overcooling
- [ ] Thermostat faulty (Chapter 3).
- [ ] Inaccurate temperature gauge sender unit (Chapter 3).

External coolant leakage
- [ ] Deteriorated or damaged hoses or hose clips (Chapter 1).
- [ ] Radiator core or heater matrix leaking (Chapter 3).
- [ ] Pressure cap faulty (Chapter 3).
- [ ] Water pump internal seal leaking (Chapter 3).
- [ ] Water pump gasket leaking (Chapter 3).
- [ ] Boiling due to overheating (Chapter 3).
- [ ] Core plug leaking (Chapter 2B).

Internal coolant leakage
- [ ] Leaking cylinder head gasket (Chapter 2A).
- [ ] Cracked cylinder head or cylinder block (Chapter 2A or 2B).

Corrosion
- [ ] Infrequent draining and flushing (Chapter 1).
- [ ] Incorrect coolant mixture or inappropriate coolant type (Chapter 1).

Fuel and exhaust systems

Excessive fuel consumption
- [ ] Air filter element dirty or clogged (Chapter 1).
- [ ] Faulty carburettor (Chapter 4A).
- [ ] Fuel injection system fault (Chapter 4B).
- [ ] Ignition timing incorrect or ignition system fault (Chapters 1 and 5).
- [ ] Brakes binding (Chapter 9).
- [ ] Tyres under-inflated ("*Weekly Checks*").

Fuel leakage and/or fuel odour
- [ ] Damaged fuel tank, pipes or connections (Chapters 1 and 4).

Excessive noise or fumes from exhaust system
- [ ] Leaking exhaust system or manifold joints (Chapters 1 and 4).
- [ ] Leaking, corroded or damaged silencers or pipe (Chapters 1 and 4).
- [ ] Broken mountings causing body or suspension contact (Chapter 4A or 4B).

Clutch

Pedal travels to floor - no pressure or very little resistance
- [ ] Leak in clutch hydraulic system (Chapter 6).
- [ ] Faulty hydraulic master or slave cylinder (Chapter 6).
- [ ] Broken clutch release bearing (Chapter 6).
- [ ] Broken diaphragm spring in clutch pressure plate (Chapter 6).

Clutch fails to disengage (unable to select gears)
- [ ] Leak or air in clutch hydraulic system (Chapter 6).
- [ ] Faulty hydraulic master or slave cylinder (Chapter 6).
- [ ] Clutch disc sticking on splines (Chapter 6).
- [ ] Clutch disc sticking to flywheel or pressure plate (Chapter 6).
- [ ] Faulty pressure plate assembly (Chapter 6).
- [ ] Clutch release mechanism worn or poorly assembled (Chapter 6).

Clutch slips (engine speed increases, with no increase in vehicle speed)
- [ ] Clutch disc linings excessively worn (Chapter 6).
- [ ] Clutch disc linings contaminated with oil or grease (Chapter 6).
- [ ] Faulty pressure plate or weak diaphragm spring (Chapter 6).

Judder as clutch is engaged
- [ ] Clutch disc linings contaminated with oil or grease (Chapter 6).
- [ ] Clutch disc linings excessively worn (Chapter 6).
- [ ] Faulty or distorted pressure plate or diaphragm spring (Chapter 6).
- [ ] Worn or loose engine mountings (Chapter 2A).
- [ ] Clutch disc hub or shaft splines worn (Chapter 6).

Noise when depressing or releasing clutch pedal
- [ ] Worn clutch release bearing (Chapter 6).
- [ ] Worn or dry clutch pedal bushes (Chapter 6).
- [ ] Faulty pressure plate assembly (Chapter 6).
- [ ] Pressure plate diaphragm spring broken (Chapter 6).
- [ ] Broken clutch disc cushioning springs (Chapter 6).

Fault Finding

Manual transmission

Noisy in neutral with engine running
- [ ] Primary gears and bearings worn (noise apparent with clutch pedal released, but not when depressed) (Chapter 7A).*
- [ ] Clutch release bearing worn (noise apparent with clutch pedal depressed, possibly less when released) (Chapter 6).

Noisy in one particular gear
- [ ] Worn, damaged or chipped gear teeth (Chapter 7A).*

Difficulty engaging gears
- [ ] Clutch fault (Chapter 6).
- [ ] Worn or damaged gear linkage (Chapter 7A).
- [ ] Incorrectly-adjusted gear linkage (Chapter 7A).
- [ ] Worn synchroniser units (Chapter 7A).*
- [ ] Seized spigot bearing in the flywheel (Chapter 2A).

Jumps out of gear
- [ ] Worn or damaged gear linkage (Chapter 7A).
- [ ] Incorrectly-adjusted gear linkage (Chapter 7A).
- [ ] Worn synchroniser units (Chapter 7A).*
- [ ] Worn selector forks (Chapter 7A).*

Vibration
- [ ] Lack of oil (Chapter 1).
- [ ] Worn bearings (Chapter 7A).*

Lubricant leaks
- [ ] Leaking oil seal (Chapter 7A).
- [ ] Leaking housing joint (Chapter 7A).*

*Although the corrective action necessary to remedy the symptoms described is beyond the scope of the home mechanic, the above information should be helpful in isolating the cause of the condition, so that the owner can communicate clearly with a professional mechanic.

Automatic transmission

Note: *Due to the complexity of the automatic transmission, it is difficult for the home mechanic to properly diagnose and service this unit. For problems other than the following, the vehicle should be taken to a dealer service department or automatic transmission specialist.*

Fluid leakage
- [ ] Automatic transmission fluid is usually deep red in colour. Fluid leaks should not be confused with engine oil, which can easily be blown onto the transmission by air flow.
- [ ] To determine the source of a leak, first remove all built-up dirt and grime from the transmission housing and surrounding areas, using a degreasing agent or by steam-cleaning. Drive the vehicle at low speed, so that air flow will not blow the leak far from its source. Raise and support the vehicle, and determine where the leak is coming from. The following are common areas of leakage:
    a) Fluid pan (Chapter 7B).
    b) Dipstick tube (Chapter 1).
    c) Transmission-to-fluid cooler fluid pipes/unions (Chapter 7B).

Transmission fluid brown, or has burned smell
- [ ] Transmission fluid level low, or fluid in need of renewal (Chapter 1).

General gear selection problems
- [ ] The most likely cause of gear selection problems is a faulty or poorly-adjusted gear selector mechanism. The following are common problems associated with a faulty selector mechanism:
    a) Engine starting in gears other than Park or Neutral.
    b) Indicator on gear selector lever pointing to a gear other than the one actually being used.
    c) Vehicle moves when in Park or Neutral.
    d) Poor gear shift quality, or erratic gear changes.
- [ ] Refer any problems to a Saab dealer, or transmission specialist.

Transmission will not downshift (kickdown) with accelerator pedal fully depressed
- [ ] Low transmission fluid level (Chapter 1).
- [ ] Incorrect selector cable adjustment (Chapter 7B).

Engine will not start in any gear, or starts in gears other than Park or Neutral
- [ ] Incorrect starter inhibitor switch adjustment - where applicable (Chapter 7B).
- [ ] Incorrect selector cable adjustment (Chapter 7B).

Transmission slips, shifts roughly, is noisy, or has no drive in forward or reverse gears
- [ ] There are many probable causes for the above problems, but the home mechanic should be concerned with only one possibility - fluid level. Before taking the vehicle to a dealer or transmission specialist, check the fluid level and condition of the fluid (see Chapter 1). Correct the fluid level as necessary, or change the fluid and filter if needed. If the problem persists, professional help will be necessary.

Driveshafts

Clicking or knocking noise on turns (at slow speed on full-lock)
- [ ] Lack of constant velocity joint lubricant, possibly due to damaged gaiter (Chapter 8).
- [ ] Worn outer constant velocity joint (Chapter 8).

Vibration when accelerating or decelerating
- [ ] Worn inner constant velocity joint (Chapter 8).
- [ ] Bent or distorted driveshaft (Chapter 8).

Fault Finding REF•17

Braking system

Note: *Before assuming that a brake problem exists, make sure that the tyres are in good condition and correctly inflated, that the front wheel alignment is correct, and that the vehicle is not loaded with weight in an unequal manner. Apart from checking the condition of all pipe and hose connections, any faults occurring on the anti-lock braking system should be referred to a Saab dealer for diagnosis.*

Vehicle pulls to one side under braking
- [ ] Worn, defective, damaged or contaminated front or rear brake pads on one side (Chapters 1 and 9).
- [ ] Seized or partly-seized front or rear brake caliper piston (Chapter 9).
- [ ] A mixture of brake pad lining materials fitted between sides (Chapter 9).
- [ ] Brake caliper mounting bolts loose (Chapter 9).
- [ ] Worn or damaged steering or suspension components (Chapters 1 and 10).

Noise (grinding or high-pitched squeal) when brakes applied
- [ ] Brake pad friction lining material worn down to metal backing (Chapters 1 and 9).
- [ ] Excessive corrosion of brake disc - may be apparent after the vehicle has been standing for some time (Chapters 1 and 9).

Excessive brake pedal travel
- [ ] Faulty master cylinder (Chapter 9).
- [ ] Air in hydraulic system (Chapter 9).
- [ ] Faulty vacuum servo unit (Chapter 9).

Brake pedal feels spongy when depressed
- [ ] Air in hydraulic system (Chapter 9).

- [ ] Deteriorated flexible rubber brake hoses (Chapters 1 and 9).
- [ ] Master cylinder mountings loose (Chapter 9).
- [ ] Faulty master cylinder (Chapter 9).

Excessive brake pedal effort required to stop vehicle
- [ ] Faulty vacuum servo unit (Chapter 9).
- [ ] Disconnected, damaged or insecure brake servo vacuum hose (Chapters 1 and 9).
- [ ] Primary or secondary hydraulic circuit failure (Chapter 9).
- [ ] Seized brake caliper piston(s) (Chapter 9).
- [ ] Brake pads incorrectly fitted (Chapter 9).
- [ ] Incorrect grade of brake pads fitted (Chapter 9).
- [ ] Brake pads contaminated (Chapter 9).

Judder felt through brake pedal or steering wheel when braking
- [ ] Excessive run-out or distortion of brake disc(s) (Chapter 9).
- [ ] Brake pad linings worn (Chapters 1 and 9).
- [ ] Brake caliper mounting bolts loose (Chapter 9).
- [ ] Wear in suspension or steering components or mountings (Chapters 1 and 10).

Brakes binding
- [ ] Seized brake caliper piston(s) (Chapter 9).
- [ ] Incorrectly-adjusted handbrake mechanism (Chapter 9).
- [ ] Faulty master cylinder (Chapter 9).

Rear wheels locking under normal braking
- [ ] Seized brake caliper piston(s) (Chapter 9).
- [ ] Faulty brake pressure regulator (Chapter 9).

Suspension and steering systems

Note: *When tracing suspension or steering faults, be sure the trouble is not due to incorrect tyre pressures, mixtures of tyre types, or binding brakes.*

Vehicle pulls to one side
- [ ] Defective tyre ("*Weekly Checks*").
- [ ] Excessive wear in suspension or steering components (Chapters 1 and 10).
- [ ] Incorrect front wheel alignment (Chapter 10).
- [ ] Accident damage to steering or suspension components (Chapters 1 and 10).

Wheel wobble and vibration
- [ ] Front roadwheels out of balance (vibration felt mainly through the steering wheel) (Chapter 10).
- [ ] Rear roadwheels out of balance (vibration felt throughout the vehicle) (Chapter 10).
- [ ] Roadwheels damaged or distorted (Chapter 10).
- [ ] Faulty or damaged tyre ("*Weekly Checks*").
- [ ] Worn steering or suspension joints, bushes or components (Chapters 1 and 10).
- [ ] Wheel nuts loose (Chapter 10).

Excessive pitching and/or rolling around corners, or during braking
- [ ] Defective shock absorbers (Chapters 1 and 10).
- [ ] Broken or weak coil spring and/or suspension component (Chapters 1 and 10).
- [ ] Worn or damaged anti-roll bar or mountings (Chapter 10).

Wandering or general instability
- [ ] Incorrect front wheel alignment (Chapter 10).
- [ ] Worn steering or suspension joints, bushes or components (Chapters 1 and 10).
- [ ] Roadwheels out of balance (Chapter 10).
- [ ] Faulty or damaged tyre ("*Weekly Checks*").
- [ ] Wheel nuts loose (Chapter 10).
- [ ] Defective shock absorbers (Chapters 1 and 10).

Excessively-stiff steering
- [ ] Lack of steering gear lubricant (Chapter 10).
- [ ] Seized track rod end balljoint or suspension balljoint (Chapters 1 and 10).
- [ ] Broken or incorrectly adjusted auxiliary drivebelt (Chapter 1).
- [ ] Incorrect front wheel alignment (Chapter 10).
- [ ] Steering rack or column bent or damaged (Chapter 10).

Excessive play in steering
- [ ] Worn steering column universal joint(s) (Chapter 10).
- [ ] Worn steering track rod end balljoints (Chapters 1 and 10).
- [ ] Worn rack-and-pinion steering gear (Chapter 10).
- [ ] Worn steering or suspension joints, bushes or components (Chapters 1 and 10).

Lack of power assistance
- [ ] Broken or incorrectly-adjusted auxiliary drivebelt (Chapter 1).
- [ ] Incorrect power steering fluid level ("*Weekly Checks*").
- [ ] Restriction in power steering fluid hoses (Chapter 1).
- [ ] Faulty power steering pump (Chapter 10).
- [ ] Faulty rack-and-pinion steering gear (Chapter 10).

Fault Finding

Suspension and steering systems - continued

Tyre wear excessive

Tyres worn on inside or outside edges
- [ ] Tyres under-inflated (wear on both edges) ("*Weekly Checks*").
- [ ] Incorrect camber or castor angles (wear on one edge only) (Chapter 10).
- [ ] Worn steering or suspension joints, bushes or components (Chapters 1 and 10).
- [ ] Excessively-hard cornering.
- [ ] Accident damage.

Tyre treads exhibit feathered edges
- [ ] Incorrect toe setting (Chapter 10).

Tyres worn in centre of tread
- [ ] Tyres over-inflated ("*Weekly Checks*").

Tyres worn on inside and outside edges
- [ ] Tyres under-inflated ("*Weekly Checks*").
- [ ] Worn shock absorbers (Chapters 1 and 10).

Tyres worn unevenly
- [ ] Tyres out of balance ("*Weekly Checks*").
- [ ] Excessive wheel or tyre run-out (Chapter 1).
- [ ] Worn shock absorbers (Chapters 1 and 10).
- [ ] Faulty tyre ("*Weekly Checks*").

Electrical system

Note: *For problems associated with the starting system, refer to the faults listed under "Engine" earlier in this Section.*

Battery will only hold a charge for a few days
- [ ] Battery defective internally (Chapter 5).
- [ ] Battery electrolyte level low - where applicable ("*Weekly Checks*").
- [ ] Battery terminal connections loose or corroded (Chapter 1).
- [ ] Auxiliary drivebelt worn - or incorrectly adjusted (Chapter 1).
- [ ] Alternator not charging at correct output (Chapter 5).
- [ ] Alternator or voltage regulator faulty (Chapter 5).
- [ ] Short-circuit causing continual battery drain (Chapters 5 and 12).

Ignition/no-charge warning light remains illuminated with engine running
- [ ] Auxiliary drivebelt broken, worn, or incorrectly adjusted (Chapter 1).
- [ ] Alternator brushes worn, sticking, or dirty (Chapter 5).
- [ ] Alternator brush springs weak or broken (Chapter 5).
- [ ] Internal fault in alternator or voltage regulator (Chapter 5).
- [ ] Broken, disconnected, or loose wiring in charging circuit (Chapter 5).

Ignition/no-charge warning light fails to come on
- [ ] Warning light bulb blown (Chapter 12).
- [ ] Broken, disconnected, or loose wiring in warning light circuit (Chapter 12).
- [ ] Alternator faulty (Chapter 5).

Lights inoperative
- [ ] Bulb blown (Chapter 12).
- [ ] Corrosion of bulb or bulbholder contacts (Chapter 12).
- [ ] Blown fuse (Chapter 12).
- [ ] Faulty relay (Chapter 12).
- [ ] Broken, loose, or disconnected wiring (Chapter 12).
- [ ] Faulty switch (Chapter 12).

Instrument readings inaccurate or erratic

Instrument readings increase with engine speed
- [ ] Faulty voltage regulator (Chapter 12).

Fuel or temperature gauges give no reading
- [ ] Faulty gauge sender unit (Chapters 3 and 4A or 4B).
- [ ] Wiring open-circuit (Chapter 12).
- [ ] Faulty gauge (Chapter 12).

Fuel or temperature gauges give continuous maximum reading
- [ ] Faulty gauge sender unit (Chapters 3 and 4A or 4B).
- [ ] Wiring short-circuit (Chapter 12).
- [ ] Faulty gauge (Chapter 12).

Horn inoperative, or unsatisfactory in operation

Horn operates all the time
- [ ] Horn contacts permanently bridged or horn push stuck down (Chapter 12).

Horn fails to operate
- [ ] Blown fuse (Chapter 12).
- [ ] Cable or cable connections loose, broken or disconnected (Chapter 12).
- [ ] Faulty horn (Chapter 12).

Horn emits intermittent or unsatisfactory sound
- [ ] Cable connections loose (Chapter 12).
- [ ] Horn mountings loose (Chapter 12).
- [ ] Faulty horn (Chapter 12).

Windscreen/tailgate wipers inoperative, or unsatisfactory in operation

Wipers fail to operate, or operate very slowly
- [ ] Wiper blades stuck to screen, or linkage seized or binding ("*Weekly Checks*" and chapter 12).
- [ ] Blown fuse (Chapter 12).
- [ ] Cable or cable connections loose, broken or disconnected (Chapter 12).
- [ ] Faulty relay (Chapter 12).
- [ ] Faulty wiper motor (Chapter 12).

Wiper blades sweep over too large or too small an area of the glass
- [ ] Wiper arms incorrectly positioned on spindles ("*Weekly Checks*").
- [ ] Excessive wear of wiper linkage (Chapter 12).
- [ ] Wiper motor or linkage mountings loose or insecure (Chapter 12).

Wiper blades fail to clean the glass effectively
- [ ] Wiper blade rubbers worn or perished ("*Weekly Checks*").
- [ ] Wiper armtension springs broken, or armpivots seized (Chapter 12).
- [ ] Insufficient windscreen washer additive to adequately remove road film ("*Weekly Checks*").

Fault Finding   REF•19

Electrical system - continued

Windscreen/tailgate washers inoperative, or unsatisfactory in operation

One or more washer jets inoperative
- [ ] Blocked washer jet ("*Weekly Checks*").
- [ ] Disconnected, kinked or restricted fluid hose (Chapter 12).
- [ ] Insufficient fluid in washer reservoir ("*Weekly Checks*").

Washer pump fails to operate
- [ ] Broken or disconnected wiring or connections (Chapter 12).
- [ ] Blown fuse (Chapter 12).
- [ ] Faulty washer switch (Chapter 12).
- [ ] Faulty washer pump (Chapter 12).

Washer pump runs for some time before fluid is emitted from jets
- [ ] Faulty one-way valve in fluid supply hose (Chapter 12).

Electric windows inoperative, or unsatisfactory in operation

Window glass will only move in one direction
- [ ] Faulty switch (Chapter 12).

Window glass slow to move
- [ ] Regulator seized or damaged, or in need of lubrication (Chapter 11).
- [ ] Door internal components or trim fouling regulator (Chapter 11).
- [ ] Faulty motor (Chapter 11).

Window glass fails to move
- [ ] Blown fuse (Chapter 12).
- [ ] Faulty relay (Chapter 12).
- [ ] Broken or disconnected wiring or connections (Chapter 12).
- [ ] Faulty motor (Chapter 11).

Central locking system inoperative, or unsatisfactory in operation

Complete system failure
- [ ] Blown fuse (Chapter 12).
- [ ] Faulty relay (Chapter 12).
- [ ] Broken or disconnected wiring or connections (Chapter 12).

Latch locks but will not unlock, or unlocks but will not lock
- [ ] Faulty switch (Chapter 12).
- [ ] Broken or disconnected latch operating rods or levers (Chapter 11).
- [ ] Faulty relay (Chapter 12).

One solenoid/motor fails to operate
- [ ] Broken or disconnected wiring or connections (Chapter 12).
- [ ] Faulty solenoid/motor (Chapter 11).
- [ ] Broken, binding or disconnected latch operating rods or levers (Chapter 11).
- [ ] Fault in door latch (Chapter 11).

Glossary of Technical Terms

A

ABS (Anti-lock brake system) A system, usually electronically controlled, that senses incipient wheel lockup during braking and relieves hydraulic pressure at wheels that are about to skid.

Air bag An inflatable bag hidden in the steering wheel (driver's side) or the dash or glovebox (passenger side). In a head-on collision, the bags inflate, preventing the driver and front passenger from being thrown forward into the steering wheel or windscreen.

Air cleaner A metal or plastic housing, containing a filter element, which removes dust and dirt from the air being drawn into the engine.

Air filter element The actual filter in an air cleaner system, usually manufactured from pleated paper and requiring renewal at regular intervals.

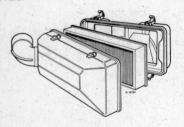

Air filter

Allen key A hexagonal wrench which fits into a recessed hexagonal hole.

Alligator clip A long-nosed spring-loaded metal clip with meshing teeth. Used to make temporary electrical connections.

Alternator A component in the electrical system which converts mechanical energy from a drivebelt into electrical energy to charge the battery and to operate the starting system, ignition system and electrical accessories.

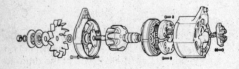

Alternator (exploded view)

Ampere (amp) A unit of measurement for the flow of electric current. One amp is the amount of current produced by one volt acting through a resistance of one ohm.

Anaerobic sealer A substance used to prevent bolts and screws from loosening. Anaerobic means that it does not require oxygen for activation. The Loctite brand is widely used.

Antifreeze A substance (usually ethylene glycol) mixed with water, and added to a vehicle's cooling system, to prevent freezing of the coolant in winter. Antifreeze also contains chemicals to inhibit corrosion and the formation of rust and other deposits that would tend to clog the radiator and coolant passages and reduce cooling efficiency.

Anti-seize compound A coating that reduces the risk of seizing on fasteners that are subjected to high temperatures, such as exhaust manifold bolts and nuts.

Anti-seize compound

Asbestos A natural fibrous mineral with great heat resistance, commonly used in the composition of brake friction materials. Asbestos is a health hazard and the dust created by brake systems should never be inhaled or ingested.

Axle A shaft on which a wheel revolves, or which revolves with a wheel. Also, a solid beam that connects the two wheels at one end of the vehicle. An axle which also transmits power to the wheels is known as a live axle.

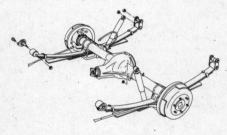

Axle assembly

Axleshaft A single rotating shaft, on either side of the differential, which delivers power from the final drive assembly to the drive wheels. Also called a driveshaft or a halfshaft.

B

Ball bearing An anti-friction bearing consisting of a hardened inner and outer race with hardened steel balls between two races.

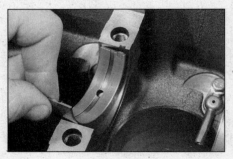

Bearing

Bearing The curved surface on a shaft or in a bore, or the part assembled into either, that permits relative motion between them with minimum wear and friction.

Big-end bearing The bearing in the end of the connecting rod that's attached to the crankshaft.

Bleed nipple A valve on a brake wheel cylinder, caliper or other hydraulic component that is opened to purge the hydraulic system of air. Also called a bleed screw.

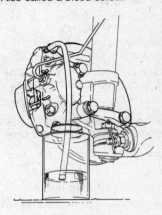

Brake bleeding

Brake bleeding Procedure for removing air from lines of a hydraulic brake system.

Brake disc The component of a disc brake that rotates with the wheels.

Brake drum The component of a drum brake that rotates with the wheels.

Brake linings The friction material which contacts the brake disc or drum to retard the vehicle's speed. The linings are bonded or riveted to the brake pads or shoes.

Brake pads The replaceable friction pads that pinch the brake disc when the brakes are applied. Brake pads consist of a friction material bonded or riveted to a rigid backing plate.

Brake shoe The crescent-shaped carrier to which the brake linings are mounted and which forces the lining against the rotating drum during braking.

Braking systems For more information on braking systems, consult the *Haynes Automotive Brake Manual*.

Breaker bar A long socket wrench handle providing greater leverage.

Bulkhead The insulated partition between the engine and the passenger compartment.

C

Caliper The non-rotating part of a disc-brake assembly that straddles the disc and carries the brake pads. The caliper also contains the hydraulic components that cause the pads to pinch the disc when the brakes are applied. A caliper is also a measuring tool that can be set to measure inside or outside dimensions of an object.

Glossary of Technical Terms

Camshaft A rotating shaft on which a series of cam lobes operate the valve mechanisms. The camshaft may be driven by gears, by sprockets and chain or by sprockets and a belt.

Canister A container in an evaporative emission control system; contains activated charcoal granules to trap vapours from the fuel system.

Canister

Carburettor A device which mixes fuel with air in the proper proportions to provide a desired power output from a spark ignition internal combustion engine.

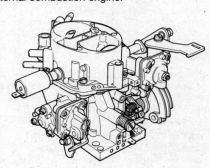
Carburettor

Castellated Resembling the parapets along the top of a castle wall. For example, a castellated balljoint stud nut.

Castellated nut

Castor In wheel alignment, the backward or forward tilt of the steering axis. Castor is positive when the steering axis is inclined rearward at the top.

Catalytic converter A silencer-like device in the exhaust system which converts certain pollutants in the exhaust gases into less harmful substances.

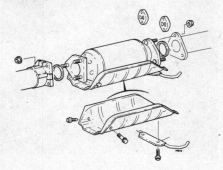
Catalytic converter

Circlip A ring-shaped clip used to prevent endwise movement of cylindrical parts and shafts. An internal circlip is installed in a groove in a housing; an external circlip fits into a groove on the outside of a cylindrical piece such as a shaft.

Clearance The amount of space between two parts. For example, between a piston and a cylinder, between a bearing and a journal, etc.

Coil spring A spiral of elastic steel found in various sizes throughout a vehicle, for example as a springing medium in the suspension and in the valve train.

Compression Reduction in volume, and increase in pressure and temperature, of a gas, caused by squeezing it into a smaller space.

Compression ratio The relationship between cylinder volume when the piston is at top dead centre and cylinder volume when the piston is at bottom dead centre.

Constant velocity (CV) joint A type of universal joint that cancels out vibrations caused by driving power being transmitted through an angle.

Core plug A disc or cup-shaped metal device inserted in a hole in a casting through which core was removed when the casting was formed. Also known as a freeze plug or expansion plug.

Crankcase The lower part of the engine block in which the crankshaft rotates.

Crankshaft The main rotating member, or shaft, running the length of the crankcase, with offset "throws" to which the connecting rods are attached.

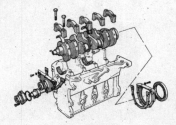
Crankshaft assembly

Crocodile clip See Alligator clip

D

Diagnostic code Code numbers obtained by accessing the diagnostic mode of an engine management computer. This code can be used to determine the area in the system where a malfunction may be located.

Disc brake A brake design incorporating a rotating disc onto which brake pads are squeezed. The resulting friction converts the energy of a moving vehicle into heat.

Double-overhead cam (DOHC) An engine that uses two overhead camshafts, usually one for the intake valves and one for the exhaust valves.

Drivebelt(s) The belt(s) used to drive accessories such as the alternator, water pump, power steering pump, air conditioning compressor, etc. off the crankshaft pulley.

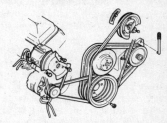
Accessory drivebelts

Driveshaft Any shaft used to transmit motion. Commonly used when referring to the axleshafts on a front wheel drive vehicle.

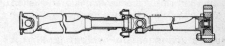
Driveshaft

Drum brake A type of brake using a drum-shaped metal cylinder attached to the inner surface of the wheel. When the brake pedal is pressed, curved brake shoes with friction linings press against the inside of the drum to slow or stop the vehicle.

Drum brake assembly

Glossary of Technical Terms

E

EGR valve A valve used to introduce exhaust gases into the intake air stream.

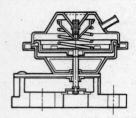
EGR valve

Electronic control unit (ECU) A computer which controls (for instance) ignition and fuel injection systems, or an anti-lock braking system. For more information refer to the Haynes Automotive Electrical and Electronic Systems Manual.

Electronic Fuel Injection (EFI) A computer controlled fuel system that distributes fuel through an injector located in each intake port of the engine.

Emergency brake A braking system, independent of the main hydraulic system, that can be used to slow or stop the vehicle if the primary brakes fail, or to hold the vehicle stationary even though the brake pedal isn't depressed. It usually consists of a hand lever that actuates either front or rear brakes mechanically through a series of cables and linkages. Also known as a handbrake or parking brake.

Endfloat The amount of lengthwise movement between two parts. As applied to a crankshaft, the distance that the crankshaft can move forward and back in the cylinder block.

Engine management system (EMS) A computer controlled system which manages the fuel injection and the ignition systems in an integrated fashion.

Exhaust manifold A part with several passages through which exhaust gases leave the engine combustion chambers and enter the exhaust pipe.

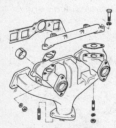
Exhaust manifold

F

Fan clutch A viscous (fluid) drive coupling device which permits variable engine fan speeds in relation to engine speeds.

Feeler blade A thin strip or blade of hardened steel, ground to an exact thickness, used to check or measure clearances between parts.

Feeler blade

Firing order The order in which the engine cylinders fire, or deliver their power strokes, beginning with the number one cylinder.

Flywheel A heavy spinning wheel in which energy is absorbed and stored by means of momentum. On cars, the flywheel is attached to the crankshaft to smooth out firing impulses.

Free play The amount of travel before any action takes place. The "looseness" in a linkage, or an assembly of parts, between the initial application of force and actual movement. For example, the distance the brake pedal moves before the pistons in the master cylinder are actuated.

Fuse An electrical device which protects a circuit against accidental overload. The typical fuse contains a soft piece of metal which is calibrated to melt at a predetermined current flow (expressed as amps) and break the circuit.

Fusible link A circuit protection device consisting of a conductor surrounded by heat-resistant insulation. The conductor is smaller than the wire it protects, so it acts as the weakest link in the circuit. Unlike a blown fuse, a failed fusible link must frequently be cut from the wire for replacement.

G

Gap The distance the spark must travel in jumping from the centre electrode to the side electrode in a spark plug. Also refers to the spacing between the points in a contact breaker assembly in a conventional points-type ignition, or to the distance between the reluctor or rotor and the pickup coil in an electronic ignition.

Gasket Any thin, soft material - usually cork, cardboard, asbestos or soft metal - installed between two metal surfaces to ensure a good seal. For instance, the cylinder head gasket seals the joint between the block and the cylinder head.

Gasket

Gauge An instrument panel display used to monitor engine conditions. A gauge with a movable pointer on a dial or a fixed scale is an analogue gauge. A gauge with a numerical readout is called a digital gauge.

H

Halfshaft A rotating shaft that transmits power from the final drive unit to a drive wheel, usually when referring to a live rear axle.

Harmonic balancer A device designed to reduce torsion or twisting vibration in the crankshaft. May be incorporated in the crankshaft pulley. Also known as a vibration damper.

Hone An abrasive tool for correcting small irregularities or differences in diameter in an engine cylinder, brake cylinder, etc.

Hydraulic tappet A tappet that utilises hydraulic pressure from the engine's lubrication system to maintain zero clearance (constant contact with both camshaft and valve stem). Automatically adjusts to variation in valve stem length. Hydraulic tappets also reduce valve noise.

I

Ignition timing The moment at which the spark plug fires, usually expressed in the number of crankshaft degrees before the piston reaches the top of its stroke.

Inlet manifold A tube or housing with passages through which flows the air-fuel mixture (carburettor vehicles and vehicles with throttle body injection) or air only (port fuel-injected vehicles) to the port openings in the cylinder head.

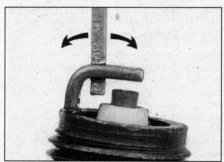
Adjusting spark plug gap

Glossary of Technical Terms

J
Jump start Starting the engine of a vehicle with a discharged or weak battery by attaching jump leads from the weak battery to a charged or helper battery.

L
Load Sensing Proportioning Valve (LSPV) A brake hydraulic system control valve that works like a proportioning valve, but also takes into consideration the amount of weight carried by the rear axle.
Locknut A nut used to lock an adjustment nut, or other threaded component, in place. For example, a locknut is employed to keep the adjusting nut on the rocker arm in position.
Lockwasher A form of washer designed to prevent an attaching nut from working loose.

M
MacPherson strut A type of front suspension system devised by Earle MacPherson at Ford of England. In its original form, a simple lateral link with the anti-roll bar creates the lower control arm. A long strut - an integral coil spring and shock absorber - is mounted between the body and the steering knuckle. Many modern so-called MacPherson strut systems use a conventional lower A-arm and don't rely on the anti-roll bar for location.
Multimeter An electrical test instrument with the capability to measure voltage, current and resistance.

N
NOx Oxides of Nitrogen. A common toxic pollutant emitted by petrol and diesel engines at higher temperatures.

O
Ohm The unit of electrical resistance. One volt applied to a resistance of one ohm will produce a current of one amp.
Ohmmeter An instrument for measuring electrical resistance.
O-ring A type of sealing ring made of a special rubber-like material; in use, the O-ring is compressed into a groove to provide the sealing action.

O-ring

Overhead cam (ohc) engine An engine with the camshaft(s) located on top of the cylinder head(s).
Overhead valve (ohv) engine An engine with the valves located in the cylinder head, but with the camshaft located in the engine block.
Oxygen sensor A device installed in the engine exhaust manifold, which senses the oxygen content in the exhaust and converts this information into an electric current. Also called a Lambda sensor.

P
Phillips screw A type of screw head having a cross instead of a slot for a corresponding type of screwdriver.
Plastigage A thin strip of plastic thread, available in different sizes, used for measuring clearances. For example, a strip of Plastigage is laid across a bearing journal. The parts are assembled and dismantled; the width of the crushed strip indicates the clearance between journal and bearing.

Plastigage

Propeller shaft The long hollow tube with universal joints at both ends that carries power from the transmission to the differential on front-engined rear wheel drive vehicles.
Proportioning valve A hydraulic control valve which limits the amount of pressure to the rear brakes during panic stops to prevent wheel lock-up.

R
Rack-and-pinion steering A steering system with a pinion gear on the end of the steering shaft that mates with a rack (think of a geared wheel opened up and laid flat). When the steering wheel is turned, the pinion turns, moving the rack to the left or right. This movement is transmitted through the track rods to the steering arms at the wheels.
Radiator A liquid-to-air heat transfer device designed to reduce the temperature of the coolant in an internal combustion engine cooling system.
Refrigerant Any substance used as a heat transfer agent in an air-conditioning system. R-12 has been the principle refrigerant for many years; recently, however, manufacturers have begun using R-134a, a non-CFC substance that is considered less harmful to the ozone in the upper atmosphere.
Rocker arm A lever arm that rocks on a shaft or pivots on a stud. In an overhead valve engine, the rocker arm converts the upward movement of the pushrod into a downward movement to open a valve.
Rotor In a distributor, the rotating device inside the cap that connects the centre electrode and the outer terminals as it turns, distributing the high voltage from the coil secondary winding to the proper spark plug. Also, that part of an alternator which rotates inside the stator. Also, the rotating assembly of a turbocharger, including the compressor wheel, shaft and turbine wheel.
Runout The amount of wobble (in-and-out movement) of a gear or wheel as it's rotated. The amount a shaft rotates "out-of-true." The out-of-round condition of a rotating part.

S
Sealant A liquid or paste used to prevent leakage at a joint. Sometimes used in conjunction with a gasket.
Sealed beam lamp An older headlight design which integrates the reflector, lens and filaments into a hermetically-sealed one-piece unit. When a filament burns out or the lens cracks, the entire unit is simply replaced.
Serpentine drivebelt A single, long, wide accessory drivebelt that's used on some newer vehicles to drive all the accessories, instead of a series of smaller, shorter belts. Serpentine drivebelts are usually tensioned by an automatic tensioner.

Serpentine drivebelt

Shim Thin spacer, commonly used to adjust the clearance or relative positions between two parts. For example, shims inserted into or under bucket tappets control valve clearances. Clearance is adjusted by changing the thickness of the shim.
Slide hammer A special puller that screws into or hooks onto a component such as a shaft or bearing; a heavy sliding handle on the shaft bottoms against the end of the shaft to knock the component free.
Sprocket A tooth or projection on the periphery of a wheel, shaped to engage with a chain or drivebelt. Commonly used to refer to the sprocket wheel itself.
Starter inhibitor switch On vehicles with an

Glossary of Technical Terms

automatic transmission, a switch that prevents starting if the vehicle is not in Neutral or Park.
Strut See MacPherson strut.

T

Tappet A cylindrical component which transmits motion from the cam to the valve stem, either directly or via a pushrod and rocker arm. Also called a cam follower.
Thermostat A heat-controlled valve that regulates the flow of coolant between the cylinder block and the radiator, so maintaining optimum engine operating temperature. A thermostat is also used in some air cleaners in which the temperature is regulated.
Thrust bearing The bearing in the clutch assembly that is moved in to the release levers by clutch pedal action to disengage the clutch. Also referred to as a release bearing.
Timing belt A toothed belt which drives the camshaft. Serious engine damage may result if it breaks in service.
Timing chain A chain which drives the camshaft.
Toe-in The amount the front wheels are closer together at the front than at the rear. On rear wheel drive vehicles, a slight amount of toe-in is usually specified to keep the front wheels running parallel on the road by offsetting other forces that tend to spread the wheels apart.
Toe-out The amount the front wheels are closer together at the rear than at the front. On front wheel drive vehicles, a slight amount of toe-out is usually specified.
Tools For full information on choosing and using tools, refer to the *Haynes Automotive Tools Manual*.
Tracer A stripe of a second colour applied to a wire insulator to distinguish that wire from another one with the same colour insulator.
Tune-up A process of accurate and careful adjustments and parts replacement to obtain the best possible engine performance.
Turbocharger A centrifugal device, driven by exhaust gases, that pressurises the intake air. Normally used to increase the power output from a given engine displacement, but can also be used primarily to reduce exhaust emissions (as on VW's "Umwelt" Diesel engine).

U

Universal joint or U-joint A double-pivoted connection for transmitting power from a driving to a driven shaft through an angle. A U-joint consists of two Y-shaped yokes and a cross-shaped member called the spider.

V

Valve A device through which the flow of liquid, gas, vacuum, or loose material in bulk may be started, stopped, or regulated by a movable part that opens, shuts, or partially obstructs one or more ports or passageways. A valve is also the movable part of such a device.
Valve clearance The clearance between the valve tip (the end of the valve stem) and the rocker arm or tappet. The valve clearance is measured when the valve is closed.
Vernier caliper A precision measuring instrument that measures inside and outside dimensions. Not quite as accurate as a micrometer, but more convenient.
Viscosity The thickness of a liquid or its resistance to flow.
Volt A unit for expressing electrical "pressure" in a circuit. One volt that will produce a current of one ampere through a resistance of one ohm.

W

Welding Various processes used to join metal items by heating the areas to be joined to a molten state and fusing them together. For more information refer to the *Haynes Automotive Welding Manual*.
Wiring diagram A drawing portraying the components and wires in a vehicle's electrical system, using standardised symbols. For more information refer to the *Haynes Automotive Electrical and Electronic Systems Manual*.

Index

Note: References throughout this index are in the form - "Chapter number" • "Page number"

A

ABS system - 9•2, 9•18, 9•19
Accelerator cable - 4A•3, 4B•4
Accelerator pedal - 4A•3, 4B•5
Acknowledgements - 0•4
Aerial - 12•11
AIC valve - 4B•14
Air cleaner - 1A•9, 1B•10, 4A•2, 4B•4
Air conditioning system - 3•8
Airflow meter - 4B•12, 4B•14
Alarm - 12•11, REF•5
Alternator - 5•4
Anti-roll bars - 10•5, 10•8
Anti-theft alarm - 12•11, REF•5
Antifreeze - 0•16, 1A•8, 1A•20, 1A•22, 1B•19, 1B•20
ATF - 0•16, 1A•17, 1B•8, 7B•1,
Automatic transmission - 7B•1 et seq
   fault diagnosis - REF•16
   fluid - 0•16, 1A•17, 1B•8, 7B•1
   kickdown cable - 7B•2
   selector cable - 7B•2
   selector lever and housing - 7B–3
   start inhibitor - 7B–4
Auxiliary air valve - 4B•12, 4B•14
Auxiliary drivebelt - 1A•13, 1B•14
Axle tube (rear) - 10•9

B

Badges - 11•12
Battery - 0•6, 0•15, 1A•8, 5•3, 5•4
Beam alignment - 1A•16, 1B•16, 12•8
Bearings (engine) - 2B•22, 2B•24
Bleeding the brakes - 9•3
Bleeding the clutch - 6•4
Blower motor (ventilation system) - 3•6
Body damage - 11•2, 11•3
Body electrical systems - 12•1 et seq
Bodywork and fittings - 11•1 et seq
Bodywork trim and mouldings - 11•12
Bonnet/lock/cable - 11•4
Boot lid /lock- 11•8, 11•9
Bosch CIS (K-Jetronic) injection system - 4B•5
Bosch LH-Jetronic injection system - 4B•6
Braking system - 9•1 et seq
   ABS - 9•2, 9•18, 9•19
   bleeding the brakes - 9•3
   brake hydraulic lines and hoses - 9•13
   brake pads - 1A•7, 1B•8, 9•5, 9•7
   brake disc - 9•11
   brake caliper - 9•8, 9•10
   brake fluid - 0•12, 0•16, 1A•20, 1B•18
   brake lights - 12•7, 12•8
   bake light switch - 9•18
   brake pedal - 9•17
   checks - 1A•17, 1B•8, 1B•16, 1B•18
   fault diagnosis - REF•17
   handbrake - 9•15 to 9•17
   master cylinder - 9•13
   road test - 1A•19, 1B•9
   vacuum servo unit - 9•14
Brushes (alternator) - 5•5
Bulbs - 12•1, 12•6, 12•8
Bumpers - 11•3, 11•4
Buying spare parts and vehicle identification numbers - REF•3

C

Cables
   accelerator - 4A•3, 4B•4
   bonnet release - 11•4
   choke - 4A•5
   handbrake - 9•15
   kickdown - 7B•2
   selector (automatic transmission) - 7B•2
   speedometer - 12•9
Caliper (brake) - 9•8, 9•10
Cam followers - 2A•5
Camber - 10•2, 10•12
Camshaft(s) - 2A•5
Capacities - 0•16, 1A•22, 1B•20
Carburettor - Pierburg - 4A•5, 4A•7
Carburettor - Zenith/Solex - 4A•5, 4A•6
Carpets - 11•2
Castor - 10•2, 10•12
Catalytic converter - 4C•2, 4C•4
Central locking - 11•10
Centre console - 11•13
Charging system - 5•4
Choke cable - 4A•5
Cigarette lighter - 12•9
Clutch - 6•1 et seq
   bleeding - 6•4
   fluid - 0•12, 0•16,
   fault diagnosis - REF•15
   hydraulic system - 6•4
   inspection - 6•6
   lining check - 6•5
   master cylinder - 6•2
   pedal - 6•1
   release bearing - 6•3
   slave cylinder - 6•3
Coil - 5•9
Coil spring (suspension) - 10•1, 10•6, 10•8
Compression test - 2A•3
Condenser - 5•11
Connecting rods - 2B•18, 2B•20, 2B•25
Contact breaker points - 1A•14
Contents - 0•2
Conversion factors - REF•2
Coolant - 0•11, 0•16, 1A•20, 1B•19, 1B•20
Cooling, heating and ventilation systems - 3•1 et seq
Cooling system
   air conditioning - 3•8
   blower motor (ventilation system) - 3•6
   fan (radiator) - 3•4
   fault diagnosis - REF•15
   heating/ventilation components- 3•6
   hoses - 3•2
   matrix - 3•7
   radiator - 3•2
   switches - 3•4
   thermostat - 3•3
   water pump - 3•5
Courtesy light switch - 12•6
Crankshaft - 2B•19, 2B•21, 2B•24
Crankshaft oil seals - 2A•16, 2B•15
Cruise control - 4B•22
Cylinder block/crankcase - 2B•19
Cylinder head - 2A•4, 2A•9, 2B•15, 2B•16, 2B•18

D

Deceleration device - 1A•12, 1B•13
Dents - 11•2, 11•3
Dimensions and weights - REF•1
Disc (brake) - 9•11
Distributor - 1A•14, 1B•15, 5•9, 5•10
Door and trim panels - 11•5
Drivebelt - 1A•13, 1B•14
Driveplate - 2B•15
Driveshafts - 8•1 et seq
   fault diagnosis - REF•16
   gaiter - 8•2
   inner joint - 8•3
   outer joint - 8•1
   test - 1A•19, 1B•9

E

Earth fault (finding) - 12•4
Electric mirror - 11•10
Electric window/switches - 11•10, 12•6
Electrical fault finding - 12•3
Electrical system (body) - 12•1 et seq
Electrical system check - 0•14
Electrical system fault diagnosis - REF•18
Electronic control unit - 4B•13
Emissions control - 4C•2
Engine (in-car repair) - 2A•1 et seq
Engine (removal and overhaul) - 2B•1 et seq
   cam followers - 2A•5
   camshaft(s) - 2A•5
   compartment - 1A•3, 1B•3
   compression test - 2A•3
   crankshaft - 2B•19, 2B•21, 2B•24
   crankshaft oil seals -2A•16, 2B•15
   cylinder block/crankcase - 2B•19
   cylinder head -2A•4, 2A•9, 2B•15, 2B•16, 2B•18
   dismantling 2B•10
   driveplate - 2B•15
   electrical systems - 5•1 et seq
   fault diagnosis - REF•13
   flywheel - 2A•17
   idler shaft - 2B•15
   in car repair - 2A•1 et seq
   main and big-end bearings - 2B•22, 2B•24
   mountings (engine/transmission) - 2A•17
   oil and filter renewal - 0•16, 1A•5, 1A•9, 1B•5, 1B•20
   oil cooler - 2A•16
   oil level - 0•11
   oil pressure warning light switch - 2A•16
   oil pump - 2A•13
   overhaul - 2B•3, 2B•23
   piston rings - 2B•23

Index

piston/connecting rod assemblies - 2B•18, 2B•20, 2B•25
removal and overhaul - 2B•1 et seq
timing chain/sprockets - 2B•10
top dead centre (locating) - 2A•4
valves - 2B•16
Engine/transmission mountings - 2A•17
Engine/transmission removal - 2B•4
Environmental considerations - REF•4
Exhaust manifold - 4A•8, 4B•25
Exhaust system - 1A•8, 1B•14, 4A•8, 4B•26
Exterior mirror - 11•10

F

Facia assembly - 11•14
Fan (radiator) - 3•4
Fault diagnosis - REF•12 et seq
  automatic transmission - REF•16
  braking system - REF•17
  clutch - REF•15
  cooling system - REF•15
  driveshafts - REF•16
  electrical system - REF•18
  engine - REF•13
  fuel and exhaust systems - REF•15
  manual transmission - REF•16
  steering - REF•17
  suspension - REF•17
Fluids - 0•16, 1A•22, 1B•8, 1B•18
Flushing (cooling system) - 1A•21, 1B•19
Flywheel - 2A•17
Fog light switch - 12•5
Front suspension wishbones - 10•4
Fuel and exhaust (carburettor) systems - 4A•1 et seq
Fuel and exhaust (fuel injection) systems - 4B•1 et seq
Fuel system
  accelerator cable - 4A•3, 4B•4
  accelerator pedal - 4A•3, 4B•5
  AIC valve - 4B•14
  air cleaner - 1A•9, 1B•10, 4A•2, 4B•4
  airflow meter - 4B•12, 4B•14
  auxiliary air valve - 4B•12, 4B•14
  Bosch CIS (K-Jetronic) fuel injection - 4B•5
  Bosch LH Jetronic fuel injection - 4B•6
  carburettor systems - 4A•1 et seq
  catalytic converter - 4C•2, 4C•4
  choke cable - 4A•5
  cruise control - 4B•22
  electronic control unit - 4B•13
  emissions control - 4C•2
  exhaust manifold - 4A•8, 4B•25
  exhaust system - 4A•8, 4B•26
  fault diagnosis - REF•15
  fuel accumulator - 4B•12
  fuel filter - 1A•6, 1B•17
  fuel gauge sender unit - 4A•4, 4B•9
  fuel pump - 4A•3, 4B•7
  fuel injection systems - 4B•1 et seq
  fuel injector(s) - 4B•12
  fuel pressure regulator - 4B•12, 4B•14
  fuel pump relay - 4B•9
  fuel supply rail/injectors - 4B•14
  fuel tank - 4A•4, 4B•10
  injector(s) - 4B•12
  inlet manifold - 4A•7, 4B•23
  intercooler - 4B•22
  Lucas CU14 fuel injection - 4B•7
  manifolds - 4A•7, 4A•8, 4B•23, 4B•25
  mixture (CO) - 1A•10, 1B•10
  Pierburg carburettor - 4A•5
  throttle cable - 4A•3, 4B•4
  throttle position switch - 4B•13, 4B•15
  turbocharger - 4B•15
  unleaded petrol - 4A•3, 4B•5
  warm-up regulator - 4B•12
  Zenith/Solex carburettor - 4A•5
Fuel tank - 4A•4, 4B•10
Fuses - 12•2, 12•3, 12•4

G

Gear lever lock - 7A•5
Gearbox - see Manual transmission or Automatic transmission
Gearbox oil - 0•16, 7A•1, 7A•6
Gearchange lever and housing - 7A•3
Gearchange linkage - 7A•4
Glass - 11•8
Glossary of technical terms - REF•20

H

Handbrake adjustment - 9•15
Handbrake - 1A•17, 1B•8, 9•17, 9•15
Handbrake warning light switch - 9•17
Handles (door) - 11•6
Hazard warning light switch - 12•5
Headlight - 12•7, 12•8
Headlight beam alignment - 1A•16, 1B•16, 12•8
Heated rear window switch - 12•5
Heated seat - 12•11
Heater matrix - 3•7
Heating systems - 3•1 et seq
Heating/ventilation components - 3•6
Hinges and locks - 1A•9
Horn and switch - 12•9
Hose and fluid leak check - 1A•10, 1B•6
Hoses (cooling system) - 3•2
Hoses brake - 9•13
HT Coil - 5•9
HT leads - 1A•14, 1B•15
Hub assembly - 10•5, 10•9
Hydraulic fluid - 0•16
Hydraulic lines/hoses (braking system) - 9•13
Hydraulic system (brakes) bleeding - 9•3
Hydraulic system (clutch) - 6•4

I

Idle speed and mixture - 1A•10, 1B•10
Idler shaft - 2B•15
Ignition - 5•1, 5•7, 5•8, 5•9, 5•11
Ignition distributor - 1A•14, 1B•15
Ignition switch - 7A•5
Ignition system specifications - 1A•22, 1B•21
Ignition timing check - 1A•15, 1B•15
Indicator/sidelight - 12•7, 12•8
Inlet manifold - 4A•7, 4B•23
Instrument panel - 12•9
Intercooler - 4B•22
Interior light switch - 12•6
Interior trim panels - 11•13
Introduction to the Saab 90, 99 & 900 - 0•4

J

Jacking and vehicle support - REF•5
Jump starting - 0•7

K

Kickdown cable - 7B•2

L

Leak checking - 0•9
Locks bonnet/door/tailgate - 11•4, 11•6, 11•9
Lubricants and fluids - 0•16, 1A•8, 1B•8, 1B•18
Lucas CU14 fuel injection system - 4B•7

M

Main and big-end bearings - 2B•22, 2B•24
Maintenance and servicing - see Routine maintenance
Manifolds - 4A•7, 4A•8, 4B•23, 4B•25
Manual transmission - 7A•1 et seq
  fault diagnosis - REF•16
  gear lever and housing - 7A•3
  gear lever lock - 7A•5
  gearbox oil - 0•16, 1A•16, 1B•16 7A•1, 7A•6
  gearchange linkage - 7A•4
  oil seals - 7A•2
  reversing light switch - 7A•3
  speedometer drive - 7A•2
Master cylinder (brake) - 9•13
Master cylinder (clutch) - 6•2
Mirror adjustment switch - 12•5
Mirrors - 11•10
Mixture (CO) adjustments - 1A•10, 1B•10
MOT test checks - REF•8
Mountings (engine/transmission) - 2A•17

O

Oil change - 1A•5, 1B•5
Oil cooler - 2A•16
Oil filter - 1A•5, 1A•9, 1B•5
Oil pressure warning light switch - 2A•16
Oil pump - 2A•13
Oil seals - 7A•2, REF•4
Oil seals crankshaft - 2A•16, 2B•15
Oil, engine - 0•16, 1A•22, 1B•20
Oil, gearbox - 0•16, 7A•1, 7A•6
Open-circuit (finding) - 12•4
Overhaul (engine) - 2B•3, 2B•23

Index

P
Pads (brake) - 1A•17, 1B•8, 9•5, 9•7
Paintwork damage - 11•2, 11•3
Panhard rod - 10•8
Parts - REF•3
Pedals
  accelerator - 4A•3, 4B•5
  brake - 9•17
  clutch - 6•1
Pierburg carburettor - 4A•5
Piston rings - 2B•23
Piston/connecting rod - 2B•18, 2B•20, 2B•25
Plastic components - 11•3
Plugs - 1A•6, 1A•23, 1B•7, 1B•21
Points - 1A•14
Pollen filter - 1A•18, 1B•9, 3•8
Power assisted steering - 10•10
Power steering fluid - 0•14, 0•16, 1A•18, 1B•8, 10•2
Pressure regulator - 4B•12, 4B•14
Punctures - 0•8

R
Radiator - 3•2
Radiator electric cooling fan - 3•4
Radiator grille - 11•4
Radio/cassette player - 12•10, REF•5
Rear lights - 12•7, 12•8
Relays - 12•3, 12•4
Release bearing (clutch) - 6•3
Repair procedures - REF•4
Respraying - 11•2
Reversing light switch - 7A•3, 7B•4
Rings (piston) - 2B•23
Road test - 1A•19, 1B•9
Roll bars - 10•5, 10•8
Routine maintenance, servicing and schedule - 1A•1 *et seq*, 1B•1 *et seq*
Routine maintenance procedures - 1A•5 et seq, 1B•5 et seq

S
Safety first! - 0•5
Scratches - 11•2, 11•3
Seat belts - 1B•9, 11•12, 12•5
Seats - 11•12
Selector cable (automatic transmission) - 7B•2
Selector lever and housing - 7B•3
Sender unit fuel gauge - 4A•4, 4B•9
Servicing - see Routine maintenance
Servo unit (braking system) - 9•14
Shock absorber - 1A•17, 1B•16
Short-circuit (finding) - 12•4
Slave cylinder - 6•3
Spare parts - REF•3
Spark plugs - 1A•6, 1A•23, 1B•7, 1B•21

Special tools - REF•4
Specifications and capacities - 1A•22, 1B•20
Speedometer drive - 7A•2
Speedometer drive cable - 12•9
Starter inhibitor switch - 7B•4
Starter lock cylinder - 5•7
Starter motor - 5•6
Starting system - 5•6
Starting/running/ignition problems - 0•6, 5•8
Steering - 10•1 *et seq*
  angles - 10•2
  camber - 10•2 10•12
  castor - 10•2, 10•12
  check - 1A•17, 1B•16
  column - 10•11, 10•12
  fault diagnosis - REF•17
  gaiters - 10•11
  gear - 10•10
  power assisted steering - 10•10
  rack - 10•10
  road test - 1A•19, 1B•9
  swivel member - 10•6
  toe setting - 10•2, 10•12
  tracking - 10•12
  wheel - 10•11
  wheel alignment - 10•2, 10•12
Sunroof - 11•11, 12•6
Support struts (tailgate) - 11•8
Suspension and steering - 10•1 *et seq*
  anti-roll bars - 10•5, 10•8
  balljoints - 10•6
  check - 1A•17, 1B•16
  coil spring - 10•1, 10•6, 10•8
  damper - 10•3, 10•7
  fault diagnosis - REF•17
  hub assembly - 10•5, 10•9
  panhard rod - 10•8
  rear axle tube - 10•9
  road test - 1A•19, 1B•9
  torque/lower arm - 10•8, 10•9
  wheel bearings - 10•5
  wishbones - 10•4
Switches
  cooling system - 3•4
  ignition - 7A•5
  brake light - 9•18
  courtesy light - 12•6
  electric window - 12•6
  fog lights - 12•5
  handbrake warning light - 9•17
  hazard light - 12•5
  heated rear window - 12•5
  horn 12•9
  interior light - 12•6
  mirror adjustment - 12•5
  oil pressure warning light - 2A•16
  reversing light - 7A•3, 7B•4
  seat belt warning - 12•5
  starter inhibitor - 7B•4
  sunroof - 12•6

T
Tail lights - 12•7, 12•8
Tailgate lock - 11•9
Tailgate supports - 11•8
TDC (locating) - 2A•4
Thermostat - 3•3
Throttle cable - 4A•3, 4B•4
Throttle position switch/sensor - 4B•13, 4B•15
Timing - 1A•15, 1B•15
Timing chain/sprockets - 2B•10
Toe setting - 10•2, 10•12
Tools and working facilities - REF•6
Top Dead Centre (locating) - 2A•4
Towing - 0•9
Tracking - 10•12
Transmission/engine removal - 2B•4
Trim panel (door) - 11•5
Turbocharger - 1A•13, 1B•13, 4B•15
Tyre checks - 0•13, 0•16

U
Underbody - 1A•4, 1B•4, 11•1
Unleaded petrol - 4A•3, 4B•5
Upholstery and carpets - 11•2

V
Vacuum servo unit (braking system) - 9•14
Valves - 1A•13, 1B•13, 2B•16
Vehicle identification numbers - REF•3
Ventilation system - 3•1 *et seq*

W
Warm-up regulator - 4B•12
Washer fluid level - 0•12
Water pump - 3•5
Weekly checks - 0•10 *et seq*
Wheel (steering) - 10•11
Wheel alignment - 10•2, 10•12
Wheel bearings - 10•5
Wheel changing - 0•8
Wheel hub assembly - 10•9
Window (electric) - 11•10
Windows/glass - 11•8, 11•11
Window regulator - 11•8
Windscreen - 11•11
Wiper arms - 12•10
Wiper blades - 0•15, 1A•23, 1B•21
Wiper motors - 12•10
Wiring diagrams - 12•12 *et seq*
Working facilities - REF•6

Z
Zenith/Solex carburettor - 4A•5

Preserving Our Motoring Heritage

The Model J Duesenberg Derham Tourster. Only eight of these magnificent cars were ever built – this is the only example to be found outside the United States of America

Almost every car you've ever loved, loathed or desired is gathered under one roof at the Haynes Motor Museum. Over 300 immaculately presented cars and motorbikes represent every aspect of our motoring heritage, from elegant reminders of bygone days, such as the superb Model J Duesenberg to curiosities like the bug-eyed BMW Isetta. There are also many old friends and flames. Perhaps you remember the 1959 Ford Popular that you did your courting in? The magnificent 'Red Collection' is a spectacle of classic sports cars including AC, Alfa Romeo, Austin Healey, Ferrari, Lamborghini, Maserati, MG, Riley, Porsche and Triumph.

A Perfect Day Out

Each and every vehicle at the Haynes Motor Museum has played its part in the history and culture of Motoring. Today, they make a wonderful spectacle and a great day out for all the family. Bring the kids, bring Mum and Dad, but above all bring your camera to capture those golden memories for ever. You will also find an impressive array of motoring memorabilia, a comfortable 70 seat video cinema and one of the most extensive transport book shops in Britain. The Pit Stop Cafe serves everything from a cup of tea to wholesome, home-made meals or, if you prefer, you can enjoy the large picnic area nestled in the beautiful rural surroundings of Somerset.

John Haynes O.B.E., Founder and Chairman of the museum at the wheel of a Haynes Light 12.

Graham Hill's Lola Cosworth Formula 1 car next to a 1934 Riley Sports.

The Museum is situated on the A359 Yeovil to Frome road at Sparkford, just off the A303 in Somerset. It is about 40 miles south of Bristol, and 25 minutes drive from the M5 intersection at Taunton.
Open 9.30am - 5.30pm (10.00am - 4.00pm Winter) 7 days a week, *except Christmas Day, Boxing Day and New Years Day*
Special rates available for schools, coach parties and outings Charitable Trust No. 292048